PREPARATION FOR ALGEBRA

Math

ADVANTAGE

HARCOURT
BRACE

Orlando • Atlanta • Austin • Boston • San Francisco • Chicago • Dallas • New York • Toronto • London

http://www.hbschool.com

Printed in the United States of America

ISBN 0-15-310697-2

2 3 4 5 6 7 8 9 10 032 2000 99 98

Senior Authors

Grace M. Burton
Chair, Department of Curricular Studies
Professor, School of Education
University of North Carolina at Wilmington
Wilmington, North Carolina

Evan M. Maletsky
Professor of Mathematics
Montclair State University
Upper Montclair, New Jersey

Authors

George W. Bright
Professor of Mathematics Education
The University of North Carolina at Greensboro
Greensboro, North Carolina

Sonia M. Helton
Professor of Childhood Education
Coordinator, College of Education
University of South Florida
St. Petersburg, Florida

Loye Y. (Mickey) Hollis
Professor of Mathematics Education
Director of Teacher Education and Under-
  graduate Programs
University of Houston
Houston, Texas

Howard C. Johnson
Dean of the Graduate School
Associate Vice Chancellor for Academic Affairs
Professor, Mathematics and
  Mathematics Education
Syracuse University
Syracuse, New York

Joyce C. McLeod
Visiting Professor
Rollins College
Winter Park, Florida

Evelyn M. Neufeld
Professor, College of Education
San Jose State University
San Jose, California

Vicki Newman
Classroom Teacher
McGaugh Elementary School
Los Alamitos Unified School District
Seal Beach, California

Terence H. Perciante
Professor of Mathematics
Wheaton College
Wheaton, Illinois

Karen A. Schultz
Associate Dean and Director of Graduate Studies
  and Research
Research Professor, Mathematics Education
College of Education
Georgia State University
Atlanta, Georgia

Muriel Burger Thatcher
Independent Mathematics Consultant
Mathematical Encounters
Pine Knoll Shores, North Carolina

Advisors

Anne R. Biggins
Speech-Language Pathologist
Fairfax County Public Schools
Fairfax, Virginia

Carolyn Gambrel
Learning Disabilities Teacher
Fairfax County Public Schools
Fairfax, Virginia

Lois Harrison-Jones
Education Consultant
Dallas, Texas

Asa G. Hilliard, III
Fuller E. Callaway Professor
  of Urban Education
Georgia State University
Atlanta, Georgia

Marsha W. Lilly
Secondary Mathematics
  Coordinator
Alief Independent School District
Alief, Texas

Judith Mayne Wallis
Elementary Language Arts/
  Social Studies/Gifted Coordinator
Alief Independent School District
Houston, Texas

CONTENTS

Focus on Problem Solving

NUMBER SENSE AND OPERATIONS   CHAPTERS 1–3

Key Skills

Key Skills

Key Skills

Modeling Decimals H11
Adding and Subtracting
   Decimals H12
Multiplying and Dividing Decimals
   by Powers of 10 .... H12
Multiplying Decimals H13
Dividing Decimals H13

Chapters 1–3 ✓Checkpoint

Key Skills

Key Skills

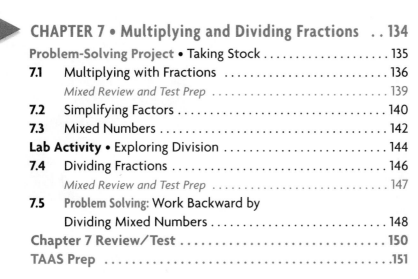

Key Skills

Mixed Numbers and Fractions . . .H16
Renaming Mixed Numbers H18
Adding Mixed Numbers H18
Subtracting Mixed Numbers H19

Key Skills

Dividing Whole Numbers H8
Simplest Form of Fractions H15
Mixed Numbers and Fractions . . .H16

Chapters 4–7 ✓ Checkpoint

GEOMETRY AND SPATIAL REASONING — CHAPTERS 8–10

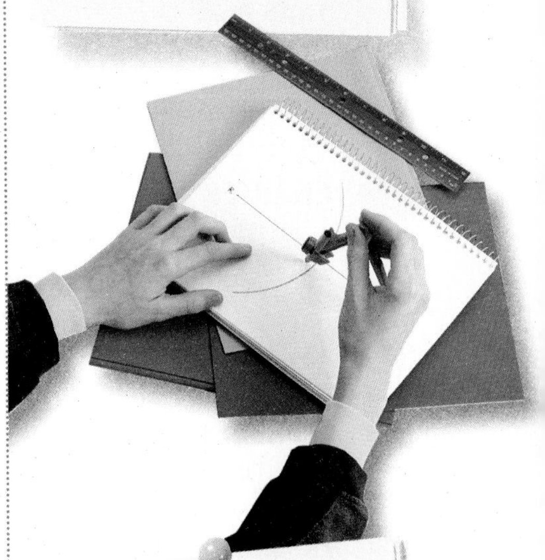

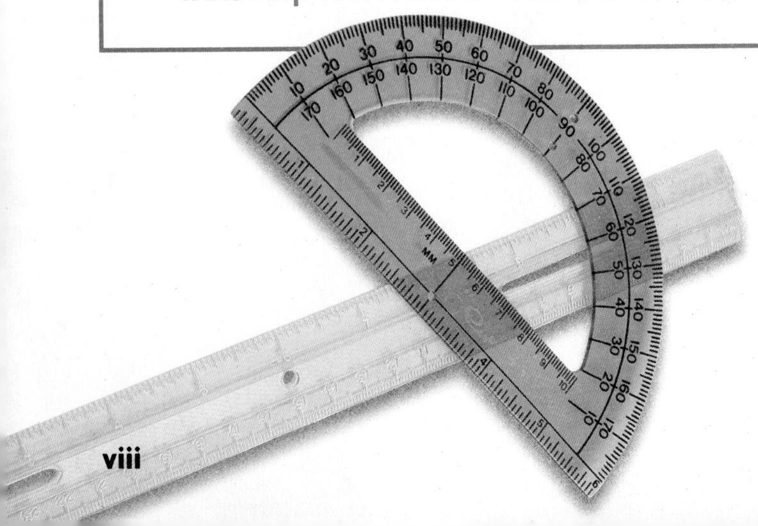

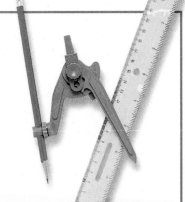

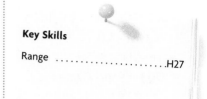

Key Skills

Range H27

Key Skills

Range H27

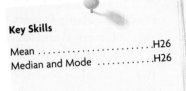

Key Skills

Mean H26
Median and Mode H26

Key Skills

Area H25

Chapters 11–14 ✓Checkpoint

Key Skills

Key Skills

Multiplying and Dividing Decimals
  by Powers of 10 ....H12
Ratios H27
Percents H28
Percents and Decimals H28

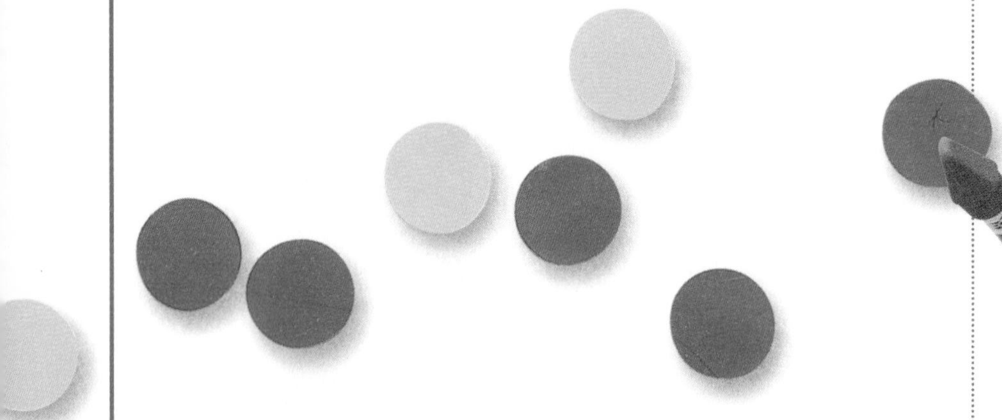

Key Skills

Percents H28
Percents and Ratios H29

Key Skills

Congruent Figures H23

Chapters 17–20 ✓Checkpoint

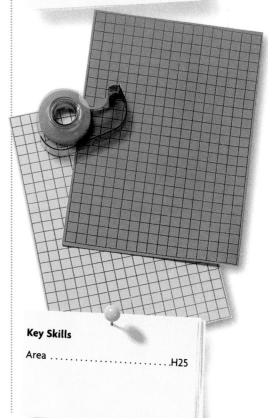

Key Skills

Solid Figures H22

Chapters 21–23 ✓ Checkpoint

Key Skills

Chapters 24–26 ✓ Checkpoint

Key Skills

Key Skills

Chapters 27–28 ✓ Checkpoint

Student Handbook

FOCUS ON PROBLEM SOLVING

Good problem solvers need to be good thinkers. They also need to know these strategies.

- Draw a Diagram
- Work Backward
- Make a Table or Graph

- Act It Out
- Find a Pattern

- Make a Model
- Guess and Check
- Write an Equation

- Use a Formula
- Solve a Simpler Problem
- Account for All Possibilities

After a strategy has been chosen, a good problem solver then decides how to solve the problem. They think about whether using paper or pencil, a calculator, manipulatives, or mental math is the best way to get the answer.

CHOOSE a strategy and a tool.

- **Draw a Diagram**
- **Use a Formula**

- **Make a Model**
- **Write an Equation**

Paper/Pencil Calculator Hands-On Mental Math

A good problem solver thinks through a problem carefully before trying to solve it. This plan can help you learn how to think through a problem.

UNDERSTAND the problem

Ask yourself...
What is the problem about?

Then try this.
Retell the problem in your own words.

What is the question?

Say the question as a fill-in-the-blank sentence.

What information is given?

List the information given in the problem.

PLAN how to solve it

Ask yourself...
What strategies might I use?

Then try this.
List some strategies you can use.

About what will the answer be?

Predict what your answer will be.
Make an estimate if it will help.

SOLVE the problem

Ask yourself...
How can I solve the problem?

Then try this.
Follow your plan and show your solution.

How can I write my answer?

Write your answer in a complete sentence.

LOOK BACK and check your answer

Ask yourself...
How can I tell if my answer is reasonable?

Then try this.
Compare your answer to your estimate.
Check your answer by redoing your work.
Match your answer to the question.

How else might I have solved the problem?

Try using another strategy to solve the problem.

On the following pages, you can practice being a good problem solver. Each page reviews a different strategy that you can use throughout the year. These pages will help you recognize the kinds of problems that can be solved with each strategy. Think through each problem you work on and ask yourself questions as you Understand, Plan, Solve, and Look Back. Then be proud of your success!

Draw a Diagram

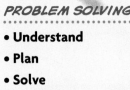

Sarah and her brother Chris visit their aunt and cousin each Saturday. From their apartment they walk 4 blocks north and 5 blocks west to their aunt's house. They then continue 2 blocks south and 3 blocks east to their cousin's house. How many ways can Sarah and Chris return home by walking four blocks?

UNDERSTAND You must mark the starting point and map out the walk. You know the number of blocks and the direction walked to arrive at each stopping place.

PLAN Draw a diagram using graph paper. Draw a line to show the number of blocks walked. Use arrows to show the direction taken to get from one place to the next.

SOLVE Draw a diagram like the one at the right to show the location of their apartment, the aunt's house, and the cousin's house.
From their cousin's, sketch the 4 remaining blocks to home.

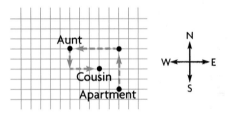

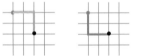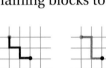

There are 6 possible ways.

LOOK BACK They are going to go south and east. The possible ways are EESS, SSEE, ESES, SESE, ESSE, and SEES.

Try These

1. Freeway poles are placed at regular intervals. The distance from the first pole to the fifth pole is 300 ft. What is the distance from the twentieth pole to the twenty-ninth pole?

2. The last Friday of a month is the 25th day of that month. What day of the week is the first day of the month?

3. There are 12 small square tables in the media center. Each table can seat only 1 person on each side. If the tables are pushed together to form one big rectangle, what arrangement will allow the most people to be seated? How many people can be seated?

Act It Out

PROBLEM SOLVING

• Understand
• Plan
• Solve
• Look Back

Eric found a toaster oven that toasts 2 slices of bread at once, but it toasts only one side at a time. Eric likes each side of his bread to be toasted for exactly one minute. How can he toast 3 slices of bread in 3 minutes?

UNDERSTAND You need to find a way to toast 3 slices of bread in three minutes. You know that each side takes 1 minute.

PLAN Place 3 pieces of paper in front of you to represent the bread. Label each side. You can use A1, A2, B1, B2, C1, and C2. Set up three steps, one for each minute.

SOLVE Step 1: Place slices A and B "in the toaster," with sides A1 and B1 facing up.

Step 2: Flip slice A1. Then replace B1 with C1.

Step 3: Replace A2 with B2. Then flip C1.

B2 C2

LOOK BACK If each step takes 1 minute, then all 3 slices have been toasted for 1 minute on each side.

Try These

1. A special matinee of a play costs $4.00 for adults and $1.00 for children under 10. If twice as many children attended as adults and the show received $72.00, how many adults attended the matinee?

2. Zachary wants to buy a comic book for $0.85. He has ten coins in his pocket that add up to exactly the correct amount. What coins does Zachary have in his pocket?

Make a Model

PROBLEM SOLVING
• Understand
• Plan
• Solve
• Look Back

The figures below show three views of a cube, revealing six faces. This cube has a different color on each face. Determine which colors are opposite each other.

B: blue   R: red
Y: yellow   G: green
W: white   P: purple

UNDERSTAND You must find the colors that are opposite each other. You know that certain colors cannot be opposites since they are touching.

PLAN Use three cubes and label each face with color dots or mark as shown in the diagram above. Focus on one particular color and determine which colors could not be opposite it.

SOLVE Given the information, place the known information on a grid. Use X's for "cannot be solutions." Use ✓'s for opposites. Opposites can occur only once in a column.

So, red is opposite purple. Green is opposite blue. Yellow is opposite white.

Make a cube with these opposites.

	Red	Blue	Green	Yellow	White	Purple
Red	X	X	X	X	X	
Blue	X	X	✓	X		
Green	X		X	X	X	X
Yellow	X	X	X	X		X
White	X		X	✓	X	
Purple	✓		X	X		X

LOOK BACK Place your cube in each of the positions shown in the diagram at the top of the page. Make sure your cube matches in each position.

Try These

1. Erica and Amanda always disagree over who sits where in their family's 7-seat minivan. Since their mother and father always sit in the front 2 seats, how many different arrangements are possible for the remaining 5 seats?

2. Sam is making a plant stand using 4 cubes. He stacks the cubes, one on top of the other, and paints the outside of the stand red (not the bottom). How many faces of the original cubes are painted?

Guess and Check

PROBLEM SOLVING
. . . .
- **Understand**
- **Plan**
- **Solve**
- **Look Back**

Handmade friendship bracelets use 20 beads. Handmade rings use 8 beads. Marne used a total of 176 beads to make 13 items. How many friendship bracelets did she make?

UNDERSTAND You know the number of beads used in a bracelet and in a ring.

PLAN Guess a number of bracelets and rings. Check to see if the total number of beads is 176 and the number of items is 13.

SOLVE Record your guess and number of beads.

Bracelets	Rings	Total Items	Number of Beads	
5	8	5 + 8 = 13	5 × 20 + 8 × 8 = 100 + 64 = 164	too low
8	5	8 + 5 = 13	8 × 20 + 5 × 8 = 160 + 40 = 200	too high
7	6	7 + 6 = 13	7 × 20 + 6 × 8 = 140 + 48 = 188	too high
6	7	6 + 7 = 13	6 × 20 + 7 × 8 = 120 + 56 = 176	correct

So, the number of bracelets is 6.

LOOK BACK Check that there are 13 items and 176 beads used.

bracelets: 6   bracelet beads: 6 × 20 = 120
rings: 7   ring beads: 7 × 8 = 56
total: 6 + 7 = 13 ✓   total beads: 120 + 56 = 176 ✓

Try These

1. Victor bought comics. He handed the clerk $20.00 and received $11.45 in change. He knows he bought more than 35 comics and each comic had the same price. How many books could he have bought and what did each cost?

2. The sum of two numbers is 49. Their difference is less than 10. What are all the possible pairs?

3. Amy bought used books for $4.95. She paid $0.50 each for some books and $0.35 each for others. She bought fewer than 8 books at each price. How many did she buy for $0.50?

Work Backward

PROBLEM SOLVING
....
• **Understand**
• **Plan**
• **Solve**
• **Look Back**

Jack, Jill, and Jodi collect baseball cards. Jack has 125 more cards in his collection than Jill. Jill has 75 cards fewer than Jodi. Jodi has 250 baseball cards. How many cards does Jack have?

UNDERSTAND   You need to find the number of cards Jack has in his collection. You know the number of cards Jodi has and how many more cards Jack has than Jill.

PLAN   Start from the end and work backward from what you know about Jodi's cards. Calculate Jill's and Jack's cards from this information.

SOLVE   Work from Jodi's collection. This is the only information you have without computing.

Jodi: 250 cards   250
Jill: 75 fewer than Jodi $250 - 75 = 175$
Jack: 125 more than Jill $175 + 125 = 300$

So, Jack has 300 baseball cards.

LOOK BACK   Check the solution with the information from the original problem.

Jack has 125 more cards than Jill: $175 + 125 = 300$

Jill has 75 cards fewer than Jodi: $250 - 75 = 175$

Try These

1. A driver travels east for 25 mi, turns north, and goes another 15 mi. To return home over the same roads, in what directions and for how far in each direction will she have to travel?

2. Mr. George ordered $50 of candy. He bought a box of deluxe caramels at $8 a box and 3 boxes of deluxe nuggets. What was the price per box for the deluxe nuggets?

3. A coin collection was distributed among four family members. Ryan received $\frac{1}{2}$ of the coins. Josephine received $\frac{1}{4}$ of the coins, and Lilly received $\frac{1}{5}$ of the coins. Beth got 100 coins. How many coins were in the collection?

Account for All Possibilities

PROBLEM SOLVING
- Understand
- Plan
- Solve
- Look Back

A palindrome reads the same forward and backward. How many palindromes are there during a 24-hr day?

UNDERSTAND You must find the number of palindromes displayed in 24 hr. You know what a palindrome is.

PLAN Make a list of the palindromes that appear within a 1-hr block of time. Be sure to account for A.M. and P.M.

SOLVE List all the possible palindromes within 1-hr blocks and look for a pattern:

1 o'clock	2 o'clock	3 o'clock	4 o'clock	5 o'clock	6 o'clock
1:01	2:02	3:03	4:04	5:05	6:06
1:11	2:12	3:13	4:14	5:15	6:16
1:21	2:22	3:23	4:24	5:25	6:26
1:31	2:32	3:33	4:34	5:35	6:36
1:41	2:42	3:43	4:44	5:45	6:46
1:51	2:52	3:53	4:54	5:55	6:56

7 o'clock	8 o'clock	9 o'clock	10 o'clock	11 o'clock	12 o'clock
7:07	8:08	9:09	10:01	11:11	12:21
7:17	8:18	9:19			
7:27	8:28	9:29			
7:37	8:38	9:39			
7:47	8:48	9:49			
7:57	8:58	9:59			

There are 6 palindromes in each single-digit hr. For 10, 11, and 12 o'clock, there is 1 palindrome each. In 12 hr, the clock displays $(9 \times 6) + (1 \times 3)$, or 57 palindromes. For 24 hr, double that amount: $2 \times 57 = 114$. So, there are 114 palindromes in a 24-hr day.

LOOK BACK Check the answer by adding the numbers of palindromes: $6 + 6 + 6 + 6 + 6 + 6 + 6 + 6 + 6 + 1 + 1 + 1 = 57$.

Try These

1. How many different 3-digit odd numbers can you make using the digits 3, 5, and 8? No digit should be used more than once in any number.

2. How many ways can you make change for a quarter using only dimes, nickels, and pennies?

Find a Pattern

PROBLEM SOLVING
....
• Understand
• Plan
• Solve
• Look Back

John made a design using hexagons and triangles. The length of each side of the hexagons and of each side of the triangles is 1 in. What is the perimeter of the next figure in his design?

UNDERSTAND You know the first 5 figures in John's design and the length of each side of the hexagons and triangles.
You need to find the perimeter of the sixth figure.

PLAN You can find a pattern in the perimeters of the first 4 figures and use the pattern to find the perimeter of the next figure.

SOLVE Find the perimeter of the figures. Then find a pattern in the perimeters.

Figure	1	2	3	4	5
Perimeter	6	7	11	12	16

Think: $6 + 1 = 7$ The rule for the pattern is add 1, add 4,
$7 + 4 = 11$ add 1, add 4, and so on.
$11 + 1 = 12$
$12 + 4 = 16$ Sixth perimeter: $16 + 1 = 17$

So the perimeter of the sixth figure will be 17 ft.

LOOK BACK Sketch a picture of the sixth figure. Check the perimeter.

Perimeter = 17 in. ✓

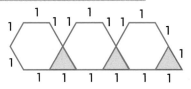

Try These

1. The Computer Connection is planning to network all 10 of its members. The company will connect each member's computer with that of each of the other 9 members' computers. How many cables will the hook-up require?

Find the pattern. Then find the next number or figure.

2. 1, 5, 9, 13, 17, . . .   3. 1, 4, 16, 64, 256, . .   4.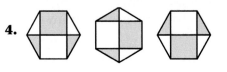

Make a Table or Graph

The Houston Public Library charges a fine of 10¢ a day for overdue books. The Fort Bend Library charges a fine of 50¢ for the first day and 5¢ for each additional day. On what day would an overdue book have the same fine at both libraries?

UNDERSTAND You need to know on which day both libraries charge the same amount. You know what each library charges and the daily rate.

PLAN Make a table to show the fines each library charges.

SOLVE Record the daily fine for each library until you find a match between the two libraries' fines.

Day	1	2	3	4	5	6	7	8	9
Houston Library	10¢	20¢	30¢	40¢	50¢	60¢	70¢	80¢	90¢
Fort Bend Library	50¢	55¢	60¢	65¢	70¢	75¢	80¢	85¢	90¢

So, on the 9th day both libraries would charge the same amount.

LOOK BACK Is it reasonable that both libraries would both charge the same amount on the 9th day? Is there another way you could solve this problem?

Try These

1. Planes leave Houston Hobby Airport for Dallas Love Field every 45 min. The first plane leaves at 5:45 A.M. What is the departure time closest to 4:30 P.M.?

2. Sixth graders have light blue, white, and green tops for their uniforms and white, navy, light blue, and tan pants. How many different combinations of tops and pants do they have for their uniforms?

3. The debate club has 10 members. Each member will debate each of the other members only once. How many debates will they have?

Technology Link

In *Data ToolKit*, you can make a table to help you solve the Try These exercises.

Solve a Simpler Problem

PROBLEM SOLVING
....
• **Understand**
• **Plan**
• **Solve**
• **Look Back**

At the end of each game, it is a tradition for each player to shake hands with every player on the opposing team. How many handshakes occur at the end of a game between two hockey teams of 20 players?

UNDERSTAND You know how many players are on each team. You need to find the total number of handshakes exchanged.

PLAN Examine the simplest case. Start with 1 player from each team. Then look at 2, 3, and 4 players from each team.

Collect and record the data in a table. Look for a pattern.

SOLVE The simplest case involves one player per team shaking hands with the opponent.

Extend the pattern in the table to include 20 players on each team.

5 → 25 handshakes
6 → 36 handshakes
7 → 49 handshakes

20 → 400 handshakes

So, the total number of handshakes exchanged is 400.

	Players on Each Team	Number of Handshakes
	1	1
	2	4
	3	9
	4	16

LOOK BACK The table shows a pattern that relates to square numbers. So, the number of handshakes between two hockey teams each with 20 players is 20^2, or 400.

Try These

1. There are 10 players in the chess tournament. Each player plays every other player in one game. How many games need to be scheduled?

2. The sum of the first 100 positive whole numbers is 5,050. What is the sum of the first 100 positive even whole numbers?

Use a Formula

PROBLEM SOLVING
· · · · · · · · · · · · · · · · ·
- **Understand**
- **Plan**
- **Solve**
- **Look Back**

The first figure below is called a tetromino. It has a perimeter of 10 units. The second shape connects two tetrominoes and has a perimeter of 18 units. You can use the formula $P = 8n + 2$, where n is the number of connected tetrominoes, to find the perimeter of any connected tetromino shape.

If the perimeter of a connected tetromino shape is 58 units, how many connected tetromino shapes are there in the figure?

UNDERSTAND   You are asked to find how many connected tetrominoes make a figure that has a perimeter of 58 units. You know the formula for finding the perimeter.

PLAN   Since you know the value of P, you can replace P with that value in the formula and solve for n.

SOLVE   Use the formula $P = 8n + 2$. Replace P with 58.

$$P = 8n + 2$$
$$58 = 8n + 2$$

Think: What number multiplied by 8, then increased by 2, gives 58?

$$8 \times 7 + 2 = 56 + 2 = 58$$

So, there are 7 connected tetrominoes in a figure with a perimeter of 58 units.

LOOK BACK   You can also use the formula to find the perimeters of different connected tetromino shapes.

Number of Connected Tetrominoes	1	2	3	4	5	6	7
Perimeter	10	18	26	34	42	50	58

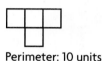

Perimeter: 10 units

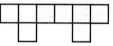

Perimeter: 18 units

NOTE: Tetrominoes are always connected the same way.

Try These

1. Suppose the input number, n, comes from the set of whole numbers and the rule is $n^2 - 1$. Which of the following are output numbers?

   80, 99, 224, 339, 481, 528

2. A side of square B is 5 times the length of a side of square A. How many times greater is the area of square B than the area of square A?

Write an Equation/Proportion

PROBLEM SOLVING
- **Understand**
- **Plan**
- **Solve**
- **Look Back**

A recipe for chocolate chip cookies uses 2 eggs to make 15 cookies. If you want to make 60 cookies, how many eggs will you need?

UNDERSTAND You know the number of eggs needed for 15 cookies. You need to know how many eggs to use to make 60 cookies.

PLAN The same two things are compared in both cases. The number of eggs should be in proportion to the number of cookies.

SOLVE Use a proportion (equation) to compare the ratio of eggs to cookies.

$$\frac{\text{eggs}}{\text{cookies}} = \frac{\text{eggs}}{\text{cookies}}$$
$$\frac{2}{15} = \frac{n}{60}$$
$$15 \times n = 60 \times 2$$
$$15n = 120$$ *Find the cross products.*
$$\frac{15n}{15} = \frac{120}{15}$$ *Divide each side by 15.*
$$n = 8$$

Use the numbers you know. Let n represent the number of eggs needed for 60 cookies.

LOOK BACK You could also solve the problem by finding how many batches of 15 cookies make 60 cookies, and then multiplying the number of batches by 2 eggs.

$60 \div 15 = 4 \rightarrow 4$ batches $4 \times 2 = 8 \rightarrow 8$ eggs ✓

Try These

1. The perimeter of the triangle is 65 m. Find the unknown length. Write an equation to solve the problem.

2. To park their car at Fun City, Savitri's parents paid $5.00 for the first hour and $1.50 for each additional hour. If they paid $14.00 for parking, how many hours were they at Fun City?

26   a

13

3. Of the 180 bicycles sold last month at Cycle Villa, 30 were Mountain Climbers. Mr. Villa is planning to make a circle graph to show this information. How many degrees should the Mountain Climbers section of the graph be?

MIXED APPLICATIONS

Use the strategy of your choice to solve.

1. A math book is lying open on a desk. What pages is it open to if the product of the page numbers of the facing pages is 7,832?

2. A rectangle has an area of 2,000 ft². One side is 50 ft long. What is the perimeter of the rectangle?

3. Jacques arranged coins on a grid and left the coins in a pattern of 4 rows with 4 coins in each row. Each row had exactly one penny, one nickel, one dime, and one quarter. No row, either horizontal or vertical, had more than one coin of each kind. How had Jacques arranged the coins?

4. Choose any three counting numbers less than 10. Write all the 2-digit numbers possible with these numbers. Do not repeat a digit. Find the sum of the 2-digit numbers. Divide by the sum of the three original numbers. Why is the quotient always 22?

5. David made the following transactions on his bank account in one month:

   Deposit $120.00
   Check $35.28
   Check $63.97
   Deposit $85.00

   The balance at the end of the month was $326.79. What was the balance at the beginning of the month?

6. A target has three circles, one inside of the other. The bull's eye is 10 points, the middle circle is 7 points, and the outer circle is 5 points. If three darts are thrown and only two hit the target, what are the possible scores?

   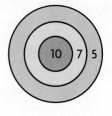

7. The bells ring at 10:00 A.M., 10:40 A.M., 11:20 A.M., and 12:00 P.M. At which of these times will the bells also ring?

   6:00 P.M.   8:00 P.M.   12:00 A.M.

8. How many cents are in d dimes? How many nickels are in q quarters?

9. A baked loaf of bread is about two times higher than the unbaked dough. A baked loaf is 5.5 in. high. How high was the unbaked dough?

10. How many ways can you add 8 odd numbers to get a sum of 20? Each number may be used more than once.

11. Aaron is designing a studio and wants to change the square window. He will make it twice as long and half as tall. What happens to its area?

12. Write a problem that can be solved with more than one strategy. Exchange with a classmate and solve.

13. David has 4 trophies. In how many different ways can he arrange them on his trophy shelf?

LOOKING AT NUMBERS

LOOK AHEAD

In this chapter you will solve problems that involve

- using place value of whole numbers and decimals

- comparing and ordering decimals

- changing decimals to fractions and fractions to decimals

- identifying and ordering integers

People have been digging up dinosaur bones for nearly 200 years, but about 50 percent of all dinosaurs we know have been identified in the last 20 years.

The smallest dinosaur, only 2 feet long, was called compsognathus. Its 3-inch head, about 12.5 percent of its body length, housed lots of small, sharp teeth.

What's the largest dinosaur? In 1986, New Mexico scientists announced they had found the bones of Seismosaurus, "earthshaker." Its tail bones suggested a creature 150 feet long, weighing 50 tons, or as much as ten large elephants.

We know by measuring their footprints that some dinosaurs could run as fast as 25 miles per hour. That's 0.416 miles per minute!

Fascinating Facts

What kinds of facts interest you?

Research and record four or five facts that interest you. Make a display of your facts. Then share and discuss with the class the type of numbers in your display, why the numbers are appropriate, and which facts could be expressed with another type of number.

PROJECT CHECKLIST

✓ Did you research at least four facts?

✓ Did you make a display of your facts?

✓ Did you share and discuss the type of numbers used in the display?

Whole Numbers and Decimals

What You'll Learn

How to use place value to express whole numbers and decimals

Why Learn This?

To understand and compare whole numbers and decimals used in population counts, prices, and sizes of bacteria

Think about the numbers you see every day in school, in the newspaper, on television, and in books and magazines. How does place value help you understand these numbers?

Tiger Woods

☐ Tiger Woods a sensation, exceeds a six-digit salary in 19 events

At the age of 3, Tiger Woods shot 48 for 9 holes. As a junior golfer, he won three consecutive U.S. Junior Amateur titles in 1991, 1992, and

1993. After winning three U.S. Amateur Championships, he turned professional at the age of 21 in August of 1996. In his first year in the PGA, he continued to break records, such as least time to earn $1,000,000. His golf earnings in 1997 exceeded $2,000,000.

ASSOCIATED PRESS

The numbers you see every day are part of the decimal system. The decimal system uses ten digits: 0, 1, 2, 3, 4, 5, 6, 7, 8, and 9. The digits and the position of each digit in a number determine the number's value.

- Read each number on the place-value chart. What is the value of the digit 3 in each number?

	Millions	Hundred Thousands	Ten Thousands	Thousands	Hundreds	Tens	Ones	Tenths	Hundredths	Thousandths	Ten - Thousandths
PLACE VALUE											
1,623,051 →	1	6	2	3	0	5	1				
0.0531 →							0	0	5	3	1
32.4 →						3	2	4			

When you read numbers, you use place value.

REMEMBER:

When reading a number with a decimal point, read the decimal point as "and." **See page H10.**

Read 8.2 as "eight and two tenths."

EXAMPLE

A. Write: 841,230 ← *standard form*

Read: eight hundred forty-one thousand, two hundred thirty

B. Write: 12.75 ← *standard form*

Read: twelve and seventy-five hundredths

- **CRITICAL THINKING** In the chart, what would be the next three places to the left of millions?

GUIDED PRACTICE

Read the number. What is the value of the digit 6?

1. 34,065   **2.** 15,425.006   **3.** 2,654,000.25   **4.** 550.76

Write the number in words.

5. 10,040   **6.** 0.605   **7.** 6.5   **8.** 7,868   **9.** 863,042.46

Write the number in standard form.

10. seven million, twenty-four thousand, two hundred one

11. eighty-four and seven hundred twenty-two thousandths

12. six hundredths

13. five thousand, forty-two

INDEPENDENT PRACTICE

Write the value of the digit 4.

1. 34.1   **2.** 400   **3.** 11.394   **4.** 940,006

Write in words the value of the underlined digit.

5. 1<u>2</u>0   **6.** 5.045<u>7</u>   **7.** <u>8</u>00,927   **8.** 345.6<u>9</u>

Write the number in words.

9. 46   **10.** 0.03   **11.** 1,500.1   **12.** 2.456   **13.** 1,037,804

Write the number in standard form.

14. one hundred twenty-five

15. fifteen hundredths

16. thirty-five and two thousand, three hundred sixty-nine ten-thousandths

17. eight hundred fifty thousand, two hundred forty-seven and fifty-six hundredths

Problem-Solving Applications

18. SCIENCE The world's smallest cut diamond is 0.0009 in. in diameter and weighs 0.0012 carat. Write both numbers in words.

19. CONSUMER MATH A movie studio announced that the box office sales for its new release reached nine million, four hundred fifty-six thousand, three hundred two in the first week. Write the number in standard form.

20. ✏️ **WRITE ABOUT IT** How do you know that 8.1 and 8.001 are not the same?

Comparing and Ordering

Do you ever draw straws to determine who will be chosen from a group? Sam is drawing straws with his friends. The person that draws the shortest straw will have to help clean up after a party.

• The first straw is 55 mm, and the second straw is 57 mm. Which straw is longer?

You can use a number line to decide whether one number is greater than or less than another number.

Since 55 is to the left of 57, 55 is less than 57.
You can write 55 < 57, or you can write 57 > 55.

     ↑   ↑
means "less than"   means "greater than"

• Copy the number line below and use it to compare 124 and 132. Which is greater?

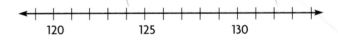

ANOTHER METHOD You can also use place value to compare numbers.

> **EXAMPLE 1** Compare 32,728 and 32,722. Use < or >.
>
> | | | |
> |---|---|---|
> | 32,728 | 32,722 | *Compare the digits. Start at the left.* |
> | 32,728 | 32,722 | ←same number of ten thousands |
> | 32,728 | 32,722 | ←same number of thousands |
> | 32,728 | 32,722 | ←same number of hundreds |
> | 32,728 | 32,722 | ←same number of tens |
> | 32,728 | 32,722 | *Compare the ones.* |
>
> Since 8 ones is greater than 2 ones, 32,728 > 32,722, or 32,722 < 32,728.
>
> • Compare 139,759 and 139,783.

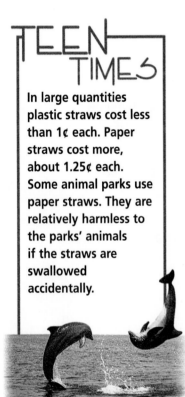

When you order two or more numbers from least to greatest or from greatest to least, you can first compare the numbers two at a time.

EXAMPLE 2 Use < to list the numbers in order from least to greatest.

$$12,010; \ 12,031; \ 12,000; \ 13,001$$

12,000 < 12,010   12,010 < 12,031   *Compare the numbers*
          12,031 < 13,001   *two at a time.*

12,000 < 12,010 < 12,031 < 13,001   *List the numbers in order.*

- Use > to list the numbers in order from greatest to least.
  $$3,812; \ 3,821; \ 3,028$$

Comparing Decimals

Comparing decimals is similar to comparing whole numbers. The number line below shows that 2.6 is to the right of 2.3, so 2.6 > 2.3.

$$
\begin{array}{ccccccccc}
2.1 & 2.2 & 2.3 & 2.4 & 2.5 & 2.6 & 2.7 & 2.8
\end{array}
$$

- Draw a number line to show 3.25 and 3.2. Which is greater?

EXAMPLE 3 Compare 7.28 and 7.2. Use < or >.

7.28	7.2	*Compare the digits. Start at the left.*
7.28	7.2	← same number of ones
7.28	7.2	← same number of tenths
7.28	7.20	*Add a zero to 7.2 so that both numbers have the same number of places.*

Since 8 hundredths is greater than 0 hundredths,
7.28 > 7.2, or 7.2 < 7.28.

Talk About It

- Which is greatest, 5.2, 5.20, or 5.200? Explain your answer.

- Explain how you would compare 0.325 and 0.335.

- Suppose you wanted to compare 745 and 8.22. Would it be easier to use a number line or to use place value and compare the digits? Explain.

REMEMBER:
You can add a zero to the right of a decimal without changing its value. For example, 7.2 and 7.20 are equivalent.
See Key Skill 21, page H12.

Knowing how to compare and order decimals is important when you need to compare prices or measurements.

EXAMPLE 4 Marcia Stone has found the same CD player at four different stores. The prices are $132.95, $132.50, $130.25, and $135.25. Order the prices from least to greatest. Which is the lowest price?

$130.25 < $132.50 $132.50 < $132.95   *Compare the numbers two at a time.*

$132.95 < 135.25

$130.25 < $132.50 < $132.95 < $135.25   *List the numbers in order.*

So, $130.25 is the lowest price.

• Order $1,250.45; $1,345.62; $1,235.78; and $1,305.45 from greatest to least.

• Order from least to greatest: 134.25 m, 130.5 m, 134.05 m

GUIDED PRACTICE

Use the number lines to compare the numbers. Write < or >.

1.
```
←+——+——+——●——+——+——●——+→
  120  121  122  123  124  125  126
```

122 ● 125

2.
```
←+——+——●——+——+——+——+——●——+→
  3.5  3.6  3.7  3.8  3.9  4.0  4.1
```

4.1 ● 3.6

Compare the numbers. Write <, >, or =.

3. 155 ● 152   **4.** 3,678 ● 3,675   **5.** 8,205 ● 8,250   **6.** 77,442 ● 77,342

7. 21.30 ● 21.3   **8.** 1.5 ● 1.7   **9.** 0.01 ● 0.12   **10.** 12.36 ● 12.3

Write the numbers in order from least to greatest. Use <.

11. 8,621; 8,612; 8,613   **12.** 1.361, 1.351, 1.363   **13.** 125.3, 124.32, 125.33

INDEPENDENT PRACTICE

Compare the numbers. Write <, >, or =.

1. 1,234 ● 1,243   **2.** 54,004 ● 54,504   **3.** 20,115 ● 20,109   **4.** 106,447 ● 106,442

5. 12.1 ● 12.2   **6.** 213 ● 223   **7.** $132.34 ● $122.34   **8.** 34.6 ● 34.60

9. 7.433 ● 7.432   **10.** 7.099 ● 7.999   **11.** 0.110 ● 0.1100   **12.** 45.678 ● 45.673

13. $644.56 ● $634.55   **14.** 99.088 ● 99.888   **15.** 133.23 ● 13.323   **16.** 457.8 ● 457.3

MORE PRACTICE Lesson 1.2, page H42

Write the numbers in order from least to greatest. Use <.

17. 299, 295, 290, 298   **18.** 4,556; 4,566; 4,555   **19.** 1.412, 1.214, 1.421, 1.124

20. 2,365; 2,305; 2,563; 2,300; 2,035   **21.** 1.45, 1.05, 1.405, 1.25, 1.125, 0.405

Write the numbers in order from greatest to least. Use >.

22. 3,245; 3,024; 3,125   **23.** $90, $90.22, $90.12   **24.** 5.004, 5.040, 5.4

25. 1,263; 1,623; 1,260; 1,063; 1,203   **26.** 125.3, 124.32, 125.33, 12.345, 120.4

Any number that is greater than 1.5 and less than 1.6 is between
1.5 and 1.6. Tell which of the given numbers are between 1.5 and 1.6.

27. 1.49   1.63   1.51   **28.** 1.58   1.505   1.605   **29.** 1.7   1.5342   1.067
     1.05   1.55   1.409   1.52   1.61   1.72   1.56   1.507   1.366

Problem-Solving Applications

For Problem 30, use the table.

Monthly Normal Rainfall for Seattle, Washington (in inches)												
Month	Jan	Feb	Mar	Apr	May	Jun	Jul	Aug	Sep	Oct	Nov	Dec
Rainfall	6.0	4.2	3.6	2.4	1.6	1.4	0.7	1.3	2.0	3.4	5.6	6.3

30. WEATHER Joey's dad is planning the family's vacation to Seattle, Washington, for next year. He knows that some months have more rain than others. He would like to make sure the chance of rain is lower while they are there on vacation.

   a. Write in order from least to greatest the rainfall per month. Use <.

   b. Which month would be the best for a visit? Why?

   c. Which month would be the worst for a visit? Why?

31. SPORTS The batting averages of 3 players are 0.268, 0.280, and 0.265. Write the averages in order from greatest to least. Use >.

32. CONSUMER MATH Robert needs a new jacket. He found the same jacket on sale at four different sports stores. The prices are $32.95, $30.50, $35.99, and $34.00. Which is the lowest price?

33. ✏️ **WRITE ABOUT IT** Explain how you would compare 5,361 and 5,316.

Decimals and Fractions

Jen ran five tenths of a mile. What is the distance Jen ran, expressed as a decimal? What is the distance Jen ran expressed as a fraction?

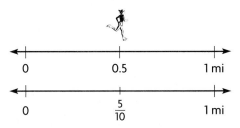

The number lines show that 0.5 and $\frac{5}{10}$ are equivalent. You can use place value to change a decimal to an equivalent fraction.

EXAMPLE 1 Use place value to change the decimal to a fraction.

A. 0.3   *Identify the place value of the last digit. The 3 is in the tenths place.*

$\frac{3}{10}$ *Use the value for the denominator.*

So, $0.3 = \frac{3}{10}$.

B. 0.42   *Identify the place value of the last digit. The 2 is in the hundredths place.*

$\frac{42}{100}$ *Use the value for the denominator.*

So, $0.42 = \frac{42}{100}$.

- Write 0.103 as a fraction.

- When changing a decimal to a fraction, what would the denominator be if the last digit in a decimal has a place value of tenths? hundredths? thousandths?

A calculator with fraction functions can be used to change a decimal to a fraction. For example, the calculator sequence below shows how to change 0.75 to a fraction.

$$0.75 \quad \boxed{\text{F} \otimes \text{D}} \quad \boxed{75/100} \leftarrow 0.75 = \frac{75}{100}$$

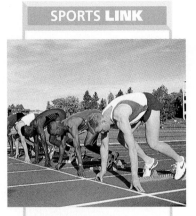

You can change a fraction to an equivalent decimal by dividing the numerator by the denominator.

EXAMPLE 2 Use division to change the fraction to a decimal.

A. $\frac{1}{5}$ *Divide the numerator by the denominator.*

$$
\begin{array}{r}
0.2 \\
5\overline{)1.0} \\
-10 \\
\hline
0
\end{array}
$$

Place the decimal point.
Then divide as with whole numbers.

So, $\frac{1}{5} = 0.2$.

B. $\frac{1}{4}$ *Divide the numerator by the denominator.*

$$
\begin{array}{r}
0.25 \\
4\overline{)1.00} \\
-8\downarrow \\
\hline
20 \\
-20 \\
\hline
0
\end{array}
$$

Place the decimal point.
Then divide as with whole numbers.

So, $\frac{1}{4} = 0.25$.

You can also use a calculator to change a fraction to a decimal.

EXAMPLE 3 Change $\frac{1}{8}$ to a decimal.

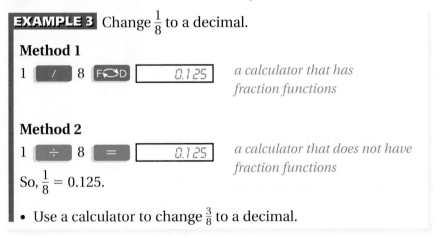

Method 1

1 [/] 8 [F⟷D] [0.125]   *a calculator that has fraction functions*

Method 2

1 [÷] 8 [=] [0.125]   *a calculator that does not have fraction functions*

So, $\frac{1}{8} = 0.125$.

• Use a calculator to change $\frac{3}{8}$ to a decimal.

GUIDED PRACTICE

Use place value to change the decimal to a fraction.

1. 0.4   **2.** 0.03   **3.** 0.32   **4.** 0.425   **5.** 0.7255

Use division to change the fraction to a decimal.

6. $\frac{3}{10}$ **7.** $\frac{1}{2}$ **8.** $\frac{23}{100}$ **9.** $\frac{3}{4}$ **10.** $\frac{3}{25}$

 Calculator Activities, page H35

23

Comparing Fractions and Decimals

You can compare fractions that have the same denominators by looking at the numerators.

For example, to compare $\frac{1}{5}$ and $\frac{2}{5}$, you compare the numerators.

$$\frac{1}{5} < \frac{2}{5} \text{ since } 1 < 2.$$

When fractions do not have the same denominator, it is easier to compare them if you change them to decimals.

EXAMPLE 4 Students at Sumter Middle School wear their school colors, blue and yellow, to show school spirit. One day $\frac{2}{5}$ of the students wore blue and $\frac{1}{4}$ of the students wore yellow. Which color was worn by more students?

Compare $\frac{2}{5}$ and $\frac{1}{4}$.

$\frac{2}{5}$ ● $\frac{1}{4}$ *To compare the fractions, write each as a*
↓   ↓   *decimal.*
0.4 ● 0.25   *Compare the decimals.*

$0.4 > 0.25$

Since $0.4 > 0.25$, $\frac{2}{5} > \frac{1}{4}$.

So, more students wore blue.

• Use decimals to compare $\frac{1}{8}$ and $\frac{2}{10}$.

Talk About It

• Suppose you were comparing $\frac{1}{4}$ and $\frac{3}{4}$. Would you need to change to decimals? Explain.

• **CRITICAL THINKING** Explain how you would compare $\frac{5}{8}$ and 0.75.

INDEPENDENT PRACTICE

Use place value to change the decimal to a fraction.

1. 0.2   **2.** 0.8   **3.** 0.17   **4.** 0.09   **5.** 0.99   **6.** 0.45

7. 0.25   **8.** 0.568   **9.** 0.325   **10.** 0.875   **11.** 0.3525   **12.** 0.0823

Write as a decimal.

13. $\frac{2}{10}$ **14.** $\frac{27}{100}$ **15.** $\frac{3}{5}$ **16.** $\frac{7}{8}$ **17.** $\frac{15}{16}$ **18.** $\frac{4}{8}$

Write as a decimal.

19. $\frac{87}{10,000}$ **20.** $\frac{7}{100}$ **21.** $\frac{1}{4}$ **22.** $\frac{3}{8}$ **23.** $\frac{7}{16}$ **24.** $\frac{2}{5}$

Write < or >.

25. $\frac{1}{2}$ ● $\frac{3}{4}$ **26.** $\frac{1}{5}$ ● $\frac{3}{10}$ **27.** $\frac{17}{20}$ ● $\frac{5}{8}$ **28.** $\frac{1}{8}$ ● $\frac{17}{100}$ **29.** $\frac{99}{100}$ ● $\frac{13}{16}$

30. $\frac{7}{16}$ ● $\frac{11}{20}$ **31.** 0.7 ● $\frac{77}{100}$ **32.** 0.18 ● $\frac{3}{50}$ **33.** 0.335 ● $\frac{5}{16}$ **34.** 0.09 ● $\frac{9}{10}$

Problem-Solving Applications

35. CONSUMER MATH Hector needs to know the decimal equivalent of $\frac{1}{8}$ to find the cost of one slice of pizza. Change the fraction to a decimal.

36. NUMBER SENSE The decimal that is equivalent to $\frac{1}{5}$ is 0.2. What is the decimal that is equivalent to $\frac{2}{5}$?

37. CONSUMER MATH Al needs $\frac{3}{4}$ lb of plant food. The plant food shows decimal weights. What decimal weight should Al buy?

38. SPORTS Jan ran $\frac{1}{8}$ mi. Karen ran 0.15 mi. Who ran farther?

39. CRITICAL THINKING Use a calculator to change $\frac{1}{3}$ to a decimal. Then use paper and pencil to change $\frac{1}{3}$ to a decimal. What do you notice about the decimal? How is this decimal different from the decimal for $\frac{1}{2}$?

40. ✏ WRITE ABOUT IT Explain two ways to change the fraction $\frac{4}{10}$ to a decimal.

Mixed Review and Test Prep

Multiply.

41. $2 \times 2 \times 2 \times 2$ **42.** 11×11 **43.** $6 \times 6 \times 6$

Find the area of the rectangle.

44. length: 2 ft
width: 2 ft

45. length: 5 in.
width: 1 in.

46. ESTIMATION The population of Houston, Texas, in 1997 was 1,702,086. What is this number rounded to the nearest hundred thousand?

    A 2,000,000   **B** 1,702,100

    C 1,702,000   **D** 1,700,000

47. ESTIMATION Allen rode his bike 27.5 miles. To the nearest mile, how far did Allen ride his bike?

    F 30 mi   **G** 28 mi

    H 27 mi   **J** 20 mi

Technology Link

In *Mighty Math Calculating Crew* the *Nautical Number Line* challenges you to find oceanic treasure by identifying equivalent fractions for decimals. Use Grow Slide Level Q.

What You'll Learn
How to represent numbers by using exponents

Why Learn This?
To be able to write large numbers, such as 10,000, in a shortened form

VOCABULARY

exponent

base

REMEMBER:

When you multiply two or more numbers to get a product, the numbers multiplied are called **factors**. See page H3.

$8 \times 3 \times 4 = 96$

The numbers 8, 3, and 4 are factors.

ALGEBRA CONNECTION

Exponents

Large numbers can be hard to understand, but you can think of them in certain ways to make sense of them. For example, to understand how much $10,000 is, first you can ask, How many $10 bills would it take to make $10,000? Then think about powers of 10.

$$10 \times 10 = 100 \qquad 10 \times 10 \times 10 = 1,000$$
$$10 \times 10 \times 10 \times 10 = 10,000$$

You can think of $10,000 as $10 \times 10 \times 10 \times 10$, or $10 \times 1,000$. So, there are one thousand $10 bills in $10,000.

Powers of numbers can be written with exponents. An **exponent** shows how many times a number called the **base** is used as a factor.

$$\overset{\text{exponent}}{\underset{\text{base}}{10^4}} = \underbrace{10 \times 10 \times 10 \times 10}_{\text{factors}} = 10,000$$

Read 10^4 as "ten to the fourth power."

You use multiplication to find the value of a number with an exponent.

EXAMPLE 1 Find the value of 2^4.

$2^4 = 2 \times 2 \times 2 \times 2 = 16$ *Use the base, 2, as a factor four times.*

So, the value of 2^4 is 16.

• Find the value of 8^2.

EXAMPLE 2 Express 81 by using an exponent and the base 3.

$81 = 9 \times 9 = 3 \times 3 \times 3 \times 3$ *Find the equal factors.*

$\qquad = 3^4$ *Write the base and the exponent.*

• Express 81 by using an exponent and the base 9.

Calculator Activities, page H30

GUIDED PRACTICE

Tell how many zeros will be in the standard form of the number.

1. 10^3 **2.** 10^{12} **3.** 10^{10} **4.** 10^1 **5.** 10^2 **6.** 10^8

Write the equal factors. Then find the value.

7. 2^3 **8.** 5^2 **9.** 3^4 **10.** 9^3 **11.** 9^4 **12.** 1^4

INDEPENDENT PRACTICE

Write in exponent form.

1. $12 \times 12 \times 12$ **2.** $1 \times 1 \times 1 \times 1 \times 1$ **3.** $4 \times 4 \times 4 \times 4$ **4.** $10 \times 10 \times 10$

5. 7×7 **6.** 14×14 **7.** $2 \times 2 \times 2 \times 2 \times 2$ **8.** $20 \times 20 \times 20 \times 20$

Find the value.

9. 4^5 **10.** 7^3 **11.** 1^{12} **12.** 5^3 **13.** 2^3 **14.** 2^5

15. 14^1 **16.** 13^2 **17.** 10^8 **18.** 20^2 **19.** 2^{10} **20.** 10^4

Express with an exponent and the given base.

21. 64, base 8   **22.** 216, base 6   **23.** 1,000; base 10

Problem-Solving Applications

24. **MEASUREMENT** Every 10 ft is equal to 1 story. How many stories is a building that is 10^2 ft tall?

25. **SOCIAL STUDIES** Texas has a population of about 18 million. Is the population greater than or less than 10^7?

26. **WRITE ABOUT IT** Which is greater, 3^2 or 2^3?

Mixed Review and Test Prep

Describe the opposite of the action.

27. Take 2 steps forward.   **28.** Turn right.   **29.** Jump up.

Write the number in standard form.

30. ten thousand, one hundred ninety

31. two hundred thousand, twenty-four

32. four hundred sixty-five

33. seventy thousand, seventy-two

34. **NUMBER SENSE** Which product is equal to 64?

     A $2 \times 2 \times 2 \times 2$ **B** $3 \times 3 \times 3 \times 2$

     C $2 \times 2 \times 16$ **D** $3 \times 2 \times 16$

35. **NUMBER SENSE** Which expression is equivalent to $(3 + 5) \times 2$?

     F $(7 - 3) + 10$ **G** $(4 \times 2) + 8$

     H $(2 \times 3) + 5$ **J** $10 + (7 \times 0)$

LAB ACTIVITY

What You'll Explore
How to use a model to find squares and square roots

What You'll Need
square tiles

ALGEBRA CONNECTION

Squares and Square Roots

A **square** is the product of a number and itself. A square can be expressed with the exponent 2. Read 3^2 as "3 squared."

You can use square arrays to model square numbers.

ACTIVITY 1

Explore

- Make a square array with 3 square tiles on each side. How many tiles did you use?

- Make a square array with 4 square tiles on each side.

- Make another square array with 5 square tiles on each side.

3

3

Think and Discuss

- How many tiles are in the square array with 4 square tiles on each side? Complete: $4^2 = $ ●.

- How many tiles are in the square array with 5 square tiles on each side? Complete: $5^2 = $ ●.

- **CRITICAL THINKING** Suppose you have a square with n square tiles on each side. How many tiles would be in the square array?

Try This

- Use square tiles to find 6^2, 7^2, and 8^2.

You can use a calculator to find squares.

Enter the value of the base, 11.

11 $\boxed{x^2}$ $\boxed{121}$ *Then press* $\boxed{x^2}$.

So, $11^2 = 121$.

- Use a calculator to find 15^2, 25^2, and 51^2.

Calculator Activities, page H38

When you find the two equal factors of a number, you are finding the **square root** of the number. The symbol for a square root is $\sqrt{}$. Finding a square root is the opposite of finding a square.

Since $5^2 = 25$, $\sqrt{25} = 5$. Read $\sqrt{25}$ as "the square root of 25."

ACTIVITY 2

Explore

You can think about the sides of a square array to help you find the square root of a number.

- Make a square array with 16 square tiles. How many tiles are on each side of your square array?

- Make a square array with 4 square tiles.

- Make another square array with 9 square tiles.

Think and Discuss

- How many tiles are on each side of your square array with 4 square tiles? Complete: $\sqrt{4} =$ ●.

- How many tiles are on each side of your square array with 9 square tiles? Complete: $\sqrt{9} =$ ●.

- How are squares and square roots different?

- CRITICAL THINKING Can you form a square with 10 square tiles? 12 square tiles? Explain.

Technology Link

You can practice squaring a number by using E-Lab, Activity 1. Available on CD-ROM and on the Internet at **www.hbschool.com/elab**

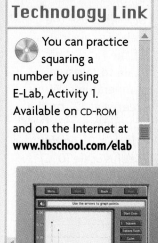

Try This

You can use a calculator to find square roots.

36 〔 2nd 〕 〔 √ 〕 ⌈ 6. ⌉   *Enter the number, 36. Then press* 〔 2nd 〕 *and* 〔 √ 〕.

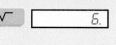

So, $\sqrt{36} = 6$.

- Use a calculator to find $\sqrt{121}$, $\sqrt{625}$, and $\sqrt{1,024}$.

ALGEBRA CONNECTION

Integers

What You'll Learn
How to order integers and how to identify opposite integers

Why Learn This?
To compare temperatures and altitudes and find opposites

VOCABULARY

integers
positive integers
negative integers
opposites

In a given year, the lowest temperature recorded for Atlantic City, New Jersey, was ⁻11°F. On a thermometer, is ⁻11°F above or below zero?

Just as a thermometer shows temperatures above and below zero, a number line can show numbers to the right and to the left of zero.

Integers can be shown on a number line. Integers greater than 0 are **positive integers**. Integers less than 0 are **negative integers**. The integer 0 is neither positive nor negative.

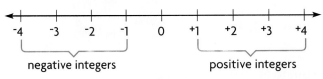

You can use a number line to compare integers. On a number line, each integer is greater than any integer to its left and less than any integer to its right.

EXAMPLES Compare the integers. Use < and >.

A. ⁺3 and ⁻4

⁻4 is to the left of ⁺3 on the number line.
So, ⁻4 < ⁺3.

⁺3 is to the right of ⁻4 on the number line.
So, ⁺3 > ⁻4.

B. ⁻4 and ⁻6

⁻6 is to the left of ⁻4 on the number line.
So, ⁻6 < ⁻4.

⁻4 is to the right of ⁻6 on the number line.
So, ⁻4 > ⁻6.

• Compare and order ⁻3, ⁺7, and 0 from least to greatest.

SCIENCE LINK

Scientists use the Celsius scale instead of the Fahrenheit scale to measure temperatures. Water freezes at 32° on the Fahrenheit scale (32°F), which is 0° on the Celsius scale (0°C). Which temperature is colder, 2°F or ⁻2°C?

For every positive integer, there is an opposite negative integer. Integers that are **opposites** are the same distance from 0 on the number line, only in opposite directions.

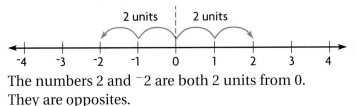

The numbers 2 and ⁻2 are both 2 units from 0.
They are opposites.

GUIDED PRACTICE

Compare the integers. Tell which is greater.

1. $^+2$ and $^-3$ **2.** $^-3$ and $^-6$ **3.** $^+3$ and $^+2$ **4.** $^-4$ and $^+3$ **5.** $^-4$ and $^-2$

Name the opposite of the given integer.

6. $^+5$ **7.** $^-10$ **8.** $^-3$ **9.** $^+15$ **10.** $^-8$

INDEPENDENT PRACTICE

Compare the integers. Use $<$ and $>$.

1. $^-3$ ● $^+3$ **2.** $^-4$ ● 0 **3.** $^-2$ ● $^-5$ **4.** $^-5$ ● $^-3$ **5.** $^+15$ ● $^+10$

6. $^-4$ ● $^-2$ **7.** $^-10$ ● $^-1$ **8.** $^+6$ ● 0 **9.** $^-6$ ● $^-10$ **10.** $^+12$ ● $^-11$

11. $^+6$ ● $^+10$ **12.** $^-17$ ● $^-3$ **13.** $^+22$ ● $^+19$ **14.** $^-2$ ● $^+20$ **15.** $^+13$ ● $^-26$

Order the integers from least to greatest. Use $<$.

16. $0, ^-4, ^+7, ^-3$ **17.** $^-6, ^-3, ^+6, ^-4$ **18.** $^+5, ^-5, ^+2, ^-3$ **19.** $^+2, 0, ^-3, ^-10$

20. $^-3, ^+3, ^+7, ^-2$ **21.** $^-8, ^-2, ^+5, ^-6$ **22.** $^+8, ^+4, ^+2, ^-2$ **23.** $^+20, ^-11, ^+2, ^-9$

Name the opposite of the given integer.

24. $^-2$ **25.** $^+9$ **26.** $^+3$ **27.** $^-11$ **28.** $^+6$ **29.** $^+14$

30. $^-7$ **31.** $^-12$ **32.** $^-19$ **33.** $^-24$ **34.** $^-54$ **35.** $^-98$

Problem-Solving Applications

36. WEATHER Juneau, Alaska, reports a temperature of $^-22°$F. Seattle, Washington, reports a temperature of $22°$F. Which city has the higher temperature?

37. GEOGRAPHY The Smiths' house is 5 ft below sea level. The Jacksons' house is 3 ft above sea level. Whose house is closer to sea level?

38. SCIENCE The countdown for the space shuttle's launch is at minus 10 sec. What integer would represent 10 sec after the launch?

39. SCIENCE On one space shuttle mission, a communications check is scheduled at $^-15$ sec. A fuel check is scheduled at $^-8$ sec. Which check will happen first?

40. ✏️ **WRITE ABOUT IT** Write *sometimes, always,* or *never* for this statement: A negative integer is less than a positive integer. Explain your answer.

Write the number in standard form. (pages 16–17)

1. one thousand, six hundred four ten-thousandths

2. two hundred fifty-five

3. eight hundred twenty-one thousand, two hundred two and four tenths

4. one million, two hundred fifty-two thousand, seven hundred four

Compare the numbers. Write <, >, or =. (pages 18–21)

5. 113.3 ● 11.33

6. 2,281 ● 2,291

7. 82.46 ● 82.461

8. 820 ● 82

9. 4.5 ● 4.50

10. 0.0821 ● 0.821

11. Use < to list the numbers in order from least to greatest:
92.01, 920.1, 92.10. (pages 18–21)

Change the decimal to a fraction. (pages 22–25)

12. 0.1

13. 0.003

14. 0.32

15. 0.097

Write as a decimal. (pages 22–25)

16. $\frac{3}{100}$

17. $\frac{3}{5}$

18. $\frac{73}{100}$

19. $\frac{9}{10}$

20. Jean has $\frac{2}{5}$ of a pizza and Joe has $\frac{1}{4}$. Who has more? (pages 22–25)

21. VOCABULARY A number that shows how many times a base is used as a factor is called a(n) __?__. (page 26)

Find the value. (pages 26–27)

22. 10^4

23. 9^2

24. 15^3

25. 7^2

26. 1^{14}

27. 4^2

28. 3^5

29. 5^4

30. VOCABULARY Integers that are an equal distance from zero on the number line are called __?__. (page 30)

Compare the integers. Use < and >. (pages 30–31)

31. $^-7$ ● $^+1$

32. 0 ● $^-2$

33. $^+4$ ● $^+3$

34. $^-5$ ● $^-10$

1. Which pair of numbers both contain the digit 5 with a value of 5 hundredths?

   A 553.621; 10,579.2
   B 87.35; 92,514.2
   C 36.157; 4,062.058
   D 6,501.35; 792.56

2. Sue has 4 bags of apples weighing 2.3 kilograms, 2.05 kilograms, 3.12 kilograms, and 2.18 kilograms. How would she put the weights in order from lightest to heaviest?

   F 2.05, 2.18, 2.3, 3.12
   G 2.3, 2.18, 3.12, 2.05
   H 3.12, 2.3, 2.18, 2.05
   J 2.18, 2.3, 2.05, 3.12
   K 2.05, 2.18, 3.12, 2.3

3. Which student ran the greatest distance?

Student	Distance
Al	0.35 mile
Betty	$\frac{3}{8}$ mile
Carl	$\frac{2}{5}$ mile
Donna	0.28 mile

   A Al
   B Betty
   C Carl
   D Donna

4. Bradley measured the length of a remote control race car course as $\frac{7}{8}$ of a mile. What is the decimal equivalent of $\frac{7}{8}$?

   F 0.78
   G 0.8
   H 0.875
   J 0.9

5. Which can be expressed as 3^4?

   A $3 \times 3 \times 3 \times 3$
   B $4 \times 4 \times 4$
   C 4×3
   D 3×3
   E Not Here

6. Jennifer's class has been saving pennies to go on a field trip to the Science and Math Museum. The class has collected 100,000 pennies. Express 100,000 using an exponent and the base 10.

   F 5^{10}
   G 10^2
   H 10^5
   J 10^{10}

7. Which letter represents the opposite of $^+3$ on the number line?

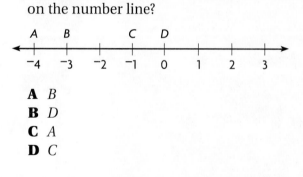

   A B
   B D
   C A
   D C

8. The average depth of the Atlantic Ocean is 11,730 feet, the average depth of the Pacific Ocean is 12,925 feet, and the average depth of the Arctic Ocean is 3,407 feet. Order the integers from least to greatest.

   F $^-11,730, \ ^-12,925, \ ^-3,407$
   G $^+11,730, \ ^+3,407, \ ^+12,925$
   H $^-12,925, \ ^-11,730, \ ^-3,407$
   J $^+12,925, \ ^+11,730, \ ^+3,407$

USING WHOLE NUMBERS

LOOK AHEAD

In this chapter you will solve problems that involve

- using mental-math strategies for addition, subtraction, and multiplication

- using the strategy *guess and check*

- multiplying and dividing whole numbers

- using the order of operations

HISTORY LINK

When was the calculator invented? The answer depends on what type you're talking about.

Study the table.

INVENTION	YEAR	INVENTOR
Slide rule	1632	Oughtred
Adding machine	1642	Pascal
Calculating machine	1833	Babbage
Electronic calculator	1972	Van Tassel

- How many years after Babbage's invention was the electronic calculator invented?

- What do you think is the difference between an adding machine and a calculating machine?

How My Calculator Works

Write a guide that shows how to work your calculator. Decide which operations to include in your guide. Then, perform each operation and record your actions in your guide. Test your completed guide.

PROJECT CHECKLIST

✓ Does your guide include examples?

✓ Does your guide include actions for different operations?

✓ Did you test your guide?

ALGEBRA CONNECTION

VOCABULARY

Commutative
 Property

Associative
 Property

compensation

Mental Math for Addition and Subtraction

How can you mentally find the total number of tickets sold by Marcus, Sandy, and Li?

One way to mentally find a sum is to use addition properties.

Number of Tickets Sold	
Marcus	64
Sandy	78
Li	56

Commutative Property of Addition	**Associative Property of Addition**
Numbers can be added in any order.	Addends can be grouped differently. The sum is always the same.
$64 + 78 + 56 = 56 + 64 + 78$	$(64 + 78) + 56 = (56 + 64) + 78$

EXAMPLE 1 Find the sum. $5 + 9 + 21 + 25$

$$
\begin{aligned}
5 + 9 + 21 + 25 &= 5 + 25 + 9 + 21 &&\leftarrow \textit{Commutative Property} \\
&= (5 + 25) + (9 + 21) &&\leftarrow \textit{Associative Property} \\
&= 30 + 30 &&\leftarrow \textit{Use mental math.} \\
&= 60
\end{aligned}
$$

So, the sum is 60.

LANGUAGE LINK

One definition of *compensation* is "something given to balance something else." If you used compensation in the following problem, what would be given and where would it come from?

$29 + 46$

A strategy you can use for addition or subtraction is **compensation**. You can change one number to a multiple of ten and then adjust the other addend to keep the balance.

EXAMPLE 2 Use compensation to solve.

A. $18 + 24$

$$
\begin{aligned}
18 + 24 &= (18 + 2) + (24 - 2) \\
&= 20 + 22 \\
&= 42
\end{aligned}
$$

Think: $18 + \blacksquare = 20$
Add 2 and subtract 2.
Use mental math.

B. $43 - 31$

$$
\begin{aligned}
43 - 31 &= (43 - 1) - (31 - 1) \\
&= 42 - 30 \\
&= 12
\end{aligned}
$$

Think: $31 - \blacksquare = 30$
Subtract 1 from each number.
Use mental math.

GUIDED PRACTICE

Find the missing addend. Name the property used. Then find the
value of the addition expression.

1. $15 + 4 + 5 = 15 + \blacksquare + 4$ **2.** $6 + 17 + 13 = 6 + (\blacksquare + 13)$ **3.** $24 + 11 + 16 = 24 + 16 + \blacksquare$

4. $12 + 11 + 8 + 9 = 12 + \blacksquare + 11 + 9$ **5.** $24 + 6 + 7 + 23 = (24 + 6) + (7 + \blacksquare)$

Use compensation to solve. Show your work.

6. $19 + 11$ **7.** $38 + 12$ **8.** $17 + 28$ **9.** $16 + 35$

10. $29 - 12$ **11.** $32 - 11$ **12.** $45 - 24$ **13.** $56 - 35$

INDEPENDENT PRACTICE

Use mental math to add.

1. $8 + 16 + 12$ **2.** $19 + 10 + 11$ **3.** $15 + 21 + 9$ **4.** $31 + 26 + 19$

5. $25 + 3 + 15 + 17$ **6.** $22 + 15 + 8 + 15$ **7.** $9 + 20 + 31 + 10$ **8.** $32 + 26 + 18 + 4$

9. $12 + 21 + 38 + 9$ **10.** $23 + 21 + 17 + 19$ **11.** $16 + 33 + 24 + 17$ **12.** $37 + 26 + 13 + 14$

Use compensation to add.

13. $17 + 24$ **14.** $19 + 23$ **15.** $14 + 27$ **16.** $18 + 26$ **17.** $16 + 28$

18. $28 + 24$ **19.** $48 + 32$ **20.** $29 + 32$ **21.** $26 + 37$ **22.** $37 + 33$

Use compensation to subtract.

23. $44 - 21$ **24.** $48 - 37$ **25.** $35 - 14$ **26.** $28 - 12$ **27.** $37 - 23$

28. $54 - 22$ **29.** $58 - 35$ **30.** $64 - 48$ **31.** $66 - 15$ **32.** $67 - 32$

Problem-Solving Applications

Use mental math to solve. For Problems 33–36, use the table below.

33. CONSUMER MATH How many CDs were bought in all?

34. NUMBER SENSE Whose purchases can be
combined to total 30 CDs?

35. If Nick gave 12 CDs to his cousin, how
many would he have left?

36. STATISTICS How many CDs would Brenda, Selena, Ricardo, and
Nick each have if they shared their CDs equally?

37. ✏ WRITE ABOUT IT Explain how to use
compensation to add.

CDs Bought			
Brenda	Selena	Ricardo	Nick
12	17	18	25

MORE PRACTICE Lesson 2.1, page H43

PROBLEM-SOLVING STRATEGY

Using Guess and Check to Add and Subtract

Sometimes it takes detective work to find the solution to a problem.

During an annual reading program, the sixth and seventh graders read 1,020 books. The sixth graders read 110 more books than the seventh graders. How many books did the sixth graders read? How many books did the seventh graders read?

PROBLEM SOLVING
• **Understand**
• **Plan**
• **Solve**
• **Look Back**

UNDERSTAND   What are you asked to find?

What facts are given?

PLAN   What strategy will you use?

You can use the strategy *guess and check*. Make a guess that satisfies the first clue. Then check your guess to see if it satisfies the second clue.

SOLVE   How will you solve the problem?

Make a table to keep track of your guesses and checks for the first and second clues. Try to guess in an organized way so that your guesses get closer to the answer.

GUESS		CHECK	
Sixth Graders	**Seventh Graders**	**Clue 1: The sum is 1,020.**	**Clue 2: The difference is 110.**
520	500	1,020	20	← *too low*
600	420	1,020	180	← *too high*
565	455	1,020	110	← *satisfies both clues*

LOOK BACK   The difference for the first guess was too low.

What changes were made for the second guess?

What if . . . the sixth graders had read 120 more books than the seventh graders? How many books would the sixth graders have read? How many books would the seventh graders have read?

PRACTICE

Guess and check to solve. Make a table to record your guesses.

1. Douglas bought a total of 50 cans of orange and grape sodas. He bought 12 more cans of grape than of orange. How many of each kind did he buy?

2. The perimeter of a rectangular garden is 40 ft. The length is 6 ft more than the width. What is the length and width of the garden?

3. Kim sold a total of 36 red and blue tickets. She sold 6 more red tickets than blue tickets. How many red and blue tickets did she sell?

4. The Brown Bears soccer team played a total of 20 games. They won 4 more than they lost, and they tied in 2 games. How many games did they win?

MIXED APPLICATIONS

Solve.

CHOOSE a strategy and a tool.
- Guess and Check
- Write an Equation
- Act It Out
- Solve a Simpler Problem
- Find a Pattern
- Make a Table

Paper/Pencil   Calculator   Hands-On   Mental Math

5. Rosalia talks on the phone with Teresa, Ronald, and Jack. If each of the four friends has one conversation with each other friend, how many calls are made?

6. Ned spent a total of $23.45. He bought a ticket to a basketball game for $6.50, food for $6.95, and some T-shirts for $5.00 each. How many T-shirts did he buy?

7. A bus travels 55 mi per hour. Abe starts out by bus at 8:00 A.M. How many miles does he travel if he arrives at his destination at 11:00 A.M.?

8. Tim has earned $2,100. To buy a used car, he needs twice that amount, plus $500. How much does the car cost?

9. Peter and his two brothers are collecting stamps. Peter has twice as many as his older brother, who has 21 stamps. Peter has three times as many as his younger brother. How many stamps do they have in all?

10. Mira started an exercise routine and jogged 4 mi each day the first week. She increased her daily routine by 2 mi each week, but every third week she added only 1 mi. How many miles did she jog each day in week 5?

11. Martha saved $0.50 from her allowance one week. Then she saved $0.50 more each week than she had the week before. How many weeks did it take to save a total of $10.50?

12. Ira has 68 baseball cards. This is twice as many as Saul has, plus 2. How many cards does Saul have?

13. ✏️ **WRITE ABOUT IT** Write a problem that can be solved by *guessing and checking*.

What You'll Learn
How to use properties to solve multiplication problems

Why Learn This?
To mentally solve multiplication problems such as determining the number of party favors

VOCABULARY

Distributive Property

Commutative Property

Associative Property

Identity Property of One

Property of Zero

ALGEBRA CONNECTION

Mental Math for Multiplication

Erica is making 32 party favors, with 12 treats in each favor. How can you use mental math to determine the number of treats she will need?

You can use the Distributive Property to make an easier problem.

DISTRIBUTIVE PROPERTY
A factor can be thought of as the sum of addends. Multiplying the sum by a number is the same as multiplying each addend by the number and then adding the products.

Use the Distributive Property to find 12×32.

$$12 \times 32 = 12 \times (30 + 2)$$
$$= (12 \times 30) + (12 \times 2)$$
$$= 360 + 24$$
$$= 384$$

So, Erica will need 384 treats.

EXAMPLE 1 Use the Distributive Property to solve 8×15.

$$8 \times 15 = 8 \times (10 + 5)$$
$$= (8 \times 10) + (8 \times 5)$$
$$= 80 + 40$$
$$= 120$$

Sometimes the Distributive Property is helpful when a factor is repeated.

EXAMPLE 2 Use the Distributive Property to solve $12 \times 2 + 12 \times 5$.

$$12 \times 2 + 12 \times 5 = 12 \times 2 + 12 \times 5 \leftarrow 12 \text{ is a repeated factor.}$$
$$= 12 \times (2 + 5)$$
$$= 12 \times (7)$$
$$= 84$$

• How could you use the Distributive Property to solve $24 \times 3 + 24 \times 7$?

GUIDED PRACTICE

Use the Distributive Property to make an easier problem.

1. 4×33 **2.** 6×25 **3.** 7×24

4. 5×42 **5.** 12×14 **6.** 9×85

Use the Distributive Property to solve.

7. 12×17 **8.** 9×36 **9.** 5×29 **10.** 11×43

11. $4 \times 33 + 4 \times 7$ **12.** $6 \times 25 + 9 \times 6$ **13.** $7 \times 24 + 4 \times 24$

Other Properties

You can use the Commutative and Associative Properties to solve multiplication problems.

COMMUTATIVE PROPERTY OF MULTIPLICATION
Factors can be multiplied in any order without changing the product.
$6 \times 7 = 7 \times 6$

• Use the Commutative Property to rewrite 22×48.

ASSOCIATIVE PROPERTY OF MULTIPLICATION
Factors can be grouped in any way without changing the product.
$(2 \times 9) \times 5 = 2 \times (9 \times 5)$

• Use the Associative Property to rewrite $16 \times (5 \times 3)$.

EXAMPLE 3 A party store has 8 party favor bags on each of 3 shelves. Each party favor bag contains 5 items. How many items are there altogether?

$$8 \times 3 \times 5 = 8 \times 5 \times 3 \quad \textit{Commutative Property}$$
$$= (8 \times 5) \times 3 \quad \textit{Associative Property}$$
$$= 40 \times 3$$
$$= 120$$

So, there are 120 items.

• Explain how you would use the Commutative and Associative Properties to find the product $12 \times 7 \times 5$.

REMEMBER:

You can use basic facts and mental math to multiply numbers that end with zeros.

70×6

Multiply nonzero numbers.

$7 \times 6 = 42$

Add the same number of zeros that are in the factors to the product.

So, $70 \times 6 = 420$.

See Key Skill 22, page H12.

Talk About It

- When you use the Commutative Property to rewrite a multiplication problem, does this change the product?

- How does using the Commutative Property help you mentally solve a multiplication problem?

- How does using the Associative Property help you mentally solve a multiplication problem?

Two other properties that may help you multiply mentally are the Identity Property of One and the Property of Zero.

IDENTITY PROPERTY OF ONE
The product of any factor and 1 is the factor.
$17 \times 1 = 17$

EXAMPLE 4 Use mental math to find the product $4 \times 25 \times 1$.

$4 \times 25 \times 1 =$

$\qquad 100 \times 1 = 100 \leftarrow$ *Identity Property of One*

- Use mental math to find the product $20 \times 40 \times 1 \times 3$.

PROPERTY OF ZERO
The product of any factor and zero is zero.
$99 \times 0 = 0$

EXAMPLE 5 Use mental math to find the product $9 \times (8 - 8)$.

$9 \times (8 - 8) =$

$\qquad 9 \times 0 = 0 \leftarrow$ *Property of Zero*

- Use mental math to find $182 \times 0 + 182 \times 1$.

INDEPENDENT PRACTICE

Find the missing factor.

1. $48 \times 5 = (40 \times 5) + (\blacksquare \times 5)$

2. $25 \times \blacksquare = (25 \times 30) + (25 \times 6)$

3. $55 \times 42 = (\blacksquare \times 40) + (55 \times 2)$

4. $6 \times 99 = (6 \times \blacksquare) + (6 \times 9)$

5. $32 \times 14 = (\blacksquare \times 10) + (32 \times 4)$

6. $78 \times 9 = (70 \times 9) + (\blacksquare \times 9)$

Use the Distributive Property and mental math to solve.

7. 17×9 **8.** 28×8 **9.** 32×6 **10.** 16×21

11. 24×7 **12.** 19×14 **13.** 45×11 **14.** 12×35

15. $17 \times 5 + 5 \times 19$ **16.** $12 \times 9 + 22 \times 12$ **17.** $6 \times 19 + 30 \times 6$

Write a number sentence that illustrates each property of multiplication.

18. Property of Zero   **19.** Commutative Property

20. Distributive Property   **21.** Associative Property

22. Identity Property of One

Find the missing factor. Name the property used.

23. $26 \times 37 = \blacksquare \times 26$ **24.** $42 \times \blacksquare = 0$ **25.** $67 \times \blacksquare = 67$

26. $(3 \times 5) \times \blacksquare = 3 \times (5 \times 2)$ **27.** $2 \times (4 \times 9) = (4 \times \blacksquare) \times 2$ **28.** $\blacksquare \times 13 = (6 \times 4) + (6 \times 9)$

Use mental math to find the product.

29. $9 \times 5 \times 2$ **30.** $4 \times 8 \times 2$ **31.** 3×21 **32.** $2 \times 6 \times 5$

33. $3 \times 10 \times 7$ **34.** 6×15 **35.** 12×6 **36.** $12 \times 4 \times 5$

37. 13×4 **38.** $25 \times 3 \times 2$ **39.** 8×28 **40.** 120×4

Problem-Solving Applications

Use mental math to solve.

41. A toy store has 6 boxes on each of 4 shelves. Each box has 10 items in it. How many items altogether are on the 4 shelves?

42. For a fundraiser, 25 students are making muffins. Each student bakes 16 muffins. How many muffins do they make altogether?

43. **LOGICAL REASONING** Raoul sells three kinds of cookies. His average sales are 24, 26, and 28 cookies for the three kinds each day. For each kind, he wants to know the total he is likely to sell in the next 14 days. Write number sentences that use the properties of multiplication and addition to simplify his calculations.

44. **WRITE ABOUT IT** What properties of multiplication can you use to help you with mental math?

Multiplication and Division

What You'll Learn
How to multiply and divide whole numbers

Why Learn This?
To solve problems such as finding the total amount earned by ticket sales

Sometimes you can use shortcuts to multiply numbers. For example, to find 32×100, you know that you can just put two zeros after 32 to get 3,200. This activity will show you a different way to multiply two whole numbers when both of the factors are numbers between 10 and 20.

ACTIVITY

Find 16×19.

• Find the sum of the first factor and the ones digit of the second factor.	$16 + 9 = 25$
• Add a zero to that sum.	250
• Find the product of the ones digits of the first and second factors.	$6 \times 9 = 54$
• Find the sum.	$250 + 54 = 304$

So, $16 \times 19 = 304$.

• Find 14×15.

CULTURAL LINK

The doubling method below is known as the Russian Peasant Method.

$26 \times 35 = 910$

Halve	Double
26	~~35~~
13	70
6	~~140~~
3	280
1	$+ 560$
	910

Find the even numbers in the left column and cross out the numbers directly across from them in the right column. Add the remaining doubles.

Use this method to find 16×22.

For most multiplication problems, you find the product by multiplying the two factors.

EXAMPLE 1 Find 132×24.

$$
\begin{array}{r}
132 \\
\times\ \ 24 \\
\hline
528 \\
+\ 2640 \\
\hline
3,168
\end{array}
$$

Multiply by the ones. 4×132
Multiply by the tens. 20×132
Use the zero as a placeholder. Add.

• Find 128×18.

You can omit the zero placeholders when you multiply. Just be careful to line up your products correctly.

	Correct:	Wrong:
	132	132
	$\times\ \ 24$	$\times\ \ 24$
	528	528
	$+264$	$+264$

Sometimes you have to use multiplication to solve a problem.

EXAMPLE 2 Season tickets to an amusement park are on sale for $125. On the first day of the sale the amusement park sold 12,383 tickets. How much money did the amusement park receive for season tickets that day?

$$
\begin{array}{r}
12,383 \\
\times\quad 125 \\
\hline
61\ 915 \\
247\ 66 \\
+1\ 238\ 3 \\
\hline
1,547,875
\end{array}
$$

Multiply by the ones. 5 × 12,383
Multiply by the tens. 20 × 12,383
Multiply by the hundreds. 100 × 12,383
Add.

So, the amusement park received $1,547,875.

- How much will the amusement park receive if it sells 15,225 season tickets?

GUIDED PRACTICE

Find the product.

1. 13
× 14

2. 380
× 52

3. 2,382
× 12

4. 962
× 40

5. 849
× 413

6. 7,658
× 111

7. 1,453
× 721

8. 16,225
× 258

9. 123
× 12

10. 918
× 27

11. 342
× 521

12. 189
× 108

Solving Division Problems

Liz is redesigning the school cafeteria to seat exactly 420 students. Each table in her design will seat 12 students. How many tables will she need?

You can use division to find the number of tables.

$$
\begin{array}{r}
35 \\
12\overline{)420} \\
-36\downarrow \\
\hline
60 \\
-60 \\
\hline
0
\end{array}
$$

Since 12 is greater than 4, the first digit of the quotient will be in the tens place. Divide the 42 tens.
Think: *12 × 3 = 36*
Bring down the 0 ones. Divide the 60 ones.
Think: *12 × 5 = 60*

So, Liz will need 35 tables in her design.

- What would happen if there were 424 students? How many tables would Liz need?

Sometimes a division problem has a zero in the quotient.

EXAMPLE 3 A school collected 2,568 newspapers. The newspapers were bundled in packages of 25. How many packages of newspapers did the school bundle?

$$\begin{array}{r} 102\ r18 \\ 25\overline{)2{,}568} \\ -25 \\ \hline 06 \\ -0 \\ \hline 68 \\ -50 \\ \hline 18 \end{array}$$

Since 25 is greater than 2, the first digit will be in the hundreds place. Divide the 25 hundreds. **Think:** $25 \times 1 = 25$

Bring down the 6 tens.
Since $25 > 6$, write 0 in the quotient.

Bring down the 8 ones.
Divide the 68 ones.
Think: $25 \times 2 = 50$

So, the school bundled 102 packages of newspapers.

Talk About It

- [CRITICAL THINKING] In Example 3, why will the school bundle 102 packages of newspapers instead of 103 packages of newspapers?

- When do you write a zero in the quotient?

ANOTHER METHOD You can use some calculators to show the whole number remainder.

145  8 $\boxed{=}$ $\boxed{= 18\ ^{R}1}$

- How does a calculator show the remainder when you divide 145 by 8 using the ÷ key?

In Example 3 the quotient had a remainder. You can express a remainder with an *r*, or you can express it as a fraction. The quotient in Example 3 can be expressed as 102 r18, or it can be expressed as $102\frac{18}{25}$.

- Find $13\overline{)254}$. Write the remainder as a fraction.

INDEPENDENT PRACTICE

Find the product.

1.
$$\begin{array}{r} 27 \\ \times\ 11 \\ \hline \end{array}$$

2.
$$\begin{array}{r} 140 \\ \times\ 25 \\ \hline \end{array}$$

3.
$$\begin{array}{r} 364 \\ \times\ 30 \\ \hline \end{array}$$

4.
$$\begin{array}{r} 3{,}473 \\ \times\ 46 \\ \hline \end{array}$$

5.
$$\begin{array}{r} 298 \\ \times\ 89 \\ \hline \end{array}$$

6.
$$\begin{array}{r} 327 \\ \times\ 123 \\ \hline \end{array}$$

7.
$$\begin{array}{r} 5{,}233 \\ \times\ 238 \\ \hline \end{array}$$

8.
$$\begin{array}{r} 462 \\ \times\ 140 \\ \hline \end{array}$$

9.
$$\begin{array}{r} 5{,}470 \\ \times\ 240 \\ \hline \end{array}$$

10.
$$\begin{array}{r} 4{,}904 \\ \times\ 196 \\ \hline \end{array}$$

Find the quotient.

11. $4\overline{)56}$ **12.** $8\overline{)432}$ **13.** $12\overline{)144}$ **14.** $4\overline{)412}$ **15.** $24\overline{)626}$

16. $16\overline{)1,664}$ **17.** $14\overline{)7,044}$ **18.** $19\overline{)1,068}$ **19.** $20\overline{)6,020}$ **20.** $34\overline{)15,912}$

21. $18\overline{)2,250}$ **22.** $22\overline{)1,829}$ **23.** $26\overline{)2,314}$ **24.** $68\overline{)24,820}$ **25.** $52\overline{)728}$

Find the quotient. Write the remainder as a fraction.

26. $5\overline{)49}$ **27.** $12\overline{)245}$ **28.** $34\overline{)855}$ **29.** $14\overline{)3,385}$ **30.** $36\overline{)7,225}$

31. $15\overline{)386}$ **32.** $32\overline{)428}$ **33.** $18\overline{)5,720}$ **34.** $20\overline{)7,349}$ **35.** $45\overline{)2,270}$

Problem-Solving Applications

36. CONSUMER MATH Bob's parents bought an entertainment center for $1,176. If they pay for it with 14 equal monthly payments, how much will each payment be?

37. CONSUMER MATH Tasha's family lives in an apartment. Tasha's parents pay $725 each month for rent. How much rent do they pay in one year?

38. BUSINESS Lincoln Middle School had a car wash to raise money for the school. They charged $3 for every car. If they washed 23 cars, how much money did they earn?

39. CALCULATOR Kristie used this calculator sequence: 298 ÷R 6 = [= 49 R4] Rewrite the answer, with the remainder expressed as a fraction.

40. CAREER Robbie's father earns $42,600 a year as an engineer. How much does he earn each month?

41. ✏ WRITE ABOUT IT Write a word problem that can be solved with multiplication or division.

Mixed Review and Test Prep

Round to the nearest thousand.

42. 3,789 **43.** 12,554 **44.** 8,432

Order the integers from least to greatest. Use $<$.

45. $^-3, 4, ^-2, 5$ **46.** $0, ^-1, 2, ^-4$ **47.** $^-3, ^-6, 7, 6$

48. CONSUMER MATH Teresa is buying perfume for her mother. The prices of the four perfumes she liked are given. She bought the least expensive one. Which one did she buy?

   A $15.89   **B** $16.00
   C $15.19   **D** $15.99

49. MEASUREMENT How many meters are in 1,500 millimeters?

   F 0.15 m   **G** 1.5 m
   H 15 m   **J** 150 m

MORE PRACTICE Lesson 2.4, page H44

Technology Link

In *Mighty Math Calculating Crew,* work with Captain Nick Knack in the *Intergalactic Trader* to multiply and divide whole numbers. Use Grow Slide Level V.

LAB ACTIVITY

What You'll Explore
How to find a solution using order of operations

What You'll Need
calculator

ALGEBRA CONNECTION

Order of Operations

Marco's older brother, Drake, was working on his homework. At the top of the page, he wrote, "Please Excuse My Dear Aunt Sally." Marco didn't understand. He said, "We don't have an aunt named Sally."

Drake said, "You'll see why I wrote this on my paper."

ACTIVITY 1

Explore

- Use paper and pencil to find the value of $3 + 2 \times 6 - 2$.

- How does your answer compare with your partner's answer?

When you find the value of an expression with more than one operation, you need to use the **order of operations**.

1. Perform operations in parentheses.
2. Clear exponents.
3. Multiply and divide from left to right.
4. Add and subtract from left to right.

- Use the order of operations to evaluate $13 - 4 \div 2 + 6$.

Think and Discuss

- Why do you think Marco's brother had "Please Excuse My Dear Aunt Sally" at the top of his homework page?

- What do the underlined letters in "Please Excuse My Dear Aunt Sally" represent?

- What order would you use to find the value of $26 + 2 - 1 \times (24 \div 2)$? Explain why the order is important.

Try This

- Tell the order in which you would perform each of the operations in the expression $120 - 14 + 4^2 \times 3$. Then find the value of the expression.

Calculator Activities, page H33

You can use a calculator to find the value of expressions with more than one operation. Some calculators use an **algebraic operating system (AOS)**. These calculators automatically follow the order of operations.

ACTIVITY 2

Explore

- Use your calculator to find the value of $8 \div 2 + 6 \times 3 - 4$.

- Following the order of operations, use paper and pencil to find the value of $8 \div 2 + 6 \times 3 - 4$.

Think and Discuss

- How does the value you calculated for $8 \div 2 + 6 \times 3 - 4$ compare with the value you got by using paper and pencil? Does your calculator use AOS?

To find the value of an expression with a calculator that does not use AOS, follow the order of operations or use the memory keys.

Follow the order of operations to find the value of the expression $2 + 6 \times 3^2 - 4$.

3 ⊠ 3 ⊠ 6 ⊞ 2 ⊟ 4 ⊜ [52.]

Use the memory keys to find the value of the expression $9^2 + 6 \div 2 \times 4$.

9 ⊠ 9 M+ 6 ⊘ 2 ⊠ 4 M+ MRC [93.]

- When you enter values into a calculator that does not have AOS, how do you know which values to enter first?

- Would you use memory keys or the order of operations to calculate the value of $4 \times 12 - 3 + 8 \times 2$?

Try This

- A calculator shows the display [9.] as the value of $12 + 15 \div 3$. Does the calculator use AOS? Explain.

Technology Link

You can practice the order of operations by evaluating expressions in E-Lab, Activity 2. Available on CD-ROM and on the Internet at **www.hbschool.com/elab**

Using Estimation

Did you know that the longest throw of a flying disc was 656 ft 2 in., made by Scott Stokely? Mark, Kevin, and Deanna are throwing their disc, trying to equal the distance thrown by Scott Stokely. They wrote the distances of their throws in a table.

You can use rounding to find a sum when you don't need an exact answer.

Name	Distance
Mark	145 ft
Kevin	148 ft
Deanna	151 ft

$$
\begin{array}{rcl}
145 & \rightarrow & 150 \\
148 & \rightarrow & 150 \\
+151 & \rightarrow & +150 \\
\hline
 & & 450
\end{array}
$$
Round each number to the nearest 10.

The estimate, 450, is not close to 656 ft, so the total distance is not about the same as the record.

You can use **clustering** to estimate a sum when all the addends are about the same.

EXAMPLE 1 Use clustering to estimate the sum.

$$
\begin{array}{r}
1,802 \\
2,182 \\
+1,999 \\
\end{array}
$$
The three addends are close to 2,000.

$3 \times 2,000 = 6,000$ *Multiply.*

So, the sum is about 6,000.

You can use rounding to estimate a difference.

EXAMPLE 2 Use rounding to estimate the difference.

$$
\begin{array}{r}
928 \\
-616 \\
\hline
\end{array}
$$

Round to the nearest hundred.   *Round to the nearest ten.*

$$
\begin{array}{r}
900 \\
-600 \\
\hline
300
\end{array}
\qquad
\begin{array}{r}
930 \\
-620 \\
\hline
310
\end{array}
$$

So, the difference is 300 when you round to the nearest hundred, and it is 310 when you round to the nearest ten.

• Use rounding to estimate $1,225 - 782$.

GUIDED PRACTICE

Use clustering to estimate the sum.

1. 2,940
3,100
+2,834

2. 1,040
1,300
+1,422

3. 4,480
4,100
+3,967

4. 5,449
4,869
+4,834

Round to the nearest ten. Then find the sum or difference.

5. 723
+819

6. 420
+388

7. 998
−114

8. 375
−132

Round to the nearest hundred. Then find the sum or difference.

9. 184
+518

10. 244
+238

11. 667
−133

12. 855
−268

Estimating Products and Quotients

You can use rounding to estimate a product.

> **EXAMPLE 3** The school is holding a talent show in the cafeteria. The students set up 21 rows of 32 seats. About how many programs should the school print for the talent show?
>
> **Think:** Are you asked to find an exact answer or an estimate?
>
> $\begin{array}{r} 32 \\ \times 21 \end{array} \rightarrow \begin{array}{r} 30 \\ \times 20 \end{array}$ *Estimate by rounding each factor to the nearest ten.*
>
> $\begin{array}{r} 30 \\ \times 20 \\ \hline 600 \end{array}$ *Multiply.*
>
> So, the school will need to print about 600 programs.
>
> • Estimate the product. 550×82

Talk About It

• When you rounded each factor in Example 3, did you round up to the next ten or down?

• **CRITICAL THINKING** Would the exact answer be a number greater than the estimated answer or less than the estimated answer? How do you know?

• Do you think it would be better for the estimated answer to be less than or greater than the exact answer? Explain.

SCIENCE LINK

In the vastness of space, distances between objects are enormous. Even within our own solar system, distances between planets are huge. Earth's orbit is about 151,000,000 km from the sun, and Mars's orbit is about 228,000,000 km from the sun. Rounding these distances to the nearest ten million, estimate the distance between Earth's orbit and Mars's orbit.

When you estimate a quotient, you can use rounding or compatible numbers. **Compatible numbers** divide without a remainder, are close to the actual numbers, and are easy to compute mentally.

EXAMPLE 4 Loughman Middle School is starting a recycling program. The school has set a goal of collecting 1,545 lb of material to be recycled. The school has 36 homerooms. About how many pounds of recyclable material should the students in each homeroom collect?

Using rounding:

$1{,}545 \div 36$

$1{,}500 \div 40$ *Round the dividend and the divisor.*

$1{,}500 \div 40 = 37 \text{ r}20$ *Divide.*

So, the students in each homeroom should collect about 38 lb.

Using compatible numbers:

$1{,}545 \div 36$

$1{,}600 \div 40$ *4 is compatible with 16.*

$1{,}600 \div 40 = 40$ *Divide.*

So, the students in each homeroom should collect about 40 lb.

- Which estimate is easier to compute with mentally?

- Is $1{,}488 \div 36$ easier to estimate using rounding or compatible numbers?

INDEPENDENT PRACTICE

Estimate the sum or difference.

1. 1,700
    2,008
    +2,324

2. 293
    348
    +343

3. 3,643
    +4,211

4. 5,765
    5,948
    +6,324

5. 6,902
    +7,219

6. 389
    − 43

7. 152
    −138

8. 3,556
    −3,339

9. 9,123
    −6,512

10. 4,687
    −1,022

Estimate the product.

11.	36 × 9	12.	43 × 7	13.	59 × 33	14.	48 × 29	15.	940 × 66
16.	364 × 12	17.	53 × 41	18.	590 × 335	19.	482 × 299	20.	846 × 562

Estimate the quotient.

21. $268 \div 5$ 22. $321 \div 4$ 23. $1,544 \div 28$ 24. $4,156 \div 64$ 25. $8,429 \div 39$

26. $1,844 \div 22$ 27. $3,575 \div 56$ 28. $4,239 \div 670$ 29. $6,435 \div 529$ 30. $9,433 \div 309$

31. $9,200 \div 8$ 32. $802 \div 23$ 33. $5,867 \div 18$ 34. $1,784 \div 178$ 35. $3,165 \div 211$

Problem-Solving Applications

36. **CONSUMER MATH** Rudy had $64 to spend. He spent $18 on a haircut, $22 on shoes, and $7 for a movie ticket. About how much does Rudy have left?

37. **BUSINESS** A garden shop sells annuals, shrubs, and trees. They have 78 annuals, 32 shrubs, and 18 trees. About how many plants do they have?

38. **HOBBIES** Maria collected souvenirs from each place she visited on her trip. She collected 156 colored leaves and 124 rocks. About how many more colored leaves did she collect than rocks?

39. **ENTERTAINMENT** The theater of the Natural Science and History Museum was filled to capacity for its 276 shows this fall. The theater holds 35 people. About how many people attended the shows?

40. The Forest Service estimates that about 3,645 people visited the Cradle of Forestry in the past 31 days. About how many people visited the attraction each day?

41. **WRITE ABOUT IT** Is it easier to use rounding or compatible numbers to estimate $756 \div 44$? Explain.

Mixed Review and Test Prep

Tell which numbers are equivalent.

42. 4.2, 4.25, 4.20   43. 1.45, 1.450, 1.405   44. 72.04, 72.40, 72.4

Write as a decimal.

45. $\frac{1}{2}$ 46. $\frac{7}{8}$ 47. $\frac{3}{4}$ 48. $\frac{1}{4}$ 49. $\frac{5}{8}$

50. **SPORTS** Joe bought 3 basketballs for $22.99 each and a net for $5.99. Which number sentence can be used to find c, the total cost?

  A $c = 3 \times (22.99 + 5.99)$
  B $c = (3 \times 22.99) + 5.99$
  C $c = (3 \times 5.99) + 22.99$
  D $c = (3 + 22.99) + 5.99$

51. **ALGEBRA** Paul wants to save $60 each year. If Paul saves $60 for n years, which expression can he use to determine the total amount of money he has saved?

  F $60 \times n$ **G** $60 \div n$
  H $60 + n$ **J** $60 - n$

1. **VOCABULARY** When you change one number to a multiple of ten and adjust the other addend, you are using __?__.
(page 36)

Find the sum or difference by using mental math. (pages 36–37)

2. $28 + 45 + 32$ 　　3. $29 + 84$ 　　4. $93 - 26$ 　　5. $77 - 23$

Solve. (pages 38–39)

6. The perimeter of a rectangular yard is 360 ft. The length is 20 ft more than the width. What are the dimensions of the yard?

7. Jane has a bag with 48 marbles. She has 12 more red marbles than yellow ones. How many does Jane have of each color?

8. Rob sold 62 game tickets. He sold 8 more student tickets than adult tickets. How many student tickets did he sell? adult tickets?

9. The sum of Dustin's and Nick's ages is 30. Nick is 6 years older. What are their ages?

Use mental math to multiply. (pages 40–43)

10. 9×510 　　11. $22 \times 2 \times 0$ 　　12. 24×5 　　13. $12 \times 6 \times 5$

14. $15 \times 6 \times 2$ 　　15. $5 \times 9 \times 3$ 　　16. $12 \times 7 \times 5$ 　　17. 5×27

Find the product. (pages 44–47)

18.
$$\begin{array}{r} 64 \\ \times\ 23 \\ \hline \end{array}$$

19.
$$\begin{array}{r} 2{,}813 \\ \times\ \ \ \ 85 \\ \hline \end{array}$$

20.
$$\begin{array}{r} 332 \\ \times\ 132 \\ \hline \end{array}$$

21.
$$\begin{array}{r} 8{,}235 \\ \times\ \ \ \ 44 \\ \hline \end{array}$$

Find the quotient. (pages 44–47)

22. $8\overline{)584}$ 　　23. $92\overline{)5{,}996}$ 　　24. $71\overline{)14{,}555}$ 　　25. $19\overline{)1{,}045}$

26. **VOCABULARY** Numbers that divide without a remainder, that are close to the actual numbers, and that are easy to compute mentally are called __?__. (page 52)

Estimate. (pages 50–53)

27.
$$\begin{array}{r} 8{,}271 \\ +\ 7{,}834 \\ \hline \end{array}$$

28.
$$\begin{array}{r} 1{,}802 \\ -\ \ \ 959 \\ \hline \end{array}$$

29.
$$\begin{array}{r} 536 \\ \times\ 54 \\ \hline \end{array}$$

30. $6\overline{)431}$

1. Ryan needs to know the decimal equivalent of $\frac{3}{8}$ to find the cost of 3 slices of pizza. What is it?

   A 0.4
   B 0.375
   C 0.67
   D 0.875

2. Three cities reported the following temperatures: ⁻10°F, 39°F, and 25°F. Which is the correct order of the temperatures from lowest to highest?

   F 39°F, 25°F, ⁻10°F
   G 25°F, ⁻10°F, 39°F,
   H ⁻10°F, 25°F, 39°F
   J ⁻10°F, 39°F, 25°F

3. The attendance for a three-day book fair was 82 students, 51 students, and 98 students. How many students attended the book fair?

   A 200
   B 220
   C 231
   D 240

4. Which expression is equivalent to $(51 + 32) + 46$?

   F $(51 + 46) + (32 + 46)$
   G $51 + (32 + 46)$
   H $88 + 51$
   J $19 + 46$

5. The Hawks soccer team played a total of 18 games. They won 5 more than they lost, and they tied in one game. How many games did they win?

   A 11
   B 13
   C 14
   D 23

6. Which expression is equivalent to 7×53?

   F $(3 \times 7) + (3 \times 50)$
   G $(50 \times 7) + (50 \times 3)$
   H $(7 \times 30) + (7 \times 50)$
   J $(7 \times 50) + (7 \times 3)$

7. Workers at a supply company are packing 540 notebooks for an order. Each carton holds 15 notebooks. How many cartons are needed to package the order?

   A 18
   B 22
   C 29
   D 32
   E 36

8. During a holiday weekend, a theater was sold out for all 18 shows. The theater holds 376 people. What is a good estimate of the number of people who attended the weekend shows?

   F Less than 4,000 people
   G Between 4,000 and 6,000 people
   H Between 6,000 and 8,000 people
   J More than 8,000 people

9. What is the best estimate of the total number of cans collected by students in Grades 3 through 6?

Grade	Number of Cans Collected
3	117
4	376
5	483
6	234

   A 800
   B 1,000
   C 1,200
   D 1,400
   E Not Here

USING DECIMALS

LOOK AHEAD

In this chapter you will solve problems that involve

• adding, subtracting, multiplying, and dividing decimals

CONSUMER LINK

The five top-selling software games in the United States in 1996 were as follows:

GAME	COMPANY
Myst	Broderbund
Top Ten Pack	Electronic Arts
Warcraft II	Davidson
Links Pro	Access
X-Wing Collector's CD	Lucas Arts

• What are your five favorite games?

Selecting Computer Programs

Suppose your class has been given $500 to buy computer programs. Make a list of computer programs and their costs. Select the ones you would buy and then find the total cost. Will you have any money left? If so, how much? Then share your list with the class. Compare lists to see who spent the most money without going over $500.

PROJECT CHECKLIST

✓ Did you list computer programs and their costs?

✓ Did you find the total cost?

✓ Did you compare lists with the class?

57

Adding Decimals

Have you ever noticed how decimals are arranged on a cash register receipt?

When you add decimals, you align the decimal points.

```
Store #16188    10-29-98

Iced Tea
Folder            1.05
Paper             0.89
Greeting Card     1.84
                  2.09
     SUBTOTAL     5.87
         TAX      0.35
     BAL DUE      6.22

    THANK YOU
```

EXAMPLE 1 Margo is buying supplies for a school project. The items she is buying and their prices are listed below. What is the total cost of the items?

poster board: $1.05
construction paper: $0.89
markers: $1.84
glue: $2.09

$$
\begin{array}{r}
\$1.05 \\
0.89 \\
1.84 \\
+\ 2.09 \\
\hline
\$5.87
\end{array}
$$

Align the decimal points.

Place the decimal point. Then add.

So, the total cost is $5.87.

• Suppose Margo also decides to buy a ruler for $1.35. What would be the total cost for supplies, including the ruler?

REMEMBER:
. . . .
You can add zeros to the right of a decimal without changing its value. **See page H12, Key Skill 21.**

$2.75 = 2.750$
$17 = 17.0$

Sometimes when you add decimals, the addends do not have the same number of decimal places. You can use zeros as placeholders after you align the decimal points.

EXAMPLE 2 Find the sum. $3.24 + 2 + 0.425$

$$
\begin{array}{r}
3.24 \\
2 \\
+\ 0.425 \\
\end{array}
\quad \text{or} \quad
\begin{array}{r}
3.240 \\
2.000 \\
+\ 0.425 \\
\hline
5.665
\end{array}
$$

Align the decimal points.
Use zeros as placeholders.

Add.

So, the sum is 5.665.

• Explain why you can write 3.24 as 3.240 and 2 as 2.000.

• Find the sum. $10.5 + 2.34 + 0.345$

GUIDED PRACTICE

Find the sum.

1. $4.25 + $3.14   **2.** $2.61 + $6.50   **3.** $10.12 + $6.59 + $8.05

Rewrite the addends so they have the same number of decimal places.

4. 4.05 + 2.356 + 8.3   **5.** 125.4 + 75 + 82.25

6. 0.64 + 1.253 + 0.8 + 2   **7.** 3 + 4.51 + 0.3755 + 2.2

Find the sum.

8. 12.4 + 6.8 + 3.2   **9.** 0.45 + 0.5 + 1.349

10. 6 + 5.43 + 1.4 + 5.755   **11.** 123.1 + 140 + 225.45

Estimating Sums

ACTIVITY

- Copy the table at the right.

- Suppose you have $50. Without finding the exact totals, decide whether you have enough money to make each of the following purchases.

  a. socks, shirt, cap
  b. watch, shirt, shorts, sunglasses
  c. shorts, socks, sunglasses, cap, shirt

- From the table, make your own list of items that you could purchase for $50. Include at least 3 items.

Item	Cost
Socks	$ 3.99
Shorts	$ 18.29
Shirt	$ 12.75
Watch	$ 20.79
Cap	$ 7.50
Sunglasses	$ 5.19
Back Pack	$ 25.15

When you don't need to know the exact answer to solve a problem, you can estimate. Rounding is a good way to estimate with decimals.

Shana has $9.50 and is eating out for dinner. She looks at her bill to see if she has enough money to order a dessert for $1.25. How can she determine whether she has enough money without finding the exact total?

She can estimate the sum by rounding.

$1.05 + $4.89 + $1.19 + $1.25
  ↓   ↓   ↓   ↓
  1 + 5 + 1 + 1 = 8 → The total with dessert
                                 will be about $8.

So, Shana has enough money to buy a dessert.

Remember that when you round decimals, you use the same rules that you use for rounding whole numbers.

EXAMPLE 3 Estimate the sum to the nearest whole number.

$$7.5 + 23.47 + 29.8$$

7.5	→	8	*Round each addend to*
23.47	→	23	*the nearest whole number.*
+ 29.8	→	+ 30	
		61	*Find the sum.*

An estimate is helpful when you want to determine whether an answer is reasonable.

EXAMPLE 4 Robert is buying 3 CDs. The prices are $12.49, $13.98, and $15.50. The salesperson tells him that the total cost is $41.97. Is this total cost reasonable?

$12.49	→	$12	*Round each addend to the*
13.98	→	14	*nearest whole number.*
+ 15.50	→	+ 16	
		$42	*Find the sum.*

Since the estimate is $42, a total cost of $41.97 is reasonable.

Talk About It

• Suppose you are buying a radio for $49.99 and a microphone for $27.85. A salesperson says the total is $104.69. Is the total reasonable? Explain.

• When you round to the nearest dollar, what decimal place do you use to determine whether to round to the next greater dollar or round to the given dollar?

• To estimate the sum $0.35 + $0.28 + $0.45, would you round to the nearest dollar? Explain.

INDEPENDENT PRACTICE

Find the sum.

1. 0.28 + 7.52   **2.** 3.95 + 4.16   **3.** 6.03 + 0.29   **4.** 5.41 + 3.20

Rewrite the addends so they have the same number of decimal places.

5. 7.398 + 14.5 + 0.75   **6.** 0.3 + 120 + 17.98   **7.** 902.7 + 1.4 + 32.886

8. $7 + $18.25 + $0.91   **9.** $13.89 + $5 + $14   **10.** 0.9871 + 3 + 14.5

Find the sum.

11. $12.5 + 8.64$ **12.** $0.94 + 1.237$ **13.** $\$21.18 + \$2$ **14.** $\$420.52 + \$35.48$

15. $2.16 + 3.5$ **16.** $\$25.16 + \$4.05$ **17.** $8.04 + 0.923$ **18.** $0.406 + 0.22$

19. $\begin{array}{r} 211.32 \\ 946.8 \\ + 63.754 \\ \hline \end{array}$
122087

20. $\begin{array}{r} 2.49 \\ 10.9521 \\ + 45.1 \\ \hline \end{array}$

21. $\begin{array}{r} 90.2 \\ 100.38 \\ 1.341 \\ + 1{,}082.6219 \\ \hline \end{array}$

22. $\begin{array}{r} 1.0981 \\ 96.235 \\ 62.1 \\ + 11.74 \\ \hline \end{array}$

23. $0.431 + 1.549 + 2.017$ **24.** $9.365 + 40.271 + 1.537$ **25.** $3.1428 + 90.6218$

26. $2.4 + 46 + 8.15$ **27.** $0.21 + 8.99 + 53.614$ **28.** $\$6 + \$118.59 + \$0.35$

Round to the nearest whole number.

29. 0.2 **30.** 1.55 **31.** 1.75 **32.** 5.42 **33.** 10.65 **34.** 20.51

Estimate the sum.

35. $0.3 + 0.6$ **36.** $1.4 + 2.7$ **37.** $7.1 + 4.5$ **38.** $0.77 + 3.24$

39. $4.09 + 5.89$ **40.** $6.95 + 2.44$ **41.** $0.392 + 1.021$ **42.** $9.356 + 6.811$

Estimate to determine if the given sum is reasonable.
Write *yes* or *no*.

43. $14.78 + 122.4 = 137.18$ **44.** $812.4 + 73.74 = 886.14$ **45.** $\$32.76 + 8.09 + \$0.49 = \$51.34$

Problem-Solving Applications

46. The computer club has raised $120.95 for new software. A math program costs $39.50, a game costs $24.75, and a reading program costs $66.29. About how much more money does the club need to pay for the software?

47. Mr. Ross made a list of computer hardware he found on sale. A mouse pad is $4.75, a 10-pack of diskettes is $5.89, and a color printer is $249.79. How much money does he need to buy the hardware?

48. **CRITICAL THINKING** Estimate $30.53 + 95.7 + 75.12$. Is the actual sum more or less than your estimate? Explain how you know.

49. Rochelle must earn a total of 37 points to advance in the gymnastics competition. She earns scores of 8.0. 8.0, 7.8, 7.9, and 7.9. Rochelle estimates that she will advance. Is her estimate reasonable? Explain.

50. ✏️ **WRITE ABOUT IT** Explain how to add two or more decimals. Show an example in your explanation.

Subtracting Decimals

What You'll Learn
How to subtract decimals

Why Learn This?
To know how much change to expect when you make a purchase

How does knowing how to subtract decimals help you determine if you received the correct amount of change from a purchase?

Andrew bought new basketball shoes for $53.49. He gave the cashier $55.00. How much change should Andrew get from the cashier?

$$\begin{array}{r} \$55.00 \\ -\ 53.49 \\ \hline \end{array}$$ *Align the decimal points.*

$$\begin{array}{r} \$55.00 \\ -\ 53.49 \\ \hline \$\ 1.51 \end{array}$$ *Place the decimal point.*
Subtract.

So, Andrew should get $1.51 in change from the cashier.

- Suppose Andrew gave the cashier $60.00. How much change should he get?

Sometimes adding zeros to the right of a decimal is helpful when subtracting.

CONSUMER LINK

When you pay for an item, you can use a strategy to avoid getting a pocketful of change or to avoid getting pennies. For example, if the total is $7.04, you can give the cashier $10.04 and get three $1 bills in change. Mark bought a magazine for $2.49. He paid the cashier $3.09. How much change should he get back?

> **EXAMPLE** Find the difference. 101.2 − 8.73
>
> $$\begin{array}{r} 101.20 \\ -\quad 8.73 \\ \hline \end{array}$$ *Align the decimal points.*
> *Add a zero to 101.2.*
>
> $$\begin{array}{r} 101.20 \\ -\quad 8.73 \\ \hline 92.47 \end{array}$$ *Place the decimal point.*
> *Subtract.*
>
> - How can you check your subtraction?

GUIDED PRACTICE

Copy the problem. Place the decimal point in the difference.

1. 9.7 − 3.01 = 669   **2.** 37.5 − 0.19 = 3731

Find the difference.

3. $4.38 − $2.26   **4.** $54.99 − $12.81   **5.** $80.93 − $60.51

6. 3.2 − 2.6   **7.** $6.18 − $5.55   **8.** 10.712 − 1.53

INDEPENDENT PRACTICE

Find the difference.

1. 9.3 − 6.1

2. 11.2 − 1.7

3. 5.95 − 3.26

4. $12.41 − $8.04

5. 4.981 − 3.235

6. 7.306 − 5.872

7. 6.0781 − 2.3195

8. 18.6902 − 9.2571

9. 12.304 − 6.92

10. 16.9278 − 5.051

11. 17.45 − 4.207

12. 38.124 − 2.3819

13. 3
    − 2.894

14. 48.28
    − 16.1790

15. 982.3024
    − 890.5196

16. 127.5
    − 3.257

17. 98.5
    − 7.21

18. 1,251.1
    − 12.7

19. 7.215
    − 0.89

20. 0.8765
    − 0.548

Find the missing number.

21. 5.262 − ■ = 2.45

22. 88.163 − ■ = 61.21

23. 7.24 − ■ = 3.22

24. 48.900 − ■ = 10.208

25. 7.560 − ■ = 4.211

26. 100.25 − ■ = 0.82

Problem-Solving Applications

27. SPORTS Jake's batting average is 0.325. Last year it was 0.235. What is the difference between his batting average last year and his batting average this year?

28. CONSUMER MATH Jan bought 3.3 lb of potatoes. Paul bought a 5-lb bag of potatoes. How many more pounds of potatoes did Paul buy than Jan?

29. MILEAGE Arnold drove 108.2 mi to his aunt's house. On the way back, he took a different route and drove 92.6 mi. How much shorter was his route back?

30. ✐ **WRITE ABOUT IT** Write a word problem that can be solved by subtracting decimals.

Mixed Review and Test Prep
Multiply.

31. 32 × 15

32. 149 × 53

33. 201 × 90

34. 698 × 712

Use the Distributive Property to find the product.

35. 25 × 8

36. 35 × 6

37. 55 × 9

38. 89 × 4

39. ESTIMATION Eric bought a game for $28.09, including tax. He paid the clerk with $50. Which is the best estimate for the amount of change he received?

  A less than $10   **B** more than $10

  C more than $15   **D** more than $20

40. CONSUMER MATH Tanya bought milk for $2.09, bread for $1.05, cheese for $4.50, and juice for $5.00. How much did Tanya pay for the groceries?

  F $12.04   **G** $12.64

  H $22.64   **J** $23.04

Multiplying Decimals

Why Learn This?
To find the total amount of money you will earn for working a number of hours

CULTURAL LINK

You can send a letter anywhere in the world. All you need is a stamp. Every country has some type of postal service. The country with the greatest number of post offices is India. The postage rate to send a letter from the U.S. to any country across an ocean is $0.60 for the first 0.5 ounce and $0.40 for each additional ounce. How much would it cost to send a letter with photos that weighs 3.5 ounces to India?

You can use a model to find the product of a decimal and a whole number.

ACTIVITY

WHAT YOU'LL NEED: colored pencils, decimal squares

- To find 3×0.14, shade 0.14, or 14 squares, three times. Use a different color each time.

- Count the number of shaded squares. What is 3×0.14?

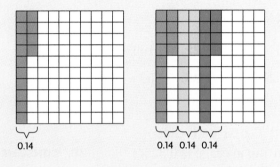

- Use a decimal square to find 2×0.16.

Sometimes, when the factors are larger, as in 8×1.52, it is easier to use paper and pencil to find the product of a whole number and a decimal.

EXAMPLE 1 Mark collects stamps. He bought 9 different stamps at $1.12 each. How much did he spend?

Estimate the products.	*Multiply as with whole numbers.*	*Since the estimate is 9, place the decimal point after the 10.*
$1.12 \times 9 \rightarrow \$1 \times 9 = \$9$	$\begin{array}{r} \tiny 1\ 1 \\ \$1.12 \\ \times\quad 9 \\ \hline \$10\,08 \end{array}$	$\begin{array}{r} \$1.12 \\ \times\quad 9 \\ \hline \$10.08 \end{array}$

So, Mark paid $10.08 for 9 stamps.

- Suppose Mark had paid $0.82 for each stamp. How much would he have paid for 9 stamps?

- **CRITICAL THINKING** How does an estimate help you determine whether the product of two factors is reasonable?

GUIDED PRACTICE

Estimate the product.

1. 3.62×7 **2.** 2.15×8 **3.** 4.04×5 **4.** 6.82×9

Copy the problem. Use an estimate to place the decimal point in the product.

5. $9 \times 5.4 = 486$ **6.** $7 \times 4.1 = 287$ **7.** $22 \times 0.55 = 121$

8. $7.32 \times 3 = 2196$ **9.** $0.82 \times 5 = 41$ **10.** $32.5 \times 6 = 195$

Find the product.

11. 0.33×3 **12.** 0.42×2 **13.** 1.25×4

14. 3.23×8 **15.** 1.3×5 **16.** 28.84×7

Multiplying a Decimal and a Decimal

You can use decimal squares or paper and pencil to find the product of two decimals.

> **EXAMPLE 2** Find 0.2×0.6.
>
>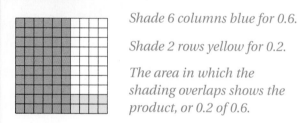
>
> *Shade 6 columns blue for 0.6.*
>
> *Shade 2 rows yellow for 0.2.*
>
> *The area in which the shading overlaps shows the product, or 0.2 of 0.6.*
>
> $$\begin{array}{r} 0.2 \\ \times\ 0.6 \\ \hline 0.12 \end{array}$$
>
> So, $0.2 \times 0.6 = 0.12$.
>
> • Find 0.5×0.3.

Talk About It

• How do you determine the number of rows and columns to shade to find the product of 2 decimals using a decimal square?

• Look at the model. Is the product greater or less than either factor?

• How is the product 0.2×0.6 different from the product 2×6?

• Look at the number of decimal places in the factors and in the product in Example 2. Write a rule for placing the decimal point in the product when multiplying decimals.

You can place the decimal point in a product by estimating. Another method of placing the decimal point is to add the numbers of decimal places in the factors and then count that total number of places from the right in the product.

$$0.8 \leftarrow \textit{1 decimal place}$$
$$\underline{\times\, 0.11} \leftarrow \textit{2 decimal places}$$
$$0.088 \leftarrow \textit{1 + 2, or 3 decimal places}$$

EXAMPLE 3 George works 37.5 hr per week at the bank. He earns $7.70 an hour. How much does he earn in a week?

$$7.70 \leftarrow \text{2 decimal places}$$
$$\underline{\times\, 37.5} \leftarrow \text{1 decimal place}$$
$$3850 \quad \textit{Multiply as with whole numbers.}$$
$$5390$$
$$\underline{2310} \qquad \textit{Place the decimal point.}$$
$$288.750 \leftarrow \text{3 decimal places}$$

So, George earns $288.75 in a week.

When you multiply decimals, sometimes you have to use zeros as placeholders.

EXAMPLE 4 Find the product 0.042×0.073.

$$0.042 \leftarrow \text{3 decimal places}$$
$$\underline{\times\, 0.073} \leftarrow \text{3 decimal places}$$
$$126 \quad \textit{Multiply as with whole numbers.}$$
$$\underline{294}$$
$$0.003066 \quad \textit{Place the decimal point.}$$
$$\textit{Since the answer must have 6 decimal places,}$$
$$\textit{you need 2 zeros to the left of 3.}$$

So, $0.042 \times 0.073 = 0.003066$.

• Find the product. 9.73×0.84

SCIENCE LINK

A day on Mars is 0.5 hr longer than a day on Earth. A year on Mars is 1.9 times as long as a year on Earth. How much longer, in hours, is a year on Mars than an average year on Earth? HINT: An average Earth year is 365.25 days.

INDEPENDENT PRACTICE

Estimate the product.

1. 0.95×4.2 **2.** 28.32×9.81 **3.** 51.42×8.16 **4.** 99.65×2.42

5. 4.82×20.19 **6.** 541.28×6.95 **7.** 603.95×9.1 **8.** 992.06×8.8

Tell the number of decimal places there will be in the product.

9. 3.79×8.2 **10.** 72×0.89 **11.** 0.876×0.2 **12.** 1.3842×0.91

Copy the problem. Place the decimal point in the product.

13. $4.01 \times 5.6 = 22456$ **14.** $6.37 \times 2.91 = 185367$ **15.** $20.4 \times 9.52 = 194208$

Find the product.

16. 2×0.6 **17.** 5×2.4 **18.** 8×1.6 **19.** $\$3.20 \times 4$

20. 2×0.12 **21.** 8×0.11 **22.** $\$1.21 \times 5$ **23.** 4×2.62

24. 6×3.35 **25.** $\$8.46 \times 2$ **26.** 9.35×7 **27.** 6.04×8

28. 0.2×0.4 **29.** 6.3×0.9 **30.** 0.21×2.1 **31.** 3.21×4.5

32. 6.15×2.4 **33.** 4.08×1.35 **34.** 6.21×0.95 **35.** 35.42×2.33

36. 24.63×1.09 **37.** 29.147×5.61 **38.** 0.189×2.09 **39.** 2.354×1.92

40. 118.001×0.37 **41.** 148.9×0.006 **42.** $1,200.5 \times 8.2$ **43.** $8,116.9 \times 1.402$

Problem-Solving Applications

44. **MEASUREMENT** Keith has a room that is 5.2 m wide. He has 3 bookcases that are each 1.6 m wide. Is there enough room for all of them? Explain.

45. **MEASUREMENT** Joanne brought 4.5 bottles of soda to the town picnic. Each bottle will fill 10 cups. How many cups can she fill with soda?

46. **BUSINESS** The middle school is having a car wash for a fund-raiser. They charge $5.25 for each car. If they wash 45 cars, how much money will they earn?

47. **CONSUMER MATH** Michael is making an apple pie. The recipe calls for 2.25 lb of apples. Apples are $1.60 per pound. How much will he pay for the apples?

48. **SAVINGS** Robbie wants 2 CDs that cost $8.99 each. He earns $5.00 a week. About how many weeks will he have to work to earn enough to buy the CDs?

49. ✏ **WRITE ABOUT IT** Explain how you know how many decimal places to put in the product.

Mixed Review and Test Prep
Divide.

50. $240 \div 6$ **51.** $320 \div 80$ **52.** $210 \div 70$ **53.** $600 \div 15$

Use mental math or compensation to solve.

54. $15 + 26 + 5$ **55.** $44 + 35 + 6$ **56.** $24 - 16$

57. **SCIENCE** Every 27 days there is a full moon. Which is the correct list for the multiples of 27?

    A 27, 47, 67, 87   **B** 27, 52, 77, 103
    C 27, 54, 81, 108   **D** Not Here

58. **FACTORS** Which list shows all the factors of 32?

    F 1, 32   **G** 1, 2, 4, 16, 32
    H 1, 2, 4, 8, 16, 32   **J** 1, 2, 3, 4, 8, 16, 32

LAB ACTIVITY

What You'll Explore
How to use a model to divide decimals

What You'll Need
decimal squares, colored pencils, scissors

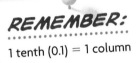

REMEMBER:
....
1 tenth (0.1) = 1 column

See page H11.

Exploring Division of Decimals

You can shade and cut apart decimal squares to divide a decimal by a whole number.

ACTIVITY 1

Explore

Find 3.6 ÷ 3.

• Shade 3.6 decimal squares.

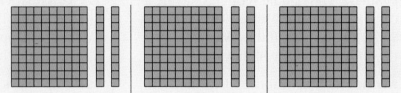

• Divide the shaded squares into 3 equal groups. Cut out each of the 6 tenths in the partially shaded decimal square.

• What decimal names each group? What is the quotient?

• Use decimal squares to find 1.2 ÷ 4. Use scissors to cut out each shaded tenth, and divide the tenths into 4 equal groups.

Think and Discuss

• Find 36 ÷ 3. How is the quotient the same as for 3.6 ÷ 3? How is it different?

• Find 12 ÷ 4. How is the quotient the same as for 1.2 ÷ 4? How is it different?

Try This

• Use decimal squares to find the quotients.

1. 4.4 ÷ 4   **2.** 3.5 ÷ 5   **3.** 1.5 ÷ 3

You can shade and cut apart decimal squares to divide a decimal by a decimal.

ACTIVITY 2

Explore

Find $3.6 \div 1.2$.

- Shade 3.6 decimal squares.

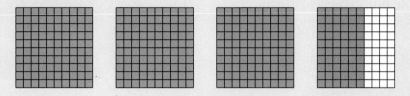

- Use scissors to cut apart the 6 tenths.

- Divide the shaded squares and shaded tenths into equal groups of 1.2. How many groups of 1.2 are in 3.6? What is the quotient?

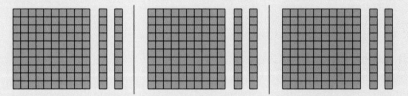

- Use decimal squares to find $3.2 \div 1.6$.

- How many groups of 1.6 are in 3.2?

Think and Discuss

- Find $36 \div 12$. How is the quotient the same as for $3.6 \div 1.2$? How is the problem different from $3.6 \div 1.2$?

- You know that $3.6 \div 12 = 0.3$ and $3.6 \div 1.2 = 3$. What do you think $3.6 \div 0.12$ equals?

Try This

Use decimal squares to find the quotients.

1. $7.8 \div 1.3$ **2.** $5.6 \div 0.8$ **3.** $1.56 \div 0.52$

4. $5.5 \div 1.1$ **5.** $3.6 \div 0.9$ **6.** $11.2 \div 1.4$

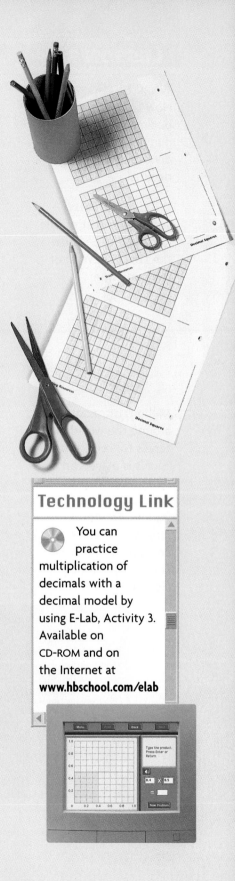

Technology Link

You can practice multiplication of decimals with a decimal model by using E-Lab, Activity 3. Available on CD-ROM and on the Internet at **www.hbschool.com/elab**

LESSON 3.4

Dividing Decimals

What You'll Learn

How to divide a decimal by a whole number and how to divide a decimal by a decimal

Why Learn This?

To find the cost of one item when you know the cost of several of the same item

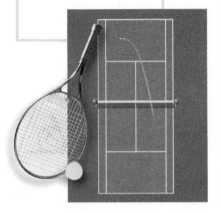

Jacob is going to take pictures on the class field trip. He bought a 3-pack of film for $9.93. What was the cost of each roll of film?

You can use division to find the cost. Dividing a decimal by a whole number is like dividing with whole numbers.

$$3\overline{)9.93}$$ *Place a decimal point above the decimal point in the dividend.*

$$
\begin{array}{r}
3.31 \\
3\overline{)9.93} \\
-9\downarrow \\
\hline
09 \\
-\ 9\downarrow \\
\hline
03 \\
-\ 3 \\
\hline
0
\end{array}
$$ *Divide as with whole numbers.*

Each roll of film cost $3.31.

You have to put a zero in the quotient when the divisor is greater than the dividend.

EXAMPLE 1 Felicia purchased a container of 3 tennis balls for $2.49. How much did she pay for each tennis ball?

$$3\overline{)2.49}$$ *Place a decimal point above the decimal point in the dividend.*

$$
\begin{array}{r}
0.83 \\
3\overline{)2.49} \\
-0\downarrow \\
\hline
2\ 4 \\
-2\ 4\downarrow \\
\hline
09 \\
-\ 9 \\
\hline
0
\end{array}
$$ *Since the divisor, 3, is greater than the 2 in the dividend, put a zero in the quotient in the ones place.*
Divide as with whole numbers.

So, Felicia paid $0.83 for each tennis ball.

• Find the quotient. 6.48 ÷ 8

Copy the problem. Place the decimal point in the quotient.

1. $4.92 \div 2 = 246$ **2.** $9.6 \div 3 = 32$ **3.** $2.96 \div 4 = 74$

4. $41.5 \div 5 = 83$ **5.** $121.2 \div 12 = 101$ **6.** $57.75 \div 15 = 385$

7. $32.5 \div 5 = 65$ **8.** $825.9 \div 3 = 2753$ **9.** $224.6 \div 4 = 5615$

Divide.

10. $7.88 \div 4$ **11.** $9.22 \div 2$ **12.** $8.93 \div 10$ **13.** $102.7 \div 5$

14. $4.44 \div 2$ **15.** $2.49 \div 3$ **16.** $30.5 \div 5$ **17.** $33.66 \div 11$

18. $5\overline{)1.61}$ **19.** $2\overline{)9.87}$ **20.** $20\overline{)7.9}$ **21.** $16\overline{)118.4}$

Dividing Decimals by Decimals

ACTIVITY

WHAT YOU'LL NEED: calculator

• Use a calculator to find the following quotients.

• Look for a pattern. Try to predict the last quotient in each set.

Set A	Set B
$0.52 \div 0.02 = $ 	$0.768 \div 0.016 = $ ▨
$5.2 \div 0.2 = $ ▨	$7.68 \div 0.16 = $ ▨
$52 \div 2 = $ ▨	$76.8 \div 1.6 = $ ▨
$520 \div 20 = $ ▨	$768 \div 16 = $ ▨

• Describe the pattern that helped you predict the last quotient in each set.

• For each set above, write another division problem that has the same quotient.

• Look at $5.2 \div 0.2$. Multiply both 5.2 and 0.2 by 10. How do the quotients of $5.2 \div 0.2$ and $52 \div 2$ compare? How does multiplying the divisor and the dividend by 10 affect the quotient?

• **CRITICAL THINKING** Look at $7.68 \div 0.16$. Multiply both 7.68 and 0.16 by 100. How do the quotients of $7.68 \div 0.16$ and $768 \div 16$ compare? How does multiplying the divisor and the dividend by 100 affect the quotient?

REMEMBER:

When you multiply a decimal by a power of 10, the decimal point moves one digit to the right for each power of 10.

$3.25 \times 10 = 32.5$
$3.25 \times 100 = 325$
$3.25 \times 1,000 = 3,250$

See page H12.

To divide a decimal by a decimal, first multiply the divisor and the dividend by a power of 10 to change the divisor to a whole number.

$$0.5\overline{)12.55} \rightarrow 5\overline{)125.5}$$

$$0.5 \times 10 = 5$$
$$12.55 \times 10 = 125.5$$

- What power of 10 would you multiply the divisor and the dividend by in the problem $425.7 \div 0.12$ to change the divisor to a whole number?

EXAMPLE 2 Divide. $22.8 \div 0.8$

$$0.8\overline{)22.8}$$

Make the divisor a whole number by multiplying the divisor and dividend by 10.

$$0.8 \times 10 = 8 \qquad 22.8 \times 10 = 228$$

$$\begin{array}{r} 28.5 \\ 8.\overline{)228.0} \\ -16\downarrow \\ \hline 68 \\ -64\downarrow \\ \hline 40 \\ -40 \\ \hline 0 \end{array}$$

Place the decimal point in the quotient.
Divide as with whole numbers.

Place a zero in the tenths place in the dividend, and continue to divide.

So, $22.8 \div 0.8 = 28.5$

- Think about $55.8 \div 0.18$. To change the divisor to a whole number, you multiply by 100. What does the dividend, 55.8, become? Explain.

You can use division of decimals to solve problems that involve money.

EXAMPLE 3 Each member of the class gave Mark $0.75 to buy a camera for the class. Mark received a total of $21.75. How many students are in the class?

$$0.75\overline{)21.75}$$

Make the divisor a whole number by multiplying the divisor and dividend by 100.

$$0.75 \times 100 = 75 \qquad 21.75 \times 100 = 2,175$$

$$\begin{array}{r} 29. \\ 75.\overline{)2175.} \\ -150\downarrow \\ \hline 675 \\ -675 \\ \hline 0 \end{array}$$

Place the decimal point in the quotient.
Divide as with whole numbers.

Since the remainder is 0, the answer is a whole number, and you do not need to show the decimal point.

So, there are 29 students in the class.

INDEPENDENT PRACTICE

Rewrite the problem so that the divisor is a whole number.

1. $9.6 \div 1.6$ **2.** $5.5 \div 1.1$ **3.** $6.3 \div 0.18$ **4.** $48.24 \div 2.4$

Complete.

5. $48.4 \div 0.4 = 484 \div \blacksquare$ **6.** $8.19 \div 0.09 = \blacksquare \div 9$ **7.** $3.57 \div 2.1 = 35.7 \div \blacksquare$

8. $3.846 \div 64.1 = \blacksquare \div 641$ **9.** $239 \div 0.075 = \blacksquare \div 75$ **10.** $84.36 \div 0.2812 = \blacksquare \div 2812$

Copy the problem. Place the decimal point in the quotient.

11. $62.44 \div 7 = 892$ **12.** $925.8 \div 3 = 3086$ **13.** $10.2 \div 2 = 51$

14. $34.178 \div 2.3 = 1486$ **15.** $45.218 \div 0.23 = 1966$ **16.** $233.58 \div 10.2 = 229$

Find the quotient.

17. $21.6 \div 3$ **18.** $12.8 \div 4$ **19.** $80.1 \div 9$ **20.** $90.3 \div 6$

21. $11\overline{)109.01}$ **22.** $12\overline{)286.8}$ **23.** $90\overline{)10.8}$ **24.** $60\overline{)12.6}$

25. $2.75 \div 0.5$ **26.** $1.26 \div 0.2$ **27.** $13.2 \div 0.06$ **28.** $42.5 \div 0.05$

29. $3.2\overline{)2.24}$ **30.** $2.8\overline{)4.48}$ **31.** $4.2\overline{)3.78}$ **32.** $8.2\overline{)229.6}$

33. $0.38\overline{)13.3}$ **34.** $0.55\overline{)2.42}$ **35.** $6.41\overline{)135.892}$ **36.** $2.48\overline{)1.3392}$

37. $49.3\overline{)201.144}$ **38.** $38.2\overline{)469.86}$ **39.** $29.1\overline{)186.24}$ **40.** $18.2\overline{)378.56}$

Problem-Solving Applications

41. ENTERTAINMENT Emilio bought 5 movie tickets for his friends. The tickets cost a total of $28.75. How much does each friend owe him for the cost of one ticket?

42. SAVINGS Jonelle is saving $4.95 every week to buy a video that costs $29.70, including tax. How many weeks will she have to save?

43. MEASUREMENT A new parking garage at the mall is 16.8 m high. Each floor is 4.2 m high. How many floors are there?

44. CONSUMER MATH Mel knows that 5 lb of mixed nuts sell for $12.50. He buys only 1 lb. How much does he pay?

45. CRITICAL THINKING Michael divided 4.25 by 0.25 and got a quotient of 0.17. Explain what he did wrong. What is the correct quotient?

46. ✏️ **WRITE ABOUT IT** When you divide a decimal by a decimal, what do you do to the divisor and the dividend?

Technology Link

💿 In *Mighty Math Number Heroes,* the game *Quizzo,* with show host Starr Brilliant, challenges you to complete decimal division problems. Use Grow Slide Level Z.

Estimate the sum. (pages 58–61)

1. $0.87 + 2.4$

2. $3.2 + 0.7 + 1.2$

3. $3.27 + 19.9$

Find the sum.

4. $3.9 + 7.2$

5. $480.7 + 0.824$

6. $28.7 + 0.45 + 3.2$

7.
$$\begin{array}{r} 125.35 \\ 2.34 \\ + \ 0.83 \\ \hline \end{array}$$

8.
$$\begin{array}{r} \$341.59 \\ + \ \$10.52 \\ \hline \end{array}$$

9.
$$\begin{array}{r} \$0.92 \\ \$7.24 \\ + \ \$0.89 \\ \hline \end{array}$$

10. Ross wants a glove that costs $28.99, a helmet that costs $21.50, and shoes that cost $35.00. How much money will he need to buy these items?

Find the difference. (pages 62–63)

11. $3.2 - 1.5$

12. $91.06 - 81.95$

13. $231.72 - 27.926$

14.
$$\begin{array}{r} 9.638 \\ -9.284 \\ \hline \end{array}$$

15.
$$\begin{array}{r} 45.2 \\ - \ 3.78 \\ \hline \end{array}$$

16.
$$\begin{array}{r} 28 \\ -17.224 \\ \hline \end{array}$$

17. Paul bought a basketball for $22.98. He gave the cashier $25.00. How much change should he get from the cashier?

Find the product. (pages 64–67)

18. 0.15×5

19. 48.23×7

20. 1.96×9

21. 9.07×75.3

22. 0.14×0.26

23. 5.7×0.98

24. Stephanie works 15 hr per week at a theme park. She earns $5.60 an hour. How much does she earn in one week?

Find the quotient. (pages 70–73)

25. $7\overline{)16.1}$

26. $9\overline{)43.65}$

27. $15\overline{)10.8}$

28. $9.2\overline{)115}$

29. $8.45 \div 0.5$

30. $7.8 \div 0.12$

31. $15.45 \div 0.3$

32. $129.585 \div 0.15$

33. Ms. Pegg bought a 25-pack of diskettes for $11.25 for her computer class. How much did she pay for each diskette?

34. A new building is 32.8 ft high. Each floor is 8.2 ft high. How many floors are there?

1. Centerville Library charges a two-cent fine for the first day a book is overdue. Then, for each day after the first, the total fine is double the preceding day's fine. What is the fine charged on a book that is 7 days overdue, expressed in exponential form?

   A 2^2 **B** 2^7
   C 7^2 **D** 7^7

2. Which points on the number line show opposite integers?

   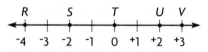

   F R and T
   G U and V
   H S and U
   J T and V
   K Not Here

3. Which expression is equivalent to $53 + 76$?

   A $50 + 79$
   B $50 + 80$
   C $53 + 72$
   D $53 + 67$

4. Carol has 87 stickers. This is twice as many as Beth has, plus 3. How many stickers does Beth have?

   F 34   **G** 40
   H 42   **J** 44

5. At the end of a trip, the odometer of a car read 3,462.1 miles. The odometer read 2,998.7 miles at the beginning of the trip. How far has the car traveled?

   A 456.2 mi
   B 463.4 mi
   C 4,670.8 mi
   D 6,460.8 mi
   E Not Here

6. Bob wants to buy a poster that costs $5.99, a desk lamp that costs $22.18, and a set of bookends that costs $18.98. What is the cost of the 3 items he wants to buy, before tax is added?

   F $46.15
   G $47.00
   H $47.15
   J $48.15

7. How many meters of fencing would it take to enclose this garden?

   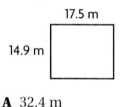

   A 32.4 m
   B 49.9 m
   C 64.8 m
   D 72.5 m

8. Seth earns $6.25 per hour. Last week he worked 18.5 hours. What is a reasonable estimate of his earnings?

   F $80   **G** $120
   H $140   **J** $200

9. Jessica bought 4 balloons. Each balloon cost $0.98, including tax. What is the total cost of the 4 balloons?

   A $3.92   **B** $3.96
   C $4.00   **D** $4.10

10. Dave bought 5 pounds of nuts for $9.95. What expression can be used to find the cost of 1 pound of nuts?

    F $5 \times \$9.95$
    G $\$9.95 + 5$
    H $\$9.95 \div 5$
    J $\$9.95 - 5$

The Numbers Are Falling!

PURPOSE To practice comparing and ordering decimals, fractions, exponents, square roots, and integers. (pages 22–31)

YOU WILL NEED blank number line

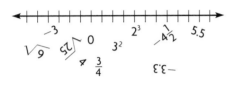

Help the numbers find their places on the number line.

Start by labeling the zero point. Find the value of any exponent or square root. Then find the location of each of the numbers.

NUMBER PUZZLES

PURPOSE To practice adding whole numbers (pages 34–37)

Copy the diagram. Then place each number listed below in one of the circles so that the sum of all three numbers in a line is the same.

21	22	23	24	25
26	27	28	29	

What is the sum?

$1,000 WINNER!

PURPOSE To practice multiplying and dividing decimals (pages 64–73)

YOU WILL NEED 6-section spinner numbered 1–6

Take turns spinning the pointer. Each player begins with $100. Use the chart to tell what to do with the number you spin. (For example, if you spin a 1, multiply your money by 5.) If you lose all your money, you're out. The first person to get $1,000, or the only one left with any money, is the winner.

If You Spin	What To Do
1	× 5
2	× 0.55
3	× 50.5
4	÷ 5
5	÷ 0.55
6	÷ 50.5

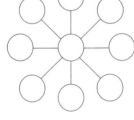 Play this game with your family.

Using Formulas to Find Sums and Differences

In this activity you will see how you can enter data in a spreadsheet and then enter formulas to find sums and differences in the data.

Look at the spreadsheet below, which has data collected from Class A and Class B about the number of hours students read.

To find the sum of Class A, enter the following formula in cell B10.

=sum(B2:B8)

To find the sum of Class B, enter the following formula in cell C10.

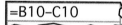
=sum(C2:C8)

To find the difference between Class A and Class B, enter the following formula in cell B11.

=B10–C10

All	A	B	C
1	Days of the Week	Class A	Class B
2	Sunday	12.5	22.5
3	Monday	22.75	24.6
4	Tuesday	27.5	23.75
5	Wednesday	30	19.45
6	Thursday	24.25	17.3
7	Friday	15	10.8
8	Saturday	9.2	11.8
9			
10	Total Hours	141.2	130.2
11	Difference	11	

For the formula to find a sum, use the first and last cell name in a row or column of data that you are adding.

For the formula to find a difference, use the cell names of the data you want to subtract.

1. Suppose you had data listed in Column D for Class C. Which cell and what formula would you use to find the sum for Class C?

2. Class C has a total of 124.5 hours. What formula would you use to find the difference between Class A and Class C? Class B and Class C?

USING A SPREADSHEET

3. Find some interesting data that can be put in a two-column format. Then insert the data into a spreadsheet.

4. Use a formula to find the sum for each column of data.

5. Use a formula to find the difference between two columns of data.

Study Guide and Review

Vocabulary Check

1. A number that shows how many times a base is used as a factor is called a(n) __?__. **(page 26)**

2. Integers that are the same distance from zero on the number line are called __?__. **(page 30)**

EXAMPLES

EXERCISES

- **Use place value to express numbers.**
  (pages 16–17)

  Write: 1,900,050.0325 *standard form*

  Read: one million, nine hundred
  thousand, fifty and three hundred
  twenty-five ten-thousandths

Write the number in standard form.

3. five hundred twenty thousand, ten and eight thousandths

4. two million, four hundred thousand, twenty-six and six tenths

---

- **Write decimals as fractions, and fractions as decimals.** **(pages 22–25)**

  0.08 *Identify the value of the last digit.
  8 is in the hundredths place.*

  $\frac{8}{100}$ *Use that value for the denominator.*

  So, $0.08 = \frac{8}{100}$.

Use place value to change the decimal to a fraction.

5. 0.2

6. 0.004

7. 0.65

8. 0.015

Write as a decimal.

9. $\frac{5}{100}$

10. $\frac{2}{5}$

- **Write numbers using exponents.** **(pages 26–27)**

  4^3 *Use 4 as a factor 3 times.*
  $4 \times 4 \times 4 = 64$ So, $4^3 = 64$

Find the value.

11. 8^2

12. 10^3

13. 1^{11}

14. 20^3

- **Multiply and divide whole numbers.**
  (pages 44–47)

  $$\begin{array}{r} 107 \text{ r}11 \\ 16\overline{)1{,}723} \\ -16 \\ \hline 12 \\ -0 \\ \hline 123 \\ -112 \\ \hline 11 \end{array}$$

  Divide.
  Multiply.
  Subtract.
  Repeat as necessary.

  ← *Remainder*

Find the product.

15. 7,654
      $\times$ 93

16. 342
      $\times$ 125

17. 8,345
      $\times$ 238

18. 10,248
      $\times$ 103

Find the quotient.

19. $7\overline{)716}$

20. $84\overline{)5{,}056}$

21. $57\overline{)325{,}398}$

22. $88\overline{)783{,}027}$

- **Estimate sums, differences, products, and quotients.** (pages 50–53)

$$\begin{array}{r} 518 \\ -329 \\ \hline \end{array}$$ *Round to the nearest hundred.* $$\begin{array}{r} 500 \\ -300 \\ \hline 200 \end{array}$$

Estimate.

23. $\begin{array}{r} 7,261 \\ +5,842 \\ \hline \end{array}$ **24.** $\begin{array}{r} 8,903 \\ + 2,342 \\ \hline \end{array}$

25. $\begin{array}{r} 2,401 \\ - 779 \\ \hline \end{array}$ **26.** $\begin{array}{r} 8,323 \\ - 289 \\ \hline \end{array}$

27. 615×49 **28.** $20,435 \times 82$

29. $895 \div 9$ **30.** $8,384 \div 22$

- **Add, subtract, multiply, and divide decimals.** (pages 58–73)

$$\begin{array}{r} 2.1 \\ 0.8\overline{)1.68} \\ \underline{1\,6} \\ 8 \\ \underline{8} \\ 0 \end{array}$$ *Make the divisor a whole number to place decimal point.*
$0.8 \times 10 = 8 \quad 1.68 \times 10 = 16.8$
Divide as with whole numbers.

Add.

31. $38.7 + 0.869$ **32.** $17.45 + 8.318$

Find the difference.

33. $85.06 - 75.98$ **34.** $224.62 - 27.948$

Find the product.

35. 63.5×9 **36.** 8.95×0.8

37. 7.81×0.24 **38.** 83.95×2.5

Find the quotient.

39. $30.45 \div 15$ **40.** $0.5\overline{)75.5}$

41. $2.8\overline{)29.12}$ **42.** $0.85\overline{)6.358}$

Problem-Solving Applications

Solve. Explain your method.

43. Roger started a worm farm with 10 worms. Each month there are 10 times as many worms as the month before. Using the exponent form, write the number of worms Roger will have in 10 mo. (pages 26–27)

44. Edward has 20 marbles. He has 10 more blue marbles than red ones. How many does Edward have of each color? (pages 38–39)

45. The PTA has $140.50 to spend on picnic supplies. They bought hot dogs for $24.75, rolls for $11.45, drinks for $48.96, and cookies for $25.29. About how much does the PTA have left? (pages 58–61)

46. In a 24-hour auto race, a car drove 3,057.6 mi. How many miles per hour is this? (pages 70–73)

Performance Assessment

Tasks: Show What You Know

1. Show how to change $\frac{3}{4}$ into a decimal. Explain your method.
(pages 22–25)

2. Choose a strategy and then explain how to solve the problem.
(pages 38–39)

Andy had 12 more pencils than pens. If Andy had a total of 48 pens and pencils, how many pens and how many pencils did Andy have?

3. Find the quotient of 34.6 ÷ 0.4. Explain your method and show your work. (pages 70–73)

Problem Solving

Solve. Explain your method.

CHOOSE a strategy and a tool.

- **Find a Pattern**
- **Make a Table**
- **Act It Out**
- **Write an Equation**
- **Make a Model**
- **Make a Graph**

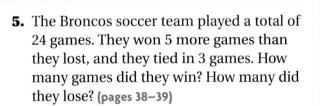

Paper/Pencil   Calculator   Hands-On   Mental Math

4. Tabby had 10 pennies. Each week she received 10 times as many pennies. If this pattern continues, how many pennies will Tabby have after five weeks? Write the total using exponent form. (pages 26–27)

5. The Broncos soccer team played a total of 24 games. They won 5 more games than they lost, and they tied in 3 games. How many games did they win? How many did they lose? (pages 38–39)

6. Shawna wants to buy film. She can buy 3 rolls of 24 exposures for $9.89, or she can buy single rolls of 36 exposures for $3.98 each. Which is the better deal? Explain why you think so. (pages 70–73)

24 Exposures .... 3 rolls for $9.89
36 Exposures .... $3.98 each roll

Cumulative Review

Solve the problem. Then write the letter of the correct answer.

1. Which is *five hundred five thousand, five hundred fifty and fifty-five ten thousandths* in standard form? (pages 16–17)

  A 55,055.55
  B 505,550.055
  C 505,550.0055
  D 505,000,550.55

2. Compare. 9.45 ● 9. 452 (pages 18–19)

  A <   **B** >
  C =   **D** ≥

3. Change 0.005 to a fraction. (pages 22–25)

  A $\dfrac{5}{10}$ **B** $\dfrac{5}{100}$
  C $\dfrac{5}{1,000}$ **D** $\dfrac{5}{10,000}$

4. Change $\frac{4}{5}$ to a decimal. (pages 22–25)

  A 0.04   **B** 0.08
  C 0.4   **D** 0.8

5. What is $3 \times 3 \times 3 \times 3$ in exponent form? (pages 26–27)

  A 2^4 **B** 3^4
  C 3^{10} **D** 4^3

6. Find the value of 10^5. (pages 26–27)

  A 50   **B** 10^5
  C 10,000   **D** 100,000

7. Compare. ⁻5 ● 0 (pages 30–31)

  A <   **B** >
  C =   **D** ≥

8. Use mental math to add.
$13 + 26 + 47 + 54$ (pages 36–37)

  A 100   **B** 130
  C 140   **D** 150

9. Use compensation to subtract.
$52 - 34$ (pages 36–37)

  A 22   **B** 20
  C 18   **D** 16

10. Use mental math to find the product.
7×18 (pages 40–43)

  A 78   **B** 126
  C 132   **D** 140

11. $6,478 \times 82$ (pages 44–47)

  A 531,196   **B** 494,486
  C 64,780   **D** 79

12. $24\overline{)4,916}$ (pages 44–47)

  A 24 r20   **B** 204
  C 204 r20   **D** 117,984

13. $42.6 + 0.968$ (pages 58–61)

  A 0.1394   **B** 4.3568
  C 43.568   **D** 4,356.8

14. $36.1 - 18.298$ (pages 62–63)

  A 1.7802   **B** 17.802
  C 17.998   **D** 1,780.2

15. 8.085×3.5 (pages 64–67)

  A 2,829.75   **B** 282.975
  C 28.2975   **D** 2.31

16. $36.12 \div 4.2$ (pages 70–73)

  A 0.086   **B** 8.6
  C 860   **D** 151.704

NUMBER THEORY AND FRACTIONS

LOOK AHEAD

In this chapter you will solve problems that involve

- multiples and factors

- prime factorization

- the least common multiple and the greatest common factor

- equivalent fractions, fractions in simplest form, and mixed numbers

SOCIAL STUDIES LINK

The table compares the wheat, rice, and corn production of the United States and China.

GRAIN PRODUCTION (METRIC TONS–1995)			
	Wheat	Rice	Corn
China	102,000,000	185,000,000	112,000,000
United States	59,000,000	8,000,000	187,000,000

- For which grain was production greatest in the United States? in China?
- The amount of rice produced in the United States is what fraction of the amount of rice produced in China?

Foods from Different Cultures

Suppose you want to prepare a lunch with foods that represent another culture.

Decide which foods you want to serve. Make a menu. Use recipes to determine the amount of each item you will need to feed your whole class. Make a shopping list of supplies. Share your menu with the class and tell what culture your foods are from. Explain how you made your supply list.

PROJECT CHECKLIST

✓ Did you make a menu?

✓ Did you make a list of the amounts of the items needed?

✓ Did you share your work with the class?

Multiples and Factors

What You'll Learn

How to find multiples and factors and how to tell whether a number is prime or composite

Why Learn This?

To use patterns of multiples in making designs and developing schedules, such as for exercise sessions

VOCABULARY

prime number

composite number

Albert has football practice every third day during November. His first day of practice is November 3. On what other dates in November will he have football practice?

S	M	T	W	Th	F	S
					1	2
3	4	5	6	7	8	9
10	11	12	13	14	15	16
17	18	19	20	21	22	23
24	25	26	27	28	29	30

The dates of Albert's practices are multiples of 3.

To find multiples of any number, multiply the number by the counting numbers 1, 2, 3, 4, and so on. The first six multiples of 3 are shown below.

```
0     3     6     9     12    15    18
├──┬──┼──┬──┼──┬──┼──┬──┼──┬──┼──┬──┼──┬──→
   3 × 1  3 × 2  3 × 3  3 × 4  3 × 5  3 × 6
```

• What are the first five multiples of 9?

Remember that when you multiply, you are using factors.

$3 \times 6 = 18 \leftarrow$ 3 and 6 are factors of 18.

REMEMBER:

Multiples of a number are the products when a number is multiplied by 1, 2, 3, 4, and so on.

When you multiply two or more numbers to get a product, the numbers multiplied are called **factors. See page H3**.

Some numbers, such as 5, have only two factors: 1 and the number itself. Numbers with only two factors are called **prime numbers**.

$5 = 5 \times 1 \leftarrow$ factors of 5: 1 and 5
$7 = 7 \times 1 \leftarrow$ factors of 7: 1 and 7

Numbers that have more than two factors are called **composite numbers**.

$6 = 1 \times 6$
$6 = 2 \times 3$ $\leftarrow$ factors of 6: 1, 2, 3, and 6

The numbers 0 and 1 are neither prime nor composite.

EXAMPLES Tell whether each number is prime or composite.

A. 12
factors: 1, 2, 3, 4, 6, 12

12 is composite.

B. 23
factors: 1, 23

23 is prime.

C. 63
factors: 1, 3, 7, 9, 21, 63

63 is composite.

• Starting with 2, what are the first ten prime numbers?

GUIDED PRACTICE

Write the first three multiples.

1. 4   **2.** 7   **3.** 9   **4.** 13   **5.** 17

6. 27   **7.** 50   **8.** 19   **9.** 14   **10.** 48

Write the factors. Tell whether the number is prime or composite.

11. 12   **12.** 11   **13.** 15   **14.** 23   **15.** 47

INDEPENDENT PRACTICE

Name the first four multiples.

1. 25   **2.** 10   **3.** 21   **4.** 15   **5.** 11   **6.** 16

Find the missing multiple or multiples.

7. 8, 16, 24, _?_ , _?_   **8.** _?_ , 24, 36, 48   **9.** _?_ , 14, 21, 28, _?_

Write the factors.

10. 9   **11.** 16   **12.** 12   **13.** 37   **14.** 34

15. 18   **16.** 23   **17.** 42   **18.** 121   **19.** 41

20. 27   **21.** 54   **22.** 31   **23.** 77   **24.** 84

Write P for *prime* or C for *composite*.

25. 31   **26.** 14   **27.** 9   **28.** 20   **29.** 33   **30.** 11

31. 16   **32.** 29   **33.** 30   **34.** 51   **35.** 48   **36.** 100

37. How do you know a number is prime?

38. What is a composite number?

39. List the prime numbers between 20 and 45.

40. List the composite numbers from 80 through 90.

Problem-Solving Applications

41. PATTERNS Jasmine's mother filled her car's gasoline tank every eighth day in June, beginning on June 8. How many times did she fill it in June? On what dates?

42. MUSIC Juan has a violin lesson every fourth day during September. His first lesson is on September 4. What are the dates of his other lessons in September?

43. LOGICAL REASONING Can a composite number have prime numbers as factors? Explain.

44. ✏️ WRITE ABOUT IT Give an example to show how a factor of a number and a multiple of that number are related.

Prime Factorization

You have learned that prime numbers have only two factors. The prime numbers less than 50 are listed below.

$$2, 3, 5, 7, 11, 13, 17, 19, 23, 29, 31, 37, 41, 43, 47$$

You have also learned that composite numbers have more than two factors. A composite number can be written as the product of prime factors. This is called the **prime factorization** of the number.

You can divide to find the prime factors of a number.

EXAMPLE 1 Find the prime factorization of 36.

$$2\overline{)36}$$
$$2\overline{)18}$$
$$3\overline{)9}$$
$$3\overline{)3}$$
$$1$$

Repeatedly divide by the smallest possible prime factor until the quotient is 1.

$2 \times 2 \times 3 \times 3$ *List the prime numbers you divided by. These are the prime factors.*

So, the prime factorization of 36 is $2 \times 2 \times 3 \times 3$, or $2^2 \times 3^2$.

• What is the prime factorization of 24?

ANOTHER METHOD You can use a factor tree to find the prime factors of a composite number. Different trees are possible but the prime factors are always the same.

EXAMPLE 2 Find the prime factorization of 100.

Choose any two factors of 100. Continue until only prime factors are left.

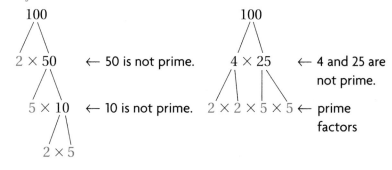

So, the prime factorization of 100 is $2 \times 2 \times 5 \times 5$, or $2^2 \times 5^2$.

• What is the prime factorization of 66?

GUIDED PRACTICE

Tell which of the prime numbers 2, 3, 5, and 7 are factors of the number.

1. 8   **2.** 24   **3.** 30   **4.** 42   **5.** 100

Draw a factor tree to show the prime factorization of the number.

6. 12   **7.** 65   **8.** 16   **9.** 66   **10.** 42

Write the prime factorization in exponent form.

11. $2 \times 2 \times 2$ **12.** $2 \times 2 \times 7$ **13.** $2 \times 3 \times 3 \times 3$ **14.** $2 \times 7 \times 7$ **15.** $3 \times 3 \times 3 \times 3$

INDEPENDENT PRACTICE

Use division to find the prime factors. Write the prime factorization.

1. 24   **2.** 36   **3.** 15   **4.** 50   **5.** 40

6. 93   **7.** 26   **8.** 38   **9.** 75   **10.** 88

Use a factor tree to find the prime factors. Write the prime factorization in exponent form.

11. 9   **12.** 18   **13.** 32   **14.** 49   **15.** 27

16. 20   **17.** 52   **18.** 45   **19.** 76   **20.** 99

21. 56   **22.** 48   **23.** 72   **24.** 64   **25.** 84

Solve for *n* to complete the prime factorization.

26. $2 \times n \times 5 = 20$ **27.** $44 = 2 \times 2 \times n$ **28.** $n \times 3 \times 7 = 42$ **29.** $75 = 3 \times 5 \times n$

30. $n \times 13 = 26$ **31.** $3 \times n \times 7 = 105$ **32.** $2^n = 32$ **33.** $78 = 2 \times 3 \times n$

Problem-Solving Applications

34. NUMBER SENSE The prime factors of a number are the three smallest prime numbers. No factor is repeated. What are the factors? What is the number?

35. CRITICAL THINKING The prime factorization of 25 is 5^2. Without dividing or using a factor tree, tell the prime factorization of 75.

36. ALGEBRA A number, *n*, is a prime factor of both 15 and 50. What is *n*?

37. ALGEBRA A number, *a*, is a prime factor of both 12 and 60. What is *a*?

38. NUMBER SENSE There are two numbers between 100 and 250 that have 3, 5, and 7 as prime factors. What are the numbers?

39. **WRITE ABOUT IT** Do the prime factors of a number differ depending on which factors you choose first? Explain.

LCM and GCF

You have learned that you can find multiples of a number by multiplying the number by 1, 2, 3, 4, and so on.

This number line shows multiples of 4 and 6.

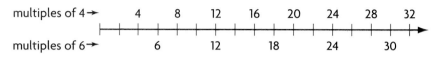

multiples of 4 → 4 8 12 16 20 24 28 32

multiples of 6 → 6 12 18 24 30

The multiples, such as 12 and 24, that are shown on each side of the number line are called common multiples. The smallest of the common multiples is called the **least common multiple**, or **LCM**.

Talk About It

• On the number line above, which multiples of 4 are also multiples of 6?

• What is the least common multiple, or LCM, of 4 and 6?

• What is another common multiple of 4 and 6?

• **CRITICAL THINKING** Is there a greatest common multiple? Explain.

You can use the LCM to solve problems.

EXAMPLE 1 Frank is buying hot dogs for a class picnic. Hot dogs are sold in packages of 10. Hot-dog buns are sold in packages of 8. What is the smallest number of hot dogs and buns Frank can buy to have an equal number of each?

10: 10, 20, 30, 40, 50, 60, 70, 80, 90   *List the multiples.*
 8: 8, 16, 24, 32, 40, 48, 56, 64, 72, 80   *Find the common*
                                              multiples.
The LCM of 10 and 8 is 40.   *Find the LCM.*

The LCM represents the smallest equal number of hot dogs and buns.
So, Frank needs 40 hot dogs and 40 buns.

• How many packages does Frank have to buy to get 40 hot dogs? to get 40 buns?

• What is the LCM of 3 and 7?

EXAMPLE 2 Country Flavor granola snacks are sold in 6-oz, 9-oz, and 18-oz packages. What is the least number of ounces you can buy to have equal amounts of the different sizes?

6: 6, 12, 18, 24, 30, 36, 42, 48, 54, 60
9: 9, 18, 27, 36, 45, 54, 63, 72, 81, 90
18: 18, 36, 54, 72, 90, 108, 126

List multiples of 6, 9, and 18. Find the common multiples.

The LCM of 6, 9, and 18 is 18.   *Find the LCM.*

So, you would need to buy 18 oz of each size.

• Since you need 18 oz, how many of each size would you need to buy?

GUIDED PRACTICE

List the first three multiples of each number in the pair.

1. 3, 6 **2.** 7, 14 **3.** 11, 8 **4.** 15, 12

Name the common multiples on the number line. Name the LCM.

5.

multiples of 5→	5	10	15	20	25	30
multiples of 10→		10		20		30

6.

multiples of 3→	3	6	9	12	15	18	21	24
multiples of 6→		6		12		18		24

Find the LCM of each pair of numbers.

7. 3, 7 **8.** 2, 3 **9.** 6, 9 **10.** 8, 20

Greatest Common Factor

Factors shared by two or more numbers are called common factors. The largest of the common factors is called the **greatest common factor**, or **GCF**.

To find the GCF of two or more numbers, list all the factors of each number, find the common factors, and then find the greatest common factor.

12: 1, 2, 3, 4, 6, 12   The common factors are 1, 2, 3, 6.
18: 1, 2, 3, 6, 9, 18   The GCF of 12 and 18 is 6.

• Why are you able to find a greatest common factor but not a greatest common multiple?

The GCF can be used to solve problems.

EXAMPLE 3 Emma is packaging items to give her friends. She has 45 pencils and 36 stickers. All packages have to contain the same number of each item. What is the greatest number of packages she can make without any items left over?

You can find the greatest number of packages by finding the GCF of 45 and 36.

45: 1, 3, 5, 9, 15, 45   *List the factors.*
36: 1, 2, 3, 4, 6, 9, 12, 18, 36   *Find the common factors.*

The GCF of 45 and 36 is 9.   *Find the GCF.*

So, Emma can make 9 packages without any items left over.

• **CRITICAL THINKING**   How many pencils and how many stickers will be in each package?

ANOTHER METHOD To find the GCF of two numbers, you can use their prime factors. List the prime factors, find the common prime factors, and then find their product.

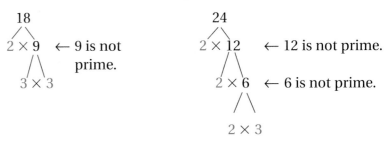

The prime factors of 18 are $2 \times 3 \times 3$.
The prime factors of 24 are $2 \times 2 \times 2 \times 3$.
The common prime factors are 2 and 3.
Find the product of the common factors: $2 \times 3 = 6$.
The GCF of 18 and 24 is 6.

EXAMPLE 4 Use prime factors to find the GCF of 54 and 72.

54: $2 \times 3 \times 3 \times 3$ *Find the prime factors.*
72: $2 \times 2 \times 2 \times 3 \times 3$

2, 3, and 3   *Find the common prime factors.*

$2 \times 3 \times 3 = 18$ *Multiply the common factors.*

So, the GCF of 54 and 72 is 18.

• What is the GCF of 36 and 60?

INDEPENDENT PRACTICE

Find the LCM for each set of numbers.

1. 3, 4   **2.** 8, 12   **3.** 1, 6   **4.** 5, 6   **5.** 4, 10   **6.** 7, 9

7. 4, 14   **8.** 9, 15   **9.** 6, 8   **10.** 1, 3, 7   **11.** 3, 5, 12   **12.** 2, 8, 10

Find the common prime factors. Then find the GCF.

13. 8, 24   **14.** 32, 128   **15.** 12, 20   **16.** 9, 15   **17.** 24, 30

18. 18, 22   **19.** 9, 12   **20.** 16, 48   **21.** 20, 48   **22.** 30, 45

Find the GCF for each set of numbers.

23. 6, 9   **24.** 4, 20   **25.** 9, 24   **26.** 12, 20   **27.** 16, 18   **28.** 5, 8

29. 9, 21   **30.** 16, 40   **31.** 3, 5, 10   **32.** 15, 20, 35   **33.** 12, 15, 24   **34.** 60, 72

Problem-Solving Applications

For Problems 35–36, use the following information.

Peter will distribute toothpaste samples with pamphlets about dental care. The toothpaste samples come in packages of 15. The pamphlets come in packages of 20.

35. BUSINESS What is the smallest number of toothpaste samples and pamphlets that he needs without having any left over?

36. How many packages of each does he need?

37. CONSUMER MATH Ruth has 36 markers and 48 erasers. She will put them in packages that are all the same. What is the greatest number of packages she can make?

38. ✏️ **WRITE ABOUT IT** What does *common* in *least common multiple* and *greatest common factor* mean? Give an example of each using 18 and 24.

Mixed Review and Test Prep

Find the value of *n*.

39. $15 \div 3 = n$ **40.** $16 \div 4 = n$ **41.** $24 \div n = 6$ **42.** $6 \times n = 30$ **43.** $n \times 8 = 56$

Solve.

44. $0.16 + 3.85$ **45.** $12.5 - 4.68$ **46.** 14.85×9.2 **47.** $70 \div 3.5$

48. COMPARE Which is the correct order of these lengths, from least to greatest?

  A 0.94 m, 1.2 m, 0.875 m
  B 1.2 m, 0.94 m, 0.875 m
  C 0.94 m, 0.875 m, 1.2 m
  D 0.875 m, 0.94 m, 1.2 m

49. EXPONENTS How is $3 \times 3 \times 3 \times 4 \times 4 \times 5 \times 5$ expressed in exponential notation?

  F $3 \times 4 \times 5^7$
  G $3^3 \times 2^4 \times 2^5$
  H $3^3 \times 4^2 \times 5^2$
  J $3^3 \times 4^4 \times 5^5$

LAB
ACTIVITY

What You'll Explore
How to use fraction bars to find equivalent fractions

What You'll Need
fraction bars

VOCABULARY

equivalent fractions

Equivalent Fractions

Fractions can be written in different ways to name the same amount or part. Fractions that name the same amount or part are called **equivalent fractions**. One way to find equivalent fractions is to use fraction bars.

ACTIVITY 1

Explore

Find how many eighths are equivalent to $\frac{1}{2}$.

• Use the $\frac{1}{2}$ fraction bar.

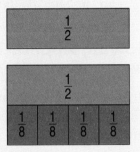

• Place $\frac{1}{8}$ bars along the $\frac{1}{2}$ bar until the lengths are equal.

Since 4 of the $\frac{1}{8}$ bars are equivalent to the $\frac{1}{2}$ bar, $\frac{1}{2} = \frac{4}{8}$.

• Use the fraction bars to find other fractions that are equivalent to $\frac{1}{2}$.

Think and Discuss

• Which other fraction bars did you use to find fractions equivalent to $\frac{1}{2}$? How many of each bar did you use?

• Explain how you would use fraction bars to find how many fourths are equivalent to $\frac{6}{8}$.

Try This

• Use fraction bars to solve.

1. How many twelfths are equivalent to $\frac{3}{4}$?

2. How many tenths are equivalent to $\frac{4}{5}$?

3. $\frac{5}{6} = \frac{\blacksquare}{12}$

4. $\frac{\blacksquare}{12} = \frac{2}{3}$

5. Find as many fractions as you can that are equivalent to $\frac{1}{4}$.

ACTIVITY 2

You can also use multiplication and division to find equivalent fractions.

Explore

- The fractions $\frac{2}{4}$ and $\frac{6}{12}$ are modeled below. Are they equivalent? Explain.

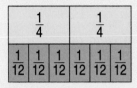

What number can you use for ■ to show that $\frac{2 \times \blacksquare}{4 \times \blacksquare} = \frac{6}{12}$?

- The fractions $\frac{5}{10}$ and $\frac{1}{2}$ are modeled below. Are they equivalent? Explain.

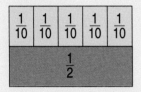

What number can you use for ■ to show that $\frac{5 \div \blacksquare}{10 \div \blacksquare} = \frac{1}{2}$?

Technology Link

You can practice finding factors to write equivalent fractions by using E-Lab • Activity 4. Available on CD-ROM and on the Internet at **www.hbschool.com/elab**

Think and Discuss

- Multiplying the numerator and denominator by 3 is like multiplying by $\frac{3}{3}$. What does $\frac{3}{3}$ equal?

- Suppose you multiply the numerator and the denominator of $\frac{2}{3}$ by 5. What equivalent fraction does that make?

- Suppose you divide both the numerator and denominator of $\frac{6}{12}$ by 2. What equivalent fraction does that make?

- What whole numbers can you use for ■ to get a fraction equivalent to $\frac{1}{3}$? $\frac{1}{3} \times \frac{\blacksquare}{\blacksquare}$

- What whole numbers can you use for ■ to get a fraction equivalent to $\frac{12}{16}$? $\frac{12}{16} \div \frac{\blacksquare}{\blacksquare}$

Try This

Copy and complete.

1. $\frac{2 \times \blacksquare}{3 \times \blacksquare} = \frac{6}{9}$ 2. $\frac{4 \times \blacksquare}{5 \times \blacksquare} = \frac{16}{20}$ 3. $\frac{30 \div \blacksquare}{35 \div \blacksquare} = \frac{6}{7}$ 4. $\frac{15 \div \blacksquare}{24 \div \blacksquare} = \frac{5}{8}$

Complete each number sentence.

5. $\frac{9}{12} = \frac{\blacksquare}{4}$ 6. $\frac{3}{24} = \frac{\blacksquare}{8}$ 7. $\frac{17}{34} = \frac{\blacksquare}{2}$ 8. $\frac{3}{5} = \frac{\blacksquare}{15}$

Fractions in Simplest Form

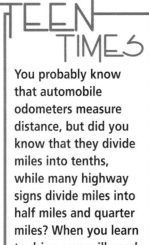

TEEN TIMES

You probably know that automobile odometers measure distance, but did you know that they divide miles into tenths, while many highway signs divide miles into half miles and quarter miles? When you learn to drive, you will need to remember that $\frac{5}{10}$ mi = $\frac{1}{2}$ mi, $\frac{2}{10}$ mi or $\frac{3}{10}$ mi is about $\frac{1}{4}$ mi, and $\frac{7}{10}$ mi or $\frac{8}{10}$ mi is about $\frac{3}{4}$ mi.

In the Lab Activity, you learned how to write equivalent fractions.

When the numerator and denominator of an equivalent fraction have no common factor other than 1, the equivalent fraction is in **simplest form**.

EXAMPLE 1 Write $\frac{6}{18}$ in simplest form.

6: 1, 2, 3, 6   *Find the common factors of 6 and 18.*
18: 1, 2, 3, 6, 9, 18

$\frac{6}{18} = \frac{6 \div 3}{18 \div 3} = \frac{2}{6}$ *Divide the numerator and denominator by a common factor until the fraction is in simplest form.*

$\frac{2}{6} = \frac{2 \div 2}{6 \div 2} = \frac{1}{3}$

So, $\frac{1}{3}$ is the simplest form of $\frac{6}{18}$.

• Explain how you know that $\frac{1}{3}$ is in simplest form.

You can use the GCF to write fractions in simplest form.

EXAMPLE 2 Write $\frac{16}{24}$ in simplest form.

16: 1, 2, 4, 8, 16   *Find the GCF of 16 and 24.*
24: 1, 2, 3, 4, 6, 8, 12, 24

$\frac{16}{24} = \frac{16 \div 8}{24 \div 8} = \frac{2}{3}$ *Divide the numerator and denominator by the GCF.*

So, $\frac{2}{3}$ is the simplest form of $\frac{16}{24}$.

• What is the simplest form of $\frac{5}{10}$?

GUIDED PRACTICE

Write the common factors for the numerator and denominator.

1. $\frac{4}{8}$ **2.** $\frac{9}{24}$ **3.** $\frac{8}{18}$ **4.** $\frac{12}{54}$

Write the fraction in simplest form.

5. $\frac{4}{32}$ **6.** $\frac{14}{21}$ **7.** $\frac{9}{54}$ **8.** $\frac{48}{54}$

9. $\frac{8}{22}$ **10.** $\frac{6}{18}$ **11.** $\frac{9}{30}$ **12.** $\frac{32}{48}$

INDEGARDPENDENT PRACTICE

INDEPENDENT PRACTICE

Write the common factors and the GCF of the numerator and denominator.

1. $\frac{1}{17}$ **2.** $\frac{9}{24}$ **3.** $\frac{6}{27}$ **4.** $\frac{9}{63}$ **5.** $\frac{10}{35}$ **6.** $\frac{16}{40}$

Write the fraction in simplest form.

7. $\frac{4}{24}$ **8.** $\frac{9}{12}$ **9.** $\frac{6}{48}$ **10.** $\frac{12}{16}$ **11.** $\frac{10}{18}$ **12.** $\frac{15}{20}$

13. $\frac{18}{90}$ **14.** $\frac{28}{42}$ **15.** $\frac{21}{33}$ **16.** $\frac{24}{30}$ **17.** $\frac{42}{60}$ **18.** $\frac{28}{98}$

Write the missing number.

19. $\frac{2}{12} = \frac{1}{\blacksquare}$ **20.** $\frac{\blacksquare}{36} = \frac{2}{9}$ **21.** $\frac{21}{24} = \frac{7}{\blacksquare}$ **22.** $\frac{40}{\blacksquare} = \frac{5}{8}$ **23.** $\frac{9}{\blacksquare} = \frac{3}{4}$

Problem-Solving Applications

24. Clint has 6 apple muffins, 2 corn muffins, and 4 bran muffins. What fraction of the muffins are bran? Write the fraction in simplest form.

25. NUMBER SENSE The numerator of a fraction is 12. The GCF of the numerator and denominator is 4. What is the denominator?

26. CALCULATOR Some calculators have a [SIMP] key that can be used to simplify fractions. What fraction would this key sequence give? 10 [/] 15 [SIMP] [=]

27. ✐ **WRITE ABOUT IT** When do you know that a fraction is in simplest form?

Mixed Review and Test Prep

Solve.

28. $(6 \div 3) + 2$ **29.** $(4 \div 2) + 1$ **30.** $(12 \div 4) + 2$

Solve.

31. 3.5×0.01 **32.** 1.1×0.2 **33.** 0.5×1.2

34. WEATHER The lowest recorded temperature in Atlanta, Georgia, is 8° below zero. What integer represents that temperature?

  A $^-18$ **B** $^-8$
  C $^+18$ **D** $^+8$

35. TRAVEL Mrs. Garcia rents a car for 5 days at $20.95 per day and $0.15 per mile. She travels 315 miles. What is her total cost?

  F $47.25 **G** $104.75
  H $147.25 **J** $152.00

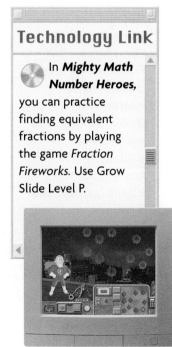

Technology Link

In *Mighty Math Number Heroes,* you can practice finding equivalent fractions by playing the game *Fraction Fireworks.* Use Grow Slide Level P.

Mixed Numbers and Fractions

A pizza store sells pizza by the slice. The pizzas are cut into fourths. You want to buy 11 slices to share with your friends. How many pizzas is this?

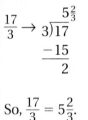

11 pieces, or $\frac{11}{4}$

Look at the model above. Since 11 pieces equal 2 whole pizzas and $\frac{3}{4}$ of another pizza, $\frac{11}{4} = 2\frac{3}{4}$.

A fraction like $\frac{11}{4}$ is greater than 1 because the numerator is greater than the denominator. A fraction greater than 1 can be written as a mixed number. A **mixed number** is a number that is made up of a whole number and a fraction.

EXAMPLE 1 Write $\frac{17}{3}$ as a mixed number.

$$\frac{17}{3} \rightarrow 3\overline{)17} \begin{array}{c} 5\frac{2}{3} \\ \hline \end{array}$$
$$\underline{-15}$$
$$2$$

Divide the numerator by the denominator.
Write the remainder as a fraction in simplest form.
Use the remainder as the numerator and the divisor as the denominator.

So, $\frac{17}{3} = 5\frac{2}{3}$.

ANOTHER METHOD You can use a calculator to change a fraction to a mixed number.

17 [b/c] 3 [SHIFT] [ab/c↔d/c] [$5\frac{2}{3}$]

You can also write a mixed number as a fraction.

EXAMPLE 2 Write $3\frac{2}{5}$ as a fraction.

$$3\frac{2}{5} \rightarrow \frac{(3 \times 5) + 2}{5} = \frac{17}{5}$$

So, $3\frac{2}{5} = \frac{17}{5}$.

Multiply the whole number by the denominator and add the numerator. Use the same denominator.

• Write $2\frac{4}{7}$ as a fraction.

Calculator Activities, page H31

GUIDED PRACTICE

Write the fraction as a mixed number.

1. $\frac{3}{2}$ **2.** $\frac{5}{3}$ **3.** $\frac{7}{2}$ **4.** $\frac{9}{5}$ **5.** $\frac{13}{4}$

Write the mixed number as a fraction.

6. $1\frac{1}{4}$ **7.** $1\frac{3}{5}$ **8.** $2\frac{2}{3}$ **9.** $3\frac{1}{3}$ **10.** $2\frac{4}{7}$

INDEPENDENT PRACTICE

Write the fraction as a mixed number or a whole number.

1. $\frac{7}{4}$ **2.** $\frac{9}{2}$ **3.** $\frac{11}{2}$ **4.** $\frac{23}{4}$ **5.** $\frac{27}{3}$ **6.** $\frac{19}{5}$

7. $\frac{31}{6}$ **8.** $\frac{18}{11}$ **9.** $\frac{29}{7}$ **10.** $\frac{32}{4}$ **11.** $\frac{75}{16}$ **12.** $\frac{50}{9}$

13. $\frac{31}{9}$ **14.** $\frac{41}{6}$ **15.** $\frac{6}{5}$ **16.** $\frac{19}{3}$ **17.** $\frac{22}{4}$ **18.** $\frac{25}{7}$

Write the mixed number as a fraction.

19. $3\frac{2}{3}$ **20.** $6\frac{1}{2}$ **21.** $5\frac{1}{3}$ **22.** $1\frac{9}{10}$ **23.** $4\frac{1}{9}$ **24.** $10\frac{2}{3}$

25. $9\frac{1}{4}$ **26.** $2\frac{3}{8}$ **27.** $4\frac{9}{11}$ **28.** $8\frac{4}{9}$ **29.** $6\frac{3}{5}$ **30.** $9\frac{7}{8}$

31. $7\frac{1}{4}$ **32.** $1\frac{3}{5}$ **33.** $7\frac{1}{2}$ **34.** $2\frac{1}{6}$ **35.** $2\frac{1}{12}$ **36.** $4\frac{3}{8}$

Tell which are equivalent numbers in each set.

37. $9\frac{1}{2}, \frac{17}{2}, \frac{19}{2}, 9\frac{1}{3}$ **38.** $\frac{13}{5}, 5\frac{1}{5}, 4\frac{1}{5}, \frac{26}{5}$ **39.** $\frac{28}{3}, \frac{13}{3}, 8\frac{1}{3}, 9\frac{1}{3}$

40. $3\frac{1}{4}, \frac{15}{4}, \frac{13}{4}, 3\frac{3}{12}$ **41.** $\frac{17}{8}, 8\frac{1}{8}, \frac{19}{9}, 2\frac{1}{8}$ **42.** $4\frac{1}{9}, \frac{49}{10}, 4\frac{9}{10}, 9\frac{4}{10}$

Write the missing number.

43. $\frac{40}{9} = 4\frac{\blacksquare}{9}$ **44.** $\frac{\blacksquare}{3} = 3\frac{2}{3}$ **45.** $\frac{\blacksquare}{6} = 9\frac{1}{6}$ **46.** $8\frac{3}{5} = \frac{43}{\blacksquare}$

Problem-Solving Applications

47. COMPARE Renee has $1\frac{3}{4}$ yd of fabric. Does she have enough for a pillow cover that requires $\frac{3}{2}$ yd of fabric? Explain.

48. COMPARE Don drinks $\frac{5}{3}$ glasses of orange juice. His sister drinks $1\frac{2}{3}$ glasses of orange juice. Do they drink the same amount? Explain.

49. ANOTHER METHOD Rick changed $3\frac{1}{4}$ to a fraction. He used this method: $3\frac{1}{4} = 3 \times \frac{4}{4} + \frac{1}{4} = \frac{12}{4} + \frac{1}{4} = \frac{13}{4}$ Explain his method.

50. ✏️ **WRITE ABOUT IT** Can any fraction be changed to a mixed number? Explain.

1. VOCABULARY Numbers with only two factors are called
___?___. **(page 84)**

Tell whether the number is prime or composite. **(pages 84–85)**

2. 7   **3.** 9   **4.** 12   **5.** 26

6. 11   **7.** 37   **8.** 49   **9.** 71

10. VOCABULARY When a composite number is written as the
product of prime factors, this is called ___?___. **(page 86)**

Find the prime factors of the number. **(pages 86–87)**

11. 9   **12.** 8   **13.** 14   **14.** 18   **15.** 80

16. 12   **17.** 33   **18.** 50   **19.** 49   **20.** 98

Find the LCM and GCF for each pair of numbers. **(pages 88–91)**

21. 3, 9   **22.** 2, 6   **23.** 6, 4   **24.** 10, 15

25. 8, 12   **26.** 9, 27   **27.** 15, 25   **28.** 12, 15

29. VOCABULARY When the numerator and denominator of an
equivalent fraction have no common factor other than 1, the
equivalent fraction is in ___?___. **(page 94)**

Write the fraction in simplest form. **(pages 94–95)**

30. $\frac{2}{8}$ **31.** $\frac{3}{9}$ **32.** $\frac{6}{21}$ **33.** $\frac{9}{24}$ **34.** $\frac{18}{32}$

35. $\frac{4}{10}$ **36.** $\frac{15}{20}$ **37.** $\frac{27}{45}$ **38.** $\frac{15}{24}$ **39.** $\frac{25}{80}$

40. VOCABULARY A number that is made up of a whole number
and a fraction is a ___?___. **(page 96)**

Write each fraction as a mixed number and each mixed number as
a fraction. **(pages 96–97)**

41. $\frac{8}{5}$ **42.** $\frac{11}{7}$ **43.** $1\frac{1}{3}$ **44.** $2\frac{3}{7}$ **45.** $7\frac{8}{11}$

46. $\frac{9}{2}$ **47.** $2\frac{1}{6}$ **48.** $3\frac{2}{5}$ **49.** $\frac{7}{3}$ **50.** $\frac{112}{3}$

1. Rosa has 4 bags of birdseed weighing 2.3 kilograms, 1.25 kilograms, 1.9 kilograms, and 2.15 kilograms. How would she put the bags in order from the lightest to the heaviest?

   A 1.9 kg, 2.15 kg, 2.3 kg, 1.25 kg
   B 2.3 kg, 2.15 kg, 1.9 kg, 1.25 kg
   C 1.25 kg, 1.9 kg, 2.15 kg, 2.3 kg
   D 2.15 kg, 1.25 kg, 2.3 kg, 1.9 kg

2. The cafeteria manager bought a total of 73 apples and oranges. She bought 11 more oranges than apples. How many of each kind of fruit did she buy?

   F 27 apples and 46 oranges
   G 31 apples and 42 oranges
   H 36 apples and 37 oranges
   J 42 apples and 31 oranges
   K 62 apples and 11 oranges

3. Which of these numbers is 550 when rounded to the nearest ten and 600 when rounded to the nearest hundred?

   A 515
   B 527
   C 554
   D 587
   E Not Here

4. Each of the 44 members of the band contributed $3.75 toward a gift for the band director. What is a reasonable estimate of the total amount collected?

   F Between $80 and $90
   G Between $90 and $110
   H About $120
   J About $160

5. Sara played basketball every fourth day in April, beginning on April 4. What was the date of the fifth day she played basketball?

   A April 5
   B April 7
   C April 10
   D April 15
   E April 20

6. Tao is buying treats for a party. Juice packs are sold in boxes of 8. Rice snacks are sold in boxes of 6. What is the least number of boxes of juice and rice snacks he should buy to have equal numbers of each?

   F 2 boxes of juice packs, 3 boxes of rice snacks
   G 3 boxes of juice packs, 3 boxes of rice snacks
   H 3 boxes of juice packs, 4 boxes of rice snacks
   J 4 boxes of juice packs, 3 boxes of rice snacks

7. Brooke has 8 mysteries, 5 biographies, and 11 adventure books. What part of her book collection is mysteries?

   A $\frac{1}{8}$

   B $\frac{1}{4}$

   C $\frac{1}{3}$

   D $\frac{1}{2}$

   E Not Here

8. Which pair contains equivalent amounts?

   F $2\frac{1}{3}$, $\frac{7}{3}$ **G** $4\frac{2}{5}$, $2\frac{1}{5}$

   H $\frac{13}{4}$, $2\frac{3}{4}$ **J** $\frac{16}{7}$, $3\frac{1}{2}$

5

ADDING AND SUBTRACTING FRACTIONS

LOOK AHEAD

In this chapter you will solve problems that involve

- adding and subtracting like and unlike fractions

- estimating sums and differences of fractions

ART **LINK**

When you sketch people, use the size of the head relative to the height to show the difference between an adult and a baby.

For an adult, the length of the oval you draw for the head should be about $\frac{1}{7}$ of the height. For a baby, on the other hand, the length of the oval you draw for the head should be about $\frac{2}{7}$ of the height.

Use these guidelines to discuss the following questions.

- An illustrator draws an oval about 2 inches long to represent the head of a person. What is the overall length of the drawing if the person is an adult? a baby?

- The oval represents a fraction of the height. What fraction can you write for the rest of an adult's height? for the rest of a baby's height?

Drawing People

Suppose you want to draw pictures of people of different ages. Use a tape measure to measure their heights and the lengths from top of the head to shoulder, shoulder to waist, and waist to floor. Make a table to record your data. Write fractions to compare the measured lengths to the height of each person. Then use the fractions to draw each person. Compare drawings and share the data you used to draw the people.

	A sixth-grader	An adult
Total height	60"	
Head + neck length	$12 \frac{12}{60} = \frac{1}{5}$	
Shoulder to waist	$15 \frac{15}{60} = \frac{1}{4}$	
Waist to knee	$15 \frac{15}{60} = \frac{1}{4}$	
Knee to ankle	$15 \frac{15}{60} = \frac{1}{4}$	
Ankle to floor	$3 \frac{3}{60} = \frac{1}{20}$	

To check:

$$\frac{1}{5} = \frac{4}{20}$$
$$\frac{1}{4} = \frac{5}{20}$$
$$\frac{1}{4} = \frac{5}{20}$$
$$\frac{1}{4} = \frac{5}{20}$$
$$+ \frac{1}{20} = + \frac{1}{20}$$
$$\frac{20}{20} = 1$$

$\frac{1}{5}$ $\frac{1}{4}$ $\frac{1}{4}$ $\frac{1}{4}$ $\frac{1}{20}$

PROJECT CHECKLIST

✓ Did you measure people of different ages?

✓ Did you use fractions to draw people of different ages?

✓ Did you share your data and compare drawings?

Adding and Subtracting Like Fractions

REMEMBER:

When the numerator and denominator of an equivalent fraction have no common factor other than 1, the equivalent fraction is in **simplest form**. See page H15.

$\frac{3}{6} = \frac{1}{2}$

The factor of 1 is 1. The factors of 2 are 1 and 2. The common factor is 1.

Kim reads his library book for $\frac{2}{6}$ hr on Tuesday and $\frac{3}{6}$ hr on Saturday. How long does Kim spend reading his library book?

You can use models to add and subtract like fractions. The model below shows $\frac{2}{6} + \frac{3}{6}$.

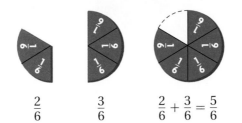

$$\frac{2}{6} \qquad \frac{3}{6} \qquad \frac{2}{6} + \frac{3}{6} = \frac{5}{6}$$

Kim spends $\frac{5}{6}$ hr reading his library book.

You can add like fractions without models.

EXAMPLE 1 Add. $\frac{7}{9} + \frac{5}{9}$

$$\frac{7}{9} + \frac{5}{9} = \frac{12}{9}$$

Add the numerators. Write the sum over the denominator.

$$= \frac{12 \div 3}{9 \div 3} = \frac{4}{3}, \text{ or } 1\frac{1}{3}$$

Write the fraction in simplest form. Write the answer as a fraction or a mixed number.

• What is $\frac{9}{15} + \frac{12}{15}$ in simplest form?

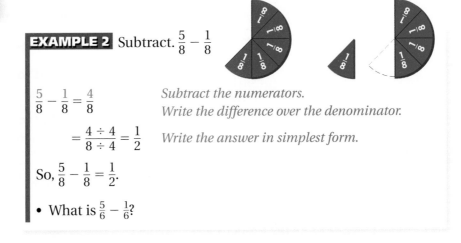

The method for subtracting like fractions is similar to the method for adding.

EXAMPLE 2 Subtract. $\frac{5}{8} - \frac{1}{8}$

$$\frac{5}{8} - \frac{1}{8} = \frac{4}{8}$$

Subtract the numerators. Write the difference over the denominator.

$$= \frac{4 \div 4}{8 \div 4} = \frac{1}{2}$$

Write the answer in simplest form.

So, $\frac{5}{8} - \frac{1}{8} = \frac{1}{2}$.

• What is $\frac{5}{6} - \frac{1}{6}$?

GUIDED PRACTICE

Find the sum or difference. Use a model to help you find the answer.

1. $\frac{1}{5} + \frac{2}{5}$ **2.** $\frac{2}{4} - \frac{1}{4}$ **3.** $\frac{1}{3} + \frac{1}{3}$ **4.** $\frac{2}{6} + \frac{3}{6}$ **5.** $\frac{3}{7} - \frac{1}{7}$

Find the sum or difference. Write the answer in simplest form.

6. $\frac{4}{5} - \frac{3}{5}$ **7.** $\frac{3}{8} + \frac{3}{8}$ **8.** $\frac{2}{3} + \frac{1}{3}$ **9.** $\frac{4}{6} - \frac{3}{6}$ **10.** $\frac{11}{12} - \frac{5}{12}$

INDEPENDENT PRACTICE

Copy and complete to make true sentences.

1. $\frac{3}{7} + \frac{2}{7} = \frac{?}{7}$ **2.** $\frac{1}{4} + \frac{2}{4} = \frac{?}{4}$ **3.** $\frac{8}{12} + \frac{3}{12} = \frac{?}{12}$ **4.** $\frac{7}{8} - \frac{2}{8} = \frac{?}{8}$ **5.** $\frac{6}{9} - \frac{5}{9} = \frac{?}{9}$

Find the sum or difference. Write the answer in simplest form.

6. $\frac{2}{5} + \frac{1}{5}$ **7.** $\frac{4}{9} + \frac{2}{9}$ **8.** $\frac{1}{4} + \frac{2}{4}$ **9.** $\frac{3}{4} + \frac{1}{4}$ **10.** $\frac{1}{6} + \frac{4}{6}$

11. $\frac{4}{7} - \frac{1}{7}$ **12.** $\frac{9}{10} - \frac{7}{10}$ **13.** $\frac{6}{10} - \frac{4}{10}$ **14.** $\frac{6}{7} - \frac{4}{7}$ **15.** $\frac{3}{8} - \frac{2}{8}$

16. $\frac{9}{12} + \frac{2}{12}$ **17.** $\frac{9}{10} - \frac{8}{10}$ **18.** $\frac{10}{20} + \frac{1}{20}$ **19.** $\frac{9}{12} - \frac{5}{12}$ **20.** $\frac{8}{9} + \frac{3}{9}$

21. $\frac{9}{17} + \frac{5}{17}$ **22.** $\frac{9}{14} - \frac{5}{14}$ **23.** $\frac{14}{20} + \frac{11}{20}$ **24.** $\frac{9}{16} - \frac{5}{16}$ **25.** $\frac{4}{15} + \frac{5}{15}$

26. $\frac{5}{8} + \frac{1}{8}$ **27.** $\frac{5}{6} - \frac{3}{6}$ **28.** $\frac{8}{15} - \frac{3}{15}$ **29.** $\frac{11}{15} - \frac{8}{15}$ **30.** $\frac{7}{10} + \frac{9}{10}$

31. $\frac{7}{12} - \frac{5}{12}$ **32.** $\frac{10}{10} - \frac{7}{10}$ **33.** $\frac{6}{8} + \frac{6}{8}$ **34.** $\frac{16}{25} + \frac{9}{25}$ **35.** $\frac{11}{12} - \frac{5}{12}$

Problem-Solving Applications

36. COMPARE Ms. Johnson measures the heights of her children on their birthdays. Lauren grew $\frac{2}{12}$ ft since her last birthday. John grew $\frac{5}{12}$ ft since his last birthday. How much more did John grow than Lauren?

37. Tori keeps pieces of lumber in his garage. Of the lumber, $\frac{1}{5}$ is pine and $\frac{3}{5}$ is oak. How much of the lumber is pine and oak?

38. MUSIC Jerry's CD collection is $\frac{5}{7}$ oldies, $\frac{1}{7}$ rock, and $\frac{1}{7}$ jazz. How much of his collection is not oldies music?

39. HOBBIES Erica has filled $\frac{3}{4}$ of an album with postcards. One fourth of the postcards are foreign postcards. What part of the album contains other postcards?

40. ✏️ **WRITE ABOUT IT** How do you add or subtract like fractions?

LAB ACTIVITY

What You'll Explore
How to use fraction bars to add and subtract fractions with unlike denominators

What You'll Need
fraction bars

REMEMBER:
....

The **LCM** is the smallest of the common multiples of two or more numbers.
See page H5.

2: 2, 4, 6, 8, 10, 12, . . .

3: 3, 6, 9, 12, 15, . . .

The LCM of 2 and 3 is 6. The LCM can be used to write common denominators of two or more fractions.

Addition and Subtraction of Unlike Fractions

You have 5 pencils and your best friend has 3 pens. How many pencils are there altogether? How many more pencils do you have than your friend?

To add or subtract, you must have things with the same name, such as 5 pencils and 3 pencils. Your sum or difference has the same name, 8 pencils or 2 pencils.

The denominators of fractions do not always have the same name. Fractions with different denominators are called **unlike fractions**. You can use fraction bars to rename the denominators before adding.

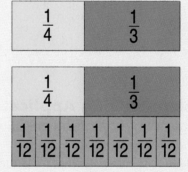

ACTIVITY 1

Explore
Find $\frac{1}{4} + \frac{1}{3}$.

- Use fraction bars to show both fractions.

- Which fraction bars fit exactly across $\frac{1}{4}$ and $\frac{1}{3}$? Think about the LCM of 4 and 3.

- What is $\frac{1}{4} + \frac{1}{3}$?

| $\frac{1}{4}$ | $\frac{1}{3}$ |

| $\frac{1}{4}$ | $\frac{1}{3}$ |
| $\frac{1}{12}$ | $\frac{1}{12}$ | $\frac{1}{12}$ | $\frac{1}{12}$ | $\frac{1}{12}$ | $\frac{1}{12}$ | $\frac{1}{12}$ |

Think and Discuss

- Look at the model for $\frac{1}{4} + \frac{1}{3}$. What do you know about $\frac{1}{4}$ and $\frac{3}{12}$? about $\frac{1}{3}$ and $\frac{4}{12}$?

- How are the denominators of $\frac{1}{4}$, $\frac{1}{3}$, and $\frac{1}{12}$ related? (HINT: Think about common multiples.)

Try This
Use fraction bars to find the sum. Draw a diagram of your model.

1. $\frac{1}{2} + \frac{1}{4}$ **2.** $\frac{1}{2} + \frac{1}{3}$ **3.** $\frac{1}{2} + \frac{2}{5}$

Fraction bars can also be used to subtract unlike fractions.

Explore

Find $\frac{1}{2} - \frac{1}{3}$.

- Use fraction bars to show $\frac{1}{2}$ and $\frac{1}{3}$.

$\frac{1}{2}$	
$\frac{1}{3}$	?

- Which fraction bars fit exactly across both $\frac{1}{2}$ and $\frac{1}{3}$? Think about the LCM.

$\frac{1}{6}$	$\frac{1}{6}$	$\frac{1}{6}$
$\frac{1}{6}$	$\frac{1}{6}$	?

- Compare $\frac{3}{6}$ and $\frac{2}{6}$. How much more is $\frac{3}{6}$ than $\frac{2}{6}$?

- What is $\frac{3}{6} - \frac{2}{6}$? What is $\frac{1}{2} - \frac{1}{3}$?

Think and Discuss

- How are the denominators of $\frac{1}{2}$, $\frac{1}{3}$, and $\frac{1}{6}$ related? (HINT: Think about common multiples.)

- Look at the model of $\frac{3}{4} - \frac{1}{3}$. Which fraction bars do you think will fit exactly across $\frac{3}{4}$ and $\frac{1}{3}$? Explain.

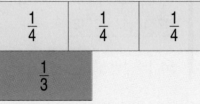

- Which fraction bars do you think would fit exactly across $\frac{1}{2} - \frac{1}{4}$?

Try This

Use fraction bars to subtract. Draw a diagram of your model.

1. $\frac{3}{4} - \frac{1}{3}$
2. $\frac{2}{5} - \frac{1}{10}$
3. $\frac{1}{3} - \frac{1}{4}$
4. $\frac{1}{2} - \frac{2}{5}$
5. $\frac{1}{2} - \frac{5}{12}$
6. $\frac{1}{4} - \frac{1}{6}$

Technology Link

You can practice finding multiples by using E-Lab, Activity 5. Available on CD-ROM and on the Internet at www.hbschool.com/elab

Adding and Subtracting Unlike Fractions

LESSON 5.2

What You'll Learn
How to use diagrams to add and subtract unlike fractions

Why Learn This?
To understand how to solve problems involving addition and subtraction of fractions by drawing a diagram

In the Lab Activity, you learned how to use fraction bars to add and subtract fractions. You can also use diagrams.

Ulises spent $\frac{1}{5}$ hr putting the clean dishes away. He spent $\frac{1}{2}$ hr cleaning the kitchen after dinner. What fractional part of an hour did Ulises spend on these chores?

Use the diagram at the right to find $\frac{1}{5} + \frac{1}{2}$.

The diagram shows that $\frac{1}{5} + \frac{1}{2} = \frac{7}{10}$.

To draw a diagram of an addition or subtraction problem, think about the LCM of the denominators and about equivalent fractions.

EXAMPLE 1 Complete the diagram to find the sum. $\frac{1}{4} + \frac{1}{6}$

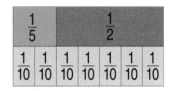

The LCM of 4 and 6 is 12. Draw twelfths under $\frac{1}{4}$ and $\frac{1}{6}$.
Think: $\frac{1}{4} = \frac{3}{12}$ and $\frac{1}{6} = \frac{2}{12}$.

So, $\frac{1}{4} + \frac{1}{6} = \frac{5}{12}$.

EXAMPLE 2 Draw a diagram to find the difference. $\frac{1}{2} - \frac{1}{5}$

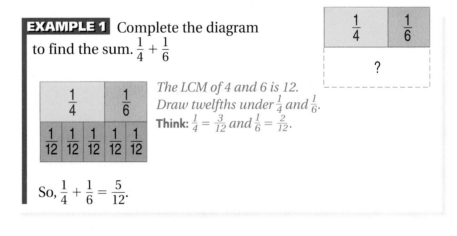

The LCM of 2 and 5 is 10.
Think: $\frac{1}{2} = \frac{5}{10}$ and $\frac{1}{5} = \frac{2}{10}$.
Draw $\frac{1}{2}$ as $\frac{5}{10}$ and $\frac{1}{5}$ as $\frac{2}{10}$.

Think: *How much greater is $\frac{5}{10}$ than $\frac{2}{10}$?*
$\frac{5}{10} - \frac{2}{10} = \frac{3}{10}$
So, $\frac{1}{2} - \frac{1}{5} = \frac{3}{10}$.

GEOGRAPHY LINK

According to the diagram below, about $\frac{1}{2}$ of all the earth's water is in the Pacific Ocean. About $\frac{1}{4}$ of all the water on earth is in the Atlantic Ocean, and about $\frac{1}{5}$ of the earth's water is in the Indian Ocean. How much of the earth's water is not part of these three oceans?

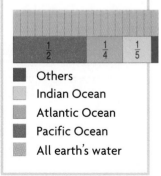

■ Others
□ Indian Ocean
▨ Atlantic Ocean
■ Pacific Ocean
▨ All earth's water

GUIDED PRACTICE

Copy and complete the diagram to find the sum or difference.

1. $\frac{1}{4} + \frac{1}{2}$

| $\frac{1}{4}$ | $\frac{1}{2}$ |

2. $\frac{1}{3} - \frac{1}{6}$

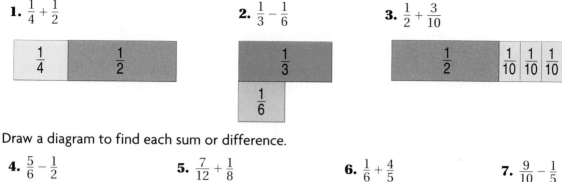

3. $\frac{1}{2} + \frac{3}{10}$

Draw a diagram to find each sum or difference.

4. $\frac{5}{6} - \frac{1}{2}$

5. $\frac{7}{12} + \frac{1}{8}$

6. $\frac{1}{6} + \frac{4}{5}$

7. $\frac{9}{10} - \frac{1}{5}$

INDEPENDENT PRACTICE

Write the addition or subtraction problem and then solve.

1.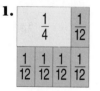

2.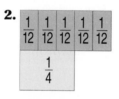

3.

Draw a diagram to find each sum or difference.

4. $\frac{7}{12} - \frac{1}{4}$

5. $\frac{1}{2} + \frac{1}{6}$

6. $\frac{3}{5} - \frac{1}{2}$

7. $\frac{2}{8} + \frac{1}{6}$

8. $\frac{3}{4} + \frac{1}{6}$

9. $\frac{2}{3} - \frac{1}{2}$

10. $\frac{5}{12} + \frac{1}{3}$

11. $\frac{7}{8} - \frac{1}{4}$

12. $\frac{7}{12} - \frac{1}{3}$

13. $\frac{9}{10} - \frac{3}{5}$

14. $\frac{3}{8} + \frac{1}{4}$

15. $\frac{7}{12} + \frac{1}{5}$

16. $\frac{1}{3} + \frac{1}{6}$

17. $\frac{5}{8} - \frac{1}{4}$

18. $\frac{5}{6} - \frac{1}{4}$

19. $\frac{7}{10} + \frac{1}{12}$

Problem-Solving Applications

20. COOKING Jackie's mom made pudding cups for the party. Of the cups she made, $\frac{5}{8}$ are chocolate and $\frac{1}{4}$ are butterscotch. How many more cups are chocolate than butterscotch?

21. HOBBIES Jane needs $\frac{1}{4}$ yd of blue fabric and $\frac{2}{3}$ yd of purple fabric for her sewing project. How much fabric does she need for her sewing project?

22. COMPARE Erin walks $\frac{1}{2}$ mi to school. Gene walks $\frac{3}{10}$ mi to school. How much farther does Erin walk than Gene?

23. ✏️ **WRITE ABOUT IT** Write a word problem that can be solved by the diagram in Exercise 2 of Independent Practice.

Adding Unlike Fractions

What You'll Learn
How to add unlike fractions

Why Learn This?
To solve problems such as finding the total distance traveled by bike

VOCABULARY

least common denominator (LCD)

When you use models or pictures to add unlike fractions, you rename the fractions so that they have a common denominator. To add fractions without models or pictures, you can write equivalent fractions by using the **least common denominator**, or **LCD**. The LCD is the LCM of the denominators.

EXAMPLE 1 You and a friend have decided to meet at the neighborhood recreation center at 4:15 P.M. Both of you will ride your bikes directly from home. You will ride $\frac{1}{2}$ mi and your friend will ride $\frac{3}{5}$ mi. What is the total distance the two of you will ride? Find $\frac{1}{2} + \frac{3}{5}$.

$$\begin{aligned}\frac{1}{2} &= \frac{1 \times 5}{2 \times 5} = \frac{5}{10}\\ +\frac{3}{5} &= \frac{3 \times 2}{5 \times 2} = \frac{6}{10}\end{aligned}$$

The LCM of 2 and 5 is 10, so the LCD of $\frac{1}{2}$ and $\frac{1}{5}$ is tenths. Multiply to write equivalent fractions using the LCD.

$$\begin{aligned}\frac{1}{2} &= \frac{5}{10}\\ +\frac{3}{5} &= \frac{6}{10}\\ \hline &\frac{11}{10}, \text{ or } 1\frac{1}{10}\end{aligned}$$

Add the numerators.
Write the sum over the denominator.

Write the answer as a fraction or as a mixed number.

So, you and your friend will ride a total of $1\frac{1}{10}$ mi.

- What if you ride $\frac{1}{2}$ mi and your friend rides $\frac{3}{4}$ mi? What is the total number of miles the two of you ride?

- **CRITICAL THINKING** When is the LCD of two fractions equal to the product of the denominators?

SPORTS LINK

In football there are four 15-min quarters, in hockey there are three 20-min periods, and in basketball there are two 24-min halves. Last Sunday you watched $\frac{1}{4}$ of a football game, $\frac{1}{3}$ of a hockey game, and $\frac{1}{2}$ of a basketball game. Is the sum of the fractions of the three events you watched more or less than one whole event?

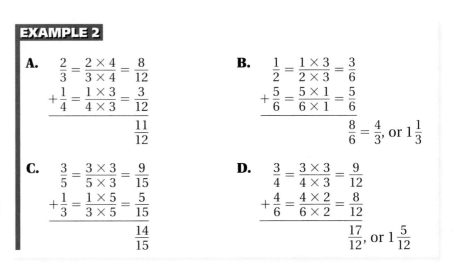

EXAMPLE 2

A.
$$\begin{aligned}\frac{2}{3} &= \frac{2 \times 4}{3 \times 4} = \frac{8}{12}\\ +\frac{1}{4} &= \frac{1 \times 3}{4 \times 3} = \frac{3}{12}\\ \hline &\frac{11}{12}\end{aligned}$$

B.
$$\begin{aligned}\frac{1}{2} &= \frac{1 \times 3}{2 \times 3} = \frac{3}{6}\\ +\frac{5}{6} &= \frac{5 \times 1}{6 \times 1} = \frac{5}{6}\\ \hline &\frac{8}{6} = \frac{4}{3}, \text{ or } 1\frac{1}{3}\end{aligned}$$

C.
$$\begin{aligned}\frac{3}{5} &= \frac{3 \times 3}{5 \times 3} = \frac{9}{15}\\ +\frac{1}{3} &= \frac{1 \times 5}{3 \times 5} = \frac{5}{15}\\ \hline &\frac{14}{15}\end{aligned}$$

D.
$$\begin{aligned}\frac{3}{4} &= \frac{3 \times 3}{4 \times 3} = \frac{9}{12}\\ +\frac{4}{6} &= \frac{4 \times 2}{6 \times 2} = \frac{8}{12}\\ \hline &\frac{17}{12}, \text{ or } 1\frac{5}{12}\end{aligned}$$

Calculator Activities, page H34

GUIDED PRACTICE

Name the LCM. Then write equivalent fractions with the LCD.

1. $\frac{1}{2} + \frac{3}{8}$ **2.** $\frac{7}{10} + \frac{1}{5}$ **3.** $\frac{1}{3} + \frac{1}{8}$ **4.** $\frac{5}{6} + \frac{1}{4}$

Find the sum. Write your answer in simplest form.

5. $\frac{1}{3} + \frac{1}{4}$ **6.** $\frac{4}{9} + \frac{1}{3}$ **7.** $\frac{2}{5} + \frac{2}{4}$ **8.** $\frac{2}{3} + \frac{3}{4}$

INDEPENDENT PRACTICE

Write equivalent fractions with the LCD.

1. $\frac{5}{6} + \frac{2}{3}$ **2.** $\frac{1}{4} + \frac{1}{2}$ **3.** $\frac{1}{5} + \frac{7}{15}$ **4.** $\frac{1}{4} + \frac{1}{3}$

5. $\frac{1}{4} + \frac{5}{8}$ **6.** $\frac{1}{2} + \frac{2}{3}$ **7.** $\frac{3}{4} + \frac{4}{5}$ **8.** $\frac{2}{3} + \frac{7}{8}$

9. $\frac{3}{8} + \frac{1}{2}$ **10.** $\frac{4}{5} + \frac{1}{7}$ **11.** $\frac{3}{4} + \frac{1}{6}$ **12.** $\frac{5}{6} + \frac{1}{3}$

Find the sum. Write your answer in simplest form.

13. $\frac{1}{6} + \frac{2}{3}$ **14.** $\frac{1}{4} + \frac{2}{3}$ **15.** $\frac{1}{2} + \frac{3}{10}$ **16.** $\frac{1}{3} + \frac{1}{2}$ **17.** $\frac{3}{4} + \frac{5}{8}$

18. $\frac{3}{8} + \frac{1}{3}$ **19.** $\frac{1}{6} + \frac{5}{12}$ **20.** $\frac{7}{12} + \frac{2}{3}$ **21.** $\frac{5}{6} + \frac{1}{4}$ **22.** $\frac{3}{8} + \frac{1}{12}$

23. $\frac{3}{5} + \frac{3}{10}$ **24.** $\frac{3}{5} + \frac{3}{4}$ **25.** $\frac{3}{4} + \frac{3}{20}$ **26.** $\frac{3}{8} + \frac{3}{20}$ **27.** $\frac{3}{5} + \frac{1}{3}$

Problem-Solving Applications

28. DATA Mary is helping her mother bake cookies. The recipe calls for $\frac{1}{2}$ c of brown sugar and $\frac{3}{4}$ c of white sugar. What is the total amount of sugar Mary's mother needs?

29. Craig feeds his pets each night. He feeds his big dog $\frac{1}{2}$ can of dog food and his small dog $\frac{1}{3}$ can of dog food. How much dog food does he need each night?

30. NUMBER SENSE Sara is landscaping her yard. She will plant trees in $\frac{1}{10}$ of the yard and flowers in $\frac{1}{5}$ of the yard. How much of Sara's yard will she plant with trees and flowers?

31. MUSIC Last week James practiced the tuba for $\frac{5}{6}$ hr on Monday, $\frac{1}{3}$ hr on Wednesday, and $\frac{1}{4}$ hr on Friday. How many hours did he practice last week?

32. ✏ **WRITE ABOUT IT** Explain how to add fractions with unlike denominators.

Subtracting Unlike Fractions

What You'll Learn
How to subtract unlike fractions

Why Learn This?
To solve problems such as finding the difference in portions of earnings spent

You have learned how to add unlike fractions. Do you think you can use a similar method to subtract unlike fractions?

Monica works part time for her uncle. She plans to spend $\frac{2}{3}$ of her earnings on school clothes and $\frac{1}{5}$ of her earnings on school supplies. What is the difference between what she spends on clothes and what she spends on school supplies?

Find $\frac{2}{3} - \frac{1}{5}$.

$$\frac{2}{3} = \frac{2 \times 5}{3 \times 5} = \frac{10}{15}$$
$$-\frac{1}{5} = \frac{1 \times 3}{5 \times 3} = \frac{3}{15}$$

The LCD of $\frac{2}{3}$ and $\frac{1}{5}$ is fifteenths. Multiply to find the equivalent fractions using the LCD.

$$\begin{array}{r} \frac{2}{3} = \frac{10}{15} \\ -\frac{1}{5} = \frac{3}{15} \\ \hline \frac{7}{15} \end{array}$$

Subtract the numerators. Write the difference over the denominator.

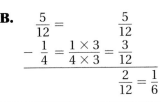

So, Monica spends $\frac{7}{15}$ more of her earnings on clothes.

- Suppose Monica spends $\frac{1}{3}$ of her earnings on clothes. What is the difference between what she spends on clothes and what she spends on school supplies?

EXAMPLES

A.
$$\begin{array}{r} \frac{5}{6} = \frac{5 \times 3}{6 \times 3} = \frac{15}{18} \\ -\frac{7}{9} = \frac{7 \times 2}{9 \times 2} = \frac{14}{18} \\ \hline \frac{1}{18} \end{array}$$

B.
$$\begin{array}{r} \frac{5}{12} = \frac{5}{12} \\ -\frac{1}{4} = \frac{1 \times 3}{4 \times 3} = \frac{3}{12} \\ \hline \frac{2}{12} = \frac{1}{6} \end{array}$$

GUIDED PRACTICE

Use the LCD to rewrite the problem with equivalent fractions.

1. $\frac{4}{5} - \frac{1}{3}$　　**2.** $\frac{2}{3} - \frac{1}{2}$　　**3.** $\frac{5}{7} - \frac{1}{2}$　　**4.** $\frac{2}{3} - \frac{2}{5}$

Subtract. Write the answer in simplest form.

5. $\frac{7}{9} - \frac{1}{6}$　　**6.** $\frac{3}{4} - \frac{1}{6}$　　**7.** $\frac{3}{4} - \frac{3}{8}$　　**8.** $\frac{2}{5} - \frac{1}{3}$

INDEPENDENT PRACTICE

Use the LCD to rewrite the problem with equivalent fractions.

1. $\frac{3}{8} - \frac{1}{4}$　　**2.** $\frac{6}{7} - \frac{3}{4}$　　**3.** $\frac{5}{9} - \frac{1}{4}$　　**4.** $\frac{9}{10} - \frac{1}{5}$　　**5.** $\frac{7}{12} - \frac{1}{5}$

Subtract. Write the answer in simplest form.

6. $\frac{1}{3} - \frac{1}{4}$　　**7.** $\frac{1}{2} - \frac{2}{5}$　　**8.** $\frac{5}{6} - \frac{1}{4}$　　**9.** $\frac{5}{9} - \frac{1}{3}$　　**10.** $\frac{7}{9} - \frac{1}{2}$

11. $\frac{5}{7} - \frac{1}{2}$　　**12.** $\frac{3}{8} - \frac{1}{4}$　　**13.** $\frac{4}{5} - \frac{1}{3}$　　**14.** $\frac{1}{4} - \frac{1}{6}$　　**15.** $\frac{7}{8} - \frac{3}{4}$

16. $\frac{8}{9} - \frac{2}{3}$　　**17.** $\frac{7}{10} - \frac{1}{5}$　　**18.** $\frac{5}{6} - \frac{1}{8}$　　**19.** $\frac{1}{3} - \frac{1}{7}$　　**20.** $\frac{3}{5} - \frac{1}{4}$

21. $\frac{5}{6} - \frac{5}{9}$　　**22.** $\frac{4}{5} - \frac{3}{10}$　　**23.** $\frac{3}{4} - \frac{1}{6}$　　**24.** $\frac{3}{10} - \frac{1}{4}$　　**25.** $\frac{7}{8} - \frac{2}{3}$

Problem-Solving Applications

26. MEASUREMENT Barnie and his family stopped to pick strawberries. Together they picked $\frac{3}{4}$ qt. Barnie's brother, Carl, ate $\frac{1}{12}$ qt before they left the strawberry field. How much was left?

27. When Barnie and his family left for their trip, they had a full tank of gas. When they reached their first stop, they had only $\frac{1}{8}$ tank. How much gas had they used?

28. ✏️ **WRITE ABOUT IT** How do you subtract fractions with unlike denominators?

Mixed Review and Test Prep

Compare the value of the numerator with the value of the denominator. Write *much less than, about the same as,* or *about $\frac{1}{2}$.*

29. $\frac{1}{5}$　　**30.** $\frac{7}{8}$　　**31.** $\frac{4}{9}$　　**32.** $\frac{2}{15}$

Write the fraction that is in simplest form.

33. $\frac{2}{4}, \frac{1}{2}, \frac{3}{6}, \frac{6}{12}$　　**34.** $\frac{4}{6}, \frac{20}{30}, \frac{6}{9}, \frac{2}{3}$　　**35.** $\frac{6}{8}, \frac{9}{12}, \frac{3}{4}, \frac{18}{24}$　　**36.** $\frac{3}{18}, \frac{7}{21}, \frac{2}{7}, \frac{3}{6}$

37. CONSUMER MATH Al buys 12 doughnuts: $\frac{1}{3}$ glazed, $\frac{1}{3}$ cake, and $\frac{1}{3}$ jelly. How many of each kind is he getting?

　　A 3　　　**B** 4　　　**C** 9　　　**D** 12

38. SPORTS Eric skated 500 meters in 37.14 seconds. Andre skated the same distance in 37.139 seconds, and Al in 37.12 seconds. What is the correct order of their times in seconds, from fastest to slowest time?

　　F 37.14, 37.139, 37.12　　　**G** 37.12, 37.14, 37.139

　　H 37.12, 37.139, 37.14　　　**J** 37.139, 37.12, 37.14

Technology Link

In *Mighty Math Number Heroes,* you can create fireworks in the game *Fraction Fireworks* by finding the correct sum or difference of fractions. Use Grow Slide Level W.

Estimating Sums and Differences

What You'll Learn
How to estimate sums and differences of fractions

Why Learn This?
To estimate answers to problems when exact answers are not needed

Sometimes when you solve problems that involve adding and subtracting fractions, you do not need to know the exact answer. One way to find the estimated answer is to use a number line.

Look at the number line below. Is $\frac{1}{8}$ closer to 0, $\frac{1}{2}$, or 1?

$$0 \quad \frac{1}{8} \quad \frac{2}{8} \quad \frac{3}{8} \quad \frac{1}{2} \quad \frac{5}{8} \quad \frac{6}{8} \quad \frac{7}{8} \quad 1$$

Since $\frac{1}{8}$ is closer to 0, you would round $\frac{1}{8}$ to 0.

To estimate with fractions less than 1, round them to 0, $\frac{1}{2}$, or 1.

Round $\frac{1}{9}$ to 0. The numerator is much less than the denominator.

Round $\frac{3}{8}$ to $\frac{1}{2}$. The numerator is about half the denominator.

Round $\frac{4}{5}$ to 1. The numerator is about the same as the denominator.

EXAMPLE 1 Gloria likes to hike and jog. She jogged $\frac{3}{4}$ mi on Monday and $\frac{4}{10}$ mi on Tuesday. What is a good estimate of the distance Gloria jogged on the two days?

Estimate. $\frac{3}{4} + \frac{4}{10}$

$$\begin{array}{r} \frac{3}{4} \rightarrow 1 \\ + \frac{4}{10} \rightarrow \frac{1}{2} \\ \hline 1\frac{1}{2} \end{array}$$
Round each fraction.

Add.

So, Gloria jogged about $1\frac{1}{2}$ mi.

• Suppose Gloria jogs the same distance every Monday and Tuesday. What is a good estimate of the number of miles she jogs in 2 weeks?

• **CRITICAL THINKING** In a fraction, if the numerator is more than half the denominator, is the fraction greater than $\frac{1}{2}$?

When you estimate the sum of more than two fractions, you follow the same rules of rounding.

EXAMPLE 2 One week Gloria couldn't hike or jog because it rained. On Monday it rained $\frac{3}{5}$ in., on Tuesday it rained $\frac{4}{10}$ in., and on Thursday it rained $\frac{4}{5}$ in. What is a good estimate of the amount of rain for the three days?

Estimate. $\frac{3}{5} + \frac{4}{10} + \frac{4}{5}$

$$\begin{array}{l} \frac{3}{5} \to \frac{1}{2} \\ \frac{4}{10} \to \frac{1}{2} \\ +\frac{4}{5} \to 1 \\ \hline \quad\quad 1\frac{2}{2}, \text{ or } 2 \end{array}$$ *Round each fraction.*

Add.
Write the answer as a whole number.

So, there was about 2 in. of rain for the three days.

- Suppose it rained $\frac{6}{10}$ in. on Friday also. What is a good estimate of the amount of rain for the four days?

- Gloria and her niece hiked $\frac{3}{4}$ mi on Wednesday, $\frac{7}{10}$ mi on Friday, $\frac{1}{5}$ mi on Saturday, and $\frac{4}{10}$ mi on Sunday. What is a good estimate for the total number of miles Gloria and her niece hiked?

GEOGRAPHY LINK

The Appalachian Trail runs 2,159 mi along the ridges of the Appalachian Mountains, from Springer Mountain, Georgia, to Mount Katahdin, Maine. The average time it takes to hike the full length of the trail is about 4–6 months. Suppose you started at the Georgia end early in May but hiked only about $\frac{1}{4}$ of the trail by the end of June. Then you hiked about $\frac{2}{5}$ of the trail in July and August. How much of the trail do you estimate you would still need to hike?

GUIDED PRACTICE

Use the number line to round each fraction to 0, $\frac{1}{2}$, or 1.

$0 \quad \frac{1}{12} \quad \frac{1}{6} \quad \frac{1}{4} \quad \frac{1}{3} \quad \frac{5}{12} \quad \frac{1}{2} \quad \frac{7}{12} \quad \frac{2}{3} \quad \frac{3}{4} \quad \frac{5}{6} \quad \frac{11}{12} \quad 1$

1. $\frac{11}{12}$ **2.** $\frac{7}{12}$ **3.** $\frac{1}{6}$ **4.** $\frac{1}{4}$

Round each fraction. Write *about 0, about $\frac{1}{2}$, or about 1.*

5. $\frac{1}{9}$ **6.** $\frac{10}{11}$ **7.** $\frac{2}{6}$ **8.** $\frac{7}{9}$ **9.** $\frac{4}{7}$

Estimate the sum.

10. $\frac{3}{8} + \frac{1}{5}$ **11.** $\frac{2}{9} + \frac{9}{11}$ **12.** $\frac{2}{3} + \frac{5}{9}$ **13.** $\frac{7}{8} + \frac{9}{10}$

14. $\frac{2}{3} + \frac{9}{10}$ **15.** $\frac{7}{8} + \frac{1}{2}$ **16.** $\frac{1}{16} + \frac{5}{8}$ **17.** $\frac{1}{2} + \frac{3}{4}$

18. $\frac{5}{6} + \frac{2}{5} + \frac{1}{8}$ **19.** $\frac{7}{9} + \frac{1}{5} + \frac{7}{8}$ **20.** $\frac{3}{10} + \frac{5}{8} + \frac{4}{9}$

You can also estimate the differences of fractions by rounding to the nearest 0, $\frac{1}{2}$, or 1.

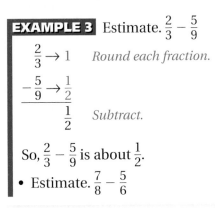

EXAMPLE 3 Estimate. $\frac{2}{3} - \frac{5}{9}$

$\frac{2}{3} \rightarrow 1$ *Round each fraction.*

$\dfrac{-\frac{5}{9} \rightarrow \frac{1}{2}}{\qquad\quad \frac{1}{2}}$ *Subtract.*

So, $\frac{2}{3} - \frac{5}{9}$ is about $\frac{1}{2}$.

• Estimate. $\frac{7}{8} - \frac{5}{6}$

EXAMPLE 4 Uriel is making a pie and a cake for a party. The cake recipe requires $\frac{5}{8}$ c of flour, and the pie recipe requires $\frac{1}{4}$ c of flour. What is a good estimate of how much more flour Uriel will need for the cake than the pie?

Estimate. $\frac{5}{8} - \frac{1}{4}$

$\frac{5}{8} \rightarrow \frac{1}{2}$ *Round each fraction.*

$\dfrac{-\frac{1}{4} \rightarrow 0}{\qquad\quad \frac{1}{2}}$ *Subtract.*

So, Uriel will need about $\frac{1}{2}$ c more flour for the cake.

• What if the cake recipe requires $\frac{2}{3}$ c of flour? Estimate how much more flour Uriel will need for the cake.

Talk About It

• Explain why $\frac{1}{4}$ could be rounded to $\frac{1}{2}$ instead of to 0. Name another fraction that could be rounded to $\frac{1}{2}$ instead of to 0.

• Look at Example 4. What would the estimate be if you rounded $\frac{1}{4}$ to $\frac{1}{2}$? Which do you think is a better estimate, this estimate or $\frac{1}{2}$? Explain.

INDEPENDENT PRACTICE

Round each fraction. Write *about 0*, *about $\frac{1}{2}$*, or *about 1*.

1. $\frac{2}{3}$ **2.** $\frac{2}{9}$ **3.** $\frac{5}{8}$ **4.** $\frac{13}{14}$ **5.** $\frac{3}{12}$ **6.** $\frac{5}{12}$

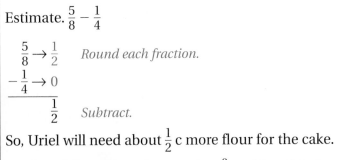

Round the fractions to 0, $\frac{1}{2}$, or 1. Then rewrite the problem.

7. $\frac{13}{14} + \frac{10}{17}$ **8.** $\frac{1}{8} + \frac{2}{3}$ **9.** $\frac{6}{10} + \frac{15}{16}$ **10.** $\frac{19}{20} - \frac{2}{50}$ **11.** $\frac{11}{12} - \frac{4}{9}$

Estimate the sum or difference.

12. $\frac{2}{3} + \frac{1}{9}$ **13.** $\frac{10}{12} - \frac{7}{10}$ **14.** $\frac{7}{9} - \frac{5}{9}$ **15.** $\frac{5}{16} + \frac{4}{9}$ **16.** $\frac{7}{8} - \frac{1}{4}$

17. $\frac{5}{9} - \frac{1}{2}$ **18.** $\frac{1}{3} + \frac{6}{7}$ **19.** $\frac{12}{15} - \frac{3}{7}$ **20.** $\frac{7}{8} + \frac{3}{4}$ **21.** $\frac{11}{12} + \frac{4}{7}$

22. $\frac{11}{12} - \frac{1}{9}$ **23.** $\frac{13}{15} + \frac{8}{9}$ **24.** $\frac{3}{5} + \frac{6}{7}$ **25.** $\frac{11}{14} - \frac{1}{3}$ **26.** $\frac{8}{10} - \frac{10}{18}$

27. $\frac{7}{8} + \frac{1}{3}$ **28.** $\frac{15}{16} - \frac{4}{5}$ **29.** $\frac{4}{5} - \frac{2}{7}$ **30.** $\frac{6}{10} + \frac{1}{5}$ **31.** $\frac{9}{12} + \frac{10}{18}$

32. $\frac{1}{9} + \frac{5}{8}$ **33.** $\frac{8}{9} - \frac{1}{6}$ **34.** $\frac{5}{6} + \frac{7}{8}$ **35.** $\frac{3}{13} + \frac{2}{15}$ **36.** $\frac{5}{6} - \frac{1}{8}$

Problem-Solving Applications

37. COOKING Lori is baking muffins. For each batch she needs $\frac{1}{2}$ c of milk, $\frac{3}{4}$ c of water, and $\frac{1}{3}$ c of apple juice. She wants to make three batches. About how many cups of liquid will she use?

38. CRAFTS Larry has three fabric samples measuring $\frac{5}{6}$ yd, $\frac{1}{4}$ yd, and $\frac{2}{3}$ yd. He estimates the total length to be about $1\frac{1}{2}$ yd. Is his estimate reasonable? Explain.

39. HEALTH Kathi bicycled $\frac{7}{8}$ mi to the post office. From there she bicycled $\frac{3}{5}$ mi to the park. She then bicycled $\frac{6}{10}$ mi home. About how many miles did she bicycle?

40. TIME Rogelio read $\frac{1}{3}$ hr on Monday, $\frac{3}{4}$ hr on Tuesday, and $\frac{1}{8}$ hr on Friday. He says he read about $1\frac{1}{2}$ hr. Is he right? Explain.

41. ✏️ **WRITE ABOUT IT** Explain how to round fractions less than 1.

Mixed Review and Test Prep

Write as a fraction.

42. $1\frac{3}{4}$ **43.** $2\frac{1}{4}$ **44.** $2\frac{5}{8}$ **45.** $3\frac{2}{3}$ **46.** $3\frac{7}{6}$ **47.** $5\frac{1}{5}$

Write the GCF of each pair of numbers.

48. 21, 9   **49.** 8, 64   **50.** 25, 10   **51.** 18, 24   **52.** 36, 16

53. REAL-LIFE The band is setting up chairs for the concert in the school auditorium. They can put 25 chairs in each row. How many rows will they need to seat 350 people?

   A 10   **B** 14
   C 18   **D** 25

54. MONEY Kayle bought 3 gifts. The prices ranged from $9 to $15. Before tax is added, which is a reasonable total cost for the 3 gifts?

   F less than $18   **G** between $18 and $27

   H between $27   **J** more than $45
       and $45

Add or subtract. Write in simplest form. (pages 102–103)

1. $\frac{7}{8} + \frac{3}{8}$
2. $\frac{8}{9} - \frac{2}{9}$
3. $\frac{13}{14} - \frac{3}{14}$
4. $\frac{6}{7} + \frac{5}{7}$

5. $\frac{2}{5} + \frac{3}{5}$
6. $\frac{7}{12} - \frac{5}{12}$
7. $\frac{3}{10} + \frac{5}{10}$
8. $\frac{11}{15} - \frac{8}{15}$

9. A blueberry-pecan muffin recipe requires $\frac{3}{4}$ c of whole-wheat flour. If Doug has $\frac{1}{4}$ c, how much more whole-wheat flour does he need?

10. **VOCABULARY** The LCM of the denominators is called the __?__. (page 108)

Add. Write in simplest form. (pages 106–109)

11. $\frac{1}{2} + \frac{1}{3}$
12. $\frac{3}{4} + \frac{1}{6}$
13. $\frac{2}{5} + \frac{2}{4}$
14. $\frac{2}{3} + \frac{3}{4}$

15. $\frac{5}{6} + \frac{2}{3}$
16. $\frac{1}{3} + \frac{5}{6}$
17. $\frac{3}{8} + \frac{3}{4}$
18. $\frac{5}{8} + \frac{1}{6}$

19. James worked on his science homework for $\frac{4}{5}$ hr on Monday and for $\frac{9}{10}$ hr on Tuesday. How long did he work on his science homework in all? Write your answer as a mixed number.

20. Sara has $\frac{5}{6}$ yd of ribbon and Mark has $\frac{3}{4}$ yd of ribbon. How much ribbon do Sara and Mark have altogether?

Subtract. Write in simplest form. (pages 106–107, 110–111)

21. $\frac{3}{4} - \frac{1}{3}$
22. $\frac{7}{8} - \frac{1}{4}$
23. $\frac{5}{6} - \frac{2}{9}$
24. $\frac{8}{9} - \frac{1}{6}$
25. $\frac{7}{8} - \frac{5}{6}$

26. On the first day of their vacation, April and her family walked $\frac{11}{12}$ mi. On the second day, they walked $\frac{1}{2}$ mi. How much farther did they walk on the first day than on the second?

27. Eric practiced $\frac{3}{4}$ hr on Monday and $\frac{2}{5}$ hr on Saturday. How much longer did Eric practice on Monday?

Estimate the sum or difference. (pages 112–115)

28. $\frac{7}{12} + \frac{1}{4}$
29. $\frac{3}{4} - \frac{1}{3}$
30. $\frac{2}{9} - \frac{1}{7}$
31. $\frac{4}{5} + \frac{3}{8}$
32. $\frac{4}{5} - \frac{1}{4}$

33. Nick has $\frac{4}{10}$ of a dollar and Narong has $\frac{4}{5}$ of a dollar. About how much money do the two boys have altogether?

34. Ed is making cookies. He needs $\frac{3}{4}$ c of flour for the recipe. He has $\frac{1}{3}$ c of flour. About how much more flour does he need?

1. The perimeter of a rectangular garden is 38 feet. The width is 5 feet more than the length. What are the length and width of the garden?

  A width: 7 ft; length: 12 ft
  B width: 8 ft; length: 15 ft
  C width: 10 ft; length: 14 ft
  D width: 12 ft; length: 7 ft

2. The cost of a dinner is \$25.50. If 6 friends share the cost equally, how much should each person contribute?

  F \$3.35
  G \$3.75
  H \$3.95
  J \$4.25
  K Not Here

3. Joe is buying items for a cookout. Hot dogs are sold 8 to a pack. Buns are sold 12 to a pack. What is the least number of packs of each he should buy to have equivalent amounts?

  A 2 packs of hot dogs and 2 packs of buns
  B 2 packs of hot dogs and 3 packs of buns
  C 3 packs of hot dogs and 2 packs of buns
  D 3 packs of hot dogs and 3 packs of buns
  E 4 packs of hot dogs and 3 packs of buns

4. Which group contains fractions that are all equivalent to $\frac{1}{4}$?

  F $\frac{2}{8}, \frac{14}{20}, \frac{11}{44}$
  G $\frac{3}{6}, \frac{20}{80}, \frac{3}{12}$
  H $\frac{6}{24}, \frac{15}{60}, \frac{50}{200}$
  J $\frac{4}{12}, \frac{25}{100}, \frac{5}{9}$
  K $\frac{12}{40}, \frac{3}{9}, \frac{30}{120}$

5. Melissa's tape collection is $\frac{4}{9}$ rock and $\frac{2}{9}$ jazz. What part of her collection is neither rock nor jazz?

  A $\frac{1}{3}$
  B $\frac{1}{2}$
  C $\frac{2}{5}$
  D $\frac{2}{3}$

6. Rick practiced the piano for $1\frac{2}{3}$ hours on Monday, $\frac{3}{4}$ hour on Wednesday, and $1\frac{1}{2}$ hour on Friday. What is the total amount of time he spent practicing the piano?

  F $3\frac{2}{5}$ hr
  G $3\frac{3}{4}$ hr
  H $3\frac{11}{12}$ hr
  J 4 hr

7. Walker had $\frac{13}{15}$ meter of wire. He used $\frac{1}{3}$ meter for a project. How much wire does he have left?

  A $\frac{1}{5}$ m
  B $\frac{1}{3}$ m
  C $\frac{8}{15}$ m
  D $1\frac{1}{3}$ m

8. Each day of a 5-day workweek, a salesman used $\frac{7}{8}$ of a tank of gas. What is a reasonable estimate of the amount of gas he used?

  F Less than 3 tanks of gas
  G Between 3 and 4 tanks of gas
  H Between 4 and 5 tanks of gas
  J Between 5 and 6 tanks of gas
  K More than 6 tanks of gas

ADDING AND SUBTRACTING MIXED NUMBERS

LOOK AHEAD

In this chapter you will solve problems that involve

- adding and subtracting mixed numbers

- estimating sums and differences of mixed numbers

HISTORY LINK

Precise measurements are important when determining the winner in a track and field event. The table shows the winning distances in the men's long jump for several recent summer Olympics.

- The winning jump in 1896 was 20 ft, 10 in., by Ellery Clark of the United States. How much greater was the 1996 winning jump?

- Study the table, and discuss any patterns you notice.

	SUMMER OLYMPICS
Year	Men's Long Jump Winning Distance
1996	27 ft $10\frac{3}{4}$ in.
1992	28 ft $5\frac{1}{2}$ in.
1988	28 ft $7\frac{1}{4}$ in.
1984	28 ft $\frac{1}{4}$ in.
1980	28 ft $\frac{1}{4}$ in.
1976	27 ft $4\frac{1}{2}$ in.

Precise Measurements

A pencil box is $8\frac{1}{2}$ in. long × $1\frac{3}{4}$ in. wide. Find six objects in the classroom and carefully measure the length and width of each to the nearest $\frac{1}{8}$ in. Then make a table to compare the length and width of each object to the pencil box. For example, suppose your first object is an 8-in. × 5-in. book. Your table should include the fact that the book is $\frac{1}{2}$ in. shorter and $3\frac{1}{4}$ in. wider than the pencil box.

PROJECT CHECKLIST

☑ Did you measure six classroom objects?

☑ Did you find the differences in both length and width compared to the pencil box?

☑ Did you make a table to compare the measurements?

	length	width	length comparison	width comparison
pencil box	$8\frac{1}{2}$ in.	$1\frac{3}{4}$ in.	—	—
floppy disk	$3\frac{3}{4}$ in.	$3\frac{1}{2}$ in.	$4\frac{3}{4}$ in. shorter	$1\frac{3}{4}$ in. wider
adult shoe	$10\frac{5}{8}$ in.	$3\frac{1}{2}$ in.	$2\frac{1}{8}$ in. longer	$1\frac{3}{4}$ in. wider
eraser	$2\frac{3}{8}$ in.	1 in.	$6\frac{1}{8}$ in. shorter	$\frac{3}{4}$ in. narrower
pencil	7 in.	$\frac{1}{4}$ in.	$1\frac{1}{2}$ in. shorter	$1\frac{1}{2}$ in. narrower
book	$8\frac{1}{2}$ in.	$5\frac{3}{4}$ in.	same	4 in. wider
paper clip box	3 in.	$2\frac{1}{4}$ in.	$5\frac{1}{2}$ in. shorter	$\frac{1}{2}$ in. wider

Adding Mixed Numbers

Have you ever assembled a model airplane? A series of diagrams showing the steps makes the assembly easier. Diagrams can also help you understand the steps for adding mixed numbers.

Mark used $2\frac{1}{4}$ yd of lumber to build a chair and $1\frac{1}{2}$ yd of lumber to build a stool. How much lumber did he use?

REMEMBER:
. . . .

A **mixed number** is a whole number and a fraction combined.
See page H16.

$4\frac{2}{5}$ is a mixed number.

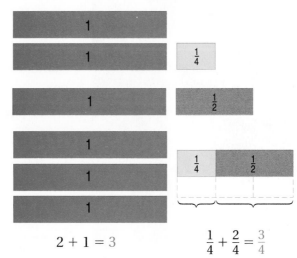

$2 + 1 = 3$ $\frac{1}{4} + \frac{2}{4} = \frac{3}{4}$

Complete the diagram to find $2\frac{1}{4} + 1\frac{1}{2}$.

Combine whole numbers. Combine fractions. Draw equivalent fractions with the LCD, fourths.

Add fractions. Add whole numbers.

So, Mark used $3\frac{3}{4}$ yd of lumber to build the chair and stool.

EXAMPLE Draw a diagram to find the sum. $1\frac{1}{5} + 1\frac{1}{2}$

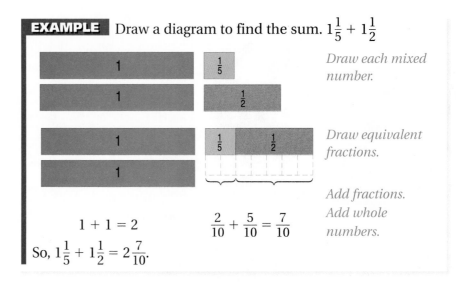

$1 + 1 = 2$ $\frac{2}{10} + \frac{5}{10} = \frac{7}{10}$

So, $1\frac{1}{5} + 1\frac{1}{2} = 2\frac{7}{10}$.

Draw each mixed number.

Draw equivalent fractions.

Add fractions. Add whole numbers.

GUIDED PRACTICE

Write the addition problem shown by the diagram. Then find the sum.

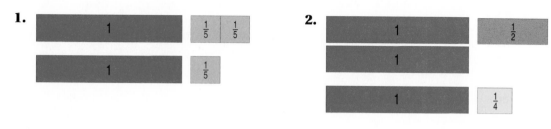

1. [diagram: 1, 1/5, 1/5 ; 1, 1/5]

2. [diagram: 1, 1/2 ; 1 ; 1, 1/4]

Draw a diagram to find each sum. Write the answer in simplest form.

3. $2\frac{3}{5} + 1\frac{1}{5}$ **4.** $1\frac{1}{3} + 1\frac{1}{3}$ **5.** $1\frac{1}{12} + 2\frac{1}{4}$ **6.** $1\frac{1}{2} + 1\frac{1}{6}$

INDEPENDENT PRACTICE

Write the addition problem shown by the diagram. Then find the sum. Write the answer in simplest form.

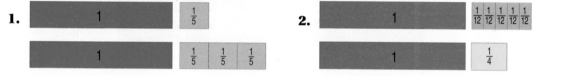

1. [diagram: 1, 1/5 ; 1, 1/5, 1/5, 1/5]

2. [diagram: 1, 1/12 1/12 1/12 1/12 1/12 ; 1, 1/4]

Draw a diagram to find each sum. Write the answer in simplest form.

3. $1\frac{1}{8} + 1\frac{5}{8}$ **4.** $2\frac{1}{8} + 2\frac{1}{4}$

5. $2\frac{1}{8} + 1\frac{3}{8}$ **6.** $1\frac{1}{3} + 1\frac{1}{6}$

Problem-Solving Applications

For Problems 7–9, use a diagram to solve.

7. ART Hank helped Madison paint scenery for the school play. He painted for $1\frac{1}{6}$ hr on Monday and $2\frac{1}{3}$ hr on Saturday. How many hours did he spend painting?

8. ENTERTAINMENT The students rehearsed for the school play for $2\frac{1}{6}$ hr on Tuesday and $3\frac{1}{4}$ hr on Saturday. How many hours did the students rehearse?

9. Joey used $2\frac{1}{8}$ yd of fabric for his costume, and Tami used $4\frac{1}{2}$ yd of fabric for her costume. How many yards of fabric did Joey and Tami use?

10. ✏️ **WRITE ABOUT IT** How is adding mixed numbers different from adding fractions?

Technology Link

💿 You can steer a submarine as you practice adding mixed numbers with Wando Wavelet of the **Mighty Math Calculating Crew** by playing the game *Nautical Number Line.* Use Grow Slide Level N.

LAB ACTIVITY

What You'll Explore
How to use fraction bars to subtract mixed numbers

What You'll Need
fraction bars

Subtracting Mixed Numbers

You can use fraction bars to subtract mixed numbers.

ACTIVITY
Explore

A. Find $2\frac{2}{5} - 1\frac{3}{10}$.

- Use fraction bars to model $2\frac{2}{5}$.

| 1 | 1 | $\frac{1}{5}$ | $\frac{1}{5}$ | $\leftarrow 2\frac{2}{5}$ |

- Since you are subtracting tenths, think about the LCD for $\frac{2}{5}$ and $\frac{3}{10}$. Change $\frac{2}{5}$ to $\frac{4}{10}$.

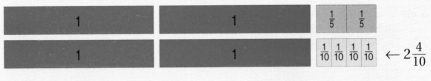

$\leftarrow 2\frac{4}{10}$

- Subtract $1\frac{3}{10}$ from $2\frac{4}{10}$.

- What is $2\frac{2}{5} - 1\frac{3}{10}$?

Sometimes you need to rename to subtract mixed numbers.

B. Find $2\frac{1}{4} - 1\frac{3}{4}$.

- Use fraction bars to model $2\frac{1}{4}$.

| 1 | 1 | $\frac{1}{4}$ | $\leftarrow 2\frac{1}{4}$ |

- Here is another way to model $2\frac{1}{4}$.

| 1 | $\frac{1}{4}$ | $\frac{1}{4}$ | $\frac{1}{4}$ | $\frac{1}{4}$ | $\frac{1}{4}$ | $\leftarrow 1\frac{5}{4}$ |

- From which model can you subtract $1\frac{3}{4}$?

- Subtract $1\frac{3}{4}$ from $1\frac{5}{4}$. What is $2\frac{1}{4} - 1\frac{3}{4}$?

C. Find $2\frac{1}{6} - 1\frac{5}{12}$.

- Use fraction bars to model $2\frac{1}{6}$.

 $\leftarrow 2\frac{1}{6}$

- Here is another way to model $2\frac{1}{6}$.

 $\leftarrow 1\frac{7}{6}$

- From which model can you subtract $1\frac{5}{12}$?

- Since you are subtracting twelfths, think of the LCD for $\frac{1}{6}$ and $\frac{5}{12}$. Change the sixths to twelfths.

$\leftarrow 1\frac{14}{12}$

- Subtract $1\frac{5}{12}$ from $1\frac{14}{12}$. What is $2\frac{1}{6} - 1\frac{5}{12}$?

Think and Discuss

- Think about $2\frac{3}{8} - 1\frac{1}{8}$. Do you need to rename before you subtract? Explain.

- Think about $3\frac{2}{5} - 1\frac{4}{5}$. Do you need to rename before you subtract? Explain.

- Think about $5\frac{1}{2} - 3\frac{5}{6}$. Do you need to rename before you subtract? Explain.

Try This

Use fraction bars to subtract. Draw a diagram of your model.

1. $2\frac{6}{10} - 1\frac{3}{10}$ **2.** $1\frac{2}{5} - \frac{1}{5}$ **3.** $1\frac{3}{4} - \frac{1}{8}$

4. $2\frac{2}{3} - 1\frac{1}{2}$ **5.** $3\frac{3}{8} - \frac{3}{4}$ **6.** $2\frac{2}{3} - 1\frac{3}{4}$

7. $3\frac{1}{2} - 1\frac{3}{5}$ **8.** $3 - 2\frac{1}{2}$ **9.** $2 - 1\frac{3}{8}$

10. $2\frac{3}{8} - 1\frac{1}{2}$ **11.** $3\frac{5}{6} - 1\frac{1}{3}$ **12.** $3\frac{1}{12} - 2\frac{5}{6}$

Subtracting Mixed Numbers

You can use diagrams to subtract mixed numbers.

Liliana tutors younger students after school. She tutored for $2\frac{1}{8}$ hr on Monday and for $1\frac{3}{8}$ hr on Wednesday. How much longer did she tutor on Monday than on Wednesday?

Complete the diagram to find $2\frac{1}{8} - 1\frac{3}{8}$.

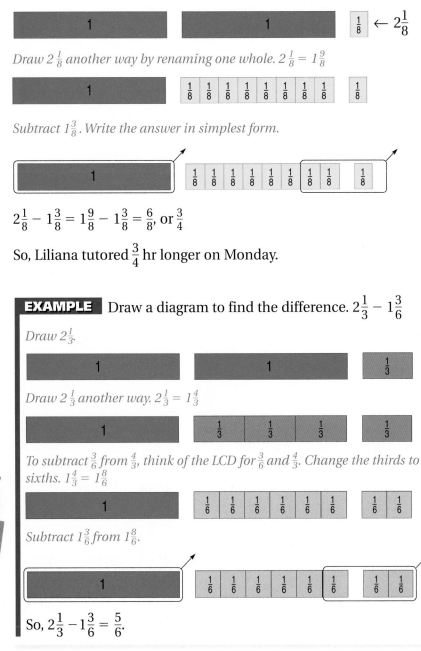

Draw $2\frac{1}{8}$ another way by renaming one whole. $2\frac{1}{8} = 1\frac{9}{8}$

Subtract $1\frac{3}{8}$. Write the answer in simplest form.

$$2\frac{1}{8} - 1\frac{3}{8} = 1\frac{9}{8} - 1\frac{3}{8} = \frac{6}{8}, \text{ or } \frac{3}{4}$$

So, Liliana tutored $\frac{3}{4}$ hr longer on Monday.

EXAMPLE Draw a diagram to find the difference. $2\frac{1}{3} - 1\frac{3}{6}$

Draw $2\frac{1}{3}$.

Draw $2\frac{1}{3}$ another way. $2\frac{1}{3} = 1\frac{4}{3}$

To subtract $\frac{3}{6}$ from $\frac{4}{3}$, think of the LCD for $\frac{3}{6}$ and $\frac{4}{3}$. Change the thirds to sixths. $1\frac{4}{3} = 1\frac{8}{6}$

Subtract $1\frac{3}{6}$ from $1\frac{8}{6}$.

So, $2\frac{1}{3} - 1\frac{3}{6} = \frac{5}{6}$.

GUIDED PRACTICE

For Exercises 1–6, match the mixed number with the diagram.

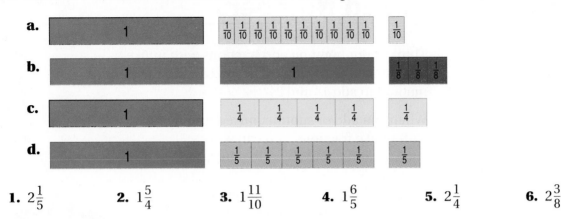

a. | 1 | | $\frac{1}{10}$ $\frac{1}{10}$ $\frac{1}{10}$ $\frac{1}{10}$ $\frac{1}{10}$ $\frac{1}{10}$ $\frac{1}{10}$ $\frac{1}{10}$ $\frac{1}{10}$ $\frac{1}{10}$ | $\frac{1}{10}$

b. | 1 | 1 | $\frac{1}{8}$ $\frac{1}{8}$ $\frac{1}{8}$

c. | 1 | $\frac{1}{4}$ $\frac{1}{4}$ $\frac{1}{4}$ $\frac{1}{4}$ | $\frac{1}{4}$

d. | 1 | $\frac{1}{5}$ $\frac{1}{5}$ $\frac{1}{5}$ $\frac{1}{5}$ $\frac{1}{5}$ | $\frac{1}{5}$

1. $2\frac{1}{5}$ **2.** $1\frac{5}{4}$ **3.** $1\frac{11}{10}$ **4.** $1\frac{6}{5}$ **5.** $2\frac{1}{4}$ **6.** $2\frac{3}{8}$

Draw a diagram to find each difference. Write the answer in simplest form.

7. $2\frac{2}{3} - 1\frac{1}{3}$ **8.** $3\frac{1}{6} - 2\frac{5}{6}$ **9.** $4\frac{1}{6} - 3\frac{1}{3}$ **10.** $5\frac{1}{4} - 1\frac{3}{8}$

INDEPENDENT PRACTICE

Write each mixed number by renaming one whole.

1. $2\frac{1}{2}$ **2.** $2\frac{3}{4}$ **3.** $3\frac{2}{3}$ **4.** $3\frac{5}{6}$ **5.** $5\frac{5}{8}$ **6.** $4\frac{5}{12}$

Draw a diagram to find each difference. Write the answer in simplest form.

7. $2\frac{3}{5} - 1\frac{1}{5}$ **8.** $2\frac{1}{3} - 1\frac{2}{3}$ **9.** $2\frac{3}{4} - 1\frac{1}{4}$ **10.** $2\frac{1}{5} - 1\frac{7}{10}$

11. $2\frac{1}{3} - 2\frac{1}{6}$ **12.** $3\frac{1}{12} - 1\frac{5}{6}$ **13.** $3\frac{5}{6} - 1\frac{1}{6}$ **14.** $3\frac{2}{3} - 2\frac{3}{4}$

Problem-Solving Applications

For Problems 15–17, draw a diagram to solve. Write the answer in simplest form.

15. COMPARE During morning break, the class drank $2\frac{3}{4}$ qt of fruit punch and $1\frac{1}{4}$ qt of apple juice. How much more of fruit punch than apple juice did the class drink?

16. Karen had $3\frac{3}{8}$ yd of cotton fabric. She used $2\frac{5}{8}$ yd for a skirt. How much fabric was left?

17. TRANSPORTATION Joel's father drives $4\frac{2}{3}$ mi to work. His mother drives $2\frac{3}{4}$ mi to work. How much farther does Joel's father drive to work than his mother?

18. ✏️ **WRITE ABOUT IT** When you draw a diagram to show subtraction of mixed numbers, how do you know when to regroup before you subtract?

Adding and Subtracting Mixed Numbers

What You'll Learn
How to add and subtract mixed numbers

Why Learn This?
To solve problems such as finding a total amount of time worked

REMEMBER:

To write a fraction as a mixed number, divide the numerator by the denominator. The quotient is the whole-number part of the mixed number. Use the remainder as the new numerator and the divisor as the denominator.

See page H16.

$$\frac{13}{3} \rightarrow 3\overline{)13} \quad \begin{array}{r} 4\frac{1}{3} \\ \phantom{3\overline{)}}-12 \\ \hline 1 \end{array}$$

Think about how you add fractions with unlike denominators. Do you think you can use the same method to add $1\frac{2}{3}$ and $2\frac{1}{4}$?

When you add or subtract mixed numbers with unlike fractions, you can write equivalent fractions by using the LCD.

Andi works with her father every Friday and Saturday. Andi's father pays her for the number of hours she works. This week Andi worked $3\frac{2}{5}$ hr on Friday and $2\frac{1}{2}$ hr on Saturday. What is the total number of hours Andi worked?

Find $3\frac{2}{5} + 2\frac{1}{2}$.

$$
\begin{array}{ll}
3\frac{2}{5} = 3\frac{4}{10} & \textit{The LCD of } \frac{2}{5} \textit{ and } \frac{1}{2} \textit{ is tenths.} \\
+2\frac{1}{2} = 2\frac{5}{10} & \textit{Write equivalent fractions, using the LCD.} \\
\hline
5\frac{9}{10} & \textit{Add fractions.} \\
& \textit{Add whole numbers.}
\end{array}
$$

So, Andi worked $5\frac{9}{10}$ hr.

• Is the time Andi worked closer to 5 hr or 6 hr? Explain.

Sometimes you need to rewrite a sum.

EXAMPLE 1 Jared, Andi's brother, likes to compete with his sister. He worked $4\frac{2}{3}$ hr on Friday and $3\frac{3}{4}$ hr on Saturday. What is the total number of hours he worked?

Find $4\frac{2}{3} + 3\frac{3}{4}$.

$$
\begin{array}{ll}
4\frac{2}{3} = 4\frac{8}{12} & \textit{Write equivalent fractions, using the LCD.} \\
+3\frac{3}{4} = 3\frac{9}{12} & \textit{Add fractions.} \\
\hline
7\frac{17}{12} = 7 + 1\frac{5}{12} = 8\frac{5}{12} & \textit{Add whole numbers.} \\
& \textit{Rename the fraction as a mixed number. Rewrite the sum.}
\end{array}
$$

So, Jared worked $8\frac{5}{12}$ hr.

Calculator Activities, page H34

Talk About It

- When should you rename the fractions before adding mixed numbers? Give an example.

- When can you rename a fraction as a mixed number?

- CRITICAL THINKING How would you rename the mixed number $2\frac{3}{2}$?

GUIDED PRACTICE

Rename the fractions, using the LCD. Rewrite each problem.

1. $1\frac{1}{3} + 1\frac{1}{2}$ **2.** $2\frac{2}{3} + 1\frac{1}{5}$ **3.** $3\frac{3}{7} + 2\frac{2}{3}$

4. $6\frac{1}{6} + 4\frac{1}{2}$ **5.** $3\frac{3}{4} + 2\frac{1}{3}$ **6.** $2\frac{4}{5} + 4\frac{1}{4}$

Rename the fraction as a mixed number. Write the new mixed number.

7. $1\frac{5}{3}$ **8.** $2\frac{4}{3}$ **9.** $2\frac{7}{4}$ **10.** $3\frac{13}{7}$

11. $2\frac{7}{5}$ **12.** $4\frac{8}{7}$ **13.** $3\frac{12}{7}$ **14.** $6\frac{15}{11}$

15. $3\frac{18}{4}$ **16.** $1\frac{12}{11}$ **17.** $5\frac{18}{3}$ **18.** $2\frac{12}{4}$

Subtracting Mixed Numbers

You can subtract mixed numbers without using diagrams or models.

EXAMPLE 2 Tyler's mother bought $3\frac{3}{4}$ yd of fabric on Wednesday and $2\frac{3}{8}$ yd on Saturday. How much more fabric did she buy on Wednesday?

Find $3\frac{3}{4} - 2\frac{3}{8}$.

$$3\frac{3}{4} = 3\frac{6}{8}$$ *Write equivalent fractions, using the LCD.*
$$-2\frac{3}{8} = 2\frac{3}{8}$$
$$\overline{\qquad 1\frac{3}{8}}$$ *Subtract the fractions.*
           Subtract the whole numbers.

So, Tyler's mother bought $1\frac{3}{8}$ yd more on Wednesday.

- What if Tyler's mother bought $1\frac{2}{3}$ yd of fabric on Saturday? How many more yards did she buy on Wednesday?

Sometimes a whole number needs to be renamed before you can subtract.

EXAMPLE 3 Find the difference. $5 - 1\frac{4}{5}$

$$5 = 4\frac{5}{5}$$ *Since you are subtracting fifths, rename 5 as $4\frac{5}{5}$.*

$$-1\frac{4}{5} = 1\frac{4}{5}$$ *Subtract the fractions.*
Subtract the whole numbers.

$$\overline{\phantom{-1\frac{4}{5}=}3\frac{1}{5}}$$

So, $5 - 1\frac{4}{5} = 3\frac{1}{5}$.

• What if you wanted to find $8 - 3\frac{4}{7}$? What would you rename 8 so you could subtract $3\frac{4}{7}$?

Sometimes a mixed number needs to be renamed before you can subtract.

EXAMPLE 4 Find the difference. $4\frac{1}{6} - 2\frac{7}{9}$

$$4\frac{1}{6} = 4\frac{3}{18}$$ *The LCD of $\frac{1}{6}$ and $\frac{7}{9}$ is eighteenths.*
Write equivalent fractions, using the LCD.

$$-2\frac{7}{9} = 2\frac{14}{18}$$

$$4\frac{1}{6} = 4\frac{3}{18} = 3\frac{21}{18}$$ *Since you can't subtract $\frac{14}{18}$ from $\frac{3}{18}$, rename $4\frac{3}{18}$.*
$$4\frac{3}{18} = 3 + \frac{18}{18} + \frac{3}{18} = 3\frac{21}{18}$$

$$-2\frac{7}{9} = 2\frac{14}{18} = 2\frac{14}{18}$$ *Subtract the fractions.*
Subtract the whole numbers.

$$\overline{\phantom{-2\frac{7}{9}=2\frac{14}{18}=}1\frac{7}{18}}$$

So, $4\frac{1}{6} - 2\frac{7}{9} = 1\frac{7}{18}$.

• **CRITICAL THINKING** What if you wanted to find $4\frac{3}{8} - 2\frac{1}{12}$? What equivalent fractions would you write using the LCD?

INDEPENDENT PRACTICE

Rename the fractions, using the LCD. Rewrite the problem.

1. $1\frac{1}{4} + 1\frac{3}{8}$ **2.** $3\frac{1}{4} + 1\frac{2}{3}$ **3.** $2\frac{1}{2} + 3\frac{2}{5}$ **4.** $2\frac{5}{6} + 4\frac{4}{9}$ **5.** $2\frac{1}{4} + 3\frac{2}{5}$

Rename the fraction as a mixed number. Write the new mixed number.

6. $2\frac{3}{2}$ **7.** $1\frac{7}{4}$ **8.** $3\frac{7}{5}$ **9.** $4\frac{11}{6}$ **10.** $2\frac{11}{7}$ **11.** $5\frac{13}{10}$

Tell whether you must rename to subtract. Write *yes* or *no*.

12. $2\frac{1}{4} - 1\frac{3}{8}$ **13.** $3\frac{1}{4} - 1\frac{2}{3}$ **14.** $7\frac{1}{2} - 3\frac{2}{5}$ **15.** $5\frac{5}{6} - 2\frac{7}{9}$ **16.** $4\frac{1}{4} - 2\frac{2}{5}$

Find the sum. Write the answer in simplest form.

17. $1\frac{1}{4} + 1\frac{1}{8}$ **18.** $2\frac{1}{4} + 4\frac{1}{3}$ **19.** $4\frac{1}{2} + 3\frac{4}{5}$ **20.** $6\frac{5}{6} + 5\frac{7}{9}$ **21.** $7\frac{3}{4} + 3\frac{2}{5}$

22. $2\frac{3}{4} + 4\frac{5}{12}$ **23.** $6\frac{1}{9} + 7\frac{1}{3}$ **24.** $3\frac{2}{7} + 8\frac{1}{3}$ **25.** $5\frac{5}{6} + 4\frac{2}{9}$ **26.** $4\frac{5}{7} + 3\frac{1}{2}$

Find the difference. Write the answer in simplest form.

27. $6\frac{1}{2} - 3\frac{1}{5}$ **28.** $7\frac{5}{6} - 2\frac{5}{9}$ **29.** $4\frac{1}{3} - 2\frac{1}{4}$ **30.** $3\frac{1}{4} - 1\frac{1}{6}$ **31.** $7\frac{1}{4} - 4\frac{3}{5}$

32. $12\frac{1}{4} - 10\frac{3}{4}$ **33.** $5\frac{1}{2} - 3\frac{7}{10}$ **34.** $12\frac{1}{9} - 7\frac{1}{3}$ **35.** $11\frac{1}{4} - 9\frac{7}{8}$ **36.** $15\frac{1}{6} - 7\frac{2}{3}$

Problem-Solving Applications

37. **INVESTMENTS** Todd's mother bought a share of stock for $25\frac{3}{8}$ dollars. A week later the stock increased $2\frac{3}{4}$ dollars. What is her share of stock worth now?

38. **MUSIC** Robin practices the guitar $5\frac{1}{4}$ hr a week. This week she has practiced $3\frac{5}{12}$ hr. How many more hours will Robin practice this week?

39. **TRANSPORTATION** Sally's father drives $19\frac{1}{3}$ mi to work in the morning. On the way home from work, he picks up Sally's little brother from day care. He drives $22\frac{4}{9}$ mi on the way home. How many miles does he drive to and from work?

40. ✏️ **WRITE ABOUT IT** Explain how subtracting or adding mixed numbers with unlike denominators differs from subtracting or adding mixed numbers with like denominators.

Mixed Review and Test Prep

Round each fraction to 0, $\frac{1}{2}$, or 1.

41. $\frac{2}{9}$ **42.** $\frac{10}{13}$ **43.** $\frac{9}{20}$ **44.** $\frac{2}{15}$ **45.** $\frac{22}{27}$ **46.** $\frac{17}{36}$

Write the equivalent fraction that is in simplest form.

47. $\frac{1}{2}, \frac{5}{10}, \frac{4}{8}, \frac{6}{12}$ **48.** $\frac{8}{12}, \frac{2}{3}, \frac{12}{18}, \frac{18}{27}$ **49.** $\frac{8}{18}, \frac{16}{36}, \frac{4}{9}, \frac{20}{45}$ **50.** $\frac{30}{36}, \frac{20}{24}, \frac{25}{30}, \frac{5}{6}$

51. **CONSUMER MATH** The scouts held a bake sale. They made $31.75 selling cake, $29.55 selling cookies, and $20.75 selling brownies. How much money did the scouts raise?

    A $88.65 **B** $89.35

    C $92.05 **D** Not Here

52. **ESTIMATION** In the United States, 1,500 aluminum cans are collected and recycled every second. Which is a reasonable estimate of the number collected in an hour?

    F 50,000 **G** 500,000

    H 5,000,000 **J** 50,000,000

Estimating Sums and Differences

What You'll Learn
How to estimate sums and differences of mixed numbers

Why Learn This?
To estimate answers to problems such as total length

Look at the ruler below. Is $2\frac{5}{8}$ in. closer to 2 in., $2\frac{1}{2}$ in., or 3 in.?

inches   1   2   3

To estimate sums and differences of mixed numbers, round each mixed number to the nearest whole number or to $\frac{1}{2}$.

REMEMBER:

To estimate with a fraction less than 1, round it to 0, $\frac{1}{2}$, or 1.
See page 112.

Round $\frac{1}{9}$ to 0.

Round $\frac{3}{8}$ to $\frac{1}{2}$.

Round $\frac{4}{5}$ to 1.

EXAMPLE 1 Arnold has $9\frac{1}{4}$ yd of rope and Sherry has $6\frac{7}{8}$ yd of rope. About how much rope do Arnold and Sherry have together?

$9\frac{1}{4} + 6\frac{7}{8}$ *Round each mixed number.*
$\downarrow$ $\downarrow$ **Think:** $\frac{1}{4}$ *rounds to 0;* $\frac{7}{8}$ *rounds to 1.*
$9 + 7 = 16$ *Add.*

So, Arnold and Sherry have about 16 yd of rope.

• What if Arnold has $10\frac{3}{4}$ yd? About how much rope do they have together?

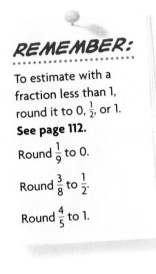

SCIENCE LINK

House cats, tigers, and lions are all closely related. There are differences among cats, especially in size. The average lion is $3\frac{1}{2}$ ft tall. The average house cat is $\frac{3}{4}$ ft tall. About how much taller is the lion?

EXAMPLE 2 Teri is going to build houses for Kevin's dog and cat. Teri needs $14\frac{4}{9}$ ft of lumber for the dog's house and $11\frac{1}{6}$ ft of lumber for the cat's house. About how many more feet of lumber does Teri need for the dog's house?

$14\frac{4}{9} - 11\frac{1}{6}$ *Round each mixed number.*
$\downarrow$ $\downarrow$ **Think:** $\frac{4}{9}$ *rounds to* $\frac{1}{2}$, $\frac{1}{6}$ *rounds to 0.*
$14\frac{1}{2} - 11 = 3\frac{1}{2}$ *Subtract.*

So, Teri needs about $3\frac{1}{2}$ more feet of lumber for the dog's house.

• What if Teri builds a birdhouse? She needs $1\frac{2}{3}$ ft of lumber to build it. About how many more feet of lumber will Teri use to build the cat's house than the birdhouse?

GUIDED PRACTICE

Round the fractions to 0, $\frac{1}{2}$, or 1. Then rewrite the problem.

1. $2\frac{1}{4} + 1\frac{1}{8}$
2. $4\frac{3}{8} - 1\frac{1}{3}$
3. $2\frac{1}{2} + 4\frac{4}{5}$
4. $5\frac{1}{6} + 2\frac{7}{9}$
5. $9\frac{7}{12} - 5\frac{1}{5}$

Estimate the sum or difference.

6. $3\frac{1}{4} + 1\frac{5}{8}$
7. $6\frac{3}{16} - 4\frac{1}{6}$
8. $4\frac{11}{12} - 3\frac{1}{5}$
9. $8\frac{7}{8} + 5\frac{5}{9}$
10. $10\frac{3}{4} - 2\frac{2}{7}$

INDEPENDENT PRACTICE

Estimate the sum or difference.

1. $4\frac{3}{4} + 1\frac{1}{6}$
2. $12\frac{1}{12} + 6\frac{7}{8}$
3. $7\frac{9}{16} - 4\frac{2}{9}$
4. $10\frac{9}{11} + 2\frac{3}{16}$
5. $8\frac{7}{12} - 6\frac{3}{14}$

6. $6\frac{5}{12} - 4\frac{7}{8}$
7. $8\frac{2}{7} + 4\frac{13}{15}$
8. $6\frac{7}{16} - 3\frac{3}{17}$
9. $8\frac{6}{11} + 8\frac{7}{16}$
10. $9\frac{2}{11} - 6\frac{7}{15}$

11. $3\frac{1}{6} + 1\frac{1}{3}$
12. $2\frac{7}{12} + 12\frac{1}{10}$
13. $8\frac{2}{9} - 2\frac{7}{18}$
14. $10\frac{1}{2} + 2\frac{7}{8}$
15. $12\frac{7}{15} - 10\frac{7}{18}$

Problem-Solving Applications

16. HOBBIES Jill used $1\frac{7}{8}$ yd of blue ribbon and $10\frac{5}{12}$ yd of purple ribbon for her craft show project. Her next project needs the same amount of ribbon but only one color. Estimate the amount of ribbon Jill will need.

17. ESTIMATION Mr. Marcos was using $22\frac{15}{16}$ gal of water a day. One faucet leaked $2\frac{3}{4}$ gal of water a day. Estimate how many gallons of water a day he will use when the faucet is fixed.

18. Mr. Kelly had $60\frac{3}{16}$ ft of plastic piping. He installed $28\frac{7}{12}$ ft in the bathroom. About how much piping does he have left?

19. ✏ **WRITE ABOUT IT** Explain how to estimate sums or differences of mixed numbers.

Mixed Review and Test Prep

Multiply.

20. 0.5×0.25
21. 0.75×0.2
22. 0.5×0.75
23. 0.125×0.2

Find the sum or difference. Write the answer in simplest form.

24. $\frac{5}{6} - \frac{1}{4}$
25. $\frac{3}{5} + \frac{2}{3}$
26. $\frac{4}{5} - \frac{1}{2}$
27. $\frac{1}{6} + \frac{3}{8}$

28. CONSUMER MATH Marci bought 3 CDs for $36.97. The first CD cost $12.99 and the second cost $10.99. How much did the third CD cost?

A $11.99
B $12.99
C $13.99
D $14.99

29. SPORTS The attendance at a Rangers game was 31,724. What is this number rounded to the nearest hundred?

F 32,000
G 31,800
H 31,700
J 30,000

Draw a diagram to find the sum or difference. Write the answer in simplest form. (pages 120–125)

1. $1\frac{3}{8} - 1\frac{1}{8}$

2. $2\frac{5}{6} + 1\frac{3}{6}$

3. $2\frac{2}{10} - 1\frac{3}{5}$

4. $1\frac{3}{6} - 1\frac{1}{3}$

5. $2\frac{2}{5} + 1\frac{1}{10}$

6. $3\frac{1}{4} - 2\frac{1}{2}$

7. $1\frac{1}{6} + 3\frac{2}{3}$

8. $2\frac{3}{4} + 3\frac{1}{8}$

9. $1\frac{1}{2} + 2\frac{1}{4}$

10. $1\frac{1}{5} + 1\frac{3}{10}$

11. $2\frac{1}{6} - 1\frac{5}{6}$

12. $3 - 1\frac{3}{8}$

13. After soccer practice, Josh drank $3\frac{1}{3}$ c of water and Travis drank $2\frac{5}{6}$ c of water. How much water did Josh and Travis drink altogether?

14. Pat lives $1\frac{2}{5}$ mi from school. Kathryn lives $2\frac{3}{10}$ mi from school. How much closer to school does Pat live than Kathryn?

Find the sum or difference. (pages 126–129)

15. $2\frac{2}{7} + 3\frac{3}{7}$

16. $7\frac{3}{4} - 5\frac{1}{3}$

17. $8\frac{1}{3} - 3\frac{1}{8}$

18. $11\frac{5}{6} - 6\frac{2}{3}$

19. $1\frac{2}{4} + 5\frac{1}{8}$

20. $6\frac{3}{5} + 2\frac{1}{3}$

21. $3\frac{1}{4} - 2\frac{1}{2}$

22. $4\frac{3}{8} + 2\frac{3}{4}$

23. $4\frac{1}{2} - 1\frac{2}{3}$

24. $6\frac{1}{3} - 3\frac{1}{4}$

25. $5\frac{2}{3} + 3\frac{1}{2}$

26. $4\frac{3}{4} + 4\frac{1}{5}$

27. Allie grew $2\frac{2}{3}$ in. during the fifth grade and $3\frac{1}{4}$ in. during the sixth grade. How much did Allie grow during the two school years?

28. Shane used $5\frac{1}{6}$ yd of rope for a rope swing. He is left with $3\frac{1}{3}$ yd of rope. How much rope did Shane have to start with?

Estimate the sum or difference. (pages 130–131)

29. $9\frac{4}{5} - 2\frac{7}{9}$

30. $6\frac{1}{4} + 3\frac{2}{9}$

31. $4\frac{9}{20} + 1\frac{4}{5}$

32. $3\frac{3}{7} + 1\frac{1}{14}$

33. $4\frac{3}{4} - 1\frac{1}{8}$

34. $3\frac{2}{4} - 1\frac{6}{7}$

35. $5\frac{1}{3} + 7\frac{2}{7}$

36. $8\frac{3}{4} - 2\frac{2}{5}$

37. $6\frac{1}{4} + 3\frac{2}{3}$

38. $8\frac{5}{8} + 5\frac{1}{6}$

39. $7\frac{5}{6} - 1\frac{3}{5}$

40. $12\frac{1}{8} - 9\frac{1}{6}$

41. Michelle practiced the piano for $1\frac{3}{7}$ hr on Monday and $2\frac{1}{6}$ hr on Wednesday. About how many hours did Michelle practice?

42. Barry is $5\frac{5}{8}$ ft tall. Kevin is $4\frac{1}{4}$ ft tall. About how much taller is Barry than Kevin?

1. A school sweatshirt costs $9 more than a school T-shirt. Together, the shirts cost $23. What is the cost of each type of shirt?

   A T-shirt: $4, sweatshirt: $19
   B T-shirt: $7, sweatshirt: $16
   C T-shirt: $12, sweatshirt: $11
   D T-shirt: $16, sweatshirt: $7
   E T-shirt: $19, sweatshirt: $4

2. With tax, a CD player costs $78.95. Ron saves $4.95 each week. What is a reasonable estimate of the number of weeks he must save for the CD player?

   F less than 15 weeks
   G between 15 and 17 weeks
   H between 18 and 20 weeks
   J more than 20 weeks

3. Leon delivered newspapers every third day in October, beginning on October 3. What was the date of the sixth day he delivered newspapers?

   A October 6   **B** October 12
   C October 15   **D** October 18

4. Rico read $\frac{1}{5}$ of a book on Saturday and $\frac{1}{8}$ of the same book on Sunday. What portion of the book does he still have to read?

   F $\frac{2}{13}$

   G $\frac{4}{15}$

   H $\frac{11}{13}$

   J $\frac{17}{40}$

   K Not Here

5. Will has 5 feet of rope. He uses $2\frac{5}{6}$ feet on a school project. How much rope does he have left?

   A $1\frac{1}{5}$ ft   **B** $2\frac{1}{6}$ ft

   C $3\frac{1}{6}$ ft   **D** $7\frac{5}{6}$ ft

6. The Drama Club rehearsed for $2\frac{1}{4}$ hours on Saturday and $3\frac{1}{6}$ hours on Sunday. Altogether, how long did the club members rehearse?

   F $3\frac{7}{12}$ hr   **G** $4\frac{1}{6}$ hr

   H $4\frac{7}{8}$ hr   **J** $5\frac{5}{12}$ hr

7. Dana has $6\frac{1}{2}$ yards of fabric. She uses $2\frac{7}{8}$ yards for a pillow. How much fabric is left?

   A $2\frac{1}{8}$ yd

   B $3\frac{5}{8}$ yd

   C $6\frac{3}{8}$ yd

   D $9\frac{3}{8}$ yd

8. Wendy rode her bicycle $1\frac{3}{8}$ miles on Monday, $2\frac{1}{4}$ miles on Wednesday, and $\frac{4}{5}$ mile on Friday. What is a reasonable estimate of the total distance Wendy rode her bicycle?

   F 3 mi

   G $4\frac{1}{2}$ mi

   H $5\frac{1}{2}$ mi

   J $6\frac{1}{2}$ mi

7

MULTIPLYING AND DIVIDING FRACTIONS

LOOK AHEAD

In this chapter you will solve problems that involve

- **multiplying fractions and mixed numbers**

- **dividing fractions and mixed numbers**

BUSINESS LINK

The price of one share of stock, in dollars, can be shown using fractions or decimals. For example, if a stock sells for $17.50 per share, its price will appear in a stock table as either

COMPANY	PRICE PER SHARE
Disney	$94\frac{3}{4}$
General Electric	$72\frac{3}{4}$
Converse	$7\frac{1}{4}$
Pepsi	38

$17\frac{1}{2}$ or 17.5. The table shows the stock prices for several familiar companies as of November 21, 1997.

- Give each share price in decimal form.

- Find the cost of 100 shares of each stock.

Problem-Solving Project

Taking Stock

Suppose that you have $1,000 to invest in stocks. Pick 4 or 5 stocks using the business section of the newspaper. Decide how many shares of each stock you could buy to be worth a total of about $1,000. Keep track of the value of each stock for two weeks. Make a table to summarize the day-to-day value of your investment for the two weeks. Explain the results.

PROJECT CHECKLIST

- ✓ Did you select 4–5 companies for your investment?
- ✓ Did you decide how many shares to buy with $1,000?
- ✓ Did you make a table to record the daily price changes?

Multiplying with Fractions

Joshua had $\frac{3}{4}$ of a candy bar. He gave his sister $\frac{1}{2}$ of what he had. To find what part of the whole candy bar Joshua gave his sister, find $\frac{1}{2}$ of $\frac{3}{4}$, or $\frac{1}{2} \times \frac{3}{4}$.

ACTIVITY

WHAT YOU'LL NEED: paper, ruler, 2 different-colored pencils or markers

Make a model to help find the product $\frac{1}{2} \times \frac{3}{4}$. Use a piece of paper to represent a whole.

- Fold the paper into 4 equal parts as shown below. Each part represents $\frac{1}{4}$. Color 3 parts to represent $\frac{3}{4}$.

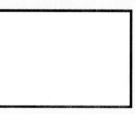 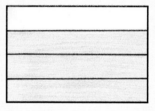

- Fold the paper in half so that you divide each of the fourths into 2 equal parts. Using the other color, shade 1 of the halves.

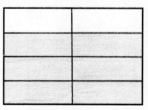 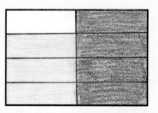

Of the 8 parts, 3 are shaded twice. These 3 parts represent $\frac{1}{2} \times \frac{3}{4}$.

The model shows that $\frac{1}{2} \times \frac{3}{4} = \frac{3}{8}$.

- Is $\frac{3}{8}$ larger or smaller than the $\frac{3}{4}$ you started with?
- Make a model to find $\frac{1}{4} \times \frac{3}{4}$.
- What is the product $\frac{1}{4} \times \frac{3}{4}$?

As you model multiplication of fractions, look for a relationship between the factors and the product.

EXAMPLE 1 Find $\frac{2}{3}$ of $\frac{1}{2}$, or $\frac{2}{3} \times \frac{1}{2}$.

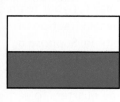

Fold a piece of paper into 2 equal parts. Color 1 part to show $\frac{1}{2}$.

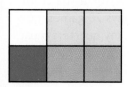

Fold the paper into thirds. Color $\frac{2}{3}$ of the paper.

Of the 6 parts, 2 are shaded twice. These parts represent $\frac{2}{3} \times \frac{1}{2}$.

So, $\frac{2}{3} \times \frac{1}{2} = \frac{2}{6}$.

- **CRITICAL THINKING** Compare the numerator and denominator of the product with the numerators and denominators of the factors. What relationship do you see?

In the Activity and Example 1, you can see this relationship:

$$\frac{\text{numerator} \times \text{numerator}}{\text{denominator} \times \text{denominator}} = \frac{\text{numerator}}{\text{denominator}}$$

$$\uparrow \qquad\qquad \uparrow \qquad\quad\; \uparrow$$

factor $\qquad$ factor $\quad = \quad$ product

You can use this relationship to multiply fractions without a model.

EXAMPLE 2 Use the relationship above to find $\frac{1}{3} \times \frac{3}{4}$. Write the product in simplest form.

$\frac{1}{3} \times \frac{3}{4} = \frac{1 \times 3}{3 \times 4}$ *Multiply the numerators.*
                 Multiply the denominators.

$\qquad = \frac{3}{12}$

$\qquad = \frac{3 \div 3}{12 \div 3}$ *Divide the numerator and the denominator by the GCF, 3.*

$\qquad = \frac{1}{4}$ *Write the product in simplest form.*

So, $\frac{1}{3} \times \frac{3}{4} = \frac{1}{4}$.

- Example 2 shows that $\frac{1}{3} \times \frac{3}{4} = \frac{1}{4}$. Explain why the product is less than the factor $\frac{3}{4}$.

LANGUAGE LINK

Fractions are an everyday part of our language. For example, if someone tells you it is quarter to two, you know that the time is 1:45, or 15 min before 2:00. Make a list of other everyday expressions that include fractions.

REMEMBER:

To write a fraction in simplest form, divide the numerator and the denominator by the greatest common factor (GCF). **See page H15.**

$\frac{4}{10} = \frac{4 \div 2}{10 \div 2}$ GCF: 2

$\qquad = \frac{2}{5}$

Multiplying Fractions and Whole Numbers

You can use the same method to multiply a whole number by a fraction.

EXAMPLE 3 Christina is on the track team at school. She practices every afternoon after school. She runs 9 times around the $\frac{3}{4}$-mi track. What is a reasonable estimate of the distance she runs?

$9 \times \frac{3}{4} \leftarrow \frac{3}{4}$ is about 1.   *Round the fraction to the nearest whole number.*

$9 \times \frac{3}{4} \approx 9 \times 1$

$9 \times 1 = 9$ *Multiply.*

So, she runs about 9 mi.

• What is a reasonable estimate of $\frac{2}{3} \times 11$?

EXAMPLE 4 Christina's coach wants to know the exact distance she runs. Find $9 \times \frac{3}{4}$.

$9 \times \frac{3}{4} = \frac{9}{1} \times \frac{3}{4}$ *Write the whole number as a fraction.*

$\phantom{9 \times \frac{3}{4}} = \frac{9 \times 3}{1 \times 4}$ *Multiply the numerators. Multiply the denominators.*

$\phantom{9 \times \frac{3}{4}} = \frac{27}{4}$, or $6\frac{3}{4}$ *Write the answer as a fraction or as a mixed number.*

So, Christina runs exactly $6\frac{3}{4}$ mi.

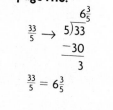

REMEMBER:

A fraction greater than 1 can be written as a mixed number.

See page H16.

$\frac{33}{5} \rightarrow$
$$5\overline{)33}$$
$$\underline{-30}$$
$$3$$
with quotient $6\frac{3}{5}$

$\frac{33}{5} = 6\frac{3}{5}$

GUIDED PRACTICE

Make a model to find the product.

1. $\frac{1}{2} \times \frac{3}{4}$ **2.** $\frac{3}{4} \times \frac{1}{2}$ **3.** $\frac{1}{3} \times \frac{5}{8}$ **4.** $\frac{1}{2} \times \frac{1}{4}$

Estimate the product.

5. $\frac{7}{9} \times 3$ **6.** $5 \times \frac{7}{8}$ **7.** $4 \times \frac{5}{7}$ **8.** $\frac{1}{2} \times \frac{3}{4}$

Find the product. Write the answer in simplest form.

9. $\frac{3}{4} \times \frac{2}{5}$ **10.** $\frac{2}{5} \times \frac{7}{8}$ **11.** $2 \times \frac{6}{7}$ **12.** $\frac{2}{3} \times 16$

13. $\frac{2}{3} \times 4$ **14.** $9 \times \frac{2}{3}$ **15.** $\frac{5}{6} \times \frac{2}{3}$ **16.** $\frac{3}{5} \times \frac{5}{6}$

INDEPENDENT PRACTICE

Make a model to help you find the product.

1. $\frac{1}{2} \times \frac{1}{4}$ **2.** $\frac{1}{2} \times 4$ **3.** $\frac{3}{5} \times \frac{1}{2}$ **4.** $\frac{1}{2} \times \frac{1}{2}$ **5.** $7 \times \frac{1}{2}$

Find the product. Write it in simplest form.

6. $\frac{1}{3} \times \frac{2}{3}$ **7.** $\frac{1}{2} \times \frac{1}{2}$ **8.** $\frac{3}{4} \times \frac{1}{4}$ **9.** $\frac{1}{5} \times \frac{2}{3}$ **10.** $\frac{1}{4} \times \frac{2}{7}$

11. $\frac{4}{5} \times \frac{7}{8}$ **12.** $\frac{2}{9} \times \frac{3}{4}$ **13.** $\frac{2}{5} \times \frac{5}{7}$ **14.** $\frac{1}{8} \times \frac{4}{5}$ **15.** $\frac{5}{9} \times \frac{3}{10}$

16. $\frac{4}{9} \times \frac{3}{5}$ **17.** $\frac{6}{7} \times \frac{7}{8}$ **18.** $\frac{2}{3} \times 21$ **19.** $24 \times \frac{1}{12}$ **20.** $\frac{1}{8} \times 16$

21. $\frac{5}{6} \times 5$ **22.** $13 \times \frac{3}{5}$ **23.** $5 \times \frac{1}{9}$ **24.** $14 \times \frac{7}{12}$ **25.** $\frac{2}{5} \times 14$

Complete the multiplication sentence.

26. $\frac{1}{5} \times \frac{\blacksquare}{3} = \frac{4}{15}$ **27.** $\frac{\blacksquare}{4} \times \frac{3}{8} = \frac{9}{32}$ **28.** $\frac{2}{\blacksquare} \times \frac{3}{5} = \frac{2}{5}$ **29.** $\frac{\blacksquare}{4} \times \frac{2}{3} \times \frac{\blacksquare}{2} = \frac{1}{2}$

Problem-Solving Applications

30. TRANSPORTATION Sandra takes $\frac{1}{2}$ hr to walk to school. She spends $\frac{1}{2}$ of that time walking down her street. What part of an hour does Sandra spend walking down her street? How many minutes is this?

31. In a stock market game, Roberto bought 150 shares of stock at $\frac{5}{8}$ per share. How much money did 150 shares cost?

32. SOCIAL STUDIES There are 144 registered voters in Booker County. In the last election, $\frac{3}{4}$ of them voted. How many voters voted in the last election?

33. ✏️ **WRITE ABOUT IT** Explain how you can multiply two fractions.

Mixed Review and Test Prep

Find the greatest common factor (GCF) of each pair of numbers.

34. 2 and 4   **35.** 4 and 8   **36.** 6 and 9   **37.** 15 and 10

Find the sum. Write it in simplest form.

38. $\frac{1}{12} + \frac{7}{12}$ **39.** $\frac{3}{4} + \frac{1}{6}$ **40.** $2\frac{2}{7} + 3\frac{3}{7}$ **41.** $1\frac{1}{2} + 2\frac{2}{5}$

42. NUMBER SENSE Cindy chose a number, added 5, multiplied the sum by 3, subtracted 10, and doubled the result. Her final number was 28. What number had she chosen?

    A 1   **B** 2
    C 3   **D** 4

43. TIME It will take Albert about $1\frac{1}{2}$ hours to decorate, and almost 45 minutes to set out food for his party. It is 3:30 P.M. now. Which is the best estimate of when he will be ready for the party?

    F about 5 P.M.   **G** about 6 P.M.
    H about 6:45 P.M.   **J** after 7 P.M.

Simplifying Factors

You can often simplify fractions before you multiply.

Cheryl has $\frac{2}{3}$ of a case of soda left from a picnic. If she gives $\frac{1}{4}$ of it away, what part of the case will she give away?

Think: What is $\frac{1}{4}$ of $\frac{2}{3}$?

Find $\frac{1}{4} \times \frac{2}{3}$.

$\frac{1}{4} \times \frac{2}{3}$ ← The GCF of 2 and 4 is 2.

$\frac{1}{\overset{4}{\underset{2}{4}}} \times \frac{\overset{1}{2}}{3}$ ← $2 \div 2 = 1$
                                    ← $4 \div 2 = 2$

$\frac{1}{\overset{4}{\underset{2}{4}}} \times \frac{\overset{1}{2}}{3} = \frac{1 \times 1}{2 \times 3} = \frac{1}{6}$

Look for a numerator and denominator with common factors. Find the greatest common factor (GCF).

Divide the numerator and denominator by the GCF, 2.

Multiply.

So, Cheryl will give away $\frac{1}{6}$ of a case of soda.

SCIENCE LINK

Fractions are used very little in science. One time when they are used, however, is when it is necessary to convert temperatures from Celsius to Fahrenheit. The formulas for these conversions are:

$°F = \frac{9}{5} \times °C + 32°$

$°C = \frac{5}{9} \times (°F - 32°)$

Suppose the temperature is 100°C. What is this in °F?

Talk About It

- Find the product of $\frac{1}{4}$ and $\frac{2}{3}$ without using the GCF.

- Would you have to simplify your answer if you didn't use the GCF? Explain.

- **CRITICAL THINKING** When you use the GCF, do you need to simplify the product after you multiply? Explain.

EXAMPLE Find $\frac{2}{3} \times \frac{3}{10}$.

$\frac{2}{3} \times \frac{3}{10} = \frac{\overset{1}{2}}{\underset{1}{3}} \times \frac{\overset{1}{3}}{\underset{5}{10}} = \frac{1}{5}$ *The GCF of 3 and 3 is 3.*
                                            The GCF of 2 and 10 is 2.

So, $\frac{2}{3} \times \frac{3}{10} = \frac{1}{5}$.

- Look at the problem $\frac{5}{8} \times \frac{4}{5}$. To completely simplify the factors, would you use one GCF or two? Explain your answer.

- Find $\frac{3}{4} \times \frac{5}{9}$.

GUIDED PRACTICE

Tell each GCF you would use to simplify the fractions.

1. $\frac{5}{8} \times \frac{8}{16}$

2. $\frac{8}{9} \times \frac{3}{16}$

3. $\frac{2}{9} \times \frac{3}{4}$

4. $\frac{4}{3} \times \frac{1}{2}$

5. $\frac{1}{8} \times \frac{4}{5}$

6. $\frac{2}{3} \times \frac{5}{6}$

7. $\frac{2}{15} \times \frac{5}{6}$

8. $6 \times \frac{2}{3}$

INDEPENDENT PRACTICE

Tell each GCF you would use to simplify the fractions.

1. $\frac{1}{2}, \frac{2}{3}$

2. $\frac{1}{5}, \frac{5}{16}$

3. $\frac{14}{27}, \frac{6}{7}$

4. $\frac{3}{4}, \frac{8}{9}$

5. $\frac{1}{4}, \frac{4}{5}$

6. $\frac{4}{20}, \frac{5}{8}$

7. $\frac{2}{9}, \frac{9}{16}$

8. $\frac{3}{5}, \frac{4}{6}$

Use GCFs to simplify the factors. Write the new problem.

9. $\frac{3}{4} \times \frac{4}{7}$

10. $\frac{1}{2} \times \frac{6}{7}$

11. $\frac{3}{5} \times \frac{5}{6}$

12. $\frac{9}{12} \times \frac{6}{18}$

13. $\frac{5}{8} \times \frac{12}{25}$

14. $\frac{4}{7} \times \frac{5}{16}$

15. $\frac{3}{5} \times \frac{1}{6}$

16. $\frac{4}{5} \times \frac{5}{16}$

Use GCFs to simplify the factors so that the answer is in simplest form.

17. $\frac{2}{3} \times \frac{1}{6}$

18. $\frac{5}{6} \times \frac{3}{8}$

19. $\frac{7}{10} \times \frac{5}{7}$

20. $\frac{4}{5} \times 5$

21. $\frac{8}{9} \times \frac{3}{4}$

22. $\frac{6}{7} \times \frac{5}{8}$

23. $\frac{4}{7} \times \frac{21}{28}$

24. $\frac{9}{10} \times \frac{5}{18}$

25. $\frac{3}{10} \times \frac{5}{9}$

26. $\frac{3}{13} \times \frac{13}{3}$

27. $\frac{5}{6} \times \frac{7}{10}$

28. $\frac{4}{5} \times \frac{5}{8}$

29. $42 \times \frac{1}{6}$

30. $\frac{8}{9} \times 27$

31. $\frac{4}{9} \times 45$

32. $81 \times \frac{1}{3}$

Problem-Solving Applications

33. SPORTS On Saturday, 39 teenagers went into the skateboard shop. Of those teens, $\frac{1}{3}$ bought skateboards. How many teens bought skateboards?

34. At Longwood School, $\frac{3}{5}$ of the students in sixth grade are boys. Of those boys, $\frac{1}{3}$ are on the honor roll. What fraction of the sixth-grade boys are on the honor roll?

35. CONSUMER MATH The Hightown Apartments are being remodeled. So far, $\frac{3}{7}$ of them have been repainted, and another $\frac{2}{7}$ have been recarpeted. How many of the 343 apartments have been repainted or recarpeted?

36. SPORTS Mark took a survey and found that $\frac{4}{5}$ of the skateboard riders wore helmets. Of those, $\frac{1}{2}$ wore knee pads. What fraction of the skateboard riders used both helmets and knee pads?

37. ✏️ **WRITE ABOUT IT** What are the advantages of simplifying before multiplying fractions?

Mixed Numbers

What You'll Learn
How to multiply fractions and mixed numbers

Why Learn This?
To solve everyday problems that involve measurements given as mixed numbers, such as $3\frac{1}{3}$ yd and $1\frac{1}{2}$ mi

You can multiply a fraction and a mixed number.

Tony took $1\frac{1}{3}$ dozen cookies to his family reunion. His relatives ate $\frac{1}{2}$ of them. What part of a dozen did they eat?

Think: What is $\frac{1}{2}$ of $1\frac{1}{3}$? Find $\frac{1}{2} \times 1\frac{1}{3}$.

$\frac{1}{2} \times 1\frac{1}{3} = \frac{1}{2} \times \frac{4}{3}$ *Write the mixed number as a fraction.*

$= \frac{1}{\underset{1}{2}} \times \overset{2}{\cancel{\frac{4}{3}}}$ *Use the GCF to simplify.*

$= \frac{1}{1} \times \frac{2}{3} = \frac{2}{3}$ *Multiply.*

So, the relatives ate $\frac{2}{3}$ dozen cookies.

- Jackie took $2\frac{1}{2}$ gallons of ice water to the reunion. The relatives drank $\frac{3}{4}$ of it. How much did they drink?

You can use the same method to multiply two mixed numbers.

EXAMPLE After lunch at the reunion, Cami rode her bike $1\frac{3}{5}$ mi. Tim rode $1\frac{1}{2}$ times as far. How far did Tim ride?

Think: Tim rode $1\frac{1}{2} \times 1\frac{3}{5}$ mi.

$1\frac{1}{2} \times 1\frac{3}{5} = \frac{3}{2} \times \frac{8}{5}$ *Write the mixed numbers as fractions.*

$= \frac{3}{2} \times \overset{4}{\cancel{\frac{8}{5}}}$ *Use the GCF to simplify.*

$= \frac{3}{1} \times \frac{4}{5}$ *Multiply.*

$= \frac{12}{5}$, or $2\frac{2}{5}$

So, Tim rode $2\frac{2}{5}$ mi.

REMEMBER:

To change a mixed number to a fraction, multiply the whole number by the denominator and add the numerator to the product. Write the sum as the new numerator. Use the same denominator.
See page H16.

$2\frac{3}{4} = \frac{(2 \times 4) + 3}{4}$

$= \frac{11}{4}$

You can use a calculator to multiply mixed numbers, such as $3\frac{3}{4} \times 1\frac{1}{2}$.

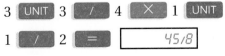

3 [UNIT] 3 [/] 4 [×] 1 [UNIT]

1 [/] 2 [=]   45/8

- What happens to the product when you press the key [Ab/c] ?

GUIDED PRACTICE

Rewrite the problem by changing each mixed number
to a fraction.

1. $2\frac{1}{4} \times \frac{1}{7}$ **2.** $\frac{2}{9} \times 2\frac{3}{8}$ **3.** $1\frac{2}{3} \times 3\frac{2}{5}$ **4.** $4\frac{1}{3} \times 1\frac{6}{11}$

Find the product.

5. $\frac{3}{4} \times 1\frac{1}{2}$ **6.** $\frac{1}{2} \times 2\frac{1}{3}$ **7.** $1\frac{1}{2} \times 1\frac{1}{2}$ **8.** $1\frac{2}{5} \times 2\frac{1}{4}$

INDEPENDENT PRACTICE

Rewrite the problem by changing each mixed number
to a fraction.

1. $4\frac{3}{4} \times \frac{1}{8}$ **2.** $\frac{2}{7} \times 6\frac{3}{5}$ **3.** $1\frac{1}{3} \times 3\frac{3}{4}$ **4.** $2\frac{1}{3} \times 1\frac{6}{7}$

Find the product. Write it in simplest form.

5. $3\frac{1}{2} \times \frac{4}{5}$ **6.** $1\frac{1}{3} \times 3\frac{3}{4}$ **7.** $3\frac{2}{3} \times 1\frac{5}{8}$ **8.** $\frac{1}{5} \times 3\frac{1}{3}$

9. $4\frac{2}{3} \times 1\frac{3}{4}$ **10.** $1\frac{3}{8} \times 4\frac{2}{3}$ **11.** $5\frac{1}{2} \times \frac{1}{6}$ **12.** $\frac{1}{2} \times 3\frac{1}{7}$

13. $4\frac{1}{6} \times 3\frac{3}{5}$ **14.** $1\frac{3}{4} \times \frac{1}{3}$ **15.** $10\frac{1}{5} \times 8\frac{1}{3}$ **16.** $\frac{1}{5} \times 5\frac{5}{6}$

17. $3\frac{1}{3} \times 2\frac{1}{7}$ **18.** $3\frac{1}{4} \times \frac{2}{5}$ **19.** $\frac{7}{8} \times 7\frac{3}{7}$ **20.** $9\frac{3}{8} \times 4\frac{4}{5}$

Problem-Solving Applications

21. TRANSPORTATION Richard and Hector walked to the picnic.
Richard walked $1\frac{3}{5}$ mi. Hector walked $2\frac{1}{2}$ times as far as
Richard. How far did Hector walk?

22. Mr. Jackson has $1\frac{2}{3}$ c of fruit. The fruit is $\frac{1}{4}$ grapes. How many
cups of grapes does he have?

23. SPORTS Lakeshia is training for a track meet. She walks $5\frac{3}{4}$ mi
every day. How many miles does she walk in one week?

24. CALCULATOR Use a fraction calculator to show the key
sequence you would use to find $\frac{1}{2} \times 2\frac{1}{4}$.

25. CALCULATOR What product would be the result of this
key sequence?

    1 [UNIT] 3 [/] 4 [×] 2 [/] 3 [=]

26. ✏️ **WRITE ABOUT IT** When you multiply two mixed
numbers, is the product less than or greater than the factors?
Give an example.

Technology Link

💿 In *Mighty Math
Number
Heroes,* the game
called *Fraction
Fireworks* challenges
you to create
fireworks to show
products of fractions.
Use Grow Slide
Level X.

LAB ACTIVITY

What You'll Explore
How to model division of fractions

What You'll Need
fraction circles

Exploring Division

Using models will help you understand division of fractions.

ACTIVITY 1

Explore

A. Use fraction circles to find $3 \div \frac{1}{2}$, or the number of halves in 3 wholes.

- Trace 3 whole circles on your paper.

- Model $3 \div \frac{1}{2}$ by tracing $\frac{1}{2}$-circle pieces on the 3 circles.

One whole equals two halves.

- How many halves are in 3 wholes? What is $3 \div \frac{1}{2}$?

B. Use fraction circles to find $\frac{1}{2} \div \frac{1}{6}$, or the number of sixths in $\frac{1}{2}$.

- Place as many $\frac{1}{6}$ pieces as you can on the $\frac{1}{2}$ piece.

- How many sixths are in $\frac{1}{2}$? What is $\frac{1}{2} \div \frac{1}{6}$?

Think and Discuss

- Why does $3 \div \frac{1}{2} = 6$?

- How can you find the number of thirds in 2 wholes?

- How would you model $3 \div \frac{1}{6}$?

Try This

Use fraction circles to model each problem. Draw a diagram of your model.

1. $3 \div \frac{1}{3}$

2. $4 \div \frac{1}{4}$

3. $\frac{1}{2} \div \frac{1}{8}$

4. $\frac{3}{4} \div \frac{1}{8}$

ACTIVITY 2

Explore

Recall that dividing a number by 2 gives you the same result as multiplying that number by $\frac{1}{2}$.

$$8 \div 2 = 4 \qquad 8 \times \frac{1}{2} = 4$$

This relationship between multiplication and division is also true for other fractions. Look at these pairs of related problems.

$$9 \div 3 = 3 \qquad 9 \times \frac{1}{3} = 3 \qquad 6 \div \frac{3}{4} = 8 \qquad 6 \times \frac{4}{3} = 8$$

$$1 \div \frac{1}{2} = 2 \qquad 1 \times \frac{2}{1} = 2 \qquad \frac{1}{3} \div \frac{1}{6} = \frac{6}{3} \qquad \frac{1}{3} \times \frac{6}{1} = \frac{6}{3}$$

- What pattern do you see in the problems of each pair?

- Use this pattern to write a related multiplication problem for $\frac{1}{3} \div \frac{2}{3} = \frac{3}{6}$.

Think and Discuss

Look at $1 \div \frac{1}{2} = 2$ and $1 \times \frac{2}{1} = 2$. Remember that dividing by $\frac{1}{2}$ is the same as multiplying by 2.

- What is the relationship between the dividend in the division problem and the first factor in the multiplication problem?

- What is the relationship between the quotient and the product?

- What is the relationship between the divisor in the division problem and the second factor in the multiplication problem?

Try This

Complete the multiplication problem.

1. $\frac{3}{4} \div \frac{1}{4} = 3$
$\frac{3}{4} \times \blacksquare = 3$

2. $\frac{1}{3} \div \frac{1}{6} = 2$
$\frac{1}{3} \times \blacksquare = 2$

3. $12 \div \frac{3}{4} = 16$
$12 \times \blacksquare = 16$

4. $15 \div \frac{1}{3} = 45$
$15 \times \blacksquare = 45$

5. $\frac{2}{3} \div \frac{1}{9} = 6$
$\frac{2}{3} \times \blacksquare = \blacksquare$

6. $\frac{3}{4} \div \frac{1}{10} = 7\frac{1}{2}$
$\frac{3}{4} \times \blacksquare = \blacksquare$

Write a multiplication problem for each. Find the quotient.

7. $\frac{5}{6} \div 6$

8. $\frac{2}{3} \div \frac{3}{4}$

9. $\frac{1}{2} \div \frac{1}{5}$

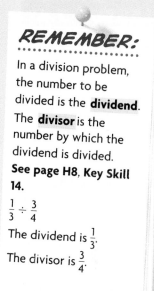

REMEMBER:

In a division problem, the number to be divided is the **dividend**. The **divisor** is the number by which the dividend is divided.
See page H8, Key Skill 14.

$\frac{1}{3} \div \frac{3}{4}$

The dividend is $\frac{1}{3}$.

The divisor is $\frac{3}{4}$.

Technology Link

You can practice dividing fractions with a computer model by using E-Lab, Activity 7. Available on CD-ROM and on the Internet at **www.hbschool.com/elab**

Dividing Fractions

In the previous activity, you saw that related multiplication and division problems give the same result. When you divide by a fraction, you can use multiplication to find the quotient.

When you rewrite the problem, you exchange the numerator and the denominator of the divisor. The new number is called the **reciprocal** of the divisor. The product of a number and its reciprocal is 1.

$$2 \div \frac{2}{3} = 2 \times \frac{3}{2} \leftarrow \frac{3}{2} \text{ is the reciprocal of } \frac{2}{3} \text{ because } \frac{2}{3} \times \frac{3}{2} = 1.$$

EXAMPLE 1 Donovan needs pieces of string for his math project. Each piece must be $\frac{1}{3}$ yd long. How many $\frac{1}{3}$-yd pieces of string can he cut from a piece $\frac{3}{4}$ yd long?

Find $\frac{3}{4} \div \frac{1}{3}$. Remember that dividing by $\frac{1}{3}$ is the same as multiplying by 3.

$\frac{3}{4} \div \frac{1}{3} = \frac{3}{4} \times \frac{3}{1}$ *Use the reciprocal of the divisor to write a multiplication problem.*

$\frac{3}{4} \times \frac{3}{1} = \frac{9}{4}$, or $2\frac{1}{4}$ *Multiply.*

So, Donovan can cut off 2 pieces of string with $\frac{1}{4}$ of a piece left.

- Suppose each piece needs to be $\frac{1}{6}$ yd long. How many $\frac{1}{6}$-yd pieces can he cut off from the original length?

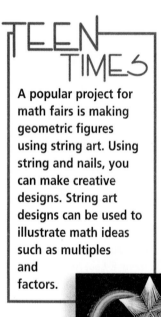

EXAMPLE 2 Find $4 \div \frac{2}{3}$.

$4 \div \frac{2}{3} = \frac{4}{1} \div \frac{2}{3}$ *Write the whole number as a fraction.*

$\frac{4}{1} \div \frac{2}{3} = \frac{4}{1} \times \frac{3}{2}$ *Use the reciprocal of the divisor to write a multiplication problem.*

$\overset{2}{\cancel{\frac{4}{1}}} \times \frac{3}{\underset{1}{\cancel{2}}} = \frac{6}{1}$, or 6 *Simplify and multiply.*

So, $4 \div \frac{2}{3} = \frac{6}{1}$, or 6.

GUIDED PRACTICE

Write the reciprocal of the number.

1. $\frac{2}{3}$ **2.** $\frac{3}{4}$ **3.** 7 **4.** $\frac{4}{7}$ **5.** $\frac{1}{9}$ **6.** 5

Find the quotient. Write it in simplest form.

7. $\frac{1}{3} \div \frac{1}{2}$ **8.** $\frac{1}{5} \div \frac{1}{4}$ **9.** $\frac{1}{4} \div \frac{1}{2}$ **10.** $\frac{1}{2} \div 3$ **11.** $4 \div \frac{2}{3}$

INDEPENDENT PRACTICE

Find the quotient. Write it in simplest form.

1. $\frac{3}{8} \div \frac{1}{2}$ **2.** $\frac{2}{3} \div \frac{4}{7}$ **3.** $\frac{7}{8} \div \frac{1}{3}$ **4.** $8 \div \frac{6}{7}$

5. $12 \div \frac{3}{5}$ **6.** $4 \div \frac{4}{5}$ **7.** $\frac{4}{9} \div \frac{3}{5}$ **8.** $\frac{3}{4} \div \frac{1}{3}$

9. $6 \div \frac{3}{4}$ **10.** $\frac{2}{5} \div 20$ **11.** $\frac{1}{8} \div \frac{4}{5}$ **12.** $\frac{5}{6} \div \frac{1}{3}$

13. $\frac{5}{8} \div 25$ **14.** $\frac{1}{9} \div 6$ **15.** $\frac{7}{12} \div \frac{2}{3}$ **16.** $\frac{1}{6} \div 2$

Problem-Solving Applications

17. COOKING How many $\frac{1}{4}$-lb hamburgers can Katie grill with 12 lb of ground beef?

18. SPORTS In a $\frac{1}{4}$-mi relay, each runner runs $\frac{1}{16}$ mi. How many runners are in the relay?

19. A recording of the current weather conditions lasts $\frac{3}{4}$ min. How many times can the recording be played in 1 hr?

20. ✏ **WRITE ABOUT IT** Explain how to use reciprocals to divide fractions.

Mixed Review and Test Prep

Write as a fraction.

21. $2\frac{1}{3}$ **22.** $1\frac{3}{8}$ **23.** $4\frac{1}{2}$ **24.** $3\frac{9}{10}$ **25.** $2\frac{4}{5}$

Find the difference. Write it in simplest form.

26. $\frac{7}{8} - \frac{3}{8}$ **27.** $\frac{7}{9} - \frac{1}{3}$ **28.** $5\frac{7}{12} - 1\frac{1}{12}$ **29.** $9\frac{5}{9} - 5\frac{1}{6}$ **30.** $3\frac{5}{7} - 1\frac{5}{7}$

31. ESTIMATION Jacki wants to buy a blouse for $12.95, a T-shirt for $15.95, a book for $10.50, and two pens for $3.75 each. What is the best estimate for the total cost?

A $40 **B** $48 **C** $55 **D** $60

32. CONSUMER MATH A 12-inch sub sandwich costs $6.95. A 6-inch sub costs $3.15. Which is a reasonable conclusion?

F A 12-in. sub costs exactly twice what a 6-in. sub costs.

G A 6-in. sub costs more than a 12-in. sub.

H Two 6-in. subs cost less than one 12-in. sub.

J Two 6-in. subs cost more than a 12-in. sub.

Work Backward by Dividing Mixed Numbers

What You'll Learn
How to work backward by dividing mixed numbers

Why Learn This?
To solve problems such as finding the length or width of a box when you know the area

PROBLEM SOLVING
- **Understand**
- **Plan**
- **Solve**
- **Look Back**

You can divide fractions that are greater than 1.

Melissa is making a rectangular box. She has $4\frac{1}{2}$ ft² of colored paper to cover the top. The width of the box must be $1\frac{1}{2}$ ft. What is the greatest length the box can have?

UNDERSTAND What are you asked to find?

What facts are given?

PLAN What strategy will you use?

You can *work backward* to find the length.

The area of a rectangle is found by multiplying the length and the width. Since you know the area and the width, you can divide to find the length.

SOLVE How will you solve the problem?

Area equals length times width: $4\frac{1}{2} = \blacksquare \times 1\frac{1}{2}$

To find the length, work backward by dividing: $4\frac{1}{2} \div 1\frac{1}{2} = \blacksquare$

$4\frac{1}{2} \div 1\frac{1}{2} = \frac{9}{2} \div \frac{3}{2}$ *Write the mixed numbers as fractions.*

$= \frac{9}{2} \times \frac{2}{3}$ *Use the reciprocal of the divisor to write a multiplication problem.*

$= \frac{\overset{3}{\cancel{9}}}{\underset{1}{\cancel{2}}} \times \frac{\overset{1}{\cancel{2}}}{\underset{1}{\cancel{3}}}$ *Simplify the factors.*

$= \frac{3}{1}$, or 3   *Multiply.*

So, she can make the box 3 ft long.

LOOK BACK How can you check your answer?

What if ... the width of the box is $\frac{3}{4}$ ft? How long can the box be?

PRACTICE

Work backward to solve.

1. Ms. Jones leaves home to pick up Amy at school. Then she travels $3\frac{1}{2}$ mi to pick up Todd and $4\frac{1}{4}$ mi to pick up Renee. Ms. Jones drives a total of $12\frac{1}{2}$ mi. What is the distance from home to school?

```
       ?          3½          4¼
   ├────────┼─────────┼─────────┤
  Home     School    Todd     Renee
```

2. Terry is trying to decrease the amount of time he spends talking on the phone. This week he talked for $2\frac{1}{2}$ hr, which is $\frac{2}{3}$ of the time he talked last week. How long did he talk on the phone last week?

3. Gerald has $8\frac{3}{4}$ yd of fabric. This is 7 times the amount he needs to make 1 costume for the school play. How much fabric does he need for each costume?

4. Marsha wants to plant a rectangular garden. She wants it to have an area of $18\frac{3}{4}$ ft². The length is 6 ft. How wide will the garden have to be?

MIXED APPLICATIONS

Solve.

CHOOSE a strategy and a tool.
- Work Backward
- Use a Formula
- Write an Equation
- Draw a Diagram
- Guess and Check
- Make a Table

Paper/Pencil   Calculator   Hands-On   Mental Math

5. A library shelf is 8 ft wide. A set of sports books fills $\frac{1}{4}$ of the shelf space. Each book in the set is $1\frac{1}{2}$ in. thick. How many books are in the set?

6. Lisa left home and walked $1\frac{1}{3}$ mi north. Then she walked $\frac{3}{4}$ mi west, $\frac{1}{3}$ mi north, $1\frac{1}{4}$ mi east, and $1\frac{2}{3}$ mi south. How far is she from her home?

7. Tim's family is driving $10\frac{1}{2}$ hr to Rompa to visit relatives. They will stop every $3\frac{1}{2}$ hr to rest. How many times will they stop to rest during the trip?

8. Caren has a $1\frac{1}{2}$-lb supply of cat treats. Each day, she gives her cat $\frac{1}{8}$ lb of the treats. How many days will her supply last?

9. Quincy went to see two one-act plays. With an intermission of $\frac{1}{4}$ hr, the evening lasted $2\frac{1}{2}$ hr. The first play was $1\frac{1}{4}$ hr long. How long did the second play last?

10. Joel chose a number and multiplied by $1\frac{1}{2}$. The product was $10\frac{1}{2}$. What was the number?

11. Tommy has a rectangular toolbox with a base 21 in. long and 9 in. wide. He wants to cover the bottom of the toolbox with felt material. How many square inches of felt does he need to cover the bottom of the toolbox?

12. 🗣 **WRITE ABOUT IT** Describe the steps needed to use the work backward strategy.

Find the product. Write it in simplest form. (pages 136–139)

1. $\frac{1}{6} \times \frac{3}{5}$ **2.** $\frac{2}{3} \times \frac{4}{7}$ **3.** $16 \times \frac{5}{12}$ **4.** $\frac{3}{8} \times 10$

5. $\frac{1}{3} \times \frac{4}{5}$ **6.** $3 \times \frac{3}{4}$ **7.** $\frac{7}{8} \times \frac{1}{2}$ **8.** $\frac{4}{7} \times 2$

Use the GCF to simplify the factors. Then find the product.
(pages 140–141)

9. $\frac{2}{3} \times \frac{1}{6}$ **10.** $\frac{8}{9} \times \frac{3}{4}$ **11.** $\frac{6}{7} \times \frac{5}{8}$ **12.** $\frac{3}{4} \times 12$

13. $6 \times \frac{2}{3}$ **14.** $16 \times \frac{5}{6}$ **15.** $\frac{3}{8} \times \frac{4}{5}$ **16.** $4 \times \frac{5}{8}$

Find the product. Write it in simplest form. (pages 142–143)

17. $1\frac{2}{3} \times \frac{3}{4}$ **18.** $\frac{4}{15} \times 4\frac{3}{8}$ **19.** $1\frac{1}{3} \times 3\frac{3}{4}$ **20.** $1\frac{7}{8} \times 4\frac{2}{3}$

21. $1\frac{1}{2} \times \frac{3}{4}$ **22.** $4\frac{1}{2} \times 2\frac{1}{3}$ **23.** $1\frac{1}{2} \times \frac{2}{3}$ **24.** $2\frac{1}{3} \times 3\frac{1}{7}$

25. $3\frac{2}{3} \times \frac{6}{7}$ **26.** $\frac{1}{2} \times 3\frac{1}{2}$ **27.** $\frac{3}{5} \times 1\frac{1}{4}$ **28.** $2\frac{1}{2} \times 3\frac{1}{7}$

Solve. (pages 142–143)

29. Susan's father had $\frac{3}{4}$ of a box of nails. He used $\frac{1}{2}$ of the nails to fix the fence. What part of the box did he use? If there were 160 nails in the box, how many nails did he use?

30. Mike rides his bicycle $6\frac{1}{2}$ min to school. Cami rides her bicycle $1\frac{1}{2}$ times as long. How long does it take Cami to ride to school?

31. VOCABULARY When the numerator and the denominator of a fraction are exchanged, the new fraction is called the ___?___ of the original fraction. (page 146)

Find the quotient. Write it in simplest form. (pages 146–147)

32. $\frac{6}{7} \div \frac{3}{5}$ **33.** $\frac{3}{4} \div \frac{1}{3}$ **34.** $\frac{1}{3} \div \frac{4}{5}$

35. $9 \div \frac{3}{8}$ **36.** $8 \div \frac{6}{7}$ **37.** $\frac{4}{5} \div 4$

38. $\frac{3}{5} \div \frac{1}{5}$ **39.** $\frac{4}{7} \div \frac{2}{3}$ **40.** $\frac{5}{8} \div 10$

Solve. (pages 148–149)

41. Terry came in third place running $1\frac{3}{4}$ mph in a three-legged race at school. After practicing, he now runs $2\frac{4}{5}$ mph. How many times faster does he run now?

42. Jami chose a number and multiplied by $2\frac{1}{4}$. The product was $7\frac{7}{8}$. What number did she choose?

TAAS Prep

1. Sheri needs $\frac{3}{4}$ pound of fish food for the next month. The fish food shows decimal weights. What decimal weight should Sheri buy?

   A 0.40 lb   **B** 0.65 lb

   C 0.75 lb   **D** 0.80 lb

2. The Seneca Middle School band had a weekend car wash to raise money for new uniforms. They charged $5 for every car. If they washed 187 cars during the weekend, how much money did they raise?

   F $890   **G** $925

   H $935   **J** $1,400

3. Jeff has $385.00 in his checking account. He writes a check for $58.99, and a check for $22.15. What is Jeff's new balance for his checking account?

   A $303.86   **B** $304.01

   C $304.86   **D** $326.01

4. Damon has a bag of fruit. He has 7 apples, 4 oranges, and 5 bananas. What fraction of the fruit are oranges?

   F $\frac{4}{18}$ **G** $\frac{1}{4}$

   H $\frac{5}{16}$ **J** $\frac{3}{4}$

5. Last week Hillary practiced the piano for $\frac{3}{5}$ hour on Monday, $\frac{5}{6}$ hour on Wednesday, and $\frac{9}{10}$ hour on Friday. How many hours did she practice during the week?

   A $1\frac{1}{3}$ hr

   B $1\frac{2}{3}$ hr

   C 2 hr

   D $2\frac{1}{3}$ hr

6. Matt used $3\frac{3}{4}$ yards of fabric for his costume. Jill used $2\frac{7}{8}$ yards of fabric for her costume. How many yards of fabric did Matt and Jill use?

   F $3\frac{2}{3}$ yd   **G** $4\frac{3}{5}$ yd

   H $5\frac{1}{8}$ yd   **J** $6\frac{5}{8}$ yd

7. Eve bought $5\frac{1}{8}$ pounds of potatoes, $1\frac{2}{3}$ pounds of carrots, and $2\frac{1}{4}$ pounds of beans. About how many pounds of vegetables did she buy?

   A 8 lb

   B 9 lb

   C 10 lb

   D 11 lb

   E Not Here

8. There are 416 students in the sixth grade at Oakdale School. During the first week of school, $\frac{1}{6}$ of the sixth grade tried out for the band. About how many students tried out?

   F 70   **G** 80

   H 90   **J** 100

9. Dan has $6\frac{2}{3}$ yards of wire. He uses $\frac{5}{6}$ yards in each birdhouse he makes. How many birdhouses can Dan make from his supply?

   A 4   **B** 8

   C 11   **D** 14

10. Jack is trying to decrease the amount of time he spends watching television. This week he spent $5\frac{2}{3}$ hours watching television. That was $\frac{2}{3}$ of the amount of time he watched television last week. How long did Jack watch television last week?

    F $4\frac{1}{2}$ hr   **G** $6\frac{1}{3}$ hr

    H 7 hr   **J** $8\frac{1}{2}$ hr

MATH FUN!

FINDING PRIMES

PURPOSE To practice finding prime numbers (pages 84–85)

The French mathematician Pierre de Fermat (1601–1665), who is called the founder of modern number theory, spent much of his time exploring prime numbers. He developed the following expression to find prime numbers.

$$2^{2n} + 1$$

A simpler expression you can use to find some prime number is $2^n - 1$.

Use the expression to find a prime number for each value of n.

1. $n = 2$ **2.** $n = 5$ **3.** $n = 7$

FRACTION RIDDLES

PURPOSE To practice finding equivalent fractions (pages 104–107)

See if you can identify the numbers from the information that is given.

1. I am a fraction in simplest form. If you multiply my numerator and denominator by the first prime number, the result is $\frac{10}{12}$. Who am I?

2. I am a fraction greater than 1. If you divide my numerator and denominator by the first composite number, the result is $\frac{7}{3}$. Who am I?

ROLL A FRACTION

PURPOSE To practice dividing fractions (pages 146–147)
YOU WILL NEED three number cubes, calculator

Work with a small group. Take turns rolling three number cubes. Use the numbers rolled to make fractions less than 1. Each student writes a division problem with two of the fractions. The person with the greatest quotient earns one point. Use a calculator to check your work. The first player to get 10 points wins the game.

EXAMPLE: Player 1 makes the problem $\frac{4}{5} \div \frac{3}{5} = 1\frac{1}{3}$. Player 2 makes $\frac{3}{4} \div \frac{4}{5} = \frac{15}{16}$.
Player 1 gets a point, because $1\frac{1}{3}$ is greater than $\frac{15}{16}$.

 Play this game at home with a family member.

Finding the LCM Using Spreadsheets

In this activity you will see how you can use a spreadsheet to list multiples of numbers quickly to find the LCM.

If you want to add $\frac{1}{8}$ and $\frac{5}{6}$, you must find the LCD. This means you must find the LCM of 8 and 6. One way to do this is to list multiples of 8 and 6.

Enter the data shown below into a spreadsheet.

All	A	B	C
1		Multiples of 8	Multiples of 6
2	1		
3	2		
4	3		
5	4		
6	5		
7	6		
8	7		
9	8		
10	9		
11	10		

To find the first multiple of 8, enter the following formula in cell B2.

```
=A2*8
```

To find the first multiple of 6, enter the following formula in cell C2.

```
=A2*6
```

The symbol * is used in spreadsheets to multiply.

Select cells B2 and C2. Then select Copy under Edit on the menu bar.

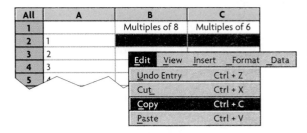

Select cells B3 through C11 by clicking and dragging the mouse until they are highlighted. Then press `Control` `V` or Paste under Edit on the menu bar. This will paste the formulas from cells B2 and C2 to all of the other cells.

All	A	B	C
1		Multiples of 8	Multiples of 6
2	1	8	6
3	2	16	12
4	3	24	18
5	4	32	24
6	5	40	30
7	6	48	36
8	7	56	42
9	8	64	48
10	9	72	54
11	10	80	60

1. Suppose the fractions were $\frac{3}{4}$ and $\frac{5}{6}$. In a column labeled Multiples of 4, what would be the new formula for cell B2?

2. Suppose you want the LCD of $\frac{1}{8}$, $\frac{5}{6}$, and $\frac{3}{7}$. What formula would you enter for multiples of 7, and in what cell?

USING A SPREADSHEET

Find the LCD by making a list of multiples.

3. $\frac{3}{7}$, $\frac{1}{6}$

4. $\frac{2}{3}$, $\frac{5}{12}$

5. $\frac{3}{8}$, $\frac{1}{3}$

6. $\frac{5}{8}$, $\frac{3}{100}$

Study Guide and Review

Vocabulary Check

1. Numbers such as 5 with only two factors 1 and 5 are __?__. **(page 84)**

2. When a composite number is written as the product of prime factors, this is called __?__. Example: $36 = 2 \times 2 \times 3 \times 3$. **(page 86)**

3. To add fractions without models or pictures, you can write equivalent fractions by using the __?__. **(page 108)**

4. When you rewrite a fraction problem as a multiplication problem to solve it, you use the __?__ of the divisor. **(page 146)**

EXAMPLES

- **Find the LCM and GCF of numbers.**
  (pages 88–91)

  Find the LCM and GCF for 8 and 12.

  LCM = 24   GCF = 4
  8: 8, 16, **24**   8: 1, 2, **4**
  12: 12, **24**   12: 1, 2, 3, **4**, 6, 12

- **Write fractions in simplest form.**
  (pages 94–95)

  Write $\frac{12}{24}$ in simplest form.

  $\frac{12}{24} = \frac{12 \div 12}{24 \div 12} = \frac{1}{2}$ *Divide by the GCF.*

- **Write mixed numbers as fractions and fractions as mixed numbers.** **(pages 96–97)**

  Write $2\frac{3}{5}$ as a fraction.

  $2\frac{3}{5} \rightarrow \frac{(5 \times 2) + 3}{5} = \frac{13}{5}$

  Write $\frac{17}{4}$ as a mixed number.

  $\begin{array}{r} 4 \\ 4\overline{)17} \\ -16 \\ \hline 1 \end{array}$ *Divide numerator by denominator.*
  Write the remainder as a fraction.
  $\frac{17}{4} = 4\frac{1}{4}$

EXERCISES

Find the LCM for each pair of numbers.

5. 5, 10   **6.** 9, 15

7. 4, 6   **8.** 5, 7

Find the GCF for each pair of numbers.

9. 5, 10   **10.** 9, 15

11. 10, 15   **12.** 8, 12

Write the fraction in simplest form.

13. $\frac{9}{21}$ **14.** $\frac{20}{40}$ **15.** $\frac{50}{100}$

16. $\frac{6}{8}$ **17.** $\frac{7}{21}$ **18.** $\frac{15}{40}$

19. $\frac{10}{15}$ **20.** $\frac{8}{10}$ **21.** $\frac{12}{48}$

Write the mixed number as a fraction.

22. $1\frac{2}{3}$ **23.** $3\frac{1}{7}$ **24.** $5\frac{2}{5}$

25. $2\frac{3}{4}$ **26.** $6\frac{1}{2}$ **27.** $4\frac{1}{3}$

Write the fraction as a mixed number.

28. $\frac{16}{7}$ **29.** $\frac{9}{5}$ **30.** $\frac{14}{3}$

31. $\frac{17}{12}$ **32.** $\frac{9}{4}$ **33.** $\frac{7}{2}$

- **Add and subtract unlike fractions.**

  (pages 106–111)

  $\dfrac{3}{4} = \dfrac{21}{28}$ *Write equivalent fractions using the LCD.*

  $-\dfrac{3}{7} = \dfrac{12}{28}$ *Subtract the numerators.*

  $\dfrac{9}{28}$ *Write the difference over the denominator.*

Add. Write the answer in simplest form.

34. $\dfrac{2}{3} + \dfrac{1}{5}$ **35.** $\dfrac{1}{3} + \dfrac{2}{6}$

36. $\dfrac{3}{4} + \dfrac{3}{8}$ **37.** $\dfrac{5}{9} + \dfrac{5}{6}$

Subtract. Write the answer in simplest form.

38. $\dfrac{3}{8} - \dfrac{1}{12}$ **39.** $\dfrac{3}{5} - \dfrac{1}{2}$

- **Add and subtract mixed numbers.**

  (pages 126–129)

  $3\dfrac{1}{6} = 3\dfrac{1}{6} = 2\dfrac{7}{6}$ *Write equivalent fractions. Rename as needed.*

  $-1\dfrac{1}{2} = 1\dfrac{3}{6} = 1\dfrac{3}{6}$ *Subtract fractions. Subtract whole numbers.*

  $1\dfrac{4}{6} = 1\dfrac{2}{3}$ *Write in simplest form.*

Add. Write the answer in simplest form.

40. $2\dfrac{3}{5} + 3\dfrac{1}{3}$ **41.** $5\dfrac{5}{6} + 4\dfrac{2}{3}$

Subtract. Write the answer in simplest form.

42. $8\dfrac{2}{3} - 3\dfrac{2}{8}$ **43.** $7\dfrac{1}{4} - 6\dfrac{3}{8}$

44. $5\dfrac{2}{9} - 2\dfrac{1}{3}$ **45.** $4\dfrac{3}{4} - 1\dfrac{5}{6}$

- **Multiply fractions and mixed numbers.**

  (pages 142–143)

  $\dfrac{1}{2} \times 1\dfrac{1}{3} = \dfrac{1}{2} \times \dfrac{4}{3} =$ *Write mixed numbers as fractions.*

  Use the GCF to simplify.

  $\dfrac{1}{2} \times \dfrac{\overset{2}{\cancel{4}}}{3} = \dfrac{2}{3}$ *Multiply numerators.*

  Multiply denominators.

Multiply. Write the answer in simplest form.

46. $\dfrac{3}{4} \times \dfrac{2}{3}$ **47.** $\dfrac{2}{9} \times 1\dfrac{1}{2}$

48. $2\dfrac{1}{3} \times 3$ **49.** $2\dfrac{3}{5} \times 2\dfrac{7}{5}$

50. $7 \times 1\dfrac{1}{3}$ **51.** $3\dfrac{1}{3} \times 2\dfrac{2}{5}$

- **Divide fractions.** (pages 146–147)

  $6 \div \dfrac{2}{3} = \dfrac{6}{1} \div \dfrac{2}{3}$ *Write the whole number as a fraction.*

  $= \dfrac{6}{1} \times \dfrac{3}{2}$ *Use the reciprocal of the divisor to write a multiplication problem.*

  $\dfrac{\overset{3}{\cancel{6}}}{1} \times \dfrac{3}{\cancel{2}} = \dfrac{9}{1} = 9$ *Simplify. Multiply.*

Divide. Write the answer in simplest form.

52. $\dfrac{1}{2} \div \dfrac{1}{3}$ **53.** $\dfrac{3}{4} \div \dfrac{9}{16}$

54. $5 \div \dfrac{10}{11}$ **55.** $\dfrac{6}{7} \div 3$

56. $\dfrac{4}{7} \div \dfrac{9}{14}$ **57.** $\dfrac{2}{5} \div \dfrac{3}{4}$

Problem-Solving Applications

Solve. Explain your method.

58. A shelf is 72 in. wide. Tom's cassettes fill up $\dfrac{1}{6}$ of the shelf. Each cassette is $\dfrac{3}{4}$ in. thick. How many cassettes does Tom have on the shelf? (pages 140–141, 146–147)

59. This week Alicia exercised for 3 hr. This is $1\dfrac{1}{2}$ times the amount she exercised last week. How long did she exercise last week? (pages 146–147)

Performance Assessment

Tasks: Show What You Know

1. Show the steps you would use to find the prime factorization of 48. Explain your method. **(pages 86–87)**

2. Show how to find the difference. Explain your method. **(pages 110–111)**

$$\frac{7}{8} - \frac{2}{3} = n$$

3. Show the steps to find the difference. $5\frac{3}{8} - 2\frac{5}{6}$ **(pages 124–125)**

4. Explain the steps and then solve the problem.

   Angela has $2\frac{1}{4}$ yd of ribbon. She plans to use it on 6 packages. She uses the same amount of ribbon on each package. How much ribbon is used for each package? **(pages 146–147)**

Problem Solving

Solve. Explain your method.

CHOOSE a strategy and a tool.

- **Draw a Diagram**
- **Make a Model**
- **Write an Equation**
- **Find a Pattern**
- **Make a Table**
- **Work Backward**

Paper/Pencil Calculator Hands-On Mental Math

5. The Olympic Summer Games are held every 4 years. They were held in 1996. How many times will they be held between the years 2000 and 2050? List the years in which they will be held. **(pages 84–85)**

6. Michael estimated that it would take him $2\frac{1}{2}$ hr to do his homework. As it turned out, he finished his homework in $1\frac{3}{4}$ hr. How much less time did it take him than expected? **(pages 126–127)**

7. Mrs. Esparza's car holds 10 gal of gas. On Friday her tank was $\frac{4}{5}$ full. On Saturday she used 5 gal. How much gas was left in the tank? **(pages 136–137)**

8. Kristin is baking bread. She uses $2\frac{1}{3}$ c of flour for $\frac{1}{4}$ recipe. How much flour does the whole recipe call for?

   (pages 148–149)

FLOUR

100% WHEAT

5 Pounds

2 CUPS

Cumulative Review

Solve the problem. Then write the letter of the correct answer.

1. Compare. 126.5 ⬤ 12.65 **(pages 18–21)**

  A. <   **B.** >
  C. =   **D.** ≥

2. Write $\dfrac{1}{1,000}$ as a decimal. **(pages 22–25)**

  A. 0.001   **B.** 0.01
  C. 0.1   **D.** 1,000

3. Find the value of 12^3. **(pages 26–27)**

  A. 36   **B.** 123
  C. 144   **D.** 1,728

4. 224×136 **(pages 44–47)**

  A. 2,240   **B.** 29,244
  C. 30,464   **D.** 31,464

Choose the best estimate. **(pages 50–53)**

5.   6,324
     + 2,731

  A. 3,000   **B.** 8,000
  C. 9,000   **D.** 10,000

6. $316.54 - 63.856$ **(pages 62–63)**

  A. 2.52684   **B.** 252.684
  C. 252.696   **D.** 25,268.4

7. 7.23×0.86 **(pages 64–67)**

  A. 6.2178   **B.** 6.3178
  C. 62.178   **D.** 621.77

8. $87.2 \div 5$ **(pages 70–73)**

  A. 154.6   **B.** 1.744
  C. 17.44   **D.** 16.25

9. $\dfrac{5}{9} + \dfrac{2}{3}$ **(pages 104–109)**

  A. $\dfrac{7}{12}$ **B.** $\dfrac{1}{9}$
  C. $\dfrac{8}{9}$ **D.** $1\dfrac{2}{9}$

10. $\dfrac{2}{5} - \dfrac{1}{3}$ **(pages 106–107, 110–111)**

  A. $\dfrac{1}{15}$ **B.** $\dfrac{2}{15}$
  C. $\dfrac{1}{3}$ **D.** $\dfrac{1}{2}$

11. $4\dfrac{3}{4} + 2\dfrac{5}{6}$ **(pages 126–129)**

  A. $6\dfrac{4}{5}$ **B.** $6\dfrac{7}{12}$
  C. $7\dfrac{7}{12}$ **D.** $7\dfrac{3}{4}$

12. $4\dfrac{3}{5} - 3\dfrac{2}{3}$ **(pages 126–129)**

  A. $\dfrac{3}{5}$ **B.** $\dfrac{14}{15}$
  C. $1\dfrac{1}{2}$ **D.** $1\dfrac{14}{15}$

13. $\dfrac{3}{4} \times \dfrac{3}{4}$ **(pages 136–139)**

  A. $\dfrac{9}{16}$ **B.** $\dfrac{3}{5}$
  C. 1   **D.** $2\dfrac{1}{9}$

14. $\dfrac{4}{7} \div 2$ **(pages 146–147)**

  A. $\dfrac{4}{7}$

  B. $\dfrac{2}{7}$

  C. $1\dfrac{1}{7}$

  D. $3\dfrac{1}{2}$

GEOMETRIC FIGURES

LOOK AHEAD

In this chapter you will solve problems that involve

- constructing more complex geometric figures from basic geometric shapes

- classifying lines

- constructing congruent line segments and angles

- recognizing a polygon by the number of sides and type of angles

ART **LINK**

The Rock and Roll Hall of Fame and Museum in Cleveland, Ohio, was built at a cost of about \$92,000,000. The total floor space is about 150,000 ft^2. The six-story central tower is about 162 ft high.

- Cleveland contributed about half the total cost. Estimate Cleveland's contribution.

- Estimate the area of your classroom floor space. Then decide how many classrooms are equivalent to the total floor space of the Rock and Roll Hall of Fame and Museum.

Catalog Geometry

Look through old catalogs to find examples of geometric figures in everyday items. Cut them out, identify the figures, and make a display. Compare your display with those of other students.

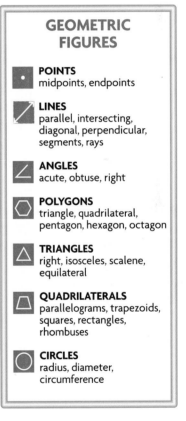

GEOMETRIC FIGURES

POINTS
midpoints, endpoints

LINES
parallel, intersecting, diagonal, perpendicular, segments, rays

ANGLES
acute, obtuse, right

POLYGONS
triangle, quadrilateral, pentagon, hexagon, octagon

TRIANGLES
right, isosceles, scalene, equilateral

QUADRILATERALS
parallelograms, trapezoids, squares, rectangles, rhombuses

CIRCLES
radius, diameter, circumference

PROJECT CHECKLIST

☑ Did you find examples of geometric figures in catalogs?

☑ Did you identify geometric figures in each item?

☑ Did you make a display of your work?

Points, Lines, and Planes

VOCABULARY

point

line

plane

line segment

ray

The building blocks of geometry are points, lines, and planes.

• Fold a sheet of paper, open it up, and mark a location on the crease. Use geometric terms to describe the paper, the crease, and the marked location.

Geometric figures are named and pictured in special ways.

A **point** is an exact location.	•P point *P* Use the name of the point.
A **line** is a straight path that goes on forever in opposite directions. It has no endpoints.	line *AB*, $\overleftrightarrow{AB}$ line *BA*, $\overleftrightarrow{BA}$ Use two points on the line to name the line.
A **plane** is a flat surface that goes on forever in all directions.	plane *LMN* Use three points that are not on a line to name the plane.

Parts of a line can be named by using named points on the line.

A **line segment** is part of a line. It has two endpoints.	line segment *XY*, $\overline{XY}$ line segment *YX*, $\overline{YX}$
A **ray** is part of a line. It begins at its endpoint and goes on forever in only one direction.	ray *JK*, $\overrightarrow{JK}$

A flagpole and a pencil are examples of line segments. A light beam from a laser can represent a ray.

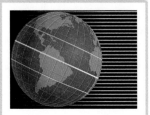

SCIENCE LINK

Because the Earth is tilted on its axis, the sun's rays at any one time are much more direct at some places on Earth than at others. Look at the diagram above. If the Earth's axis were not tilted, where would the direct rays of the sun strike the Earth all year?

GUIDED PRACTICE

Name the geometric figure.

1.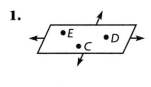

2.

3.

4.

5.

6.

INDEPENDENT PRACTICE

For Exercises 1–4, use the figure below.

1. Name three different segments.

2. Name six different rays.

3. Two names for the line are $\overleftrightarrow{PR}$ and $\overleftrightarrow{RP}$. Give four more names for the line.

4. Give another name for ray *RQ*.

Name the geometric figure that is suggested.

5. railroad tracks through a town

6. a path between bus stops on a road

7. a star in the sky

8. the short or long hand on a clock

Problem-Solving Applications

For Problems 9–12, name the geometric figure suggested by each part of the map.

9. the three towns on the map

10. the two interstate highways

11. the route from Pompton Lakes to Parsippany

12. **LOGICAL REASONING** Is a line a good model to use to describe Route 202? Explain.

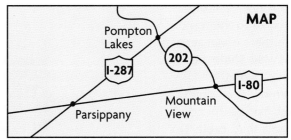

13. **MAPS** Draw a map of your neighborhood. Label your house with point *A*. Include other landmarks, such as parks and buildings. Use geometric terms to describe the landmarks and routes from one landmark to another.

14. **WRITE ABOUT IT** Can a line segment be part of a ray? Explain.

Classifying Lines

What You'll Learn
How to name the different types of lines

Why Learn This?
To help you understand directions to a place

VOCABULARY

parallel lines
intersecting lines
perpendicular lines

A line is a straight path that goes on forever in opposite directions. The chart below shows some line relationships.

GEOMETRIC FIGURES

Parallel lines are lines in a plane that are always the same distance apart. They never intersect and have no common points.	*B* *A* *L* *M*	Line *AB* is parallel to line *ML*. $\overleftrightarrow{AB} \parallel \overleftrightarrow{ML}$
Intersecting lines are lines that cross at exactly one point.	*E* *C* *D* *H* *F*	Line *EF* intersects line *CD* at point *H*.
Perpendicular lines are lines that intersect to form 90° angles, or right angles.	*R* *T* *U* *S*	Line *RS* is perpendicular to line *TU*. $\overleftrightarrow{RS} \perp \overleftrightarrow{TU}$

Line segments can also be parallel, intersecting, or perpendicular.

Examples of lines and line segments can be found in this picture of a covered bridge.

Stone Mountain, located in Georgia, is 825 ft high. Visitors can hike to the top or ride there on an aerial tramway. There are two cars on the tramway at all times. Each car is suspended from a thick cable. Do you think the cables are parallel lines or intersecting lines? Explain.

Talk About It

• In the picture of the covered bridge, what type of line segments are represented by $\overline{AB}$ and $\overline{CD}$?

• In the same picture, which figures represent intersecting line segments?

GUIDED PRACTICE

Name the type of lines.

1.

2.

3.

Make and label a drawing for Exercises 4–5.

4. Line *KL* is perpendicular to line *MN*. **5.** Line *CD* is parallel to line *EF*.

INDEPENDENT PRACTICE

The figure at the right shows 12 lines drawn on the edges of a cube.

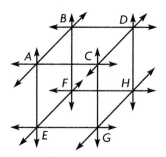

1. Name all the lines that are parallel to $\overleftrightarrow{AC}$.

2. Name all the lines that intersect $\overleftrightarrow{CD}$.

3. Name all the lines that are perpendicular to and intersect $\overleftrightarrow{DB}$.

4. Name all the lines that are parallel to $\overleftrightarrow{EF}$.

Problem-Solving Applications

5. MAPS Tina's friend knows 3rd Street intersects Oak Street. Tina tells her that she lives on 4th Street, which is parallel to 3rd Street and intersects Oak Street. Draw a map showing 3rd Street, 4th Street, and Oak Street.

6. Can two lines be both parallel and intersecting? Can two lines be both parallel and perpendicular?

7. ▭ **WRITE ABOUT IT** How are perpendicular and intersecting lines the same? different?

Mixed Review and Test Prep

Tell whether each angle is greater than or less than 90°.

8.

9.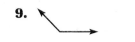

10.

11.

Multiply.

12. $2\frac{1}{2} \times 3\frac{1}{3}$

13. $4\frac{3}{8} \times 1\frac{1}{5}$

14. $6\frac{2}{3} \times 2\frac{1}{4}$

15. CONSUMER MATH Damon Clark is paid $350 a week. This week he spent $60 on groceries, saved $110, and paid a bill. He has $100 left. How much was his bill?

A $110 **B** $100 **C** $90 **D** $80

16. HOBBIES Maureen has 3 stamps to trade. She trades each stamp for 3 coins. How many coins does she get? Express the answer with an exponent.

F 2^2 **G** 2^3 **H** 3^2 **J** 3^3

Angles

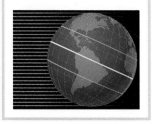
You know that a ray begins at a point and goes on forever in only one direction. An angle is formed by two rays with a common endpoint, the **vertex** of the angle. An angle can be named by three letters, a point from each side and the vertex as the middle letter. It can also be named with a single letter, its vertex.

$\angle DEF$, $\angle FED$, or $\angle E$

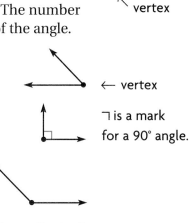

Angles are measured in degrees. The number of degrees determines the type of the angle.

The measurement of an **acute angle** is less than 90°.

← vertex

The measurement of a **right angle** is 90°.

⌐ is a mark for a 90° angle.

The measurement of an **obtuse angle** is more than 90° and less than 180°.

The measurement of a **straight angle** is 180°.

You can find the number of degrees in an angle by using a protractor. A protractor has a top and bottom scale. Either scale can be used to measure angles.

EXAMPLE Find the measure of $\angle ZYX$. Since the angle opens from the left, use the top scale of the protractor, so that 0° is on $\overrightarrow{YZ}$.

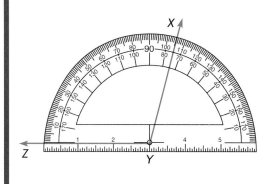

Place the center mark of the protractor on the vertex of the angle.

Place the base of the protractor along ray YZ.

Read the top scale where $\overrightarrow{YX}$ crosses to find the number of degrees in the angle.

• There are 105° in the angle. What kind of angle is it?

GUIDED PRACTICE

Write *right, acute, obtuse,* or *straight* for each angle measure.

1. 90°   **2.** 36°   **3.** 180°   **4.** 100°   **5.** 89°

Name each angle. Tell whether it is *acute, right,* or *obtuse.*

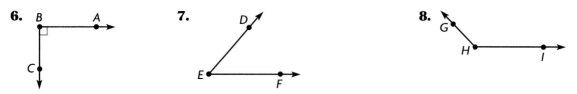

6.   **7.**   **8.**

INDEPENDENT PRACTICE

Write *right, acute, obtuse,* or *straight* for each angle measure.

1. 82°   **2.** 174°   **3.** 5°   **4.** 41°   **5.** 96°

Measure the angle. Write the measure and *acute, right, obtuse,* or *straight.*

6.   **7.**   **8.**

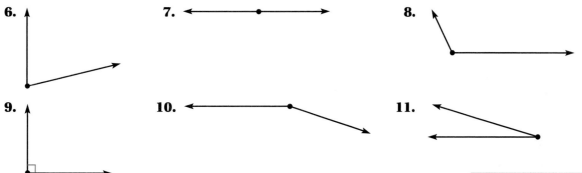

9.   **10.**   **11.**

Problem-Solving Applications

At 3:00, the hands of a clock form a right angle. For Problems 12–19, name the type of angle formed at each of the given times.

12. 2:00   **13.** 3:00   **14.** 6:00   **15.** 5:30

16. 10:25   **17.** 11:45   **18.** 9:00   **19.** 11:08

20. CRITICAL THINKING One acute angle and one obtuse angle share the vertex, *C.* The two angles form a straight line. What is the sum of the measures of the two angles?

21. CRITICAL THINKING Does the measure of an angle depend on the lengths of the rays that are drawn for its sides? Explain.

22. ✏️ **WRITE ABOUT IT** Describe the difference between an acute angle and an obtuse angle.

MORE PRACTICE Lesson 8.3, page H53

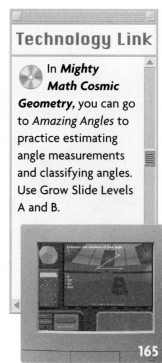

Technology Link

In *Mighty Math Cosmic Geometry,* you can go to *Amazing Angles* to practice estimating angle measurements and classifying angles. Use Grow Slide Levels A and B.

Constructing Congruent Segments and Angles

What You'll Learn
How to construct congruent line segments and angles

Why Learn This?
To make drawings for construction projects or art projects in school or in future jobs

VOCABULARY
congruent

All line segments of the same length are said to be **congruent**.

We usually compare by measuring with a ruler, but we can also use a compass to see if two line segments are congruent.

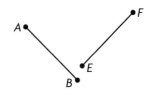

ACTIVITY

WHAT YOU'LL NEED: compass

• Trace $\overline{EF}$ and $\overline{AB}$ on your paper.

• Place the compass point on point E. Open the compass to the length of $\overline{EF}$.

• Use the compass to show that $\overline{AB}$ has the same length.

• What is the relationship between the length of $\overline{AB}$ and the length of $\overline{EF}$?

You can use a compass and a straightedge to construct congruent line segments.

The compass that you use to draw circles and arcs is a tool used in occupations such as architecture, carpentry, and engineering.

EXAMPLE 1

Trace $\overline{CD}$. Construct a line segment congruent to $\overline{CD}$.

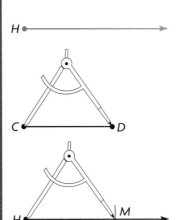

Draw a ray that is longer than $\overline{CD}$. Label the endpoint H.

Place the compass point on point C. Open the compass to the length of $\overline{CD}$.

Use the same opening, but with the compass point on H.

Draw an arc that intersects the ray. Label the intersection point M.

$\overline{CD} \cong \overline{HM} \leftarrow$ segment CD is congruent to segment HM.

Talk About It

• In Example 1, why must you draw a figure longer than $\overline{CD}$?

• What are some examples of congruent line segments in your classroom?

• What are some careers in which people must know how to construct congruent figures?

GUIDED PRACTICE

Use a compass to determine if the line segments in each pair are congruent. Write *yes* or *no*.

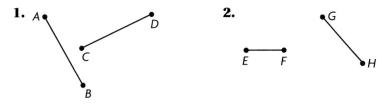

1. A, D, C, B

2. G, E, F, H

Use a compass and a straightedge to construct a line segment congruent to each figure.

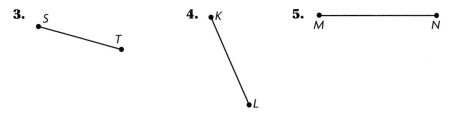

3. S, T

4. K, L

5. M, N

6. Use a ruler to draw a line segment *XY* that is 6 cm long. Then construct a line segment congruent to *XY*.

Construct Congruent Angles

Remember that angles are measured in degrees. All angles that have the same measure in degrees are congruent angles.

Look at the picture of the handmade quilt, at the right. When the quilt was made, pieces of fabric were cut by using congruent line segments and congruent angles. The pieces were then sewn together to make the pattern that you see.

• In what other handmade items might congruent angles be used?

You used a compass and a straightedge to construct congruent line segments. You can use the same tools to construct congruent angles.

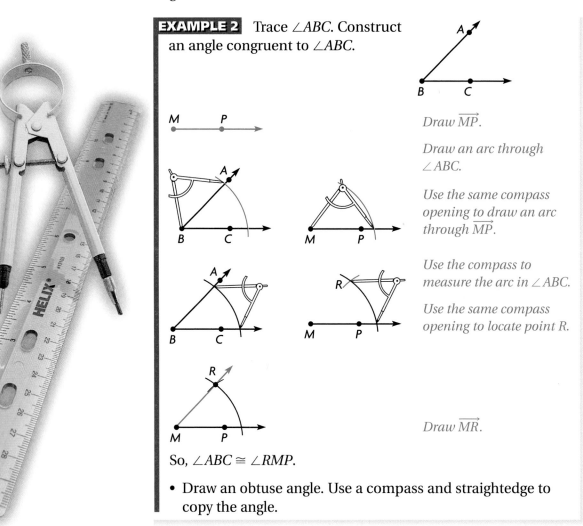

EXAMPLE 2 Trace ∠*ABC*. Construct an angle congruent to ∠*ABC*.

Draw $\overrightarrow{MP}$.

Draw an arc through ∠ABC.

Use the same compass opening to draw an arc through $\overrightarrow{MP}$.

Use the compass to measure the arc in ∠ABC.

Use the same compass opening to locate point R.

Draw $\overrightarrow{MR}$.

So, ∠*ABC* ≅ ∠*RMP*.

• Draw an obtuse angle. Use a compass and straightedge to copy the angle.

INDEPENDENT PRACTICE

Use a compass to determine if the line segments in each pair are congruent. Write *yes* or *no*.

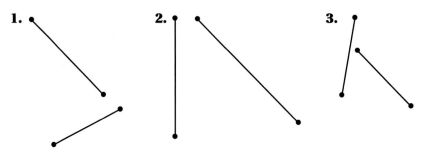

1. **2.** **3.**

Use a protractor to measure the angles in each pair. Tell whether they are congruent. Write each measure and *yes* or *no*.

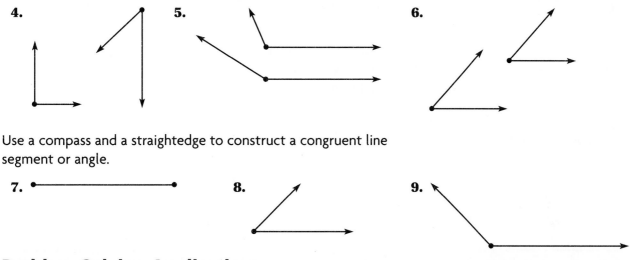

4. 5. 6.

Use a compass and a straightedge to construct a congruent line segment or angle.

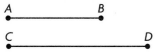

7. 8. 9.

Problem-Solving Applications

10. **GEOMETRY** Construct a line segment that has a length equal to the length of $\overline{AB}$ plus the length of $\overline{CD}$.

A   B

C   D

11. **GEOMETRY** Construct an angle that has a measure equal to the measure of $\angle ABC$ plus the measure of $\angle DEF$.

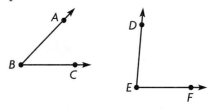

12. ✏ **WRITE ABOUT IT** Explain how to construct congruent angles.

Mixed Review and Test Prep

Tell how many sides each figure has. Name the figure.

13.   14.   15.   16.

Divide. Write the answer in simplest form.

17. $3\frac{1}{2} \div \frac{3}{4}$ 18. $2\frac{1}{4} \div 1\frac{1}{2}$ 19. $1\frac{1}{4} \div 1\frac{1}{8}$ 20. $3\frac{1}{5} \div 2\frac{2}{3}$

21. **TRANSPORTATION** Ben's car gets 16 miles per gallon of gasoline. He used about 9 gallons of gasoline driving to Austin. Which is the best estimate of the distance Ben drove?

  A 2 mi   **B** 15 mi
  C 150 mi   **D** 1,500 mi

22. **HOBBIES** Juanita has $\frac{1}{2}$ yd of fabric. She uses only $\frac{1}{3}$ yd. How much is left?

  F $\frac{1}{6}$ yd   **G** $\frac{1}{3}$ yd
  H $\frac{1}{2}$ yd   **J** $\frac{5}{6}$ yd

What You'll Explore
How to bisect a line segment

What You'll Need
compass
straightedge

VOCABULARY
bisect

Bisecting a Line Segment

You can use a compass and straightedge to bisect a line segment. When you **bisect** a line segment, you divide it into two equal parts.

ACTIVITY

Explore

Draw a segment $\overline{RS}$.

R •————————• S

- Place the compass point on point R. Open the compass to a little more than half the distance from R to S. Draw an arc through $\overline{RS}$ as shown.

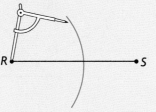

- Keep the same compass opening. Place the compass point on point S. Draw an arc as shown. Label the points T and U where the arcs intersect.

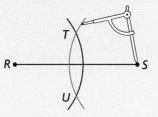

- Use a straightedge to draw a line through T and U. Label the point P where $\overleftrightarrow{TU}$ intersects $\overline{RS}$.

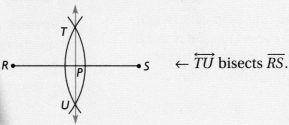

← $\overleftrightarrow{TU}$ bisects $\overline{RS}$.

Think and Discuss

- $\overleftrightarrow{TU}$ bisects $\overline{RS}$. What does that mean?

- Look at $\overleftrightarrow{TU}$ and $\overleftrightarrow{RS}$. Would you describe them as parallel or intersecting?

- $\angle TPS$, $\angle SPU$, $\angle UPR$ and $\angle RPT$ are right angles. How many degrees are in each angle?

- **CRITICAL THINKING** $\overrightarrow{TU}$ and $\overline{RS}$ intersect at point P to form four right angles. Besides intersection, what other relationship do you see between $\overleftrightarrow{TU}$ and $\overline{RS}$?

- Look at the figure at the right. How could you determine whether $\overline{SD}$ bisects $\overline{WR}$?

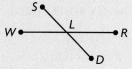

Try This

Tell whether $\overline{JK}$ bisects the other line segment. Write *yes* or *no*.

1.

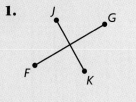

2.

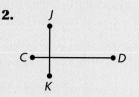

3.

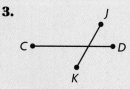

4.

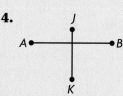

5. Draw a line segment. Label it *CD*. Then construct a segment *AB* that bisects $\overline{CD}$.

6. Draw a line segment. Label it *JK*. Then construct a segment *LM* that bisects $\overline{JK}$.

Polygons

What You'll Learn
How to name polygons by the number of sides and angles

Why Learn This?
To recognize different polygons that you see in patterns and designs

VOCABULARY

polygon

regular polygon

Line segments are used to form other geometric figures. A **polygon** is a closed plane figure formed by three or more line segments. Polygons are named by the number of their sides and angles.

COMMON POLYGONS	
Name	**Sides and Angles**
Triangle	3
Quadrilateral	4
Pentagon	5
Hexagon	6
Octagon	8

ACTIVITY

WHAT YOU'LL NEED: geoboard and dot paper

- Use the geoboard to show as many different triangles as you can. Include isosceles, scalene, right, acute, and obtuse triangles. Record your triangles on dot paper.

- Use the geoboard to show as many different quadrilaterals as you can. Include a square, rectangle, parallelogram, and trapezoid. Record your quadrilaterals on dot paper.

- Look at your quadrilaterals. Which have right angles? Which have both pairs of opposite sides parallel? In which are both pairs of opposite sides congruent?

- **CRITICAL THINKING** Show other types of polygons such as pentagons, hexagons, and octagons. Record them on dot paper.

REMEMBER:

Triangles are classified by their angles and by their sides.
See page H21.

Classified by Sides
Equilateral
Isosceles
Scalene

Classified by Angles
Acute
Right
Obtuse

Some of the polygons you made may be regular polygons. A **regular polygon** has all sides congruent and all angles congruent.

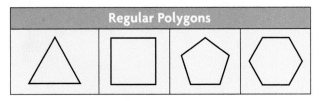

Regular Polygons

- Look at the regular polygons above. What types of angles do you see in each polygon?

GUIDED PRACTICE

Cut out a triangular piece of paper.

1. How many sides does the triangle have? How many vertices?

2. Cut off one vertex. Tell how many sides and vertices the polygon has. Name the polygon.

3. Cut another vertex. Tell how many sides and vertices the polygon has. Name the polygon.

4. Predict the number of sides and vertices if another vertex is cut.

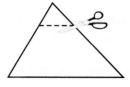

INDEPENDENT PRACTICE

Name the quadrilateral. Tell whether both pairs of opposite sides are parallel. Write *yes* or *no*.

1. 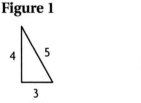　　2. 　　3. 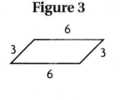　　4.

For Exercises 5–8, use Figures 1–4. All measurements are in inches.

Figure 1　　**Figure 2**　　**Figure 3**　　**Figure 4**

5. Which figure is a hexagon?

6. Which figures have only obtuse angles?

7. Is Figure 1 an equilateral, scalene, or isosceles triangle?

8. Which figures appear to be regular polygons?

Problem-Solving Applications

9. **GEOMETRY** Tell how many triangles and how many quadrilaterals are in the figure at the right.

10. **LOGICAL REASONING** Draw a square. Then divide it into two congruent triangles. What type of triangles are they?

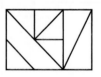

11. **CONSUMER MATH** Janice's mother bought a table that seats six people, one at each side. What shape is the table?

12. **GEOMETRY** All parallelograms have opposite sides parallel. Are squares and rectangles parallelograms? Explain.

13. ✏️ **WRITE ABOUT IT** Explain the difference between a regular polygon and a polygon that is not regular.

1. **VOCABULARY** Part of a line that has two endpoints is a(n) __?__. (page 160)

Identify each figure. Give the symbol. (pages 160–161)

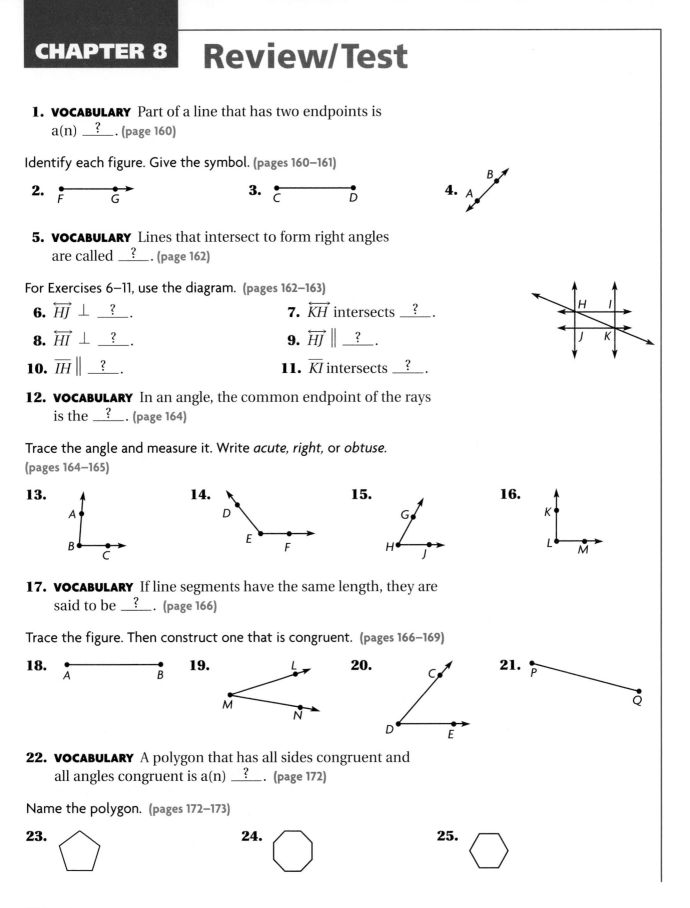

2.
3.
4.

5. **VOCABULARY** Lines that intersect to form right angles are called __?__. (page 162)

For Exercises 6–11, use the diagram. (pages 162–163)

6. $\overleftrightarrow{HJ} \perp$ __?__.

7. $\overleftrightarrow{KH}$ intersects __?__.

8. $\overleftrightarrow{HI} \perp$ __?__.

9. $\overleftrightarrow{HJ} \parallel$ __?__.

10. $\overleftrightarrow{IH} \parallel$ __?__.

11. $\overline{KI}$ intersects __?__.

12. **VOCABULARY** In an angle, the common endpoint of the rays is the __?__. (page 164)

Trace the angle and measure it. Write *acute*, *right*, or *obtuse*. (pages 164–165)

13.
14.
15.
16.

17. **VOCABULARY** If line segments have the same length, they are said to be __?__. (page 166)

Trace the figure. Then construct one that is congruent. (pages 166–169)

18.
19.
20.
21.

22. **VOCABULARY** A polygon that has all sides congruent and all angles congruent is a(n) __?__. (page 172)

Name the polygon. (pages 172–173)

23.
24.
25.

1. Which of the following is equal to 28 + 42?

   A 25 + 50   **B** 30 + 40
   C 28 + 40 + 12   **D** 20 + 60

2. Jessica drove 156.5 miles on Saturday and 204.7 miles on Sunday. What is the total distance she drove both days?

   F 361.2 mi
   G 389.3 mi
   H 409.1 mi
   J 461.2 mi
   K Not Here

3. Which of the following pairs of fractions are equal to $\frac{3}{7}$?

   A $\frac{1}{3}, \frac{7}{10}$ **B** $\frac{12}{28}, \frac{15}{30}$
   C $\frac{7}{3}, \frac{14}{6}$ **D** $\frac{9}{21}, \frac{36}{84}$

4. A road is $\frac{5}{8}$ mile long. There is construction on $\frac{1}{4}$ mile. How much of the road is free of construction?

   F $\frac{3}{8}$ mi   **G** $\frac{1}{2}$ mi
   H $\frac{5}{8}$ mi   **J** $\frac{3}{4}$ mi

5. Shawna is 5 feet $3\frac{1}{2}$ inches tall. In the last year, she grew $1\frac{1}{2}$ inches. How tall was she last year?

   A 5 ft 5 in.   **B** 5 ft $3\frac{1}{2}$ in.
   C 5 ft 2 in.   **D** 5 ft 1 in.

6. At which of the following times would the hands of a clock form a right angle?

   F 1:00   **G** 6:00
   H 9:00   **J** 12:00

7. A tree stands $8\frac{3}{4}$ feet high. It grew $1\frac{1}{4}$ feet last year, and $1\frac{1}{2}$ feet the year before. How tall was it two years ago?

   A 5 ft

   B 6 ft

   C $6\frac{1}{4}$ ft

   D $7\frac{1}{4}$ ft

8. What construction do the figures below represent?

   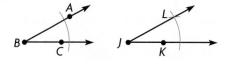

   F Parallel line segments
   G Congruent line segments
   H Bisected line segments
   J Congruent angles

9. Jack cut a string in half. Then he cut one of the pieces in half again. The smallest piece is now 7 inches long. How long was the string to start with?

   A 28 in.
   B 35 in.
   C 56 in.
   D 60 in.

10. What type of polygon is shown below?

    F Quadrilateral
    G Pentagon
    H Triangle
    J Hexagon

SYMMETRY AND TRANSFORMATIONS

LOOK AHEAD

In this chapter you will solve problems that involve

- symmetry and congruence of figures

- transformations of plane figures

- tessellations

ART LINK

M.C. Escher, a Dutch artist, created the art on these pages by using repeated patterns. The dimensions of several of Escher's works are given in the table below.

- Which work is nearly a square?

TITLE	DIMENSIONS	YEAR
Another World	31.5 cm × 26 cm	1947
High and Low	50.5 cm × 20.5 cm	1947
Curl-Up	17 cm × 23.5 cm	1951
House of Stairs	47 cm × 24 cm	1951
Relativity	28 cm × 29 cm	1953

- Which two works have dimensions that make them almost congruent rectangles?

Letter Chains

Can you form letter chains using your initials?

Use scissors to cut letter chains from fan-folded paper. As you cut, be sure to leave part of each fold so the sections will hold together. Are some letters easier to form into letter chains than others? Write a paragraph explaining your results.

PROJECT CHECKLIST

☑ Did you use fan-folded paper and cutout letters?

☑ Did you determine which letters can be cut easily in a chain and which cannot?

☑ Did you write a paragraph about your conclusions?

Symmetry and Congruence

What You'll Learn
How to identify line symmetry and rotational symmetry

Why Learn This?
To recognize and make symmetric and congruent designs as used in art and clothing

VOCABULARY

line symmetry

rotational symmetry

point of rotation

Symmetry is an interesting part of geometry that can be found all around you. You can find many examples of symmetry in nature and in manufactured objects.

ACTIVITY

WHAT YOU'LL NEED: paper, scissors, dark crayon

- Fold the paper in half. Use the crayon to write your name in cursive along the fold line.

- Fold the paper along the fold line so your name is inside. Use the handle of the scissors to make a rubbing of your name.

- Unfold the paper. Your name appears on the other half of the paper.

The design you made has line symmetry. A figure has **line symmetry** if it can be folded or reflected so that the two parts of the figure match, or are congruent.

- Where is the line of symmetry on the design you made above?

- How many lines of symmetry does the design have?

Some figures have several lines of symmetry.

REMEMBER:
....
Regular polygons have all sides congruent and all angles congruent.

See page H21.

EXAMPLE 1 Find all the lines of symmetry in the regular pentagon.

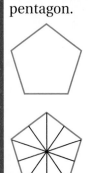

Trace the figure and cut it out.

Fold it in half in different ways.

If the halves match, then the fold is a line of symmetry.

Count the different lines of symmetry.

The figure has five lines of symmetry.

- Do all pentagons have five lines of symmetry? Explain.

If a figure has line symmetry, its two halves are congruent, or the same size and same shape, when the figure is folded.

EXAMPLE 2 Look at the figure. Are the halves on either side of the line congruent?

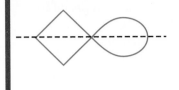

Trace the figure.

Fold along the dashed line.

The two halves are the same size and shape.

So, the two halves are congruent.

GUIDED PRACTICE

Trace the figure. Draw the lines of symmetry.

1. 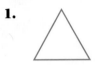   2.   3. 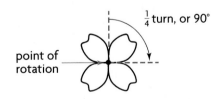   4.

Is the dashed line a line of symmetry? Write *yes* or *no*.

5.   6.    7.   8.

Rotational Symmetry

Another type of symmetry involves turning the figure instead of folding. A figure has **rotational symmetry** when it can be rotated less than 360° around a central point, or **point of rotation**, and still match the original figure.

The figure at the right has rotational symmetry. If you turn it $\frac{1}{4}$ turn, or 90°, around the point of rotation, it will match the original figure.

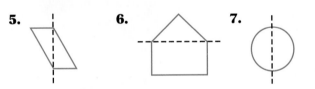

point of rotation

$\frac{1}{4}$ turn, or 90°

• CRITICAL THINKING How many times would you have to turn the figure 90° to get it back to its original position?

EXAMPLE 3 Does the figure below have rotational symmetry?

Trace the figure.

$\frac{1}{8}$, or 45°

Place your pencil point at the point of rotation.

point of rotation

Rotate the figure until it looks like the original figure.

The figure has rotational symmetry.

- Does the figure above have line symmetry?

EXAMPLE 4 The figure below has rotational symmetry. What are the fraction and the angle measure of each turn?

Trace the figure.

Place your pencil point in the center of the figure.

$\frac{1}{2}$, or 180°

Rotate the figure until it looks like the original figure.

Identify the fraction and the angle measure of the turn.

So, the figure matches itself when turned $\frac{1}{2}$ turn, or 180°.

- What if you rotated the figure $\frac{1}{4}$ turn? What would it look like? Would it match the original figure?

INDEPENDENT PRACTICE

Trace the figure. Draw all lines of symmetry.

1.

2.

3.

4.

Is the dashed line a line of symmetry? Write *yes* or *no*.

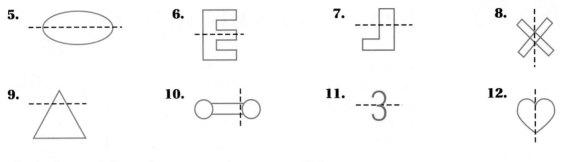

5.

6.

7.

8.

9.

10.

11.

12.

Tell whether each figure has rotational symmetry. Write *yes* or *no*.

13.

14.

15.

16.

Each figure has rotational symmetry. Tell the fraction and angle measure of each turn.

17.

18.

19.

20.

Problem-Solving Applications

21. GEOMETRY Draw a figure that has at least two lines of symmetry.

22. Does a circle have line symmetry? rotational symmetry? Explain.

23. CRITICAL THINKING Brenda has a square cake. She cuts it into pieces along all the lines of symmetry of the square. How many pieces does she have?

24. ✏ WRITE ABOUT IT Why must a rotation be less than 360° to show rotational symmetry?

Mixed Review and Test Prep

Trace the letter. Then flip it over from top to bottom and draw the new figure.

25. A

26. T

27. ∩

28. M

Identify the type of lines.

29.

30.

31.

32. CONSUMER MATH The Harpers traveled 982.6 miles. How far did they drive, rounded to the nearest hundred?

    A 100 mi   **B** 900 mi

    C 983 mi   **D** 1,000 mi

33. MEASUREMENT Ed bought 32 ounces of juice. He knows that 8 ounces = 1 cup. How many cups of juice did Ed buy?

    F 2 c   **G** 3 c

    H 4 c   **J** 8 c

LAB ACTIVITY

What You'll Explore
How to use tangrams to build symmetric figures

What You'll Need
tangram pieces

VOCABULARY
tangram

Symmetry and Tangrams

Many people enjoy solving puzzles. The **tangram** is a puzzle made of seven polygons. The seven pieces are cut from a square. The square is not the only shape that can be made with the tangram pieces.

Explore

ACTIVITY 1

- Use the tangram pieces to build the shape at the right.

- Trace the outline of the shape you made.

- Draw the line or lines of symmetry. How many lines of symmetry are there?

ACTIVITY 2

- Use five tangram pieces to build the shape at the right.

- Trace the outline of the shape you made.

- Draw the line or lines of symmetry. How many lines of symmetry are there?

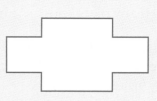

- Use all seven of the tangram pieces to build another shape with line symmetry.

- Trace the outline of your shape, and exchange with a classmate. Build your classmate's shape with tangram pieces.

- Trace the new shape you built, and draw the line or lines of symmetry.

Think and Discuss

- Which of the tangram pieces are congruent?

- Can the two small triangles be arranged to form a shape congruent to another piece? Which piece?

- Which of the tangram pieces has more than one line of symmetry? How many lines of symmetry are there?

- Which of the tangram pieces have rotational symmetry?

- Do any of the shapes you built have rotational symmetry? If so, which ones?

Try This

Use more than one tangram piece to make each shape. Trace the outline of each shape you make, and draw the line or lines of symmetry. If a shape has no lines of symmetry, write *none*.

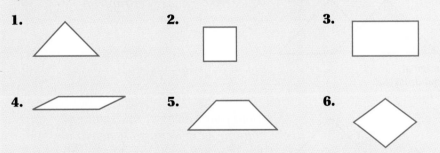

1.

2.

3.

4.

5.

6.

Use two or more tangram pieces to make each shape. Tell whether the shape has *line symmetry*, *rotational symmetry*, or *both*.

7.

8.

9.

Technology Link

You can practice transforming a figure by playing a game using E-Lab, Activity 9. Available on CD-ROM and on the Internet at **www.hbschool.com/elab**

183

Transformations

When you turn a figure around a point of rotation, it doesn't change in size or shape. A movement that doesn't change the size or shape of a figure is a rigid **transformation**. Three types of these transformations are described below.

A **translation** is the movement of a figure along a straight line. Only the location of the figure changes.

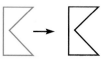

Turning a figure around a point is called a **rotation**. Both the position and the location of the figure change.

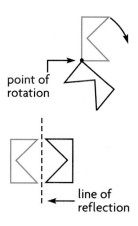

point of rotation

Flipping a figure over a line is called a **reflection**. Both the position and the location of the figure change.

line of reflection

TEEN TIMES

Have you ever heard the old expression "The mirror doesn't lie"? It means that the way you see yourself in a mirror is the way others see you. The next time you look at your reflection in a mirror, raise your right hand and wave. The reflected "you" will be waving its left hand.

EXAMPLE 1 Tell whether the transformation is a *translation*, a *rotation*, or a *reflection*.

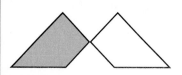

original quadrilateral

Trace the original quadrilateral on the left.

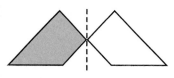

Move your tracing by translating, rotating, or reflecting it to determine how it was moved.

This transformation of the figure is a reflection.

• What position would the original quadrilateral have after being transformed with a translation?

A point of rotation can be on or outside a figure.

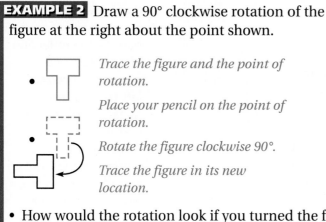

EXAMPLE 2 Draw a 90° clockwise rotation of the figure at the right about the point shown.

Trace the figure and the point of rotation.

Place your pencil on the point of rotation.

Rotate the figure clockwise 90°.

Trace the figure in its new location.

• How would the rotation look if you turned the figure 180° clockwise?

Talk About It

• Look at Examples 1 and 2. Is the new figure similar or congruent to the original figure? Explain.

• **CRITICAL THINKING** If a figure is rotated 360°, how does its new position compare to its old position?

GUIDED PRACTICE

Tell which type of transformation the second figure is of the first figure. Write *translation, rotation,* or *reflection.*

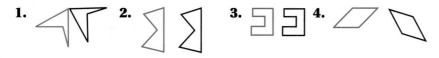

1.   2.   3.   4.

Trace the figure, line, and point. Draw a reflection about the line and then rotate that figure 90° clockwise.

5.   6.   7.   8.

Read the label, and write *true* or *false.*

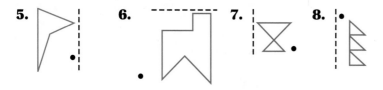

9.   10.   11.   12.

translation   rotation   reflection   translation

You can use more than one transformation to change a figure's position and location. Patterns in material, wallpaper, and draperies are examples of figures being transformed more than one way.

EXAMPLE 3 What moves were made to transform the block letter *E* into each new position?

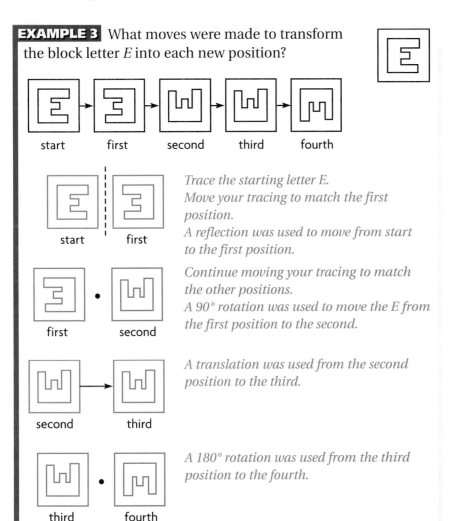

So, the transformations made were a reflection, a rotation, a translation, and a rotation.

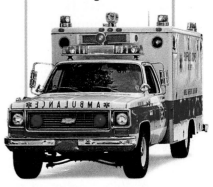

INDEPENDENT PRACTICE

Tell whether the second figure is a *translation* or a *rotation* of the first figure.

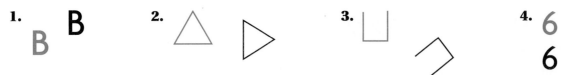

1. B B

2. △ ▷

3. ⊔ ◇

4. 6 6

Trace the figure, line, and point. Draw a reflection about the line and then rotate that figure 90° clockwise.

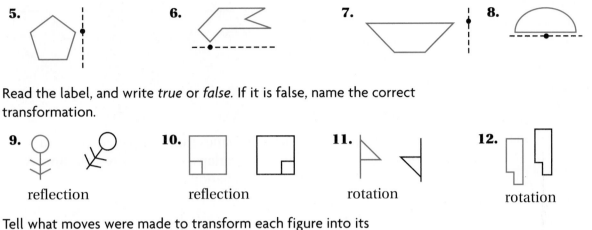

5.   **6.**   **7.**   **8.**

Read the label, and write *true* or *false*. If it is false, name the correct transformation.

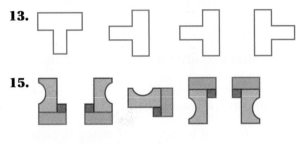

9.   **10.**   **11.**   **12.**

reflection   reflection   rotation   rotation

Tell what moves were made to transform each figure into its next position.

13.   **14.**

15.   **16.**

Problem-Solving Applications

17. LOGICAL REASONING Simone was looking in the mirror. She blinked her left eye. Which eye did her reflection blink?

18. Draw a figure. Then draw a translation, a rotation, and a reflection of the figure.

19. PATTERNS Reflect the word *bob* vertically. What word appears?

20. WRITE ABOUT IT Can a reflection also be a rotation? Explain.

Mixed Review and Test Prep

Use a protractor to measure the angle.

21.    **22.**   **23.**   **24.**

Tell whether the angle is *acute*, *right*, or *obtuse*.

25. 100°   **26.** 45°   **27.** 90°   **28.** 160°

29. TIME A 2-hour 24-minute movie ends at 4:54. What time did the movie start?
  A 2:00   **B** 2:30
  C 6:58   **D** 7:18

30. GEOMETRY Which figure always has line symmetry?
  F a triangle   **G** a circle
  H a pentagon   **J** a hexagon

LESSON 9.3

What You'll Learn
How to use polygons to make a tessellation

Why Learn This?
To understand which polygons will form tessellations so you can use them in art and construction projects

VOCABULARY
tessellation

Tessellations

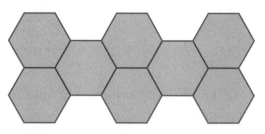

The design at the right can be seen in nature and in manufactured objects. Where have you seen this design before?

A repeating arrangement of shapes that completely covers a plane, with no gaps and no overlaps, is called a **tessellation**. A hexagon was used in the tessellation above.

ACTIVITY

WHAT YOU'LL NEED: pattern blocks, colored pencils or markers, protractor

Make a tessellation.

• Choose a pattern block shape to use for your tessellation.

• Design your tessellation. Remember that the shapes must fit together without overlapping or leaving gaps.

• Record your tessellation. Color it to make a pleasing design.

• Choose a vertex inside your design. Measure each of the angles around the vertex. What is the sum of the angle measures?

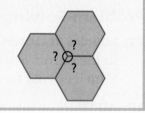

Talk About It

• Will each of the six pattern block shapes form a tessellation? Explain.

• Look at the designs you and your classmates made. What is the sum of the angles where vertices meet?

GUIDED PRACTICE

Trace each polygon. Then cut out several of each shape. Tell whether the polygon forms a tessellation. Write *yes* or *no*.

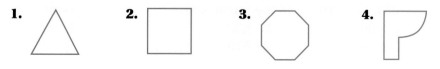

1.   2.   3.   4.

INDEPENDENT PRACTICE

Trace the polygon several times, and cut out your tracings. Tell whether the polygon forms a tessellation. Write *yes* or *no*.

1.

2.

3.

4.

Find the measures of the angles that surround the circled vertex. Then find the sum of the measures.

5.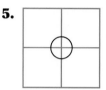

6.

7.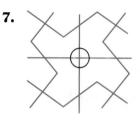

For Exercises 8–10, use the figure at the right.

8. What polygons are at the circled vertex?

9. What is the sum of the measures of the angles at the circled vertex?

10. Trace the design. Use the design to make a tessellation by connecting several more of the same pattern.

For Exercises 11–13, use the figure at the right.

11. Which polygons are at the circled vertex?

12. What is the sum of the measures of the angles at the circled vertex?

13. Trace the design. Then continue the pattern. Does it form a tessellation?

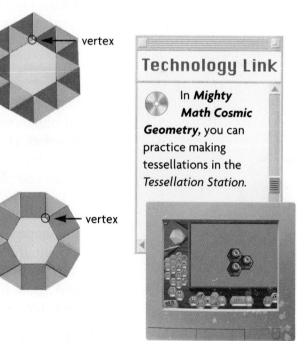

vertex

vertex

Technology Link

In *Mighty Math Cosmic Geometry,* you can practice making tessellations in the *Tessellation Station.*

Problem-Solving Applications

14. ART Draw a design of your own that forms a tessellation.

15. PATTERNS Use two equilateral triangles to form a diamond. Then draw a tessellation using diamonds and half-diamonds.

16. CRITICAL THINKING Can any parallelogram be used to make a tessellation? Explain.

17. ✏ WRITE ABOUT IT Explain how you know when shapes form a tessellation.

How to use the strategy *make a model* to solve problems that involve tessellations

Why Learn This?
To solve problems that involve woodworking, decorating, or craft projects

PROBLEM SOLVING
. . . .
- **Understand**
- **Plan**
- **Solve**
- **Look Back**

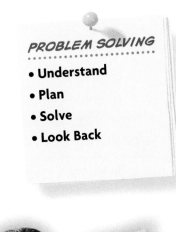

PROBLEM-SOLVING STRATEGY

Making a Model

Nikki is designing a tile mosaic for the top of her table. The shape Nikki uses must tessellate a plane. She wants to use the shape at the right. Can she use this shape for her design?

You can make a model to solve problems like this.

UNDERSTAND What are you asked to find?

What information is given?

PLAN What strategy will you use?

You can use the strategy *make a model*. Trace the shape Nikki wants to use. Cut out several copies of the shape. Use the paper shapes to see whether the shape will tessellate a plane.

SOLVE How will you solve the problem?

Begin by moving the shapes around to see if they fit together. The pieces must not overlap or leave any gaps when you place them together.

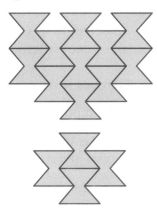

You may also check to see if the shapes will cluster around a point. If they do, check to see if the cluster can be drawn again and again. If it can, the shape will tessellate a plane.

LOOK BACK If the figures cluster around a point, the sum of the measures of the angles is 360°. What is the sum of the measures of the angles of the figure Nikki is using?

What if . . . Nikki wants to use right triangles to make her mosaic? Will a right triangle tessellate a plane? Make a model to test your ideas.

PRACTICE

Make a model to solve.

1. Denise is making a design from the shape below. She wants the design to tessellate a plane. Can she use this figure?

2. Will this shape tessellate a plane? Explain.

3. For a tile design, Bill used squares and equilateral triangles. Will his design tessellate a plane?

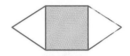

4. Draw a figure that does not tessellate a plane.

MIXED APPLICATIONS

Solve.

CHOOSE a strategy and a tool.
- **Work Backward**   • **Guess and Check**
- **Write an Equation** • **Use a Formula**
- **Make a Model**   • **Act It Out**

Paper/Pencil   Calculator   Hands-On   Mental Math

5. Colin sat to the left of Gregory. Gregory sat across from June. Heather did not sit next to June, but next to Howie, who sat across from Colin. Did Heather sit to the right or left of Howie?

6. Maria spent $40.00 at the mall. She bought a present for $12.45, a concert ticket for $15.95, and two hats that were the same price. How much did each hat cost?

7. Mr. Rodriguez bought a washing machine for $450. He made a down payment of $130 and paid the rest in equal payments of $80 a month. How many months did it take to pay for the washing machine?

8. Pedro is stocking toys on the shelves in the storeroom. The shelves are 0.5 m apart, and the bottom shelf is 1.5 m from the floor. There are 6 shelves. How far from the floor is the top shelf?

9. Marissa wants to use stones shaped like the one below to make a walk around her garden. She wants to make sure the shape will tessellate. Can she use this shape?

10. ✏️ **WRITE ABOUT IT** Explain how you can decide whether a figure will tessellate a plane.

1. **VOCABULARY** When a figure can be folded or reflected so that the two parts of the figure match or are congruent, it has __?__ (page 178)

Trace the figure. Draw the lines of symmetry. (pages 178–181)

2.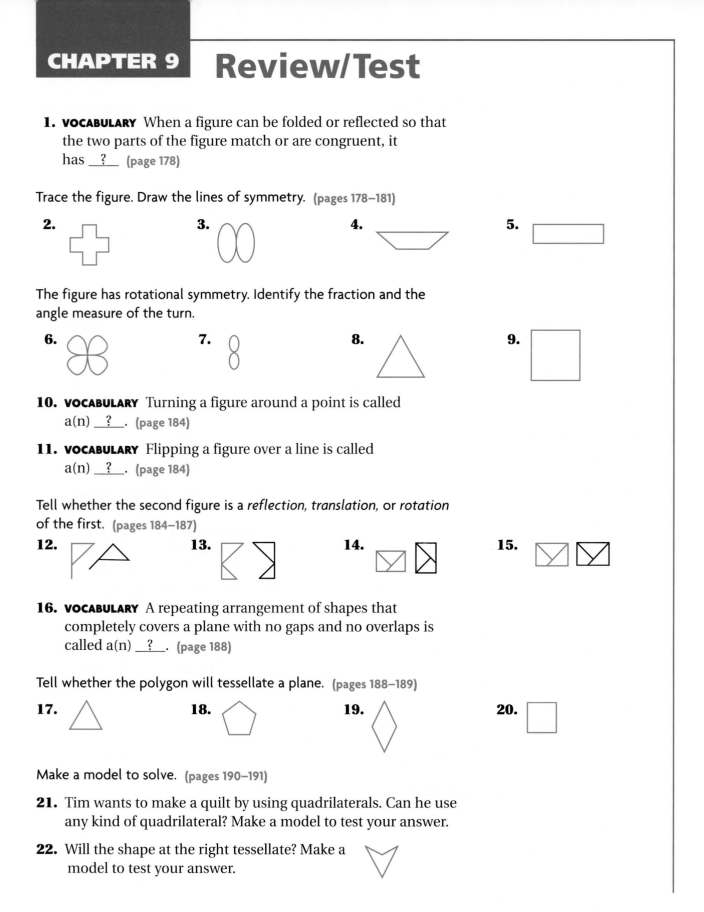

3.

4.

5.

The figure has rotational symmetry. Identify the fraction and the angle measure of the turn.

6.

7.

8.

9.

10. **VOCABULARY** Turning a figure around a point is called a(n) __?__. (page 184)

11. **VOCABULARY** Flipping a figure over a line is called a(n) __?__. (page 184)

Tell whether the second figure is a *reflection, translation,* or *rotation* of the first. (pages 184–187)

12.

13.

14.

15.

16. **VOCABULARY** A repeating arrangement of shapes that completely covers a plane with no gaps and no overlaps is called a(n) __?__. (page 188)

Tell whether the polygon will tessellate a plane. (pages 188–189)

17.

18.

19.

20.

Make a model to solve. (pages 190–191)

21. Tim wants to make a quilt by using quadrilaterals. Can he use any kind of quadrilateral? Make a model to test your answer.

22. Will the shape at the right tessellate? Make a model to test your answer.

1. Luke bought a stereo that cost $684 with tax. He is paying for it in 18 monthly payments. How much is each payment?

   A $18
   B $26
   C $32
   D $38
   E $43

2. Jon is asked to find the product of 0.029 and 0.005. How many decimal places will his answer contain?

   F 2
   G 3
   H 4
   J 5
   K 6

3. What geometric figure is modeled by the beam of light from a flashlight?

   A point   **B** line segment
   C ray   **D** line

4. Which fraction is equivalent to $\frac{3}{5}$?

   F $\frac{5}{7}$ **G** $\frac{4}{6}$

   H $\frac{6}{10}$ **J** $\frac{6}{15}$

5. What is a reasonable estimate of the sum of the fractions $\frac{1}{10}$, $\frac{4}{9}$, $\frac{5}{8}$, and $\frac{9}{10}$?

   A 1   **B** 2
   C 4   **D** 5

6. What is a reasonable estimate of the answer for $15\frac{5}{8} \div 3$?

   F 4   **G** 5
   H 6   **J** 7

7. Which type of transformation of the first figure is shown by the second figure below?

   A translation
   B rotation
   C reflection
   D Not Here

8. The sign read, "Town Line: $5\frac{1}{2}$ miles." If Bart drives $2\frac{1}{4}$ miles farther, how far is he from the town line?

   F $2\frac{1}{4}$ mi   **G** $2\frac{1}{2}$ mi

   H 3 mi   **J** $3\frac{1}{4}$ mi

9. What is the measure of each angle that surrounds the circled vertex?

   A 30°
   B 45°
   C 50°
   D 60°

10. Four baseball players are standing in a line. The outfielder is standing between the pitcher and the infielder. The catcher is in front of the pitcher. The infielder is last. How are they arranged in the line?

    F Pitcher, catcher, outfielder, infielder
    G Catcher, pitcher, infielder, outfielder
    H Outfielder, catcher, infielder, pitcher
    J Catcher, pitcher, outfielder, infielder

10

SOLID FIGURES

LOOK AHEAD

In this chapter you will solve problems that involve

- classifying solid figures

- naming and counting edges, faces, and vertices in solid figures

- identifying views of solid figures

- drawing two-dimensional views of solid figures

CONSUMER **LINK**

Some boxes for take-out food have a rectangular base and a larger rectangular opening at the top. Each of the four sides of the box is a trapezoid. Suppose one of these boxes has the following dimensions:

base of container: a 3 in. × 2.5 in. rectangle
top of container: a 4 in. × 3.5 in. rectangle
height of each trapezoidal side: 4 in.

- Are all four sides of the container the same shape?

Box Patterns

Locate an empty cardboard box—any size, any shape. Record the length, width, and height of the box. To learn more about the box, pull it apart to find the flat pattern that was used to make the box. Then measure each part of the flat pattern. Make a poster to summarize what you learned.

PROJECT CHECKLIST

☑ Did you measure the length, width, and height of the box?

☑ Did you find the measurements for each section of the flattened box?

☑ Did you make a poster to show the pattern of the box and the three-dimensional shape of the box?

Solid Figures

Solid figures come in many sizes and shapes. What shapes are used most often in packaging food?

At a grocery store, many items are sold in boxes. Most boxes have flat, rectangular faces.

Talk About It
• Name items likely to be packaged in the four boxes shown.

• How many rectangular faces does each box have?

• Which faces on each box are congruent?

Another familiar shape in a grocery store is a can.

• Name items likely to be packaged in the four cans shown.

• Which two parts of each can are congruent?

• How would you describe the parts that are not circles?

Some foods are packaged in other shapes of containers.

• Name a food for each of these shapes.

• Describe the shape of any package you have seen that is different from any shown here.

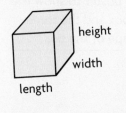

REMEMBER:

A **solid figure** is a three-dimensional figure. **See page H22.**

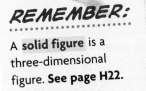

In geometry many of these shapes are given special names.

A **polyhedron** is a solid figure with flat faces that are polygons.

A **prism** is a polyhedron with two congruent, parallel bases. Its **lateral faces** are rectangles. A prism is named for the shape of its **bases**.

Boxes are usually rectangular prisms that have rectangular bases. Other polygons can be used as bases as well.

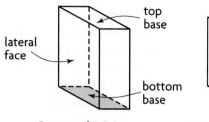

top base

lateral face

bottom base

Rectangular Prism

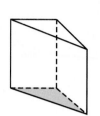

Triangular Prism

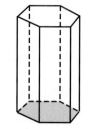

Hexagonal Prism

In science the word *prism* refers to a triangular prism that can separate rays of white light into the colors of the visible spectrum. Violet has the shortest wavelength (about 0.0000001 m). Just outside the visible spectrum is ultraviolet, the rays that cause sunburn. Ultraviolet has a wavelength of about 0.00000001 m. Is this wavelength longer or shorter than that of violet?

EXAMPLE 1 Classify this solid figure.

All the faces are flat and are polygons, so the figure is a polyhedron.

The rectangular lateral faces indicate that the figure is a prism.

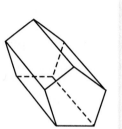

Two of the faces are congruent pentagons, so they must be the bases.

Since the lateral faces are rectangles, the figure is a pentagonal prism.

• What is the name of a prism that has squares as its bases?

A **cylinder** has two flat, circular bases and a curved lateral surface. Most cans are examples of cylinders.

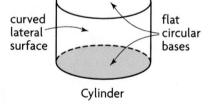

curved lateral surface

flat circular bases

Cylinder

• **CRITICAL THINKING** How are a cylinder and a prism similar? How are they different?

If you connect a single point on the top base of a cylinder to all the points around the bottom base, you form a cone.

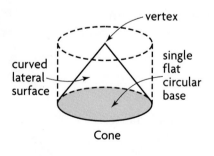

vertex

curved lateral surface

single flat circular base

Cone

A **cone** has one flat circular base, a curved lateral surface, and a **vertex**.

• **CRITICAL THINKING** How are a cone and a cylinder similar? How are they different?

A pyramid is related to a prism as a cone is related to a cylinder.

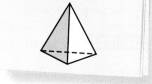

triangular lateral face

single flat square base

Pyramid

EXAMPLE 2 Classify this solid figure.

All the faces are flat and are polygons, so the figure is a polyhedron.

The triangular lateral faces with a common vertex indicate that it is a pyramid.

The one square face is the base.

The solid figure is a square pyramid.

• What is the name of a pyramid that has a pentagon as its base?

REMEMBER:

A **pyramid** is a solid figure whose one base is a polygon and whose other faces are triangles that have a common vertex. **See page H23.**

GUIDED PRACTICE

Is the figure a polyhedron? Explain your thinking.

1.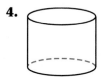

2.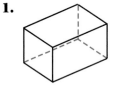

3.

4.

5.

6.

7. Which figures above are prisms? Name them.

8. Which figures above are pyramids? Name them.

9. Name the other figures.

INDEPENDENT PRACTICE

Write *base* or *lateral face* to identify the shaded face.

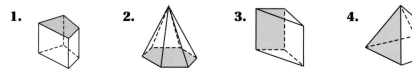

1.   **2.**   **3.**   **4.**

Name each figure.

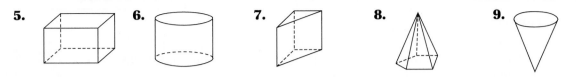

5. **6.** **7.** **8.** **9.**

Write *true* or *false* for each statement. Change each false statement into a true statement.

10. A cone has two vertices.

11. A cylinder has two bases.

12. A cone has no flat surfaces.

13. The bases of a cylinder are congruent.

14. The lateral surface of a cylinder is curved.

15. A cone can be cut from a solid cylinder.

16. If a cylinder is placed on its lateral surface, it can roll in a straight line.

17. A cylinder is a polyhedron.

Problem-Solving Applications

18. **HOBBIES** Jerome made a clay paperweight in the shape of a pentagonal prism. How many faces does the paperweight have?

19. Jessica carved a wooden pyramid with seven faces. How many sides does the base of the pyramid have?

20. **LOGICAL REASONING** Beth made two square pyramids and glued the congruent bases together to make an ornament. How many faces does her ornament have?

21. **WRITE ABOUT IT** Name three household items that are shaped like rectangular prisms.

Mixed Review and Test Prep

Tell how many sides and how many vertices each polygon has. Then name the polygon.

22. **23.** **24.** **25.** **26.**

Tell whether the two figures are similar. Write *yes* or *no*.

27.  **28.** **29.**

30. **MENTAL MATH** Mark has completed $\frac{2}{3}$ of an 18-week course. How many weeks has he completed?

   A 6   **B** 8

   C 10   **D** 12

31. **ALGEBRA** Let h represent Ed's height in inches. What expression represents Joe's height if he is 6 inches taller than Ed?

   F $h \times 6$ **G** $h - 6$

   H $h + 6$ **J** $h \div 6$

Faces, Edges, and Vertices

REMEMBER:

Vertices are named as points.
Edges are named as line segments.
Faces are named as polygons.
See page H23.

Jamie wants to paint this box. No two faces that touch will have the same color. What is the least number of colors she needs?

Talk About It

- How many faces does the box have?

- The faces meet to form edges. How many edges does the box have?

- The edges meet to form vertices. How many vertices does the box have?

- **CRITICAL THINKING** Which pairs of faces have no edges or vertices in common?

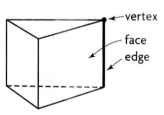

There are three pairs of opposite faces on the box. So, Jamie needs three colors.

EXAMPLE 1 Name and count the vertices, edges, and faces of this triangular pyramid.

Vertices: points *A*, *B*, *C*, and *P*; 4 vertices

Edges: line segments *AB*, *BC*, *AC*, *PA*, *PB*, and *PC*; 6 edges

Faces: triangles *APB*, *BPC*, *APC*, and *ABC*; 4 faces

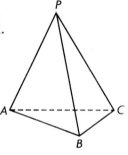

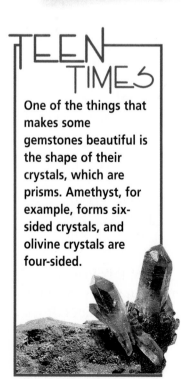

EXAMPLE 2 Count the vertices, edges, and faces of this triangular prism.

Vertices: 3 on each base, 6 vertices in all

Edges: 3 on each base, 3 lateral edges, 9 edges in all

Faces: 2 bases, 3 lateral faces, 5 faces in all

Triangular Prism

vertex
face
edge

GUIDED PRACTICE

For Exercises 1–7, use the figure at the right.

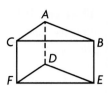

1. Name the vertices.

2. Name the edges.

3. Name the faces.

4. Count the vertices.

5. Count the edges.

6. Count the faces.

7. Which of the words below describe the figure?
prism   pyramid   polyhedron

INDEPENDENT PRACTICE

For Exercises 1–5, use the figure at the right.

1. Name the vertices.

2. Name the edges.

3. Name the faces.

4. Name the figure.

5. Write *polyhedron* or *not a polyhedron* to describe the figure.

Copy and complete the table.

	Triangular Pyramid	Triangular Prism	Rectangular Pyramid	Rectangular Prism	Pentagonal Pyramid	Pentagonal Prism
6. Number of faces	?	?	?	?	?	?
7. Number of vertices	?	?	?	?	?	?
8. Number of edges	?	?	?	?	?	?

Problem-Solving Applications

9. ART Maria wants to paint a triangular pyramid, with no two touching faces the same color. What is the least number of colors she needs?

10. PATTERNS Jon paints a pentagonal pyramid so that no two touching faces are the same color. What is the least number of colors he needs?

11. GEOMETRY Mark made a prism and a pyramid that have 7 faces each. What figures did he make?

12. ✏ **WRITE ABOUT IT** What is the least number of faces a prism can have? What is the least number for a pyramid?

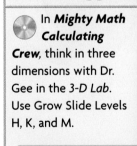

Technology Link

In *Mighty Math Calculating Crew*, think in three dimensions with Dr. Gee in the *3-D Lab*. Use Grow Slide Levels H, K, and M.

Building Solids

What You'll Learn
How to build models of prisms

Why Learn This?
To build patterns for boxes that will have the dimensions you need

VOCABULARY

net

You can build solid figures by joining faces. The faces can be cut from paper, taped together, and then folded to form the solid.

ACTIVITY

WHAT YOU'LL NEED: 4-in. × 6-in. index card, inch ruler, scissors, tape

Look at the rectangular prism at the right.

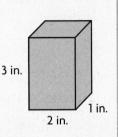

3 in.

2 in.

1 in.

• How many faces does the prism have?

• What are the dimensions of the faces?

Follow these steps to make a pattern for the prism.

Step 1: Draw the faces on the index card.

2 × 3	2 × 3
1 × 3	1 × 3
1 × 2	1 × 2	waste

Step 2: Cut out the six rectangles.

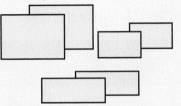

Step 3: Tape the pieces together to form the prism.

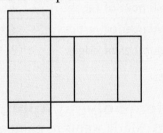

Step 4: Remove the tape from some of the edges so that the pattern lies flat.

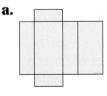

An arrangement of the rectangles that folds to form the prism is called a **net** for the prism.

Which arrangement is a net for the prism?

a.

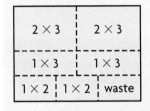

b.

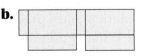

GUIDED PRACTICE

Cut six 2-in. × 2-in. squares from an index card. Arrange the squares as shown below. Is the arrangement a net for a cube? Write *yes* or *no*.

1.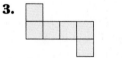

2.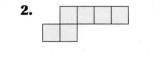

3.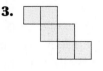

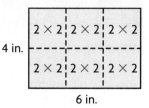

INDEPENDENT PRACTICE

1. Use the rectangular prism at the right. Draw the faces on an index card.

2. Cut out the faces, and arrange them to form a net for the prism. Tape the pieces together to form a prism.

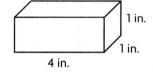

1 in.

1 in.

4 in.

Will the arrangement of squares fold to form a cube? Write *yes* or *no*.

3.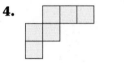

4.

5.

6.

Problem-Solving Applications

Melanie made a net for a prism that is 2 in. high and has two of these triangles as the bases.

3 in.   3 in.

3 in.

7. GEOMETRY How many faces does her prism have?

8. LOGICAL REASONING Draw a net she could have made.

9. What are the dimensions of the lateral faces?

10. ✏️ **WRITE ABOUT IT** How would the net change if Melanie used two 3-in. squares for the bases?

Mixed Review and Test Prep

Draw each polygon.

11. right triangle

12. square

13. rectangle

14. pentagon

Draw a rectangle. Then draw its image after each transformation.

15. reflected horizontally

16. rotated 90°

17. DATA A survey showed that $\frac{1}{3}$ of the 120 people at a movie were between 12 and 16 years old. How many is this?

A 40   **B** 60
C 80   **D** 100

18. NUMBER SENSE Which number sentence relates to 3 × 9 = 27?

F 6 × 9 = 54   **G** 27 ÷ 9 = 3
H 9 ÷ 3 = 3   **J** 9 × 9 = 81

Two-Dimensional Views of Solids

Kris wants to show how objects appear from different views. She decides to show how her pet, Cleo, appears when you look at him from the top, side, and front.

top view   side view   front view

You can draw different views of a solid.

ACTIVITY

WHAT YOU'LL NEED: cylinder

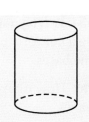

- Look at the top of the cylinder. Draw the top view.

- Look at the front of the cylinder. Draw the front view.

- Look at a side of the cylinder. Draw the side view.

EXAMPLE Identify the solid that has these views.

 top view   *The top view shows that the base is square and that the sides come together at a point.*

front view   *The front and side views show that the solid has triangular sides.*

side view

So, this solid is a square pyramid.

GUIDED PRACTICE

Name each solid that has the given top view. Refer to the solids in the box.

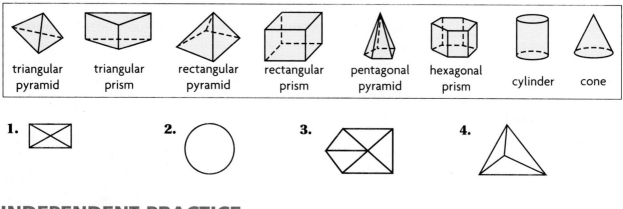

triangular pyramid   triangular prism   rectangular pyramid   rectangular prism   pentagonal pyramid   hexagonal prism   cylinder   cone

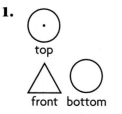

1.   2.   3.   4.

INDEPENDENT PRACTICE

Name the solid figure that has the given views.

1.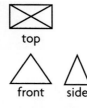
top   front   bottom

2. top   front   side

3. top   front   side

For Exercises 4–7, use the solids at the top of the page.

4. Which solid has rectangles in all of its views?

5. Which solid has triangles in all of its views?

6. Which solid has a pentagon in one of its views?

7. Which solids have circles in some of their views?

Draw the front, top, and bottom views of each solid.

8. 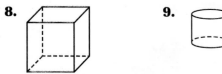   9.    10.    11.

Problem-Solving Applications

12. PATTERNS Draw the top, front, and side views of an object in your classroom.

13. CRITICAL THINKING Describe the top and side views of a cylinder.

14. PATTERNS Every side view of a triangular prism shows what shape?

15. ✏️ WRITE ABOUT IT Name objects at home that have a rectangle when you view the object from the top.

LAB ACTIVITY

What You'll Explore
How the views of a solid figure differ

What You'll Need
centimeter cubes
centimeter graph paper

Different Views of Solids

How do you think this solid would look if you viewed it from the top?

ACTIVITY 1

Explore

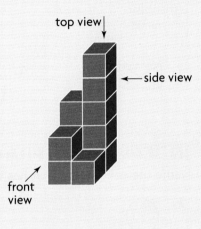

top view

side view

front view

- Use centimeter cubes to build the solid at the right.

- The top view of the solid is shown. Draw the front view and the side view on graph paper.

top view

Think and Discuss

- How many cubes did it take to build the solid?

- How many cubes do you see in the top view? the front view? the side view?

- Which views show how high the solid is?

Try This

Each solid is made with 10 cubes. On graph paper, draw a top view, a front view, and a side view for each solid.

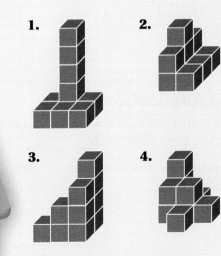

1.

2.

3.

4.

Explore

• Build this 2-cm cube.

• Draw a top view, a front view, and a side view on graph paper.

• All three views are the same 2 × 2 squares. Build a different solid whose three views are the same.

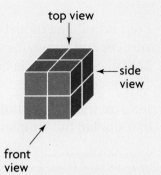

top view

side view

front view

Think and Discuss

You can make other figures by removing one, two, or three of the cubes from the whole figure. Here are some possibilities.

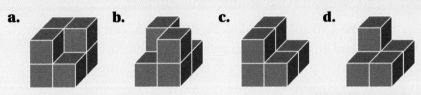

a.   b.   c.   d.

Which of the solids above match the three views given?

1.

top   front   side

2.

top   front   side

3.

top   front   side

Try This

• Build another 2-cm cube.

• Make a different solid by removing just three of the eight cubes.

• Draw the three views for your figure.

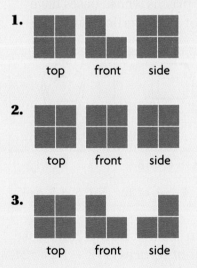

Technology Link

You can practice identifying views of solid figures by using E-Lab, Activity 10. Available on CD-ROM and on the Internet at **www.hbschool.com/elab**

PROBLEM-SOLVING STRATEGY

Solve a Simpler Problem

What You'll Learn
How to solve a difficult problem by looking at a related simpler problem

Why Learn This?
To find a simpler problem in tasks that seem too difficult to solve

PROBLEM SOLVING
....
• **Understand**
• **Plan**
• **Solve**
• **Look Back**

Brett is building models of prisms, using balls of clay for the vertices and straws for the edges. How many balls of clay and how many straws will he need to make a prism whose bases have 12 sides each?

If a problem seems difficult to solve, it sometimes helps to think of a similar, but simpler, problem.

UNDERSTAND What are you asked to find?

What facts are given?

PLAN What strategy will you use?

You can *solve a simpler problem* by thinking about the numbers of vertices and edges on prisms whose bases have 3, 4, and 5 sides. Then use these numbers to try to find a pattern.

SOLVE How will you solve the problem?

Make a table to organize the data for prisms whose bases have 3, 4, and 5 sides. Show the numbers of sides, vertices, and edges.

Sides on base	3	4	5
Vertices	3 + 3, or 6	4 + 4, or 8	5 + 5, or 10
Edges	3 + 3 + 3, or 9	4 + 4 + 4, or 12	5 + 5 + 5, or 15

From the pattern in the table, you can see that for a prism whose bases have 12 sides each, the number of vertices is 12 + 12, or 24, and the number of edges is 12 + 12 + 12, or 36.

So, Brett will need 24 balls of clay for the vertices and 36 straws for the edges.

LOOK BACK What other strategy could you use to solve this problem?

What if . . . the prism had bases with 20 sides each? How many vertices and how many edges would it have?

PRACTICE

Solve by first solving a simpler problem.

1. Aaron wants to make a model of a prism whose bases have 10 sides each. He will use balls of clay for the vertices and straws for the edges. How many balls of clay will he need?

2. Look back at Problem 1. How many straws will Aaron need? How many faces will his prism have?

3. Amy wants to make a model of a pyramid whose base has 10 sides. She will use balls of clay for the vertices and toothpicks for the edges. How many balls of clay will she need?

4. Look back at Problem 3. How many toothpicks will Amy need? How many faces will her pyramid have?

MIXED APPLICATIONS

Solve.

CHOOSE a strategy and a tool.
- **Work Backward**
- **Use a Formula**
- **Write an Equation**
- **Draw a Diagram**
- **Guess and Check**
- **Make a Table**

Paper/Pencil   Calculator   Hands-On   Mental Math

5. Mr. Dressel went to a garden shop. He spent $8.59 for a rosebush, $4.95 for fertilizer, and $9.95 for a sprinkler. He had $6.51 when he returned home. How much money did he take with him when he went to the garden shop?

6. T.J. used 5 gal of gasoline to make a 130-mi trip. How much gasoline will he need to make a 780-mi trip?

7. After the party $\frac{3}{4}$ of the cake was left. Sarah divided the cake equally among the 6 guests. How much of the cake did each guest get?

8. Mike wants to put a $3\frac{1}{2}$-ft table against a 13-ft wall. If he centers the table, how far will it be from each end of the wall?

9. Heidi baked cookies after school. She gave half to her neighbor and divided the rest equally among the 4 people in her family. Each person in her family got 3 cookies. How many cookies did Heidi bake in all?

10. Joni bought 3 model airplanes and a tube of glue at the hobby shop. The glue cost $2.59. The total bill was $20.59. What was the average price Joni paid for each of the models?

11. Jason is building a model of a prism. He used 24 toothpicks as edges to make his model. How many sides did the base of his prism have?

12. ✏️ **WRITE ABOUT IT** Look back at Problem 1. Explain how you used the strategy *solve a simpler problem*.

1. VOCABULARY A solid figure with flat faces that are polygons is called a(n) ___?___ . **(page 196)**

Name the figure. Then write *polyhedron* or *not a polyhedron.* **(pages 196–199)**

2.   **3.**   **4.**   **5.**

Name and count the faces, edges, and vertices. **(pages 200–201)**

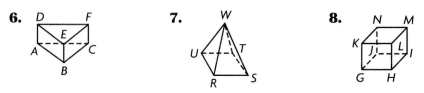

6.   **7.**   **8.**   **9.**

10. VOCABULARY An arrangement of rectangles that folds to form a prism is called a(n) ___?___ for the prism. **(page 202)**

Will the net fold to form a cube? Write *yes* or *no.* **(pages 202–203)**

11.   **12.**   **13.**   **14.**

15. What are the dimensions of the faces of this prism?

2 in.   1 in.   3 in.

Draw the top, front, and bottom views of the solid. **(pages 204–205)**

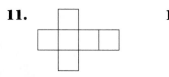 **16.**   **17.**   **18.**   **19.**

20. Name a solid figure that has circles in some of its views.

Solve by first solving a simpler problem. **(pages 208–209)**

21. Mari made a model of a prism whose base has 9 sides. How many edges and how many faces did her model have?

22. Matt made a model of a pyramid with 16 edges. How many sides did the base of his pyramid have?

TAAS Prep

1. The 38 members of an environmental club shared the cost of some plants equally. The plants cost $198. What is a reasonable estimate of the amount each club member paid?

   A Between $3 and $4
   B Between $5 and $6
   C Between $7 and $8
   D Not Here

2. Carol wants to express $\frac{17}{5}$ as a mixed number. What should she do first?

   F Divide 5 by 17
   G Multiply 17 by 5
   H Divide 17 by 5
   J Add 17 and 5
   K Not Here

3. Ben mows lawns to earn money. On Monday he spent $1\frac{2}{3}$ hours mowing. On Thursday he spent $2\frac{1}{2}$ hours mowing. What was the total time he spent mowing lawns?

   A $3\frac{1}{2}$ hr

   B $3\frac{3}{5}$ hr

   C $4\frac{1}{6}$ hr

   D $4\frac{1}{3}$ hr

4. What type of polygon has more than 5 sides and 5 angles?

   F Pentagon
   G Hexagon
   H Triangle
   J Quadrilateral
   K Both F and G

5. What type of solid figure is a textbook an example of?

   A Triangular prism
   B Rectangular prism
   C Square pyramid
   D Pentagonal prism
   E Not Here

6. Classify the solid figure shown below.

   F Square pyramid
   G Triangular prism
   H Pentagonal prism
   J Hexagonal prism
   K Rectangular prism

7. What solid figure can you make from the net shown below?

   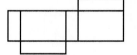

   A Rectangular prism
   B Triangular prism
   C Pentagonal prism
   D Not Here

8. Which kind of figure always has line symmetry?

   F Triangle
   G Rectangle
   H Trapezoid
   J Pentagon

MATH FUN!

IT'S ALL IN THE ALPHABET

PURPOSE To practice using geometry vocabulary (pages 160–169)

YOU WILL NEED set of 2-in. letters, construction paper

A B C D E F G H I J K L M N O P Q R S T U V W X Y Z

Classify the letters into categories using vocabulary from Chapter 8. Here are some ideas to get you started.

- Which letters consist of 2 line segments? 3 line segments?
- Which letters have parallel lines?
- Do some letters have acute angles? right angles?

Be sure every letter fits at least one category.

PUZZLE-GRAMS

PURPOSE To practice using translations, rotations, and reflections (pages 182–187)

YOU WILL NEED tangram pattern, tagboard or construction paper

Use the tangram pattern to make your own set of tangram pieces. Decide how many polygons will be in your puzzle-gram. Use translations, rotations, and reflections of your pieces to make a new shape such as a house, a person, or an animal. Trace the shape. Give the outline and the puzzle pieces you used to a classmate to solve.

HOME NOTE Take your new puzzle-gram home. Challenge your family to create new puzzles.

iMAGiNE THAT!

PURPOSE To practice thinking in three dimensions (pages 196–199)

Does the shape of some buildings suggest a rectangular prism? Does a tall tree suggest a cylinder?

List all the solid figures discussed in Chapter 10: cylinders, prisms, pyramids, and so on. For each shape, think of as many real-life objects as you can that suggest that shape. See how long a list you can make.

Geometric Translations

In this activity you will see how you can make geometric patterns and translate them by using a computer drawing program.

From the Draw tool bar, select the Freeform draw tool. Draw the shape shown.

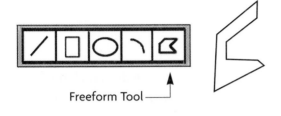

Freeform Tool

Click on the shape. On the menu bar, choose Edit. Then click on Copy.

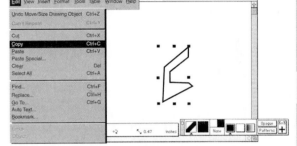

On the menu bar, choose Paste. The shape appears selected.

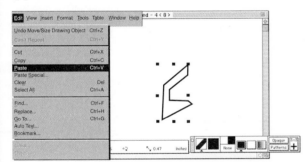

Then click and hold onto the shape and drag it to the right.

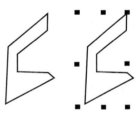

This is a translation of the shape along a line.

1. How does the shape change when you move it to the right? Explain.

2. What would the shape look like if you moved it down?

USING THE COMPUTER

Draw each shape on the left and then translate it to form the one on the right.

3.

4.

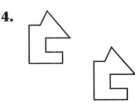

Study Guide and Review

Vocabulary Check

1. Lines that intersect to form 90° angles, or right angles, are ___?___ to each other. **(page 162)**

2. If a figure can be folded or reflected so that its two parts match, or are congruent, that figure has ___?___. **(page 178)**

3. A cone has one flat circular base, a curved lateral surface, and a(n) ___?___. **(page 197)**

EXAMPLES

EXERCISES

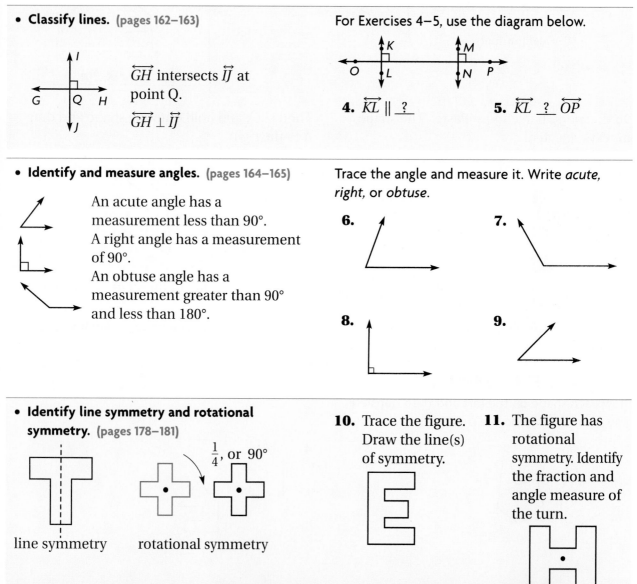

- **Classify lines.** (pages 162–163)

$\overleftrightarrow{GH}$ intersects $\overleftrightarrow{IJ}$ at point Q.

$\overleftrightarrow{GH} \perp \overleftrightarrow{IJ}$

For Exercises 4–5, use the diagram below.

4. $\overleftrightarrow{KL} \parallel$ ___?___

5. $\overleftrightarrow{KL}$ ___?___ $\overleftrightarrow{OP}$

- **Identify and measure angles.** (pages 164–165)

An acute angle has a measurement less than 90°.
A right angle has a measurement of 90°.
An obtuse angle has a measurement greater than 90° and less than 180°.

Trace the angle and measure it. Write *acute*, *right*, or *obtuse*.

6.

7.

8.

9.

- **Identify line symmetry and rotational symmetry.** (pages 178–181)

$\frac{1}{4}$, or 90°

line symmetry   rotational symmetry

10. Trace the figure. Draw the line(s) of symmetry.

11. The figure has rotational symmetry. Identify the fraction and angle measure of the turn.

- **Use polygons to make a tessellation.**
  (pages 188–189)

A hexagon will
tessellate.

Tell whether the polygon will tessellate a plane.
Write *yes* or *no*.

12. 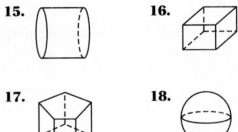 **13.** **14.**

- **Classify solid figures.** (pages 196–199)

Name the figure. Then write *polyhedron*
or *not a polyhedron*.

rectangular pyramid; polyhedron

Name the figure. Then write *polyhedron* or *not
a polyhedron.*

15. **16.**

17. **18.**

- **Name and count faces, edges, and vertices of
  solid figures.** (pages 200–201)

5 faces: *DEF, ABC, EBCF, FCAD,
DABE*

9 edges: *DE, EF, FD, AB, BC, CA,
BE, CF, AD*

6 vertices: *A, B, C, D, E, F*

Name and count the faces, edges, and vertices.

19. 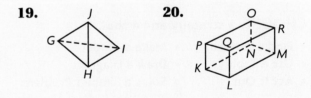 **20.**

Problem-Solving Applications

Choose a strategy to solve.

21. Name the geometric figure that is
suggested by the small or large hand of a
clock. Can a line segment be part of this
geometric figure? Explain. (pages 160–161)

22. Monica wanted to order a table that could
seat eight people, but wasn't a
quadrilateral. She also wanted each
person to have equal space. What shape
should she order? (pages 172–173)

23. Jervey wants to make a quilt entirely from
octagonal pieces of cloth. Will he be able
to do this? Why or why not? (pages 190–191)

24. Susan made a model of a prism. Its base
has 5 sides. What kind of prism is her
model? How many vertices, edges, and
faces does it have? (pages 200–201)

Performance Assessment

Tasks: Show What You Know

1. Explain how a regular pentagon is different from any other pentagon. Draw a regular pentagon. **(pages 172–173)**

2. Write the letters of the alphabet in all capital letters. Draw all the lines of symmetry on each letter. Explain. **(pages 178–181)**

3. Which of the nets below can be folded into a cube? Explain. **(pages 202–203)**

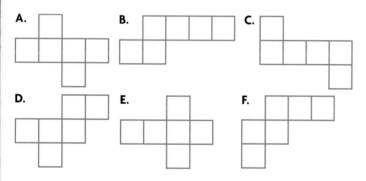

A.   B.   C.

D.   E.   F.

Problem Solving

Solve. Explain your method.

CHOOSE a strategy and a tool.	
• **Find a Pattern** • **Make a Model** • **Use a Table** • **Draw a Diagram** • **Act It Out** • **Solve a Simpler Problem**	Paper/Pencil Calculator Hands-On Mental Math

4. Draw $\overline{AB}$ parallel to $\overline{CD}$. Then draw $\overline{EF}$ perpendicular to $\overline{AB}$. What is the relationship between $\overline{EF}$ and $\overline{CD}$?
   (pages 166–169)

5. Amy wants to design a tessellation screen saver for her computer. Choose a polygon that Amy could use. Then tessellate a plane using that polygon and draw at least 3 rows of the tessellation. **(pages 188–189)**

6. Tony is creating a model of a pyramid on his computer. If the base has 8 sides, how many vertices will his pyramid have? How many edges? How many faces?
   (pages 200–201)

Cumulative Review

Solve the problem. Then write the letter of the correct answer.

1. Which of the following is the standard form for two hundred thirty and seventy-one ten-thousandths? (pages 16–17)

    A. 230.0071   **B.** 230.071
    C. 230.71   **D.** 2,300.0071

2. 7.306×4.8 (pages 64–67)

    A. 35.0688
    B. 350.688
    C. 3,506.88
    D. 350,688

For Exercises 3–6, choose the correct answer that is in simplest form.

3. $\frac{1}{9} + \frac{2}{3}$ (pages 106–107)

    A. $\frac{1}{4}$ **B.** $\frac{7}{9}$

    C. $\frac{14}{18}$ **D.** $1\frac{1}{9}$

4. $3\frac{5}{7} - 1\frac{4}{5}$ (pages 126–129)

    A. $\frac{60}{35}$ **B.** $1\frac{32}{35}$

    C. $2\frac{1}{7}$ **D.** $2\frac{32}{35}$

5. $\frac{4}{9} \times \frac{3}{7}$ (pages 136–139)

    A. $\frac{12}{72}$ **B.** $\frac{12}{56}$

    C. $\frac{4}{21}$ **D.** $1\frac{1}{27}$

6. $\frac{6}{8} \div 2$ (pages 146–147)

    A. $\frac{3}{8}$ **B.** $\frac{7}{16}$

    C. $\frac{3}{4}$ **D.** $1\frac{1}{2}$

7. Identify the figure. (pages 160–161)

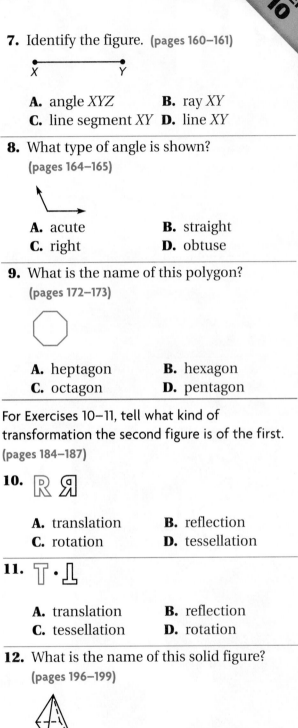

    A. angle XYZ **B.** ray XY
    C. line segment XY **D.** line XY

8. What type of angle is shown? (pages 164–165)

    A. acute   **B.** straight
    C. right   **D.** obtuse

9. What is the name of this polygon? (pages 172–173)

    A. heptagon   **B.** hexagon
    C. octagon   **D.** pentagon

For Exercises 10–11, tell what kind of transformation the second figure is of the first. (pages 184–187)

10.

    A. translation   **B.** reflection
    C. rotation   **D.** tessellation

11.

    A. translation   **B.** reflection
    C. tessellation   **D.** rotation

12. What is the name of this solid figure? (pages 196–199)

    A. rectangular pyramid   **B.** cone
    C. triangular pyramid   **D.** cube

ORGANIZING DATA

LOOK AHEAD

In this chapter you will solve problems that involve

- defining a problem
- choosing a sample
- writing a survey and determining whether it is biased
- collecting and organizing data

SPORTS LINK

Major League Soccer's first season was 1996. Here are the final standings for the Western Conference that year.

SOCCER TEAM	WINS	SHOOTOUT WINS	LOSSES
Los Angeles	15	4	13
Dallas	12	5	15
Kansas City	12	5	15
San Jose	12	3	17
Colorado	9	2	21

- How many games did each team play?

- The final standings are based on total points: 3 points for each win in regulation time, and 1 point for each shootout win. Find the total points for each team.

- Suppose one of San Jose's shootout wins had been a win in regulation time. Would San Jose have finished ahead of Kansas City? Assume Kansas City's record did not change.

What Was the Question?

Newspapers and magazines often show the results of surveys. Look for survey results. Write questions that might have  been asked to get those results. Then write three questions you can answer using the data. Display your questions and the survey results.

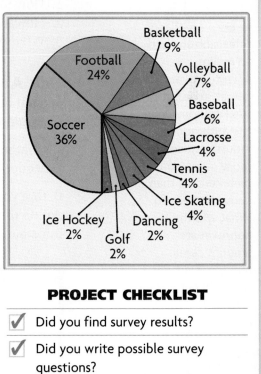

- Basketball 9%
- Volleyball 7%
- Football 24%
- Baseball 6%
- Soccer 36%
- Lacrosse 4%
- Tennis 4%
- Ice Skating 4%
- Ice Hockey 2%
- Golf 2%
- Dancing 2%

PROJECT CHECKLIST

☑ Did you find survey results?

☑ Did you write possible survey questions?

☑ Did you write questions you can answer from the data?

☑ Did you make a display of your results?

Defining the Problem

What You'll Learn
How to define a problem by deciding what information you need

Why Learn This?
To make decisions from information you get from a survey

A sixth-grade class is planning an overnight camping trip. The students in charge of the food need to buy breakfast cereal for the trip. They plan to survey everyone going on the trip, because they have to make some decisions. What decisions do they have to make? What information do they need?

The students wrote down the decisions they have to make.

a. the amount of cereal to take
b. the type of cereal to take

To make the decisions, the students need this information:

1. How many people are going on the trip?
2. How much cereal does the average person eat at breakfast?
3. What type of cereal do the people going on the trip like?

Talk About It
- What other decisions can you think of that the students should make?

- What method would you use to gather the information needed?

Before you take a survey to gather information, you have to define the problem. You can do this by deciding what information you need to make decisions.

CONSUMER LINK

In 1993, American television showed 1.3 million advertisements for cereal, at a cost of $762 million. Only the automobile industry spends more money on commercials than cereal makers do. What was the approximate amount cereal makers spent per day in 1993 for their television advertisements?

EXAMPLE Another group of students is in charge of transportation for the camping trip. What is one decision they have to make? What information do they need?

Here is a possible decision for them to make:

whether to use cars or buses for transportation

The students need to get this information:

1. How many people are going on the trip?
2. How many people can ride in each car or in each bus?
3. How much money will be needed to buy gasoline for the cars or to rent the buses?

GUIDED PRACTICE

Solve.

1. Wendy and Jo are in charge of buying beverages for a sixth-grade party. Which of these decisions will they have to make?
  a. which beverages to buy
  b. the location of the party
  c. the amount of beverages to buy

2. For the decisions you chose in Problem 1, tell what information is needed to make the decisions.

INDEPENDENT PRACTICE

Solve.

1. The Student Council is going to sell popcorn at the next basketball game. Which of these decisions will they have to make?
  a. the amount of popcorn to make
  b. the price they should charge for the popcorn
  c. the name of the team their school is playing

2. Which of the following questions should the Student Council answer before they sell popcorn?
  a. How many games did their school win this season?
  b. What is the average number of people that attend each game?
  c. How much are people willing to pay for popcorn?

Problem-Solving Applications

3. **MUSIC** A group of students is in charge of music for a dance. What decisions do they need to make? What information do they need to make the decisions?

4. 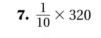 **WRITE ABOUT IT** Describe a situation in which you had to gather information in order to make a decision. Tell what information you needed.

Mixed Review and Test Prep

Find the product.

5. $\frac{1}{10} \times 30$ **6.** $\frac{1}{10} \times 70$ **7.** $\frac{1}{10} \times 320$ **8.** $\frac{1}{10} \times 240$

Name the figure.

9.   **10.** 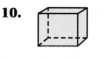   **11.**   **12.**

13. **CHOOSE A STRATEGY** Walt has 160 stamps. He has 3 times as many U.S. stamps as international stamps. How many U.S. stamps does he have?

  A 40   **B** 80   **C** 100   **D** 120

14. **NUMBER SENSE** Suzanne bought 3 cans of lemonade, 4 cans of punch, 3 cans of soda, and 10 cans of juice. What fraction of the cans were punch?

  F $\frac{4}{7}$ **G** $\frac{3}{10}$ **H** $\frac{4}{20}$ **J** $\frac{3}{20}$

Choosing a Sample

What You'll Learn
How to identify samples

Why Learn This?
To know how to choose a sample when conducting a survey

VOCABULARY

population
sample
random sample

For the sixth-grade camping trip, the game committee is going to take a survey to find out which game most students want to play. How many students should they ask?

First, you need to look at the population. A **population** is a particular group of people, such as sixth graders. If the population is large, you might survey a **sample**, or a part of the population, to represent the population.

EXAMPLE 1 There are 340 people going on the camping trip. The game committee wants to survey a sample of this population. How many people should the committee survey?

The size of a sample depends on the size of the population. For a population of 340 students, a good sample is about 1 out of every 10, or 34 students.

- Suppose you are surveying 1 person out of every 10 for a sample. How many students would you survey out of 521?

A sample can be chosen randomly. It is a **random sample** if every individual in the given population had an equal chance of being selected.

EXAMPLE 2 Ann is surveying the students at her school to find out their favorite sport. Will each of the following methods give a random sample of all the students? Explain.

A. At a basketball game, Ann can survey 1 out of every 10 students who are from her school.

B. Ann can use a computer to select 92 students from a mixed-up list of the 920 students in her school. She can then survey those students.

Method A will not produce a random sample since some of the students at Ann's school probably don't go to basketball games. These students would not have an equal chance of being selected for the sample.

Method B will give a random sample since every student will have an equal chance of being selected.

GUIDED PRACTICE

Tell how many you would survey from each group if you survey 1 out of every 10 people.

1. 420 students

2. 110 teachers

3. 280 boys

4. 630 girls

5. 390 voters

6. 860 doctors

7. 510 drivers

8. 330 athletes

Tell whether each is a random sample. Write *yes* or *no*. Then explain.

9. Kerry wanted to know the most popular movies in the sixth grade. She surveyed 30 boys out of the 300 students in the sixth grade.

10. Albert wanted to find out what snack foods students in his school prefer. He surveyed 1 out of every 10 students entering the school.

INDEPENDENT PRACTICE

Eric wants to find out if Park School cafeteria has a good selection of school lunches. There are 570 sixth, seventh, and eighth graders in the school.

1. How many people should he survey?

2. Whom should Eric survey?

3. How can Eric get a random sample?

4. Is a random sample of only sixth graders fair? Explain.

Tell whether a random sample was chosen. Explain.

5. The news station wanted to survey a large number of voters about the election. As voters left the voting booth, 4 out of every 50 were surveyed.

6. Vanessa wanted to ask shoppers to name their favorite store. She surveyed 1 out of every 10 shoppers in Tony's Marketplace.

Problem-Solving Applications

For Problems 7–9, use the table at the right.

7. DATA The table shows the results of a survey Rob took at school. How many students did he survey?

8. Rob surveyed 1 out of every 10 students as they left school. Was this a random sample of the students at school?

9. NUMBER SENSE Suppose Rob surveyed 200 students. About how many would you expect to choose rock music?

10. ✏ **WRITE ABOUT IT** Choose a topic you could find out about by conducting a survey. Then tell how you would choose a random sample.

Favorite Music	
Type of Music	Number of Students
Pop	30
Rock	28
Country	27
Rap	15

Bias in Surveys

What You'll Learn

How to determine whether a sample or a question in a survey is biased

Why Learn This?

To understand what is a good sample and a good question for a poll or survey

VOCABULARY

biased

A group of students wants to find out the favorite sport at North Middle School. Dwayne surveyed 15 students on the hockey team. Do you think the sample is a good one?

When you collect data from a survey, your sample should be large enough to represent the whole population, and every individual in that population should have an equal chance of being selected.

A sample is **biased** if individuals in the population are not represented in the sample. Dwayne's sample is biased since it only contains students on the hockey team.

EXAMPLE 1 Regina wants to survey sixth-grade students to find out the number of weekend hours students play video games. There are 450 students in the sixth grade. Which of the following sampling methods would be biased? Tell how you know.

A. Randomly survey 45 sixth-grade boys.

B. Randomly survey 45 sixth graders who do not play video games.

C. Randomly survey 45 sixth-grade students.

D. Randomly survey 4 sixth-grade students.

Choices A, B, and D are biased. Choice A excludes girls, choice B excludes students who play video games, and choice D calls for too few students to be representative of the population of 450.

Sometimes, the questions are biased. Questions that lead to a specific response or exclude a group are biased.

EXAMPLE 2 Is the following question biased?

Do you agree with the well-known and recognized Shoe Association that expensive shoes are more comfortable?

This question is biased since it leads you to agree with the Shoe Association.

• Rewrite the question so that it is not biased.

TEEN TIMES

Some magazines, such as *Zillions*® include surveys. In a recent issue, these results were given for the number of hours kids play video games in a week. (The percent is based on the number of kids surveyed.)

Hours	Percent
0	28%
1–2	27%
3–4	17%
5–6	8%
7–8	6%
9+	14%

GUIDED PRACTICE

Whitney is surveying his math class of 40 students to find out if more students pack a lunch or get the school lunch. Tell whether the sampling method is biased or not biased. Explain.

1. Randomly survey 10 girls.

2. Randomly survey students at lunch.

3. Randomly survey 10 students in the class.

4. Randomly survey 2 students in the class.

Determine whether the question is biased.

5. Do you agree with the fruit industry spokesperson that green apples are good?

6. Did you like the movie *James and the Giant Peach?*

INDEPENDENT PRACTICE

Tell whether the sampling method is biased or not biased. Explain.

The 500 Orange County teachers are doing a survey to find out how much time is spent correcting papers.

1. Randomly survey 50 male teachers.

2. Randomly survey English teachers.

3. Randomly survey 10 teachers.

4. Randomly survey 50 teachers.

The Good Soup Company is conducting a survey to find out what kind of soup consumers like best.

5. Randomly survey 2 out of 100 shoppers.

6. Randomly survey 10 out of every 100 shoppers.

7. Randomly survey 20 out of 100 women shoppers.

8. Randomly survey people who buy Good Soup.

Determine whether the question is biased.

9. What is your favorite song?

10. Is rock music your favorite type of music?

Problem-Solving Applications

For Problems 11–12, the 600 students at Major Middle School are being surveyed to find out their choice of a mascot.

11. CRITICAL THINKING If 60 students are surveyed, do equal numbers of boys and girls have to be chosen? Explain.

12. DATA Write a survey question that is biased. Then rewrite the question so it is not biased.

13. CRITICAL THINKING Suppose you want an unbiased sample of voters. Why is it unwise to choose them from a list of homeowners?

14. ✏️ **WRITE ABOUT IT** Why is this question biased? *I think Martin is the best candidate, don't you?*

LAB
ACTIVITY

What You'll Explore
How to write survey questions and how to conduct a survey

Questions and Surveys

You have learned how to define a problem, identify random samples, and determine whether a sample or question is biased. Now you can use what you know to write survey questions and conduct a survey.

ACTIVITY 1

Explore

Many companies use surveys to identify consumers' preferences. A lot of time is spent writing the questions for these surveys so that the data gathered are accurate and consistent. When you write a survey question, you should keep the following in mind.

Questions should

1. use vocabulary that is easily understood.
2. be clear and concise.
3. have the same meaning for everyone.
4. result in one clear response per person.

- The questions below were written for a survey about cafeteria food. Review the questions and decide how they could be improved.

   Do you eat broccoli and peas, do you prefer salad without dressing, and do you like to drink milk for lunch?

   Since hamburgers have more fat than turkey sandwiches, and mayonnaise has a lot of calories, wouldn't a turkey sandwich with mayonnaise be more healthful?

- A software company is thinking about opening a new store. The company wants to conduct a survey to find out if people would shop at this store, and how often. Write two survey questions the company can use.

- Check your questions. Then rewrite the questions, if necessary.

ACTIVITY 2

Explore

• Select one of the following topics for a survey:

1. favorite sport
2. favorite television show
3. favorite type of snack food

• Decide what your target group, or population, is, such as middle-school students, teachers, or parents and neighbors. What is the size of the population? How many people do you need to survey to have a good sample?

• Prepare three questions for your survey. Check and revise the survey questions so that they are not biased and are written according to the list on page 226.

• Make a recording sheet for your data. Here is a sample recording sheet for collecting data.

Person Number	Question 1	Question 2	Question 3
1			
2			
3			
4			

• Survey a random sample of the population you chose.

Think and Discuss

• What other questions could you have written?

• Compare your survey results with a classmate who chose the same topic. Are your results about the same? How do your questions compare?

Try This

Write a paragraph describing how you carried out your survey. Do you think your sample was random? Explain.

Technology Link

You can analyze experiment results by using E-Lab, Activity 11. Available on CD-ROM and on the Internet at **www.hbschool.com/elab**

Collecting and Organizing Data

What You'll Learn
How to record and organize data collected in a survey

Why Learn This?
To make it easy to understand the data you collect

VOCABULARY

tally table

frequency table

cumulative frequency

range

Mike conducted a survey to find out his classmates' favorite type of snack food. He recorded the data in a tally table. A **tally table** is a table that has categories that allow you to record each piece of data as it is collected.

FAVORITE SNACK FOODS	
Snack	**Tally**
Fruit	ЖЖ ЖЖ
Cereal	ЖЖ
Chips	ЖЖ IIII
Cookies	ЖЖ III

• How can you determine the number of students who prefer each type of snack food?

Another method for recording data is by using a line plot.

EXAMPLE 1 A movie theater surveyed 50 students to find out if they go to the movies at least once a month. The age of each student who said *yes* was recorded. Use the data below to make a line plot.

Ages of Students Who Said *Yes*								
12	15	11	16	15	16	14	12	14
16	16	15	13	16	15	13	15	14

Step 1: Draw a horizontal line.

Step 2: On your line, write the numerical values for the ages, using vertical tick marks.

Step 3: Plot the data by placing an X on your line plot for each value, or student's age, in the table.

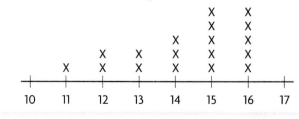

A frequency table helps you organize the data from a tally table or a line plot. A **frequency table** gives you the total for each category or group. You can add a cumulative frequency column to the table. **Cumulative frequency** keeps a running total of the number of people surveyed.

EXAMPLE 2 A new radio station surveyed people in a community to determine the type of music they preferred. Use the data below to make a frequency table. What type of music do you think this radio station will play?

TYPES OF MUSIC PREFERRED	
Music Type	**Tally**
Classical	ɪɪɪɪ ɪɪɪɪ ɪɪɪɪ
Country	ɪɪɪɪ ɪɪɪɪ ɪɪɪɪ ɪɪɪɪ ɪɪɪɪ ɪɪɪɪ
Rock and roll	ɪɪɪɪ ɪɪɪɪ
Pop	ɪɪɪɪ ɪɪɪɪ ɪɪɪɪ ɪɪɪɪ

List the music categories in one column. Put the total for each category in the frequency column and the running total in the cumulative frequency column.

Music Type	Frequency	Cumulative Frequency	
Classical	15	15	
Country	30	45	← 15 + 30 = 45
Rock and roll	9	54	← 45 + 9 = 54
Pop	20	74	← 54 + 20 = 74

So, they will probably play country music since 30 out of 74 people surveyed chose this music type.

- **CRITICAL THINKING** What part of the frequency table tells you the size of the sample surveyed?

MUSIC LINK

One of the best-known composers of classical music is Wolfgang Amadeus Mozart. A musical prodigy, he started composing at five years of age. His first major opera was performed 9 years later, in 1770. How old was he when his first opera was performed?

GUIDED PRACTICE

Use the data in the table.

1. Make a tally table.

2. Make a line plot.

3. Make a cumulative frequency table.

Miles Jogged in One Week						
20	21	26	35	38	26	40
28	33	44	32	42	28	20
42	32	24	28	35	28	20

Range and Intervals

A country music radio station took a survey to find out the ages of its listeners. The survey results are shown below.

Ages of Country Music Listeners									
14	16	35	38	43	28	36	43	41	27
21	12	27	33	18	24	19	30	29	35

The age of the youngest listener is 12. The age of the oldest listener is 43. Instead of listing each age, you can use the range to determine intervals for the data. The **range** is the difference between the greatest number and the least number in a set of data.

EXAMPLE 3 Use the data above to make a frequency table with intervals.

$43 - 12 = 31$ *Find the range.*

Since $4 \times 8 = 32$, which is close to 31, make 4 intervals that include 8 consecutive ages.   *Use the range to determine intervals.*

From the data, tally each age in the appropriate interval and record the value for each row in the frequency column.

COUNTRY MUSIC LISTENERS		
Age Group	**Tally**	**Frequency**
12–19	卌	5
20–27	||||	4
28–35	卌		6
36–43	卌	5

- **CRITICAL THINKING** What age groups could you have if you made 2 intervals?

INDEPENDENT PRACTICE

For Exercises 1–2, use the table at the right.

1. Copy and complete the table.

2. What is the size of the sample?

Ability	Frequency	Cumulative Frequency
Beginner	20	?
Intermediate	26	?
Advanced	34	?

Norma asked the students in her math class about their favorite meal. For Exercises 3–4, use the tally table.

3. Organize the data into a frequency table. Include a cumulative frequency column.

4. What is the size of the sample?

For Exercises 5–8, use the data in the box at the right.

5. Make a line plot.

6. Find the range.

7. How many heights would be in each interval if you made 4 intervals?

8. Make a cumulative frequency table using intervals of 9 for the heights.

FAVORITE MEALS	
Meals	**Tally**
Pizza	ⵏ⵰ ⵏ⵰
Tuna	ⵏ⵰	
Spaghetti	ⵏ⵰ |||
Meatloaf	||||

Students' Heights (cm)					
160	130	142	153	164	160
161	162	132	155	140	130
150	145	140	138	166	155
154	155	160	160	155	158

Problem-Solving Applications

9. CONSUMER MATH Sheri surveyed her classmates about their cereal preferences and collected the following data: 4 classmates like Hi-Fiber, 6 like Hi-Sugar, 12 like Munchy, and 8 like Crunchy. Organize this information in a tally table and a frequency table.

10. RATES Make a frequency table of each vowel using the following sentence: *When I go to school, I like math class the most.*

11. ✏️ **WRITE ABOUT IT** Explain why information on a frequency table can be more helpful than information on a tally table.

Technology Link

💾 In *Data ToolKit,* you can use data to make a frequency table and then show this data as a line plot.

Mixed Review and Test Prep

Tell whether you should use a line graph, circle graph, or bar graph.

12. high temperatures for one week

13. soccer team's heights

Tell how many faces each figure has.

14. rectangular prism

15. triangular pyramid

16. RATES Sharon rode her bike a total of 36 kilometers at a rate of 9 kilometers per hour. How long did she ride?
  A 2 hr   **B** 4 hr
  C 6 hr   **D** 9 hr

17. TIME Manuel spent $4\frac{1}{2}$ hours mowing lawns. He finished at 4:45 P.M. When did he start?
  F 11:45 A.M.   **G** 12:15 P.M.
  H 1:15 P.M.   **J** 9:15 P.M.

Solve. (pages 220–221)

1. Ted's class is in charge of planning a sixth-grade field trip. Which of these decisions will the students have to make?
   a. the type of transportation they will use
   b. where the seventh graders want to go

2. What information might you need to decide how many chaperones to have on a field trip?

3. Holly is planning a party. Which of these decisions will she have to make?
   a. when to have the party
   b. who to invite

4. What information might you need to decide how much food to buy for a party?

If you survey 1 out of every 10 people, how many would you survey from each group? (pages 222–223)

5. 350 students

6. 520 residents

7. 2,000 voters

8. 840 children

9. To determine the favorite snack at school, Al surveyed the students in his class. Is this a random sample of the school? Explain.

10. To predict who would win the election, a reporter surveyed the students at a local college. Does this sample represent voters in that city?

11. **VOCABULARY** In a survey, if any type of individual in the population is not represented by the sample, it is __?__. (page 224)

Solve. (pages 224–225)

12. A store randomly surveys 10 out of every 100 customers about the quality of its service. Is this sample biased or not biased?

13. A teacher surveys girls about the best day to give a test. Is this sample biased or not biased?

Determine whether each of the following questions is biased.

14. Which do you prefer: hot, delicious apple pie or cold pudding?

15. Are you going to her house or will you join me for ice cream at my house?

For Exercises 16–20, use the data to the right. (pages 228–231)

16. What is the sample size?

17. Find the range.

18. Make a line plot.

19. Make a cumulative frequency table with two intervals.

20. Make a cumulative frequency table with three intervals.

Students' Heights (in inches)				
62	72	63	62	69
70	60	64	66	63
71	62	65	68	63
70	64	62	67	70

1. How is $3 \times 3 \times 4 \times 4 \times 4$ expressed in scientific notation?

   A $2^3 \times 3^4$
   B $3^2 \times 4^3$
   C $3^3 \times 4^4$
   D $(3 \times 4)^5$

2. Jill was comparing the lengths of four different automobiles. Which is the correct order of the lengths of the automobiles from shortest to longest?

   F 176.5 cm, 182 cm, 180.2 cm, 180.8 cm
   G 182 cm, 180.8 cm, 180.2 cm, 176.5 cm
   H 176.5 cm, 180.2 cm, 180.8 cm, 182 cm
   J 176.5 cm, 180.8 cm, 180.2 cm, 182 cm

3. Albert bought a greeting card for $2.50, a music CD for $14.99, and wrapping paper for $3.29. All of the amounts included tax. What else do you need to know to find how much change Albert should get back from the cashier?

   A Whether the music CD was on sale
   B Whether the gift wrap included a bow
   C The number of square feet in the gift wrap
   D The amount of money Albert paid the cashier

4. Mike played football for $2\frac{1}{4}$ hours on Wednesday and $1\frac{2}{3}$ hours on Friday. How long did he play football on those days?

   F $3\frac{1}{2}$ hr

   G $3\frac{11}{12}$ hr

   H 4 hr

   J Not Here

5. Which expression can be used to find $\frac{3}{4} \div \frac{1}{3}$?

   A $\frac{3}{4} + \frac{1}{3}$

   B $\frac{3}{4} - \frac{1}{3}$

   C $\frac{3}{4} \times \frac{3}{1}$

   D Not Here

6. Classify the solid figure shown below.

   F cylinder
   G cone
   H cube
   J pyramid
   K Not Here

7. Rob is surveying 1 out of every 10 people who live in an apartment building. He spoke with 47 people. What is the population of the apartment building?

   A 94
   B 147
   C 200
   D 407
   E 470

8. What is the range of the data shown below?

Age of Theatergoers							
21	27	31	26	22	27	20	19
28	31	25	32	20	28	25	20

   F 8
   G 13
   H 15
   J 17
   K Not Here

DISPLAYING DATA

LOOK AHEAD

In this chapter you will solve problems that involve

- displaying data in bar graphs, stem-and-leaf plots, line graphs, histograms, and circle graphs

- displaying two or more sets of data in one graph

SCIENCE LINK

While a roller coaster maintains a constant speed through most of the ride, its velocity is constantly changing. Velocity changes when the roller coaster's direction changes. The *Vortex* at Paramount's Kings Island climbs to a height of 148 feet at a velocity of 5 kilometers per hour. It reaches a velocity of 83 kilometers per hour at the base of its 138-foot drop.

This ride can accommodate up to 1,600 riders per hour. The ride itself lasts just 2.5 minutes.

- Using the data above, at what point does the velocity of a roller coaster reach its maximum?

- Suppose it takes 2 minutes to unload and load the riders in the *Vortex*. About how many runs can the *Vortex* make in 1 hour?

Create the Data

Survey results don't always show all the data. For example, the graph below shows ideal vacation choices of sixth graders. But how many were surveyed? How many actually chose "amusement park"?

In newspapers and magazines, find several survey results which show percents but not totals. For each, make up reasonable data. Display your data in a table with the survey results.

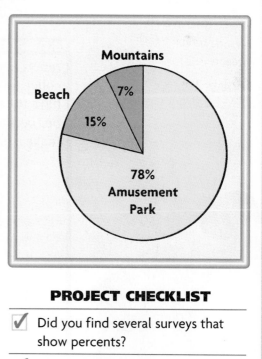

Mountains
Beach
7%
15%
78%
Amusement Park

PROJECT CHECKLIST

✓ Did you find several surveys that show percents?

✓ Did you make up your own data for the survey?

✓ Did you display your data in a table with the survey results?

Using Graphs to Display Data

What You'll Learn
How to display data in a bar graph, a line graph, and a stem-and-leaf plot

Why Learn This?
To be able to organize and visually display data for a report or project

VOCABULARY
stem-and-leaf plot

REMEMBER:

A **frequency table** shows a total in each row, one total for each category.

See page 229.

Hailey's class is conducting a survey on the favorite type of vacation. Hailey organized the data in a frequency table. She wants to make a graph for a class presentation.

FAVORITE VACATION	
Vacation Spot	**Frequency**
Amusement Park	12
Beach	10
Mountains	4
Camping	10

ACTIVITY

• Make a graph for the data above.

• Tell which type of graph you made and why you chose it.

A bar graph is a good graph to show data grouped by category.

EXAMPLE 1 Leslie is working on a report about transportation. She collected data about places that have the highest and lowest gas cost per person for a year. Use her data to make a bar graph.

YEARLY GAS COST	
Place	**Cost**
Wyoming	$734
Montana	$675
New York	$354
Washington, D.C.	$365

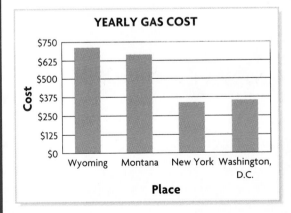

Use a scale from $0 to $750 with equal intervals.

Use bars of equal width, and leave equal space between them.

Title the graph and both axes.

• Why do you think the data sets for favorite vacation and yearly cost for gas are appropriate as bar graphs?

A bar graph and a line graph are often used to display the same data. However, a line graph is best used when the data show change over time.

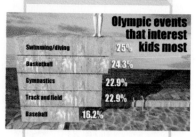

EXAMPLE 2 Mr. Henry Phillips works for a small business. He has to report to the business owners about the amount of profit the company has made over the last 6 months. Use Mr. Phillips' data to make a line graph.

PROFIT REPORT						
Month	**Nov**	**Dec**	**Jan**	**Feb**	**Mar**	**Apr**
Profit	$4,500	$5,750	$6,000	$7,500	$8,000	$8,400

Use a scale from $0 to $9,000 with equal intervals of $1,500.

Mark a point for each month.

Connect points.

Title the graph and both axes.

- **CRITICAL THINKING** Suppose the data included a seventh month with a profit of $12,000. How would you change the vertical scale on the line graph?

GUIDED PRACTICE

1. Use the data in the table below to determine the numerical scale for a bar graph.

AMOUNT OF SPORTS PARTICIPATION								
Student	A	B	C	D	E	F	G	H
Weeks	32	28	8	16	5	2	20	12

Tell whether a bar graph or a line graph is more appropriate.

2. a city's temperature readings for one week

3. sales made by several salespersons on April 1

Stem-and-Leaf Plots

You can use a **stem-and-leaf plot** to organize data when you want to see each item in the data. For a stem-and-leaf plot, choose the stems first and then write the leaves. In the example below, the tens digits appear vertically in order from least to greatest as stems. The ones digits appear horizontally in order from least to greatest as leaves.

EXAMPLE 3 In a recent tower-building competition with cards, the following numbers of levels were reached without the cards falling. Use the data to make a stem-and-leaf plot.

CARD-STACKING COMPETITION					
21	18	32	47	50	33
19	21	11	54	31	18
33	42	21	29	16	12

11	12	16	18	18	19	*First, group data by tens digits.*
21	21	21	29			*Then, order data from least to*
31	32	33	33			*greatest.*
42	47					
50	54					

Card-Stacking Competition

Stem	Leaves
1	1 2 6 8 8 9
2	1 1 1 9
3	1 2 3 3
4	2 7
5	0 4

Use tens digits as stems.
Use ones digits as leaves.
Write leaves in increasing order.

The entry 4|2 means 42 levels.

- How is this type of display useful in determining each of the levels achieved in the competition?

Talk About It

- How many levels are shown by the second stem and its third leaf?

- How many levels are shown by the fifth stem and its first leaf?

- Suppose you added 55 to the data. Where would it appear on your stem-and-leaf plot?

- In your own words, describe how to make a stem-and-leaf plot.

INDEPENDENT PRACTICE

Choose the appropriate graph for the data. Then make the graph.

1.

NAME	Mary	Tom	Carol	Jim	Tony	Patty
Height (in inches)	35	48	58	60	66	40

2.

MONTH	Jan	Feb	Mar	Apr	May	Jun
Average Temperature (in °F)	65	72	75	77	80	85

3.

MATH TEST SCORES											
60	65	66	70	71	80	81	92	59	77	75	71
80	82	84	93	95	98	80	82	75	75	77	84

Problem-Solving Applications

A school record shows that six classes have the following numbers of students: 24, 28, 32, 33, 34, and 26.

4. Which graph, a bar or a line graph, would be better for the data?

5. Make a graph of the data.

Joe's bank balances for June through December were $210, $350, $600, $400, $1,000, $750, and $900.

6. DATA What kind of graph could Joe make to see the increases and decreases in his account over six months?

7. CONSUMER MATH Make a graph of Joe's data.

8. Make a stem-and-leaf plot for Mrs. Green's golf scores: 95, 92, 88, 88, 85, 90, 92, 90, 88, 85, 82, 82, 85, 84, 80, 82, 83, 77.

9. ✏️ **WRITE ABOUT IT** Make up a problem in which you could use a stem-and-leaf plot to show the data.

Mixed Review and Test Prep

10. Make a frequency table for the data at the right.

STUDENT AGES							
10	15	18	20	15	11	13	14
9	10	15	14	18	19	15	11

If you survey 1 out of 10, how many would you survey for each group?

11. 100 teachers

12. 240 boys

13. 620 students

14. 530 girls

15. MEASUREMENT Natalie uses 2 cups of flour to bake a loaf of bread. How much flour will she need for 8 loaves?

A 4 c **B** 8 c **C** 12 c **D** 16 c

16. CONSUMER MATH If oranges cost $3 a dozen, find the cost of 36 oranges.

F $108 **G** $12 **H** $9 **J** $1

Histograms

What You'll Learn
How to make histograms

Why Learn This?
To graph groups of data when you want to show the number of times data occur within intervals

VOCABULARY

histogram

A **histogram** is a bar graph that shows the frequency, or the number of times, data occur within intervals. The bars in a histogram are connected, rather than separated.

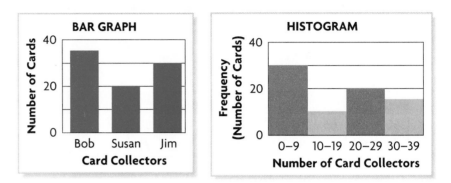

• In which graph do you find more information about the individual card collectors?

• In which graph do you find more information about card collectors in general?

REMEMBER:
....

You can use the range to find intervals.

See page 230.

2	8	12	15
15	20	25	30
35	40	42	44

Range: 44 − 2 = 42
5 intervals: 42 ÷ 5 ≈ 9
Use 5 intervals of 9 or 10 values.

0–9, 10–19, 20–29, 30–39, 40–49

EXAMPLE The table below shows the number of runs each player scored in one season. Use the data to make a histogram.

RUNS SCORED BY TEXAS RANGERS						
108	115	110	76	21	46	58
24	78	24	25	37	36	38

First, make a frequency table with intervals of 25. Start with 0.

Interval	0–24	25–49	50–74	75–99	100–124
Frequency	3	5	1	2	3

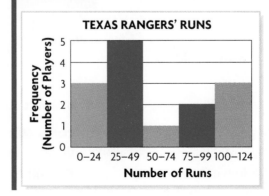

Title the graph and label the scales and axes.

Graph the number of players for each interval.

GUIDED PRACTICE

Tell whether a bar graph or a histogram is more appropriate.

1. number of voters at different intervals of time

2. heights of tallest buildings in the United States

3. favorite rock groups of sixth-grade students

For Exercises 4–5, use the frequency table at the right.

4. Make a histogram.

5. Tell how you could change the way the histogram looks.

10-MILE RACE	
Minutes	**Runners**
0–49	10
50–99	40
100–149	20

INDEPENDENT PRACTICE

Tell whether a bar graph or a histogram is more appropriate.

1. 100 scores on a social studies test

2. ages of 75 ice-skating competitors

3. frequency of certain girls' names

For Exercises 4–5, use the table at the right.

4. Make a histogram.

5. How would the number of members change in each age group if you changed the histogram to show four age groups?

MEMBERS IN THE COMMUNITY BAND					
Age	20–29	30–39	40–49	50–59	60–69
Members	10	12	18	6	8

Problem-Solving Applications

For Problems 6–8, use the histogram at the right.

6. LOGICAL REASONING What type of table might have been used before making the histogram?

7. Make a frequency table that corresponds to the histogram.

8. LOGICAL REASONING How would the histogram change if the intervals were 7–9:59, 10–12:59, and 1–3:59?

9. Write a problem based on data that can be displayed in a histogram.

10. ✏️ **WRITE ABOUT IT** Explain the difference between a bar graph and a histogram.

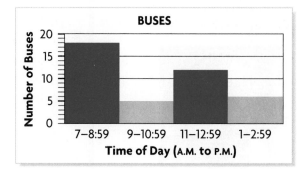

Graphing Two or More Sets of Data

VOCABULARY

multiple-bar graph
multiple-line graph

CULTURAL LINK

Temperature readings for cities on either side of the United States-Mexico border are often given in both Fahrenheit and Celsius. Suppose the temperature in both Laredo, Texas, and Nuevo Laredo, Mexico, on the United States-Mexico border, is given as 30° on a nice summer day. Is this Celsius or Fahrenheit?

The **multiple-bar graph** and the **multiple-line graph** both show two or more different sets of data on one graph. The multiple-bar graph below shows two sets of data.

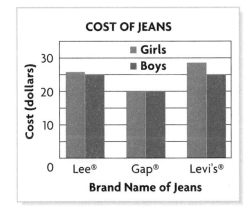

• Why does the graph need a key?

EXAMPLE Use the data below to make a multiple-line graph.

AVERAGE DAILY TEMPERATURES					
	Mon	**Tue**	**Wed**	**Thu**	**Fri**
High	82°F	78°F	75°F	64°F	70°F
Low	70°F	67°F	65°F	54°F	62°F

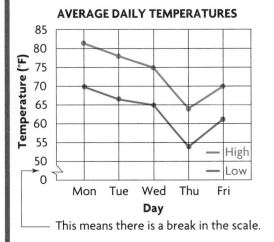

Determine an appropriate scale.

Mark a point for each high temperature, and connect the points.

Mark a point for each low temperature, and connect the points.

Title the graph and both axes. Include a key.

This means there is a break in the scale.

• CRITICAL THINKING When you make a multiple-line graph, why should you graph one set of data at a time?

GUIDED PRACTICE

1. Make a multiple-bar graph for the following data, from a survey of 100.

NUMBERS OF PETS				
	Dogs	Cats	Birds	Fish
Adult Only Household	24	10	2	1
Children in Household	38	16	4	5

2. Make a multiple-line graph for the following data, which compare scores from two seasons.

BASKETBALL SCORES					
Game	1	2	3	4	5
1997	60	66	45	50	55
1998	64	75	82	80	71

INDEPENDENT PRACTICE

1. Make a multiple-bar graph showing the following price comparisons for some regular and nonfat dairy products.

COSTS OF DAIRY PRODUCTS				
	Milk	Yogurt	Cottage Cheese	Ice Cream
Regular	$1.20	$0.89	$1.75	$1.99
Nonfat	$1.40	$0.99	$2.05	$2.20

2. Make a multiple-line graph for the following data, which compare some average stock prices.

STOCK PRICES				
	Jan	Feb	Mar	Apr
1997	$44	$53	$64	$38
1998	$48	$55	$62	$38

Problem-Solving Applications

Lisette is recording band membership in her school for the past three years. For Problems 3–5, use the table below.

3. COMPARE Which type of graph would you use for the data? Explain.

4. Make a graph for Lisette's data.

BAND MEMBERSHIP									
	Sep	Oct	Nov	Dec	Jan	Feb	Mar	Apr	May
1996	48	48	45	44	44	44	44	42	38
1997	50	51	53	53	52	52	51	50	51
1998	55	55	54	51	50	48	48	44	44

5. CRITICAL THINKING How many sets of data does your graph show? Why?

6. A survey shows how many boys and girls from each classroom are in the band. What type of graph would you use? Explain.

7. WEATHER Suppose you find the predicted high and low temperatures for your city for one week. What type of graph would you make?

8. ✏️ **WRITE ABOUT IT** Tell when you would use a multiple-bar or multiple-line graph.

LAB
ACTIVITY

What You'll Explore
How to make a circle graph from a bar graph

What You'll Need
$8\frac{1}{2}$-in. × 11-in. or larger paper, markers, scissors, tape

Exploring Circle Graphs

Data can be displayed in more than one kind of graph. In this lab you will show data in a bar graph. Then you will use the bar graph to make a circle graph.

Explore

A. Use the following data to make a bar graph.

SCHOOL LUNCHES		
Lunch Preference	Number of Students	Cumulative Frequency
Hot lunch	75	75
Salad	50	125
Pack a lunch	75	200

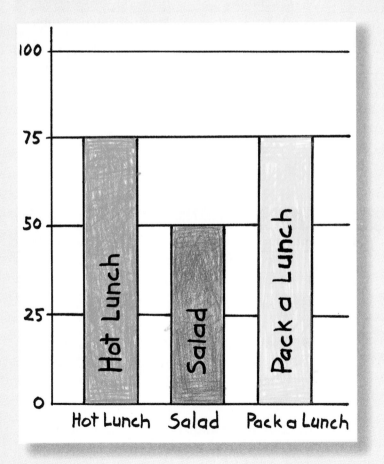

- Make each bar in the graph a different color.

- Write the lunch preference on the corresponding bar in the graph.

- Cut out each bar from your graph.

- Tape the ends of the bars together, without overlapping, to form a circle.

B. You can use your circle of bars to make a circle graph.

- Place your circle on a piece of paper, and trace around it. Mark where each bar begins and ends around the circle.

- Mark the center of the traced circle.

- Draw a radius from each of the lines you marked on the circle.

- Color the sections of the circle to match the colors of the bars. Label each region, and title the graph.

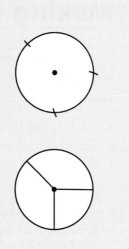

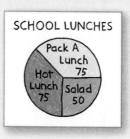

Technology Link

You can practice making a circle graph by using E-Lab Activity 12. Available on CD-ROM and on the Internet at **www.hbschool.com/elab**

Think and Discuss

- What does the whole circle graph represent?

- How is the circle graph different from the bar graph on page 244?

- What fraction could you write on your circle graph in each section?

SCHOOL LUNCHES

Pack A Lunch 75 · Hot Lunch 75 · Salad 50

Try This

- Use the data below to make a bar graph. Then use the bar graph to make a circle graph.

CAROL'S DAY		
Activity	Number of Hours	Cumulative Frequency
School	6	6
Study	3	9
TV	3	12
Sleep	8	20
Other	4	24

Making Circle Graphs

What You'll Learn
How to read and make a circle graph

Why Learn This?
To display data that are parts of a whole, such as the different ways an allowance is spent

VOCABULARY

circle graph

A **circle graph** shows parts of a whole. You can use decimals or fractions to divide the circle.

- What is the total weekly allowance? What decimal part of the circle do snacks represent?

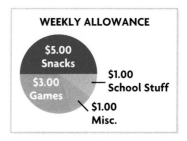

WEEKLY ALLOWANCE

$5.00 Snacks
$3.00 Games
$1.00 School Stuff
$1.00 Misc.

To make a circle graph, you need to find the number of degrees represented by each part. Since there are 360° in a circle, multiply the fraction or the decimal for each part by 360°.

EXAMPLE   Mr. Collins' class earned $400 at the school fair. Use the data below to make a circle graph.

bake sale: $200; crafts: $100; games: $50; drinks: $50

$200: $\frac{200}{400} = \frac{1}{2}$, $\frac{1}{2} \times 360° = 180°$

$100: $\frac{100}{400} = \frac{1}{4}$, $\frac{1}{4} \times 360° = 90°$

 $50: $\frac{50}{400} = \frac{1}{8}$, $\frac{1}{8} \times 360° = 45°$

Write each amount as a part of the whole. Then multiply by 360°.

Use a compass to draw a circle. Draw a radius.

Bake sale $200
Crafts $100
Drinks $50
Games $50

Use a protractor to draw a 180° angle. Label it Bake sale $200.

Draw a 90° angle. Label it Crafts $100.

Draw two 45° angles. Label them Games $50 *and* Drinks $50.

GUIDED PRACTICE

For Exercises 1–2, use the circle graph in the Example.

1. What decimal and fraction does the Crafts section represent?

2. Suppose only bake sale, games, and drinks are graphed. What angle measure would each section be?

INDEPENDENT PRACTICE

For Exercises 1–2, use the circle graph at the right.

1. How many people were surveyed about their favorite color?

2. What fraction of the circle does blue represent?

For Exercises 3–6, use the following data.

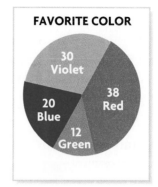

FAVORITE COLOR

30 Violet
38 Red
20 Blue
12 Green

SCOTT'S DAILY ACTIVITIES (IN HOURS)				
Sleep	School	Play	Homework	Other
8	6	3	3	4

3. Into how many sections would you divide a circle graph for the data?

4. For a circle graph, what angle measure would you use for the Sleep section?

5. Make a circle graph of the data.

6. What is the total measure of all the angles in a circle graph of the data?

Problem-Solving Applications

The chorus at Shoreline Middle School has 60 members. There are 15 sixth graders, 21 seventh graders, and 24 eighth graders. Use this information for Problems 7–9.

7. Make a circle graph to show how the chorus is divided.

8. Make a bar graph to show how the chorus is divided.

9. ✎ **WRITE ABOUT IT** Compare the circle graph you made for Problem 7 with the bar graph you made for Problem 8. Which display is easier for you to understand? Explain.

Technology Link

💾 In *Data ToolKit,* you can use data to make a table and then show the data as a circle graph.

Mixed Review and Test Prep

Tell the type of data you might graph with each.

10. bar graph
11. line graph
12. circle graph

For Exercises 13–14, tell if the sample is biased. Write *yes* or *no*.

13. Ask car makers their favorite car.

14. Ask people their favorite TV show.

15. **ALGEBRA** Daniel is 3 inches shorter than Marian. Suppose *h* represents Marian's height. What is Daniel's height?

    A $h + 3$ **B** $h - 3$
    C $h \times 3$ **D** $h \div 3$

16. **SPORTS** Joe spent $17.50 bowling. It cost $2.50 for shoes and $3.00 for each game. How many games did Joe bowl?

    F 7   **G** 6
    H 5   **J** 4

Solve. (pages 236–239)

1. Ed found that there were 40 cars, 6 vans, 5 bikes, 1 bus, and 7 trucks on a street. Would a bar graph or a line graph be a better choice for Ed's data?

2. The temperature was taken each hour from 6 A.M. to 6 P.M. Would a bar graph or a line graph be a better choice to show the data?

3. Make a stem-and-leaf plot for the data.

POINTS SCORED					
33	52	45	47	34	52
34	58	48	52	46	59

4. Make a line graph for the data.

AGE/HEIGHT TABLE				
Age of baby (months)	1	2	3	4
Height (inches)	21	23	24	25

5. **VOCABULARY** A bar graph that shows frequencies within intervals is a(n) __?__. (page 240)

Solve. (pages 240–241)

6. Would a histogram or a bar graph be more appropriate for graphing the numbers of runners in different age groups?

7. Would a histogram or a bar graph be more appropriate for graphing the number of cars sold last month for each of three models?

Use the data to make a histogram. (pages 240–241)

8.

HEIGHTS OF BUILDINGS (IN FEET)						
20	50	80	20	40	45	85
25	30	80	60	10	15	55

9.

AGE OF STUDENTS (IN MONTHS)					
60	70	63	62	60	67
69	61	67	68	64	63
61	69	65	65	66	64

Solve. (pages 242–243)

10. What type of graph would best show high and low temperatures for a week?

11. What type of graph would best show the highest and lowest temperatures for each of the last four years?

12. Make a multiple-line graph with the data at the right.

13. Make a multiple-bar graph with the data at the right.

STOCK PRICES					
	Sep	Oct	Nov	Dec	Jan
Stock A	$800	$740	$450	$500	$525
Stock B	$500	$525	$525	$500	$450

Use the data to make a circle graph. (pages 246–247)

14.

FAVORITE ICE-CREAM FLAVORS	
Flavor	Number of Students
Vanilla	30
Chocolate	50
Strawberry	20

15.

FAVORITE CANDIDATE	
Candidate	Number of Votes Sampled
A	20
B	24
C	6

⭐ TAAS Prep

1. Rose had $5\frac{1}{4}$ cups of sugar. She used $1\frac{3}{8}$ cups in a recipe. How much sugar is left?

  A $1\frac{3}{4}$ c

  B $2\frac{3}{8}$ c

  C $3\frac{1}{8}$ c

  D $3\frac{7}{8}$ c

  E Not Here

2. A can of soup is an example of what type of solid figure?

  F Cylinder

  G Cone

  H Triangular prism

  J Rectangular prism

3. Kerry is surveying 1 out of every 8 people. How many people would he speak with if the population were 664?

  A 67

  B 83

  C 94

  D 103

4. Darla blends $\frac{3}{4}$ cup of water, $\frac{5}{8}$ cup of milk, and $\frac{1}{3}$ cup of ice. What is a reasonable estimate of the total?

  F Less than 1 c

  G Between $1\frac{1}{2}$ and 2 c

  H More than 2 c

  J About $2\frac{1}{2}$ c

5. An angle which measures 120° is—

  A Acute

  B Right

  C Obtuse

  D Straight

6. Tom is making a circle graph to display the data shown below. What angle measure should he use for the pizza section?

STUDENTS' FAVORITE SCHOOL LUNCH			
Pizza	**Salad**	**Burger**	**Turkey**
30	4	18	8

  F 30°

  G 45°

  H 90°

  J 180°

7. Which of the following data is best suited for display in a multiple-bar graph?

  A Average amount of rain a town received during a 2-year period

  B Number of CDs owned by sixth-grade students

  C Number of hours sixth- and seventh-grade students spend reading each week

  D Changes in water temperature over a 24-hour period

8. Chloe made a multiple-line graph to show changes in her height each month during the past 4 years. How many lines are on her graph?

  F 2   **G** 4

  H 12   **J** Not Here

INTERPRETING DATA AND PREDICTING

LOOK AHEAD

In this chapter you will solve problems that involve

- analyzing graphs and making predictions from graphs

- identifying misleading graphs

- finding the mean, median, and mode for a set of data

- displaying and analyzing data in a box-and-whisker graph

SOCIAL STUDIES **LINK**

The population of Texas has grown throughout the 20th century. Study the table.

POPULATION OF TEXAS	
Year	Population (nearest 100,000)
1900	3,000,000
1920	4,700,000
1940	6,400,000
1960	9,600,000
1980	14,200,000

- Estimate the population of Texas in 1910.

- Estimate the population of Texas in 1950.

Population Prediction

Find historical and present-day population data for your city or state. Make a table or graph to display the data. Then use the data to predict what the population will be 10 years from now. Write a paragraph to explain your prediction.

Dallas

Houston

PROJECT CHECKLIST

✓ Did you find population data for your city or state?

✓ Did you make a table or graph?

✓ Did you predict the population for 10 years from now?

✓ Did you explain how you made your prediction?

ALGEBRA CONNECTION

Analyzing Graphs

Graphs are widely used in business reports, newspapers, and magazines. Advertisers sometimes use graphs to show how their products compare with competitors' products.

OUR CHIPS ARE CHEAPER

■ Potato Chips A
■ Potato Chips B

Cost (in dollars) / Size of Package

Talk About It

• Look at the graph.
  Describe the relationship between the height of the bars and the cost of the potato chips.

• How much greater is the cost of brand A than the cost of brand B?

• Which brand of potato chips is cheaper? How do you know?

To analyze a graph, you have to know what relationship to look for. In a bar or line graph, identify the relationships shown by the axes. In a circle graph, compare the parts to the whole. Then, look for greater-than or less-than relationships.

COMPUTER LINK

Colorful graphs may be a part of interactive Internet pages. Multimedia authoring tools and computer languages like Java allow attention-grabbing animated graphs and graphs that are updated instantly as new data become available. Make a list of World Wide Web sites that include graphs.

EXAMPLE A survey of 100 people was done to find out how many people drive alone to work and how many use other types of transportation. Analyze the graph.

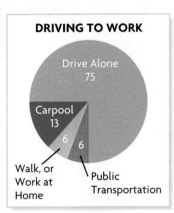

DRIVING TO WORK

Drive Alone 75
Carpool 13
Walk, or Work at Home 6
Public Transportation 6

Compare parts with the whole.

75 is $\frac{3}{4}$ of 100.

$13 + 6 + 6 = 25$;

25 is $\frac{1}{4}$ of 100.

Compare parts with other parts.
75 is 3 times as great as 25.

So, the number of people who drive alone to work is 3 times as great as the number of people who use other types of transportation.

GUIDED PRACTICE

The newspaper's movie critic conducted a survey to find out how teens enjoyed the latest teen movie released. The results are displayed in the circle graph.

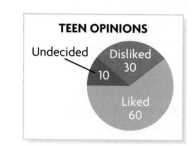

TEEN OPINIONS

1. Which opinion forms the largest part of the graph? the smallest?

2. What did the movie critic learn from this survey?

3. How many more teens liked the movie than disliked the movie?

INDEPENDENT PRACTICE

For Exercises 1–3, use the line graph at the right.

AUTO SALES

1. Which month had the best sales? Which month had the worst sales? What is the relationship between the two months' sales?

2. What kinds of factors could have affected sales in July?

3. What is the overall trend for sales for the seven months?

For Exercises 4–7, use the bar graph below at the right.

4. Which flavor was the most popular? the least popular? How do you know?

5. How does the number of gallons of Coco Delight sold compare with the number of gallons of Stellar Kiwi sold?

6. How does the number of gallons of Good Fruit sold compare with the number of gallons of Stellar Kiwi sold?

7. For next year's production of the new flavors, what would you recommend?

NEW ICE-CREAM-FLAVOR SALES

Problem-Solving Applications

8. **DATA** Write a question that can be answered by using a line graph. You may want to use the information in the graph for Exercises 1–3.

9. **DATA** Find a bar graph, line graph, or circle graph in a magazine or newspaper. Write a question about the graph.

10. **WRITE ABOUT IT** Suppose you want to compare the prices of two products. Would you use a bar graph or a circle graph? Explain.

Misleading Graphs

Sometimes advertisements in newspapers or magazines show a
graph that is misleading. The data in a misleading graph may be
factual, but the presentation of the data is misleading.

OUR SHOES LAST LONGER

Our shoes
will last 3
times as
long as other
brands of shoes!!!

Talk About It

• How does the life of Our Brand shoes compare with the life of
Brand C?

• How is this graph misleading?

Sometimes graphs can be misleading when two similar sets of
data are compared on graphs that have different scales.

EXAMPLE How are these graphs, taken together, misleading?

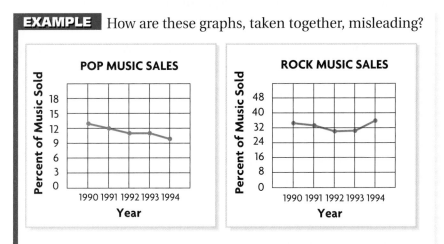

Together, the graphs make it appear as if the sales of pop music
and rock music were about the same. However, if you read the
scales and find the values of the points, you find that rock
music had greater sales.

GUIDED PRACTICE

Tell whether the graph is misleading. Write *yes* or *no*. If *yes*, explain.

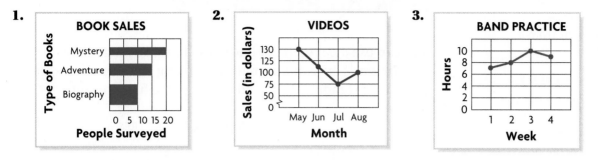

1.

BOOK SALES

2.

VIDEOS

3.

BAND PRACTICE

INDEPENDENT PRACTICE

For Exercises 1–4, use the bar graph at the right.

1. Who sold more boxes of cookies?

2. About how many times as high is the bar for Lucy's sales compared to the bar for Terri's sales?

3. Did Lucy sell twice as many boxes of cookies as Terri? Explain.

4. Make a new graph, one that is not misleading, to show Lucy's and Terri's cookie sales.

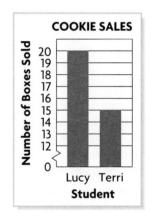

For Exercises 5–7, use the line graphs at the right.

5. In Graph A, about how many times as great as the sales in Week 4 do the sales in Week 1 appear?

6. In Graph B, about how many times as great as the sales in Week 4 do the sales in Week 1 appear?

7. What is the actual difference in sales between Week 1 and Week 4? Were the Week 1 sales 4 times as great?

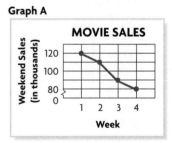

Problem-Solving Applications

8. Look at Graph A and Graph B. Which graph gives a better picture of the data? Explain.

9. **CONSUMER MATH** Share with a classmate graphs you have found in newspapers, magazines, or other sources. Discuss which are misleading and which are not.

10. **WRITE ABOUT IT** How can you determine whether a graph presents an accurate picture of a set of data?

What You'll Learn

How to make predictions from a graph

Why Learn This?

To go beyond the given data, as in predicting what the population will be in the year 2020

VOCABULARY

prediction

Making Predictions

You can use a bar graph or line graph to make predictions. A **prediction** is an estimate made by looking at a trend over time and then extending that trend to describe a future event.

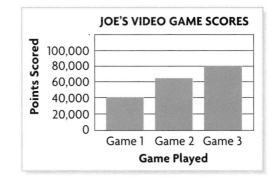

JOE'S VIDEO GAME SCORES

You can look at a graph and observe changes. On this graph, for example, there is a steady growth with no declines. Each increase in the score is about 20,000 points.

• How many points do you think Joe will score in Game 4? Do you think his score will increase indefinitely? Why or why not?

Since a line graph shows change over time, it is used most often to make predictions.

BUSINESS LINK

Stock prices for shares traded in most U.S. markets are graphed to show day-to-day trends. The Monday-to-Thursday prices for Company X stock are shown in the graph below. What do you predict the price will be on Friday?

CLOSING PRICES FOR COMPANY X

EXAMPLE The line graph below shows the population of the United States every 10 years from 1930 to 1990. Use the graph to predict the population for 2000.

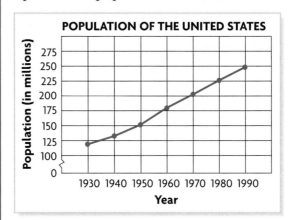

POPULATION OF THE UNITED STATES

The trend is an increase with no declines.

From 1950 to 1990, the population increased about 25 million every 10 years.

Since the population increased about 25 million every 10 years from 1950 to 1990, a good prediction for the year 2000 would be about 250 million + 25 million, or 275 million.

GUIDED PRACTICE

Use the graph at the right.

1. What was Regis's height at age 11? at age 12?

2. How much did Regis's height change from age 11 to age 12?

3. What has been the pattern for Regis's growth?

4. What do you predict Regis's height will be at age 15?

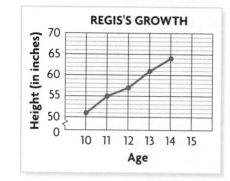

INDEPENDENT PRACTICE

For Exercises 1–5, use the graph at the right.

1. How many CDs did Lucinda have in 1994? in 1995?

2. How many more CDs did Lucinda have in 1997 than in 1996?

3. What has been the pattern for the number of CDs Lucinda buys each year?

4. How many CDs would you predict Lucinda will have in 1999?

5. How many CDs would you predict Lucinda will have in 2000?

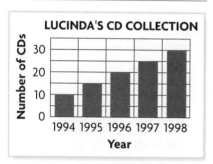

For Exercises 6–7, use the graph at the right.

6. What has been the pattern for the amount of money Andrew saves each year?

7. What do you predict Andrew's yearly savings will be in 2000?

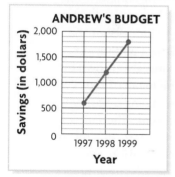

Problem-Solving Applications

8. **RATES** The electric bills for a business over the past four months were $300, $280, $250, and $210. What has the trend been for the electric bills?

9. **PREDICTION** Use the information in Problem 8 to make a line graph. Look at the graph. What do you think the electric bill will be in the fifth month?

10. **PATTERNS** The price of a particular car sold was $14,800. After one year, the car was worth $10,600. After two years, the car was worth $6,200. What has the trend been for the value of the car?

11. Use the information in Problem 10 to make a line graph. Look at the graph. What do you think the car will be worth after 3 years?

12. **✐ WRITE ABOUT IT** What must a graph display for you to make a prediction?

Mean, Median, and Mode

What You'll Learn
How to find the mean, median, and mode of a set of data

Why Learn This?
To summarize sets of data, such as test scores or sports scores

VOCABULARY

mean

median

mode

Suppose a gymnast received these scores:

8.3   7.9   8.3   8.0

There are three measures of central tendency that can be used to describe a set of data as one value: mean, median, and mode.

The **mean**, or average, is the sum of a group of numbers, divided by the number of addends.

The mean of the gymnast's scores is
$(8.3 + 7.9 + 8.3 + 8.0) \div 4 = 32.5 \div 4 = 8.125$.

The **median** is the middle number in a group of numbers arranged in numerical order. When there are two middle numbers, the median is the mean of the two middle numbers.

7.9   8.0   8.3   8.3   ← numerical order

The median is between 8.0 and 8.3.

$(8.0 + 8.3) \div 2 = 16.3 \div 2 = 8.15$

The median of the gymnast's scores is 8.15.

The **mode** of a group of numbers is the number that occurs most often. There may be one mode, more than one mode, or no mode at all.

The mode of the gymnast's scores is 8.3.

SPORTS LINK

In international figure-skating competitions, each skater is scored by 8 judges, but the highest and lowest scores are thrown out. The mean of the remaining 6 scores becomes the skater's final score. In a recent competition, the judges posted the following scores: 5.6, 5.6, 5.4, 5.3, 5.7, 5.7, 5.5, and 5.7. What would the skater's final score be, to the nearest hundredth?

EXAMPLE 1 Find the mean, median, and mode for the data.

Basketball Points				
22	18	8	34	18

Mean:
$(22 + 18 + 8 + 34 + 18) \div 5$ *Add the scores, and divide by 5.*
$100 \div 5 = 20$

Median:
8   18   18   22   34   *Order the data. The middle number is 18.*

Mode:
18   *18 occurs twice.*

 Calculator Activities, page H41

For some sets of data, you can use a line plot or a stem-and-leaf plot to help you find the median and the mode.

EXAMPLE 2 Use Ms. Jones's class test-scores data below to make a line plot. Find the mode and the median.

Ms. Jones's Class Test Scores										
72	82	83	78	81	78	73	74	75	73	76
71	75	80	83	72	72	78	81	79	82	76

```
         X                 X
         X  X        X  X  X           X  X  X
      X  X  X  X  X  X     X  X  X  X  X  X
      +--+--+--+--+--+--+--+--+--+--+--+--+--+--+
      70 71 72 73 74 75 76 77 78 79 80 81 82 83 84
```

72 and 78 each occur three times.

There are 22 scores. The median is between the 11th and 12th scores.

Modes: 72 and 78
Median: $(76 + 78) \div 2 = 77$

EXAMPLE 3 Use the fitness test-scores data to make a stem-and-leaf plot. Find the mode and the median.

Fitness Test Scores									
97	88	74	96	98	58	68	90	80	90
72	86	69	78	93	84	99	92	85	

Stem	Leaves
5	8
6	8 9
7	2 4 8
8	0 4 5 6 8
9	0 0 2 3 6 7 8 9

90 occurs more than any other number.

There are 19 scores. The median is the 10th score.

Mode: 90
Median: 86

REMEMBER:

A **stem-and-leaf plot** is a way to organize data.

Stem	Leaves
2	5 7 9
3	2 2 4 6 8
4	0 1 3

In this example, the tens digits are the **stems**. The ones digits are the **leaves. See page 238.**

GUIDED PRACTICE

For Exercises 1–3, use the table below.

Day	Sun	Mon	Tue	Wed	Thu	Fri	Sat
Miles Jogged	4	4	5	5	6	7	4

1. Find the mean.　　　　**2.** Find the mode.　　　　**3.** Find the median.

4. Find the mean, median, and mode: 10, 12, 16, 11, 10, 13.

When you want to summarize a set of data as one value, you can use one of the three measures of central tendency.

EXAMPLE 4 As a T-ball coach, you kept track of the number of runs each of your players scored for the season. Use the data in the graph to find the mean, median, and mode. Then tell which measure of central tendency best represents the data.

Mean:
$(16 + 15 + 15 + 15 + 15 + 18 + 16 + 17 + 30 + 30)$
$\div 10 = 187 \div 10 = 18.7$

So, the mean is 18.7.

Median:

15	15	15	15	16
16	17	18	30	30

So, the median is 16.

Mode:
The mode is 15. It occurs four times in the data.

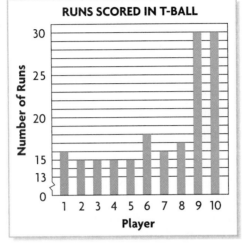

RUNS SCORED IN T-BALL

Because of the two high scores of 30, the mean is much larger than the mode and the median. So, the mean is not a good measure of the data. Either the median or the mode represents the data better.

- **CRITICAL THINKING** Which part of the data would have to be different for the mean to be a good measure of central tendency?

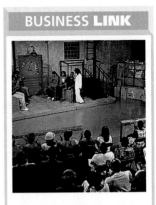
INDEPENDENT PRACTICE

Copy and complete the table.

	Data	Mean	Median	Mode
1.	7, 6, 11, 7, 9	?	?	?
2.	84, 73, 92, 77, 89	?	?	?
3.	16, 32, 24, 10, 48, 32	?	?	?
4.	9.5, 8.2, 7.1, 8.0	?	?	?
5.	45.7, 45.9, 45.7, 48.3	?	?	?
6.	27, 48, 83, 76, 48, 27	?	?	?
7.	1.3, 7.8, 0.7, 7.8, 3.4	?	?	?

Write *true* or *false* for each.

8. The mean is always one of the numbers in a set of data.

9. The median may be the average of the two middle numbers in the data.

For Exercises 10–13, use the table below.

Test Scores									
78	82	96	65	100	100	94	61	70	88

10. Make a line plot.

11. Use your line plot to find the median and the mode.

12. Make a stem-and-leaf plot.

13. Use your stem-and-leaf plot to find the median and the mode.

14. In Exercises 10–13, which display made it easier to find the median and the mode? Explain.

15. Can you tell what the mean is for this set of data by looking at the stem-and-leaf plot? Explain.

Problem-Solving Applications

16. SPORTS Use the data shown in the graph to find the mean, median, and mode. Then tell which measure of central tendency best represents the data.

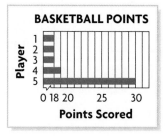

17. Eight joggers ran the following number of miles: 8, 5, 6, 4, 8, 8, 7, and 10. Determine the mean, median, and mode of the miles jogged.

18. ✏️ **WRITE ABOUT IT** Explain the differences between the mean, median, and mode of a group of numbers.

Mixed Review and Test Prep

For Exercises 19–21, use the data 10, 12, 28, 9, and 17.

19. Find the lowest value.

20. Find the highest value.

21. Find the mean.

NUMBER OF BOOKS READ		
Grade 6	Grade 7	Grade 8
59	70	65

22. Use the data in the table to the right to make a bar graph.

23. CONSUMER MATH A company claims that its 8-year warranty lasts 4 times as long as any other warranty. What is the longest of the other warranties?

A 32 yr   **B** 12 yr   **C** 2 yr   **D** $\frac{1}{2}$ yr

24. NUMBER SENSE There are 180 players in a baseball league. Each team has 12 players. How many teams are in the league?

F 12   **G** 15   **H** 16   **J** 18

LAB ACTIVITY

What You'll Explore
How to make a box-and-whisker graph and understand its parts

What You'll Need
at least eleven
3-in. × 5-in. cards,
marker

VOCABULARY

box-and-whisker graph
lower extreme
upper extreme
lower quartile
upper quartile

Exploring Box-and-Whisker Graphs

A **box-and-whisker graph** shows how far apart and how evenly data are distributed.

Explore

- Write each of the hours worked, shown in the table below, on a separate card.

Number of Hours Worked									
30	16	19	27	31	15	19	24	22	23

- Order the data from least to greatest.

- What is the least number of hours worked? What is the greatest number of hours worked?

- The least number is called the **lower extreme**. The greatest number is called the **upper extreme**. Draw a star on the cards with the lower extreme and the upper extreme.

- What is the median of the data? If the median is not one of the numbers already written, write it on a card, put it in the middle of the data, and circle it.

- The lower half of the data is to the left of the median. What is the median of the lower half? Circle it. This median is called the **lower quartile**. Separate the data to the left of the lower quartile from the rest of the data.

- The upper half of the data is to the right of the overall median. What is the median of the upper half? Circle it. This median is called the **upper quartile**. Separate the data to the right of the upper quartile from the rest of the data.

Think and Discuss

- Into how many parts do the lower quartile, the median, and the upper quartile separate the data?

- What fraction of the data are to the left of the lower quartile?

- What fraction of the data are to the right of the upper quartile?

- What fraction of the data are between the lower quartile and the upper quartile?

You have found all you need to make a box-and-whisker graph. Below is the box-and-whisker graph of the data, labeled with the parts.

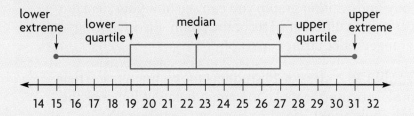

- How do you know where to start and end the box in the graph?

Try This

- For the data below, find the lower and upper extremes, the median, and the lower and upper quartiles.

2	5	11	7	9	8
8	3	4	8	5	6

- Make a box-and-whisker graph of the data.

263

Box-and-Whisker Graphs

Angela made a box-and-whisker graph to represent the number of cookies she sold each day for one week.

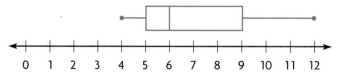

• What is the least number of cookies Angela sold in one day? What is the greatest number?

• Look at the graph. Which of the following can you determine: mean, median, mode, range?

In a box-and-whisker graph, you can see how your data are distributed. In different parts of the graph, the values may be closer together or farther apart.

If you have ever taken a standardized test, you may have noticed that quartiles were reported. These can tell you how your performance compared with that of others who took the test.

EXAMPLE A class of 20 students took a math test. Their scores were used to make the graph below. What does the graph show about how the scores are distributed?

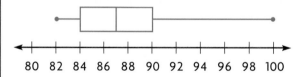

The scores in the lowest $\frac{1}{4}$ of the data are very close together. The scores in each part of the middle $\frac{1}{2}$ are farther apart. These scores are closer to the lower extreme than to the upper extreme. The scores in the highest $\frac{1}{4}$ of the data are even farther apart than those in the middle. At least one student scored 100.

• **CRITICAL THINKING** What fraction of the class had scores between 90 and 100? What fraction had scores between 80 and 90? If 90 was needed for an A, how many students got an A?

In a box-and-whisker graph, the only actual values you can identify from the data set are the extremes, the highest and lowest values.

• What are the two actual scores you can identify in the box-and-whisker graph for the example?

GUIDED PRACTICE

Use the box-and-whisker graph at the right.

1. What is the median?

2. What are the lower and upper quartiles?

3. What are the lower and upper extremes?

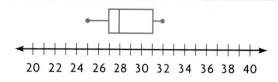

20 22 24 26 28 30 32 34 36 38 40

4. What is the range?

INDEPENDENT PRACTICE

For Exercises 1–4, use the table at the right.

1. What is the median?

2. What are the lower and upper quartiles?

3. What are the lower and upper extremes?

4. Make a box-and-whisker graph.

Lengths of Phone Calls (in min)					
20	24	21	16	15	26
17	32	30	28	16	23

Problem-Solving Applications

This box-and-whisker graph shows the number of runs scored by a baseball team in one season's games.

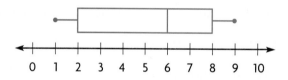

0 1 2 3 4 5 6 7 8 9 10

5. SPORTS What was the least number of runs? What was the greatest number of runs?

6. DATA Describe how the data are distributed.

7. ✏️ **WRITE ABOUT IT** Explain how a box-and-whisker graph is divided into four parts.

Mixed Review and Test Prep

Write the ratio as a fraction.

8. 3 out of 4

9. 7 out of 8

10. 1 out of 2

11. Use the data in the table to make a multiple-bar graph.

Club Name	Grade 6	Grade 7	Grade 8
Computer	19	9	7
Math	10	10	9
Yearbook	4	7	10

12. NUMBER SENSE Yolonda gave a clerk $10.03 for frozen yogurt and received $7.00 in change. How much was the yogurt?
A $3.00   **B** $3.03   **C** $3.97   **D** $7.03

13. CRITICAL THINKING The mean of 4 numbers is 13.5. What is the sum of the numbers?
F 54   **G** 27   **H** 13.5   **J** 9.5

Technology Link

💾 In *Data ToolKit*, you can use data to make a table and then show these data as a box-and-whisker graph.

For Exercises 1–3, use the graph at the right. (pages 252–253)

COMEDY FANS

1. Describe the relationship between the height of the bars and the number of students who like comedies.

2. Which grade likes comedies the least? the most?

3. The number of comedy fans in Grade 6 is how many times the number in Grade 8?

For Exercises 4–6, use the graph at the right. (pages 254–255)

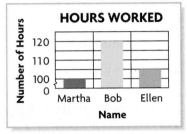

HOURS WORKED

4. The bar for Bob is how many times as tall as the bar for Martha?

5. It appears from the heights of the bars that Ellen worked about twice as many hours as Martha. Is this accurate?

6. How can you change the graph so that it is not misleading?

For Exercises 7–10, use the graph at the right. (pages 256–257)

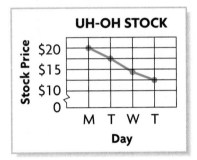

UH-OH STOCK

7. About how much did the stock price decrease from Monday to Tuesday?

8. About how much did the stock price decrease from Tuesday to Wednesday?

9. About how much did the stock price decrease from Monday to Thursday?

10. What do you think the stock price will be on Friday?

11. **VOCABULARY** The sum of a group of numbers, divided by the number of addends, is the __?__. (page 258)

Find the mean, median, and mode. (pages 258–261)

12. 17, 12, 23, 5, 19, 23

13. 81, 76, 92, 98, 87

14. 2, 2, 6, 5, 1, 2

15. 6.2, 5.5, 8.4, 5.5

16. 19, 20, 20, 19, 16

17. 265, 235, 171, 253

For Exercises 18–20, use the graph at the right. (pages 264–265)

18. What is the median?

19. What is the lower extreme? upper extreme?

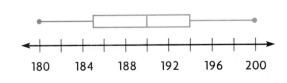

20. What is the lower quartile? upper quartile?

1. The batting averages of four players are 0.309, 0.234, 0.312, and 0.289. Which shows these averages arranged from least to greatest?

A 0.234, 0.289, 0.312, 0.309
B 0.312, 0.309, 0.289, 0.234
C 0.234, 0.289, 0.309, 0.312
D 0.289, 0.312, 0.234, 0.309
E Not Here

2. Jane brought 3.2 pounds of cold cuts to the picnic. Maria brought 2.9 pounds. How much did they bring altogether?

F 0.3 lb   **G** 5.1 lb
H 6.1 lb   **J** 7.1 lb

3. Which is the prime factorization of 24 using exponents?

A 8×3 **B** 6×2^2
C 2×3^2 **D** $2^3 \times 3$

4. Dan hiked $\frac{7}{12}$ of a mile on Saturday and $\frac{15}{16}$ of a mile on Sunday. Estimate how far Dan hiked in all.

F about 2 mi
G about $1\frac{1}{2}$ mi
H about 1 mi
J about $\frac{1}{2}$ mi

5. Zack is in charge of decorations for a school dance. Which of these decisions does Zack have to make?

A Date of the dance
B Type of decorations needed
C Cost of admission
D Who will chaperone the dance
E Not Here

6. The principal of Central Middle School has data on the number of tardy students and the number of absent students during the past three months. She wants to display the data on a graph. What type of graph should she make?

F histogram
G stem-and-leaf plot
H circle graph
J multiple-line graph

7. A food critic asked 100 people their opinion of a new restaurant. The results are shown in the circle graph below. Which opinion forms the largest part of the graph?

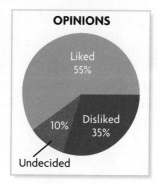

A Liked
B Disliked
C Undecided
D Not Here

8. Jon had the following scores for 5 rounds of golf: 78, 80, 69, 75, 73. What is his mean score?

F 68
G 71
H 74
J 75

PROBABILITY

LOOK AHEAD

In this chapter you will solve problems that involve

- the problem-solving strategy *account for all possibilities*

- using mathematical and experimental probability

Chewing gum was invented by Thomas Adams in 1869. He originally purchased the rubbery substance chicle to make tires for the automobile industry. It wasn't until 1906 that bubble gum was first produced.

CHEWING GUM SALES IN 1996	
Brand	Sales (in millions)
Trident	$334
Clorets	$211
Chiclets	$181

The table above shows the sales for some chewing gum brands.

- How did the sales for Chiclets compare to the sales for Trident?
- Suppose a box of Chiclets has the following pieces of gum: 4 red, 2 blue, and 1 yellow. Without looking, which color are you most likely to pull out of the box of chewing gum?

Which Item?

If you randomly select one item from a mixture, does each item have the same chance of being selected?

Find a mixture. Then count and make a tally table for the number of each type of item in the mixture. Find the likelihood of selecting each item. Decide if each item has the same chance of being randomly selected.

Gumballs		
blue	𝍸𝍸𝍸𝍸 𝍸𝍸𝍸𝍸 𝍸𝍸𝍸𝍸 𝍸𝍸𝍸𝍸 III	= 23
green	𝍸𝍸𝍸𝍸 𝍸𝍸𝍸𝍸 𝍸𝍸𝍸𝍸 𝍸𝍸𝍸𝍸 𝍸𝍸𝍸𝍸	= 25
orange	𝍸𝍸𝍸𝍸 𝍸𝍸𝍸𝍸 𝍸𝍸𝍸𝍸 𝍸𝍸𝍸𝍸 𝍸𝍸𝍸𝍸 𝍸𝍸𝍸𝍸 𝍸𝍸𝍸𝍸 𝍸𝍸𝍸𝍸 II	= 42
red	𝍸𝍸𝍸𝍸 𝍸𝍸𝍸𝍸 I	= 11
purple	𝍸𝍸𝍸𝍸 𝍸𝍸𝍸𝍸 𝍸𝍸𝍸𝍸 IIII	= 19

PROJECT CHECKLIST

✓ Did you list each item in the mixture?

✓ Did you make a tally table?

✓ Did you decide if each possibility was equally likely?

What You'll Learn
How to use the strategy *account for all possibilities* to solve problems

Why Learn This?
To find the number of possible choices, such as the number of choices for a yogurt cone

VOCABULARY
sample space

PROBLEM SOLVING
....

• **Understand**

• **Plan**

• **Solve**

• **Look Back**

BUSINESS LINK

When Ben & Jerry's® Ice Cream Company began making and selling its own Low Fat Frozen Yogurt, company sales increased from $77,000,000 to $97,000,000 in one year. How much did the company's sales increase?

PROBLEM-SOLVING STRATEGY
Account for All Possibilities

At the Frozen Treat yogurt shop, Jorge wants to buy a yogurt cone. He can have chocolate, vanilla, or swirl yogurt, and sprinkles, peanuts, or candy pieces for the topping. How many different yogurt cones are possible?

UNDERSTAND   What are you asked to find?

What information is given?

PLAN   What strategy will you use?

You can use the strategy *account for all possibilities* to show the possible choices, or **sample space**.

SOLVE   How will you solve the problem?

One way to find the number of possible choices is to make a tree diagram.

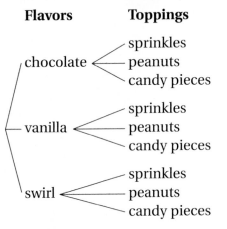

Flavors	Toppings	Choices (Sample Space)
chocolate	sprinkles	*chocolate with sprinkles*
	peanuts	*chocolate with peanuts*
	candy pieces	*chocolate with candy pieces*
vanilla	sprinkles	*vanilla with sprinkles*
	peanuts	*vanilla with peanuts*
	candy pieces	*vanilla with candy pieces*
swirl	sprinkles	*swirl with sprinkles*
	peanuts	*swirl with peanuts*
	candy pieces	*swirl with candy pieces*

There are 9 choices.

Another way to find the number of choices Jorge has is to multiply the number of flavors and the number of toppings.

Flavors		**Toppings**	
3	×	3	= 9 choices

LOOK BACK   What other strategy could you use to solve the problem?

What if . . . Jorge could have one of 4 toppings with any of the 3 flavors? How many different cones would be possible?

PRACTICE

Use the strategy c...

1. Matthew is...
   a white or...
   or gray i...
   he have...

2. Ms. Clark is making
   a dental appointment.
   It can be on Monday, Tuesday,
   or Wednesday, at 10:00 A.M., 10:30 A.M., or
   2:00 P.M. Find the total number of choices.

3. Miche...
   buy ...ipe,
   or a... ...ong,
   or v... ...s does
   sh...

4. The sixth-grade debate team is having
   lunch. The students can have pizza, a
   garden salad, macaroni, a tuna sub, or a
   taco, with milk, fruit juice, or soda. Find
   the total number of choices.

MI...

Solve.

CHOOSE a strateg... ...ool.		
• **Make a Table** • **Draw a Diagram** • **Account for All Possibilities**	• **Write a Proportion** • **Find a Pattern** • **Guess and Check**	Paper/Pencil Calculator Hands-On Mental Math

5. Julia has been working on a jigsaw puzzle
   for an hour and has put 90 of the 360
   pieces together. How much longer will it
   take her to finish the puzzle if she
   continues to put 90 pieces together every
   hour?

6. Amy is going to buy a music CD. She can
   buy either a single or an album of either
   rock-and-roll or country music. How
   many choices does she have?

7. The dance company performs in groups
   of 8 and 10. There are 64 dancers in all.
   There are more groups of 10 than groups
   of 8. How many groups of each size are
   there?

8. For 30 days Leon works every third day.
   He starts on Saturday. What day of the
   week is his last day?

9. Melissa will take 1-hr swimming lessons 3
   times a week. She can pay $12 in advance
   each week, or she can pay $5 an hour.
   Which payment plan will cost her less?

10. Elena is planting a garden. She knows
    that 50 plants cost $20. She is waiting for
    a sale so she can pay only $\frac{3}{4}$ of that price.
    How much is she willing to pay?

11. On Saturday the museum sold 100 more
    than twice the number of tickets it sold
    on Friday. On Saturday 500 tickets were
    sold. How many tickets were sold on
    Friday?

12. ✏️ **WRITE ABOUT IT** Write a problem
    using the strategy *account for all
    possibilities*. Explain how to solve the
    problem.

Probability

You use the ideas of probability in everyday life. When you have a lengthy chore to do, you may think, "I probably won't finish this today." While playing basketball, you may say, "She has a 50-50 chance of making that basket."

These expressions indicate your understanding of probability. **Probability**, P, is a comparison of the number of favorable outcomes to the number of possible outcomes.

The **mathematical probability** of an event can be written as a fraction.

$$P = \frac{\text{number of favorable outcomes}}{\text{number of possible outcomes}}$$

EXAMPLE 1 Each letter of the word *MATH* is written on a card and placed in a bag. Find P(A), the probability of choosing an *A*.

$\boxed{M}\ \boxed{A}\ \boxed{T}\ \boxed{H}$

1 favorable outcome: *A*   *List the favorable outcomes.*

4 possible outcomes: *M, A, T, H*   *List all the possible outcomes.*

$P(A) = \dfrac{\text{number of favorable outcomes}}{\text{number of possible outcomes}} = \dfrac{1}{4}$ *Write the probability as a fraction.*

So, the probability of choosing an *A* is $\frac{1}{4}$.

• How many favorable outcomes are there for choosing *T* or *H*?

Many of the things you do have a high probability of success. Every time you get on a bus, the probability is high that you will arrive safely at your destination. The probability is high that you will graduate from high school. In fact, most of the routine things you do have high probabilities of success.

You can use a number cube to find favorable outcomes.

EXAMPLE 2 You roll a number cube numbered 1 to 6. Find P(1), P(4), and P(1 or 4).

$P(1) = \dfrac{1}{6}$ ← 1 choice out of 6

$P(4) = \dfrac{1}{6}$ ← 1 choice out of 6

$P(1 \text{ or } 4) = \dfrac{1+1}{6}$ ← 2 choices out of 6

$\qquad\qquad = \dfrac{2}{6} = \dfrac{1}{3}$ *Write in simplest form.*

• What is the probability of rolling a number greater than 2?

Sometimes outcomes, or events, cannot occur. For example, what is the probability there will be eight days in the week to come?

The diagram shows that the probability of an event ranges from 0, or impossible, to 1, or certain. A probability is always 0, 1, or a fraction between 0 and 1.

impossible   possible   certain

0            $\frac{1}{2}$ 1

EXAMPLE 3 Look at the spinner at the right to find each probability.

A. $P(\text{white}) = \frac{0}{5}$. None of the sections are white.

B. $P(\text{blue or green}) = \frac{1+1}{5} = \frac{2}{5}$

C. $P(\text{not green}) = \frac{1+1+1+1}{5} = \frac{4}{5}$

• What relationship does the probability of spinning green have with the probability of not spinning green?

SPORTS LINK

There are many areas in sports where probability plays a big part. Even in darts no one is good enough to fully overcome the probability factor. If a blindfolded person is good enough only to hit the dartboard somewhere, other than the bull's-eye, what would be the probability of hitting a certain one of 20 equal sections?

GUIDED PRACTICE

For Exercises 1–4, use the numbered cards.

| 4 | 6 | 1 | 7 | 2 | 8 | 3 | 9 | 0 |

1. List the favorable outcomes for choosing an odd number.

2. How many possible outcomes are there?

3. Find P(odd).

4. Find P(6 or 0).

5. You are given a choice of three answers to a test question. If you don't know the answer, what is the probability of guessing the correct answer?

Geometric Probability

You can use the areas of geometric figures to find probabilities.

In a certain dart game, the target looks like the square at the right. The total area of the target is 144 in.² To find the probability of hitting the blue section with a dart, compare the areas. Assume the dart hits the target.

6 in.   6 in.

6 in.

6 in.

$P(\text{blue}) = \frac{\text{area of blue section}}{\text{area of target}} = \frac{6 \times 6}{144} = \frac{36}{144} = \frac{1}{4}$

REMEMBER:

The **area** is the number of square units needed to cover a surface.
See page H25.

The area of this figure is 4 × 4, or 16 square units.

EXAMPLE 4 If a randomly thrown dart hits the target at the right, what is the probability of it hitting the red section?

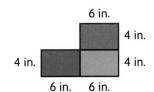

6 in.

4 in.

4 in.

4 in.

4 in.

6 in.   6 in.

$4 \times 6 = 24$
area of red section: 24 in.^{2}
$3 \times (4 \times 6) = 72$

area of target: 72 in.^{2}

$P(red) = \dfrac{24}{72} = \dfrac{1}{3}$

Find the areas of the red section and the target.

Find the probability of hitting the red section.

So, the probability of hitting the red section is $\frac{1}{3}$.

Talk About It

• Look at the target in Example 4. If you threw a dart and it hit the target, would you have a greater chance of hitting the red section than the green section? Explain.

• Suppose you wanted the probability of a dart hitting any section to be $\frac{1}{5}$. How many sections would you have on the target?

• **CRITICAL THINKING** What if the probability of a randomly thrown dart hitting red is $\frac{1}{2}$ and the probability of hitting green is $\frac{1}{4}$? What do you know about the areas of those two sections on the target?

INDEPENDENT PRACTICE

For Exercises 1–3, use the spinner at the right.

1. How many favorable outcomes are there for choosing 1?

2. How many possible outcomes are there? Name them.

3. What is the probability of choosing 1?

A number cube is numbered 2, 4, 6, 8, 10, and 12. Find each probability.

4. P(2)   **5.** P(not 8)   **6.** P(4 or 10)   **7.** P(a number less than 8)

The letters *E, D, U, C, A, T, I, O,* and *N* are put in a bag. Find each probability.

8. P(*A*)   **9.** P(*E* or *T*)   **10.** P(*S*)   **11.** P(*A, C, D, E, O,* or *U*)

For Exercises 12–14, use the rectangle at the right. A point on the figure is chosen at random. What is the probability that it has this color?

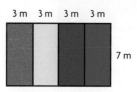

3 m 3 m 3 m 3 m

7 m

12. red   **13.** green   **14.** blue or green

For Exercises 15–18, use the figure at the right. Find each probability.

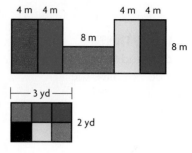

4 m 4 m   4 m 4 m

8 m

8 m

15. P(white)   **16.** P(not yellow)   **17.** P(red or green)

18. P(blue, green, red, purple, or yellow)

For Exercises 19–21, use the figure at the right. Find each probability.

3 yd

2 yd

19. P(red)   **20.** P(blue or green)   **21.** P(yellow, black, or green)

Problem-Solving Applications

22. There are 3 desserts on the menu, but one is not available. What is the probability that in making a random choice you choose that dessert?

23. CRITICAL THINKING In a new board game, the probability of landing on each section of the board is $\frac{1}{12}$. How many sections are on the board?

24. PROBABILITY Katie has a choice of 6 chores. She draws to see which chore she has for this week. What is the probability she will draw the same chore as last week?

25. PROBABILITY If you have a cube numbered 1 to 6, what is the probability you will roll a number less than 7?

26. ART Georgia is coloring a poster. She has 1 blue, 1 green, 1 purple, and 1 yellow marker in a box. She chooses a marker without looking. What is the probability of choosing a green marker?

27. WRITE ABOUT IT Explain the difference between a favorable outcome and a possible outcome. Give an example.

Mixed Review and Test Prep

Write a fraction that represents the part of the figure that is

28. green   **29.** yellow   **30.** red   **31.** blue

Determine the median, mode, or mean.

32. median: 1, 2, 3, 4, 5, 6   **33.** mode: 4, 5, 10, 4, 2, 11   **34.** mean: 2, 4, 6, 8, 10

35. NUMBER SENSE There are 24 students in the concert band. There are twice as many girls as boys. How many girls are there?

A 8   **B** 12
C 16   **D** 20

36. FRACTIONS Margaret spent $\frac{1}{3}$ of the dimes she saved. She gave her brother $\frac{1}{4}$ of the remaining ones. Then she had 45 dimes left. How many dimes had she saved?

F 100   **G** 90
H 80   **J** 60

More on Probability

What You'll Learn
How to find the probability of events that are not equally likely

Why Learn This?
To understand that all events do not have the same probability of happening

You have learned that probability is a comparison of the number of favorable outcomes and the number of possible outcomes. Sometimes there is a greater chance of one event occurring than another.

On this spinner, there are twice as many orange sections as yellow or green sections.

EXAMPLE 1 Find the probability of the pointer stopping on each color. Which color do you think the pointer will stop on most often?

P(yellow) $= \frac{1}{4}$ *Find each probability.*

P(green) $= \frac{1}{4}$

P(orange) $= \frac{2}{4} = \frac{1}{2}$

$\frac{1}{2} > \frac{1}{4}$ *Compare the probabilities.*

So, the pointer will stop on orange most often.

CONSUMER LINK

The probability of a pepper plant growing red peppers is the same as the probability of a pepper plant growing green peppers. However, consumers believe red peppers are less common, so farmers grow fewer red peppers and charge more for them. If green peppers are $0.29 per pound and red peppers are $0.89 per pound, how much more would 11 lb of red peppers cost than 11 lb of green peppers?

EXAMPLE 2 Tamara has a bag with 8 cubes all the same size: 1 red, 4 yellow, and 3 blue. Without looking, she chooses a cube from the bag. After recording the color, she places the cube back in the bag and chooses again. What is the probability of Tamara choosing a yellow cube each time?

P(yellow) $= \frac{4}{8} = \frac{1}{2}$ *4 yellow cubes; 8 cubes in all*

So, the probability of a yellow cube is $\frac{1}{2}$ each time.

• **CRITICAL THINKING** Tamara chose a blue cube and did not return it to the bag. What is the probability of choosing a yellow cube once a blue cube has been removed?

GUIDED PRACTICE

Use the spinner at the right to find each probability.

1. P(red)   2. P(blue)

3. P(yellow)   4. P(blue or yellow)

5. P(white)   6. P(not red)

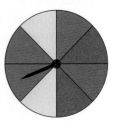

INDEPENDENT PRACTICE

For Exercises 1–6, use the spinner at the right.
Find each probability on a single spin.

1. P(*F*)   **2.** P(*T*)   **3.** P(*B*)   **4.** P(*T, D,* or *O*)

5. P(*O* or *D*)   **6.** P(*F, A, S,* or *T*)

A bag contains 4 red, 2 blue, 1 yellow, and 3 green pencils. You
choose one pencil without looking. Find each probability.

7. P(yellow)   **8.** P(red)   **9.** P(blue or green)   **10.** P(yellow, green, or red)

Cards numbered 1, 1, 2, 2, 3, 4, 5, and 6 are placed in a hat. You
choose one card without looking. Find each probability.

11. P(1)   **12.** P(1 or 2)   **13.** P(1, 2, or 3)   **14.** P(even numbers)

15. P(1, 2, 3, or 4)   **16.** P(1, 2, 3, or 6)   **17.** P(8)

For Exercises 18–23, use the figure at the right. A point in the figure
is chosen at random. Find each probability. Each measurement is
in feet.

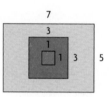

18. P(red)   **19.** P(blue)   **20.** P(green)

21. P(blue or red)   **22.** P(green or blue)   **23.** P(blue, red, or green)

Problem-Solving Applications

24. MUSIC Tasha wants to borrow a CD from her brother, Carlos.
Carlos has 6 country music CDs and 2 rock CDs. Tasha is in a
hurry, so she picks one without looking. What is the
probability she chose a country music CD?

25. Brenda can't decide which movie to pick for her party. There
are 2 adventure movies, 3 comedies, and 1 mystery. She
writes the titles on separate pieces of paper and puts them
in a bag. Brenda will choose one piece of paper at random.
What is the probability that she will choose a comedy?

26. PROBABILITY Rafael is going to his grandmother's house for
dinner and must wear a tie. He has 4 ties in his closet: 1 blue,
2 print, and 1 striped. He takes one at random and puts it
on. What is the probability that he chose the striped tie?

27. ✏️ **WRITE ABOUT IT** Write a problem that involves finding
the probability of events that are not equally likely to
happen.

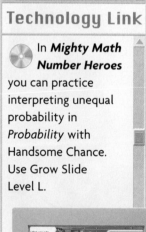

LAB
ACTIVITY

What You'll Explore
How to use a simulation to model an experiment

What You'll Need
number cube numbered 1 to 6, calculator

Simulations

A juice company is having a contest. To win a prize, you have to collect six bottle caps that spell out *ORANGE*. One of the six letters is put under each bottle cap when the cap is produced. The letters are divided equally among the juice bottles.

You can conduct an experiment to simulate how many bottles of juice you have to buy to get all six letters.

ACTIVITY 1

Explore

- Use a number cube to generate random numbers. Each of the numbers 1 to 6 will represent one of the letters in the word *ORANGE*.

$$O\ R\ A\ N\ G\ E$$
$$1\ 2\ 3\ 4\ 5\ 6$$

- Roll the number cube, and tally the numbers you get.

- Continue to roll the number cube until you get all the numbers at least once.

- Repeat the experiment.

Number	Times Rolled
1			
2		
3					
4				
5		
6		

Think and Discuss

- How many rolls did it take in the first experiment to get all six numbers?

- How many rolls did it take in the second experiment?

- What is the average of the rolls in your two experiments?

- How many bottles of juice do you expect you will have to buy? If you bought this many bottles, would you be sure to win?

Try This

- Repeat the experiment three more times.

- Guess how many times each number will be rolled. Record your guess and compare it with the results.

- Find the average of the rolls in all five experiments.

 Calculator Activities, page H39

Instead of rolling a number cube, you can produce random numbers using a calculator. Use a calculator to do the following activity.

ACTIVITY 2

Explore

- Use this key sequence to produce random numbers from 1 to 10.

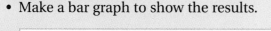

1   2nd   RAND   10   =

- Record the number you get each time.

- Repeat the key sequence until you have gotten each of the numbers 1 to 10.

- Make a bar graph to show the results.

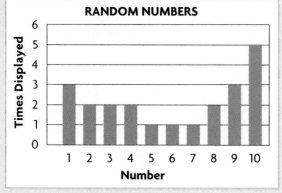

RANDOM NUMBERS

Think and Discuss

- How many times did you have to use the key sequence to get all ten numbers?

- How do your results compare with those of your classmates?

- Look at the results in the graph. What was the last random number produced?

Try This

- Use the calculator to produce random numbers from 1 to 20. Record the number of times you get each number. Continue until you get each number.

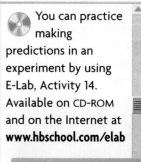

Technology Link

You can practice making predictions in an experiment by using E-Lab, Activity 14. Available on CD-ROM and on the Internet at **www.hbschool.com/elab**

279

Experimental Probability

With a number cube numbered 1 to 6, the mathematical probability of rolling each number is $\frac{1}{6}$. By performing an experiment, you can find the experimental probability of rolling each number. The **experimental probability** of an event is the number of times a success occurs compared with the total number of times you do the activity.

Bailey rolled a cube numbered 1 to 6. The table below shows the results of rolling the cube 50 times.

Number	1	2	3	4	5	6
Times rolled	4	9	11	6	10	10

$$\text{experimental probability} = \frac{\text{number of times success occurs}}{\text{total number of trials}}$$

EXAMPLE 1 Use Bailey's results to find the experimental probability of rolling each number. Write each fraction in simplest form.

$P(1) = \frac{4}{50} = \frac{2}{25}$ $P(2) = \frac{9}{50}$ $P(3) = \frac{11}{50}$

$P(4) = \frac{6}{50} = \frac{3}{25}$ $P(5) = \frac{10}{50} = \frac{1}{5}$ $P(6) = \frac{10}{50} = \frac{1}{5}$

Talk About It

- What if Bailey rolled the number cube 50 more times? Do you think the results would be the same as in Example 1? Explain.

- How does the experimental probability of each number compare with the mathematical probability of $\frac{1}{6}$?

You can use the experimental probability to predict future events.

EXAMPLE 2 Based upon his experimental results, how many times can Bailey expect to roll a 6 in his next 10 rolls?

$P(6) = \frac{1}{5}$ *Use the experimental probability.*

$\frac{1}{5} \times 10 = 2$ *Multiply 10 rolls by $\frac{1}{5}$.*

So, he can expect to roll a 6 two times in the next 10 rolls, based on his past rolls.

GUIDED PRACTICE

Rickie tossed a coin 40 times and got these results. For Exercises 1–2, use the table at the right to find the experimental probability.

Coin	Heads	Tails
Toss	28	12

1. P(heads) **2.** P(tails) **3.** What is the mathematical probability of getting tails?

4. Is the experimental probability of getting tails, based on these results, less than the mathematical probability? Explain.

INDEPENDENT PRACTICE

Tania spins the pointer of the spinner 50 times. For Exercises 1–4, use her results in the table below to find the experimental probability.

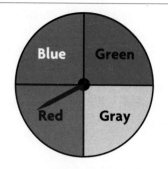

Color	Blue	Green	Red	Gray
Times lands on	20	10	12	8

1. P(blue) **2.** P(green) **3.** P(red) **4.** P(gray)

5. What is the mathematical probability for each color?

Problem-Solving Applications

For problems 6–7, use the data from Exercises 1–5.

6. Based on her experimental results, how many times can Tania expect the pointer to land on blue in the next 20 spins?

7. Based on her experimental results, how many times can Tania expect the pointer to land on red in the next 100 spins?

8. **SPORTS** Fran was in a batting cage. Out of 100 balls, she hit 30 of them. How many balls can she expect to hit in her next 30 tries?

9. **WRITE ABOUT IT** Explain how you get the experimental probability of rolling a 4 when rolling a number cube labeled 1 to 6.

Mixed Review and Test Prep

Find the value.

10. 6 + 8 + 4 **11.** 6 × 3 + 5 **12.** 18 − 3 × 2 **13.** 16 ÷ 4 + 5

A bar graph showed that more children than adults have pets.

14. Would the taller bar represent children or adults?

15. What does a taller bar show?

16. **DATA** Emily wants to get a 90 average in math. She will have a total of 8 tests. How many total points will she need for all 8 tests?

A 90 **B** 360 **C** 720 **D** 900

17. **NUMBER SENSE** A giant tortoise travels at a speed of 0.2 kilometer per hour. How far does it travel in 3 hours?

F 0.6 km **G** 0.5 km
H 0.4 km **J** 0.3 km

Review/Test

Solve. **(pages 270–271)**

1. The A.M. Cafe serves a choice of cereal, a bagel, or pancakes and a choice of milk or juice for breakfast. Find the number of possible food and drink choices.

2. For dessert, Cary has a choice of 6 different fruits and 4 different pies. He will choose one of each. Find the number of possible choices.

3. Bill can walk or take the subway to Central Station. From there, he rides either a bus or a train to work. Find the number of different ways Bill can get to work.

4. Elisa has a choice of 3 blouses and 5 skirts to wear. Find the number of possible outfits.

5. **VOCABULARY** A comparison of the number of favorable outcomes and the number of possible outcomes is ___?___. **(page 272)**

A bag has slips of paper numbered 2–10. Find the probability of randomly drawing the number from the bag. **(pages 272–275)**

6. P(5)

7. P(7 or 9)

8. P(1)

9. P(2, 5, or 9)

10. P(even numbers)

11. P(not 8)

12. P(5 or 6)

13. P(odd numbers)

A bag has 10 new pencils: 3 red, 4 yellow, 1 blue, and 2 green. Find the probability of randomly choosing the given color from the bag. **(pages 276–277)**

14. P(green)

15. P(yellow)

16. P(brown)

17. P(blue or red)

18. P(red or green)

19. P(red, yellow, or blue)

20. **VOCABULARY** The number of times an event occurs compared with the total number of times the event was done is called the ___?___. **(page 280)**

The table shows the results of rolling a number cube 100 times. Use the table to answer Exercises 21–25. **(pages 280–281)**

Number	1	2	3	4	5	6
Times rolled	8	12	10	21	15	34

21. P(3)

22. P(6)

23. P(4)

24. P(1)

25. P(2 or 5)

TAAS Prep

1. A frozen yogurt cone costs $1.75. Sprinkles cost $0.55 more. If Diane buys a cone with sprinkles and pays with a $5 bill, how much change does she receive?

   A $2.70
   B $2.90
   C $3.25
   D $3.55

2. Denise has three pieces of fabric measuring $\frac{1}{4}$ yard, $\frac{7}{8}$ yard, and $\frac{9}{10}$ yard. What is a reasonable estimate of the total length of fabric?

   F 1 yd
   G 2 yd
   H 3 yd
   J 4 yd

3. The area of a rectangular rug is 108 square feet. The rug is 8 feet wide. How long is the rug?

   A 9.6 ft
   B 13.5 ft
   C 14.2 ft
   D 15 ft
   E Not Here

4. Which is the measure of an acute angle?

   F 125°
   G 100°
   H 90°
   J 75°
   K Not Here

5. Which solid figure has 5 faces, 6 vertices, and 9 edges?

   A Pentagonal prism
   B Rectangular prism
   C Triangular pyramid
   D Rectangular pyramid
   E Triangular prism

6. The graph shows rainfall for 5 days. How many inches did it rain in all?

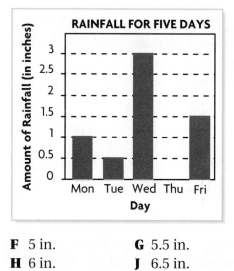

   F 5 in.   **G** 5.5 in.
   H 6 in.   **J** 6.5 in.

7. Rhea wants to find the mean of her math test scores. What is the first thing she should do?

   A Find the sum of the scores.
   B Arrange the scores from greatest to least.
   C Determine which score appears most often.
   D Find the difference between the greatest and least scores.

8. Ron wants to make an appointment for a haircut. He can go to the shop on either Friday, Saturday, or Sunday, at 10:00, 1:00, or 3:00. How many choices does Ron have?

   F 3   **G** 6
   H 9   **J** 12

9. Dan has a spinner with 5 equal sections labeled 1 to 5. He knows the probability of spinning a 2 is $\frac{1}{5}$. What is the probability of not spinning a 2?

   A $\frac{1}{5}$ **B** $\frac{2}{5}$
   C $\frac{4}{5}$ **D** 1

MATH FUN!

What kind of pet do you have?

PURPOSE To practice recording and organizing data (pages 228–231)
YOU WILL NEED poster paper, markers, ruler

Work in a small group. Conduct a pet survey. Ask each person what kind of pet he or she has. Make a frequency table and a cumulative frequency table from the data you collect.

Share your results with the class. Compare tables. Are they exactly alike? Why or why not?

Make up some fun pet questions of your own that you can ask in a survey.

WHAT'S WRONG WITH THIS PICTURE?

PURPOSE To practice identifying misleading graphs (pages 254–255)
YOU WILL NEED graph paper

Lynn and Linda are amazingly alike. They were born on the same day in the same year by the same parents. You think they are twins but they are not. How can you explain that?

Like words, graphs can be designed to mislead you. Look at the graph. Why is it misleading?

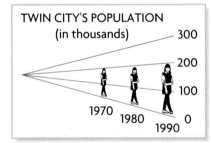

TWIN CITY'S POPULATION
(in thousands)
300
200
100
0
1970 1980 1990

How many graphs in newspapers or magazines can you find that present data in misleading ways?

YOU'LL FLIP!

PURPOSE To practice working with probability (pages 272–277)
YOU WILL NEED four coins

Flip four coins at once. How many come up heads? None? one? two? three? all four? Try this experiment 20 times and keep track of the results. Copy and complete the tally table at the right.

HOW MANY HEADS?	
Number of heads	**Tally**
0	
1	
2	
3	
4	

Finding the Mean Using Spreadsheets

In this activity you will see how you can enter data into a spreadsheet and then enter a formula to find the mean. Look at the spreadsheet below with the data collected about test scores for Class A, Class B, and Class C.

All	A	B	C	D
1		Class A	Class B	Class C
2		84	67	82
3		78	84	70
4		94	83	72
5		86	86	74
6		75	81	76
7		73	98	73
8		84	89	84
9		92	88	88
10	MEAN			

To find the mean for Class A, enter the formula shown in cell B10.

All	B
1	Class A
2	84
3	78
4	94
5	86
6	75
7	73
8	84
9	92
10	=mean(B2:B9)

To find the mean for Class B, enter the formula shown in cell C10.

All	C
1	Class B
2	67
3	84
4	83
5	86
6	81
7	98
8	89
9	88
10	=mean(C2:C9)

1. Suppose you add a new class, Class D, by entering the scores in the next column. Which cell and what formula would you use to find the mean for Class D?

2. What formula would you use to find the mean for the three classes combined?

USING A SPREADSHEET

3. List all of your test and homework grades in a spreadsheet.

4. Use a formula to find the mean for your test grades.

5. Use a formula to find the mean for your homework grades.

6. Use a formula to find the mean for all your work combined.

Study Guide and Review

Vocabulary Check

1. The difference between the greatest number and the least number in a set of data is the ___?___. **(page 230)**

2. The sum of a group of numbers, divided by the number of addends, is the average, or ___?___. **(page 258)**

3. A comparison of the number of favorable outcomes and the number of possible outcomes is called ___?___. **(page 272)**

EXAMPLES

- **Determine whether a sample or a question is biased.** (pages 224–225)

Bill surveys 5 out of 15 boys in his class about class elections. Is the sample biased? Explain.

Yes. His sample should include girls.

- **Collect and organize data.** (pages 228–231)

A cumulative frequency table keeps a running total of the number of people.

AGES OF ZOO VISITORS		
Age Group	**Frequency**	**Cumulative Frequency**
1–20	10	10
21–40	3	13
41–60	2	15
61–80	1	16

- **Graph two or more sets of data on one graph.** (pages 242–243)

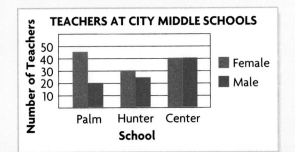

EXERCISES

4. Connie surveys 10 out of 30 sixth-grade students on what the voting age should be in city elections. Is this sample biased or not biased? Explain.

For Exercises 5–6, use the following data.

Ages of Neighborhood Cats							
3	9	1	1	4	12	7	1
6	5	4	2	10	6	3	4

5. Make a line plot.

6. Make a cumulative frequency table with intervals.

7. Make a multiple-line graph with the data.

High and Low Temperatures					
	Mon	**Tue**	**Wed**	**Thu**	**Fri**
Highs	75°	80°	75°	85°	80°
Lows	60°	70°	65°	70°	65°

- **Make predictions from graphs.**
  (pages 256–257)

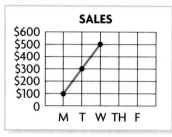

For Exercises 8–10, use the graph at the left.

8. How much did sales rise from Monday to Tuesday?

9. How much did sales rise from Tuesday to Wednesday?

10. What do you predict sales will be on Friday?

- **Find the mean, median, and mode.**
  (pages 258–261)

data: 15, 15, 20, 25, 30
mean: (15 + 15 + 20 + 25 + 30) ÷ 5 = 21
median: 20   mode: 15

Find the mean, median, and mode.

11. 4, 6, 7, 5, 8, 9, 3

12. 4.2, 5.5, 8.3, 4.2, 7.8

13. 90, 100, 84, 75, 110, 90, 95

- **Find the probability of events that are not equally likely.** (pages 276–277)

A bag has 10 marbles in it: 4 blue, 3 red, 2 green, and 1 yellow. One marble is drawn.

Find P(blue). $\frac{4}{10} = \frac{2}{5}$

For Exercises 14–17, use the description of the marbles at the left. Find each probability.

14. P(purple)   **15.** P(yellow)

16. P(red or green)   **17.** P(blue, red, or green)

- **Find the experimental probability of an event.**
  (pages 280–281)

The table shows the results of rolling a number cube 30 times.

Number	1	2	3	4	5	6
Times Rolled	2	3	10	6	5	4

experimental probability of rolling 1 = $\frac{2}{30} = \frac{1}{15}$

Use the table at the left to find each experimental probability.

18. P(2)   **19.** P(3 or 4)

20. P(even number)   **21.** P(2, 3, or 5)

22. P(4, 5, or 6)   **23.** P(odd number)

Problem Solving and Reasoning

Solve. Explain your method.

24. The school band has 60 members. There are 36 girls and 24 boys. If you used a circle graph to display the data, what would be the angle measures?
(pages 246–247)

25. The mean of eight numbers is 100. What is the sum of the numbers? (pages 258–261)

Performance Assessment

CHAPTERS 11–14

Tasks: Show What You Know

1. There are 750 students in your school. A dance is planned. The dance committee wants to find out about how many people will attend. How will you survey the students? **(pages 222–223)**

2. The table at the right shows the number of home runs hit by a Little League team. Use the data to make a histogram. **(pages 240–241)**

Runs					
24	3	8	47	19	28
21	40	36	15	2	16

3. Use the information in the table at the right to find the mean, median, and mode. Explain your method. **(pages 258–261)**

David's Test Scores				
97	88	96	93	93
98	92	97	99	97

4. You roll 1 number cube numbered 1 to 6. Find P(even number). **(pages 272–275)**

Problem Solving

Solve. Explain your method.

CHOOSE a strategy and a tool.

- **Find a Pattern**
- **Make a Table**
- **Draw a Diagram**
- **Write an Equation**
- **Make a Model**
- **Make a Graph**

Paper/Pencil   Calculator   Hands-On   Mental Math

5. The scores on a science test were 100, 83, 88, 61, 95, 85, 51, 89, 51, 75, 63, 58, 99, 67, 74, 72, 94, 88, 72, 70, 83, 89, 80, 85, and 87. Find the range. Then make a tally table and a frequency table using intervals. **(pages 228–231)**

6. Use the data below to make a circle graph. **(pages 246–247)**

Activities			
Reading	Math	Science	Homeroom
90 min	36 min	36 min	18 min

7. Jen bought a car for $5,100. It was worth $4,650 after 6 months, $4,200 after 12 months, and $3,750 after 18 months. What will the car be worth after 3 years? **(pages 256–257)**

8. Pizzas come in small, medium, and large. Toppings include pepperoni, mushrooms, onions, and green peppers. How many different types of one-topping pizzas are possible? **(pages 270–271)**

Cumulative Review

Solve the problem. Then write the letter of the correct answer.

1. Compare. $^-70 \bullet ^+1$ **(pages 30–31)**

    A. $<$ **B.** $>$
    C. $=$ **D.** $-$

2. Find the quotient. $8\overline{)818}$ **(pages 44–47)**

    A. 12 r2   **B.** 101
    C. 102   **D.** 102 r2

3. Find the LCM for 8 and 10. **(pages 88–91)**

    A. 2   **B.** 10
    C. 40   **D.** 80

4. Estimate the sum. $2\frac{3}{4} + 2\frac{1}{3}$ **(pages 130–131)**

    A. $^-3$ **B.** 0
    C. 5   **D.** 6

For Exercises 5–6, use the diagram below.
(pages 162–163)

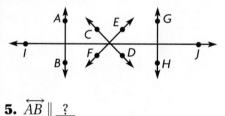

5. $\overleftrightarrow{AB} \parallel \underline{\ ?\ }$

    A. $\overleftrightarrow{CD}$ **B.** $\overleftrightarrow{EF}$
    C. $\overleftrightarrow{GH}$ **D.** $\overleftrightarrow{IJ}$

6. $\overleftrightarrow{GH} \perp \underline{\ ?\ }$

    A. $\overleftrightarrow{AB}$ **B.** $\overleftrightarrow{CD}$
    C. $\overleftrightarrow{EF}$ **D.** $\overleftrightarrow{IJ}$

7. If you survey 1 out of every 10 students, how many would you survey out of a group of 350 students? **(pages 222–223)**

    A. 10   **B.** 35
    C. 70   **D.** 3,500

8. You want to graph some data that show favorite subjects of a sample of students. Which type of graph would be the best choice? **(pages 236–239)**

    A. bar graph   **B.** histogram
    C. line graph   **D.** stem-and-leaf plot

9. You want to make a circle graph. A period of 15 min out of a 60-min workout is spent in warm-up exercises. What angle measurement should you use to represent 15 min? **(pages 246–247)**

    A. 15°   **B.** 25°
    C. 90°   **D.** 180°

For Exercises 10–11, use the graph below.
(pages 252–253)

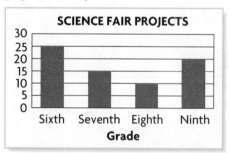

10. Which grade had the most projects?

    A. sixth   **B.** seventh
    C. eighth   **D.** ninth

11. How many more projects did the ninth grade have than the eighth grade?

    A. 0   **B.** 5
    C. 10   **D.** 20

12. For dessert, there are choices of four different pies and three flavors of ice cream. How many possible choices of pie with ice cream are there? **(pages 270–271)**

    A. 3   **B.** 4
    C. 7   **D.** 12

ALGEBRA: EXPRESSIONS AND EQUATIONS

LOOK AHEAD

In this chapter you will solve problems that involve

- writing and evaluating numerical and algebraic expressions

- using input-output tables to evaluate algebraic expressions and to solve real-world problems

- solving addition and subtraction equations

HEALTH **LINK**

A box of raisin bran displays the following nutrition facts.

Serving size: 1 cup   Carbohydrates: 47 grams
Calories: 200   Dietary fiber: 7 grams
Fat: 1.5 grams   Protein: 6 grams

- If your total daily intake is about 2,000 calories, what part of this total is one serving of raisin bran?

- In a daily diet of 2,000 calories, it is recommended that 25 grams of dietary fiber be included. Based on the data above, is raisin bran a high-fiber food? Explain.

Equations for Nutrition Facts

Equations can help you see the relationships among the contents of food. Select a nutrition label from a container. Analyze the nutrition facts. Write at least three equations that show relationships among the contents. Then, make a display of your nutrition label and equations.

PROJECT CHECKLIST

☑ Did you select a label?

☑ Did you analyze the nutrition facts?

☑ Did you write equations that show relationships among the contents?

☑ Did you make a display of your nutrition label and equations?

NUTRITION FACTS

Serving Size 3/4 cup (29g)
Servings Per Container: about 18

Amount Per Serving	Cereal	Cereal with 1/2 cup Skim Milk
Calories	100	140
Calories from Fat	10	10
	% Daily Value	
Total Fat 1g	1%	1%
Saturated Fat 0g	0%	0%
Cholesterol 0mg	0%	0%
Sodium 140mg	6%	8%
Potassium 80mg	2%	8%
Total Carbohydrate 24g	8%	10%
Dietary Fiber 3g	11%	11%
Sugars 5g		
Other Carbohydrate 16g		
Protein 3g		
Vitamin A	15%	20%
Vitamin C	0%	2%
Calcium	0%	15%
Iron	45%	45%
Vitamin D	10%	25%
Thiamin	25%	30%
Riboflavin	25%	35%
Niacin	25%	25%
Vitamin B$_6$	25%	25%
Folate	25%	25%
Vitamin B$_{12}$	25%	35%
Phosphorus	8%	20%
Magnesium	8%	10%
Zinc	8%	10%
Copper	8%	8%

Numerical and Algebraic Expressions

What You'll Learn
How to write, interpret, and identify numerical and algebraic expressions

Why Learn This?
To write expressions representing game scores, budgets, and recipes

VOCABULARY

numerical expression

variable

algebraic expression

In a basketball game, the Chicago Bulls scored 57 points in the first half and 46 points in the second half. To find the total points they scored, you could use a numerical expression.

$$57 + 46 \leftarrow \text{ total points scored}$$

A **numerical expression** is a mathematical phrase that includes only numbers and operation symbols. Numerical expressions are used daily for recipes, point totals, and family budgets, among other things. Here are some numerical expressions.

$$60 + 25 \qquad\qquad 42 \div 7 \qquad\qquad 4^2 - 3 \qquad\qquad \frac{1}{2} \times \frac{3}{4}$$

Suppose you knew that the Bulls scored 60 points in the first half of a game but didn't know how many points they scored in the second half. You could use a variable to represent the points scored in the second half. A **variable** is a letter or symbol that stands for one or more numbers.

$$60 + p \leftarrow \text{ total points scored}$$

Use p to represent points in second half.

An expression that includes a variable is called an **algebraic expression**. Here are some algebraic expressions.

$$5 + n \qquad\qquad 7 \times a \qquad\qquad k - 3 \qquad\qquad y \div 2$$

> **EXAMPLE 1** Tell whether each expression is numerical or algebraic. Give a reason for your answer.
>
> **A.** $17 + 5^3$ *This is a numerical expression, since it includes only numbers and operation symbols.*
>
> **B.** $d \div 5$ *This is an algebraic expression, since it includes a variable.*

Word expressions can be translated into numerical or algebraic expressions.

> **EXAMPLE 2** Write a numerical or algebraic expression for each word expression.
>
> **A.** three less than 5 $\qquad\qquad 5 - 3$
>
> **B.** 12 times a number, y $\qquad\qquad 12 \times y$

SPORTS LINK

In the movie *Space Jam* Michael Jordan and his animated teammates play against slam-dunking aliens. If Jordan scores 45 points and the aliens score 77 points, what numerical expression represents the difference in their scores?

GUIDED PRACTICE

Determine if the expression is numerical or algebraic.
Explain your answer.

1. $7 + 12$

2. $x - 12$

3. $p \times 2.5$

4. $\frac{3}{4} \times 5\frac{2}{3}$

Write a numerical or algebraic expression for the word expression.

5. six less than twelve

6. seven more than a number, y

7. y divided by 15

INDEPENDENT PRACTICE

Write a numerical or algebraic expression for the word expression.

1. the product of one half and two fifths

2. 15 divided by 3

3. 87 decreased by a number, k

4. q more than five ninths

5. 32.7 times a number, p

6. the sum of six squared and 8

Write a word expression for the numerical or algebraic expression.

7. $42 + 2.5$

8. $30 \div 5$

9. $c - 12$

10. $45 \times a$

11. $52 + x$

12. $9 \times m$

13. $98 \div 4.1$

14. $1,733 - 18.5$

Problem-Solving Applications

15. HEALTH Dan had 2 servings of cereal. Let s represent the number of servings in the box. Write an algebraic expression for the number of servings left in the box.

16. Let x represent Jose's height, in inches. If Carla were 5 in. taller, she would be just as tall as Jose. Write an algebraic expression for Carla's height.

17. ✏️ **WRITE ABOUT IT** What is the difference between numerical and algebraic expressions?

Mixed Review and Test Prep

Perform the indicated operations.

18. $17 + 3^2$

19. $16 - 2 \times 3$

20. $2^3 - 5.4$

21. $44 - 18 \div 2$

A number cube is labeled 0, 1, 3, 5, 7, and 9. Find each probability.

22. P(3)

23. P(4)

24. P(3 or 9)

25. P(0 or 5)

26. P(1, 3, or 7)

27. CONSUMER MATH Joe buys a pen for $2.85, a CD for $9.95, and a plant for $8.75. What is a reasonable estimate of the total cost?

A $18   **B** $22   **C** $24   **D** $26

28. NUMBER SENSE Mindy wants to average 15 points per game. She scored 12, 17, 14, and 9 points in her first four games. What must she score in her fifth game?

F 24   **G** 23   **H** 22   **J** 21

Evaluating Numerical and Algebraic Expressions

What You'll Learn
How to evaluate numerical and algebraic expressions

Why Learn This?
To compute total points, family budgets, and amounts for recipes

VOCABULARY

evaluate

REMEMBER:
....
Recall the **order of operations**. See page 48.

1. Operate inside parentheses.
2. Clear exponents.
3. Multiply and divide from left to right.
4. Add and subtract from left to right.

Vanessa is baking two kinds of cookies for the holiday season. She needs $2\frac{1}{2}$ cups of flour for one recipe and $5\frac{1}{2}$ cups for the other. To find the total amount of flour she needs, she could evaluate the numerical expression $2\frac{1}{2} + 5\frac{1}{2}$.

$$2\frac{1}{2} + 5\frac{1}{2} = 8$$

To **evaluate** a numerical expression, perform the operations and write the expression as one number.

EXAMPLE 1 Evaluate each numerical expression.

A. $5^2 \times (5 - 2)$ *Operate inside parentheses.*   **B.** $9 + 8 \div 2$ *Divide first.*
    25×3 *Clear exponent.* $9 + 4$ *Add.*
    75 *Multiply.* 13

Every time Vanessa bakes cookies, she eats 3 to test them. You can use the variable c to represent the number of cookies Vanessa has baked. To find the number of cookies left after Vanessa tests them, you could evaluate the algebraic expression $c - 3$. To evaluate an algebraic expression, replace the variable with a number and perform the operation in the expression.

EXAMPLE 2 Vanessa baked 70 chocolate chip cookies. After testing them, how many did she have left?

     $c - 3$ *Write the algebraic expression $c - 3$.*
    $70 - 3$ *Replace c with 70.*
     67 *Perform the operation.*

So, she had 67 cookies left.

EXAMPLE 3 Evaluate.
A. $x + 7$, for $x = 12$ **B.** $\sqrt{81} \div k$, for $k = 9$

     $x + 7$ $\sqrt{81} \div k$
    $12 + 7$ $\sqrt{81} \div 9$
     19 $9 \div 9$
                         1

So, for $x = 12$, $x + 7 = 19$.   So, for $k = 9$, $\sqrt{81} \div k = 1$.

GUIDED PRACTICE

Tell which operation you would perform first to evaluate the numerical expression.

1. $7 + 5 \times 2$ **2.** $4.7 + 6.5 - 2.9$ **3.** $10 \div 2 \times 8$ **4.** $(17.4 - 5.7) \div 2$

Evaluate the algebraic expression for $x = 4$.

5. $x - 3$ **6.** $17 + x$ **7.** x^3 **8.** $92 \div x$

INDEPENDENT PRACTICE

Evaluate the numerical expression.

1. $7 + 5 \times 2$ **2.** $4.7 + 6.5 - 2.9$ **3.** $10 \div 2 \times 8$ **4.** $(17.4 - 5.7) \div 2$

5. $3^2 - 2.9$ **6.** $(35 + 10) \div 9$ **7.** $\frac{3}{4} \times \frac{1}{3} + \frac{1}{2}$ **8.** $4\frac{1}{5} \times \left(8 - 2\frac{1}{2}\right)$

Evaluate the algebraic expression for the given value of the variable.

9. $x - 7$, for $x = 15$ **10.** $17 + k$, for $k = 2.1$ **11.** $x^2 - 4$, for $x = 3$

12. $y^3 - 12$, for $y = 3$ **13.** $42 + k - 8$, for $k = 18$ **14.** $25 + a^4$, for $a = 2$

15. $7 \times p$, for $p = 2.5$ **16.** $m \times \frac{1}{3}$, for $m = \frac{9}{14}$

17. $j \div 2$, for $j = 28$ **18.** $18 \div y$, for $y = 2\frac{1}{4}$

19. $(w - 8) \div 2$, for $w = 46$ **20.** $a^2 \div (3 + 7)$, for $a = 9$

Problem-Solving Applications

21. SPORTS A ticket in the upper section of the stadium is $12 less than a ticket in the lower section. A ticket in the lower section costs $30. Write and evaluate a numerical expression representing the cost of a ticket in the upper section.

22. BUSINESS At Barry's Burger Barn, burgers cost $1.25 each. If b represents the number of burgers sold, what algebraic expression represents the amount of burger sales? If Barry sold 229 burgers, how much were burger sales?

23. Serena has 45 pieces of bubble gum to give to her friends. Let f represent the number of friends. Write an algebraic expression for dividing the gum evenly. If there are 10 friends, how much gum does each one get?

24. ✏️ **WRITE ABOUT IT** Suppose you are evaluating the numerical expression $17 + 25$ and the algebraic expression $x + 19$, for $x = 12$. How are your methods different? How are they the same?

Input-Output Tables

What You'll Learn
How to use input-output tables to evaluate algebraic expressions

Why Learn This?
To solve real-world problems that involve wages or sports statistics

TEEN TIMES

The first commercial video game—Pong—was developed in 1972. Since then, video games have gone from little blips of light bouncing around a TV screen to incredibly realistic simulations that can make you feel as if you're part of a space-shuttle mission. Video gaming is also a huge business that makes more money annually than all the Hollywood movie studios or Nashville recording studios.

Calvin is trying to earn $75.00 to purchase a video game. He earns $4.25 per hour delivering newspapers.

• What numerical expression describes how much Calvin would earn if he worked 4 hr?

• What algebraic expression describes how much Calvin would earn if he worked h hr?

One way Calvin can find out how many hours he needs to work to earn $75.00 is to use an input-output table. To use an input-output table, replace the variable in the algebraic expression with each input value. Then evaluate the expressions to get the output values.

INPUT	ALGEBRAIC EXPRESSION	OUTPUT
Hours Worked	**$4.25 × h**	**Money Earned**
14	$4.25 × 14	$59.50
15	$4.25 × 15	$63.75
16	$4.25 × 16	$68.00
17	$4.25 × 17	$72.25
18	$4.25 × 18	$76.50

So, Calvin needs to work 18 hr to earn enough money to buy the video game.

EXAMPLE Make an input-output table for the algebraic expression $k + 17$. Use 4, 4.5, and 5 as the input.

INPUT	ALGEBRAIC EXPRESSION	OUTPUT
k	$k + 17$	
4	4 + 17	21
4.5	4.5 + 17	21.5
5	5 + 17	22

GUIDED PRACTICE

Find the output, using 3, 4, and 5 as the input.

1. $x + 4$ **2.** $h - 2$ **3.** $y × 5$ **4.** $m ÷ 2$

5. $6.4 + p$ **6.** $k - 1.2$ **7.** $r × 0.5$ **8.** $w ÷ 4$

INDEPENDENT PRACTICE

Copy and complete the input-output table.

1.

INPUT	ALGEBRAIC EXPRESSION	OUTPUT
x	x − 6	
6	?	?
7	?	?
8	?	?

2.

INPUT	ALGEBRAIC EXPRESSION	OUTPUT
y	y + 14.7	
12	?	?
12.5	?	?
13	?	?

3.

INPUT	ALGEBRAIC EXPRESSION	OUTPUT
w	w ÷ 4	
100	?	?
102	?	?
104	?	?

4.

INPUT	ALGEBRAIC EXPRESSION	OUTPUT
a	a × 15	
10	?	?
14	?	?
23	?	?

Make an input-output table for the algebraic expression. Evaluate the expression for 2, 3, 4, and 5.

5. $x + 10$ **6.** $24 - y$ **7.** $20 \div z$ **8.** $8.9 \times a$

Determine the input for the given output. Make a table if necessary.

9. $x + 5$
output = 10
input = ?

10. $x - 7$
output = 20
input = ?

11. $x \div 2$
output = 4
input = ?

12. $t \times 8$
output = 48
input = ?

Problem-Solving Applications

13. CONSUMER MATH Sara wants to purchase a CD player for $115.99. She earns $5.25 per hour baby-sitting. If h represents the number of hours she baby-sits, what algebraic expression represents her earnings? Make an input-output table to find out how many hours she must baby-sit to buy the CD player.

14. TABLES An input-output table shows some entries, but the algebraic expression is missing. The input column includes 5, 6, and 7. The corresponding output is 12, 13, and 14. What is the missing expression?

15. TABLES An input-output table has $1 + y^2$ for its algebraic expression. What is the output if 7 is the input?

16. ✏️ **WRITE ABOUT IT** At Bill's Birdhouses, Bill can make 3 birdhouses every hour he works. Describe how Bill could use an input-output table to find out how many birdhouses he can make in a certain number of hours.

ACTIVITY

What You'll Explore
How to model the solutions to one-step equations

What You'll Need
algebra tiles or green paper rectangles and red and yellow squares

---

VOCABULARY

equation

solve

solution

Modeling and Solving One-Step Equations

An **equation** is an algebraic or numerical sentence that shows two quantities are equal. Here are some examples of equations.

$$x + 3 = 4 \qquad 2 + 5 = 7 \qquad k - 3 = 1$$

ACTIVITY

Explore

- You can model an equation by using algebra tiles. To model an algebraic equation, use a green rectangle to represent the variable. Use a yellow square to represent addition of 1. Here is a model of the equation $x + 3 = 5$.

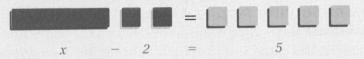

- Use algebra tiles or paper rectangles to model $x + 2 = 5$.

- You can also use algebra tiles to model an equation involving subtraction. A green rectangle still represents the variable. Use a red square to represent subtraction of 1. Copy this model of the equation $x - 2 = 5$. Then model $x - 4 = 3$.

Think and Discuss

- What is the difference between the red and the yellow squares?

- How would you model the equation $2x + 3 = 5$ differently from $x + 3 = 5$?

- Is it possible to model an equation that has the variable on the right side, as in $8 = x - 6$? If so, how?

Try This

- Model each equation by using algebra tiles or paper rectangles.
$$x + 3 = 7 \qquad 4 = x + 3 \qquad x - 7 = 10 \qquad 3x - 1 = 5$$

To **solve** an equation, you find the value of the variable that makes the equation true.

$x + 4 = 7$ Since $3 + 4 = 7$, then $x = 3$.

You can model solving an equation by using algebra tiles.

Technology Link

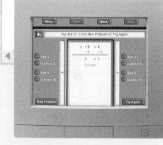

You can practice solving addition and subtraction equations by using E-Lab, Activity 15. Available on CD-ROM and on the Internet at **www.hbschool.com/elab**

ACTIVITY 2

Explore

Solve the equation $x + 3 = 5$.

- Model $x + 3 = 5$.

- To solve an equation, you must get the variable alone on one side. To do this, take 3 ones away from each side.

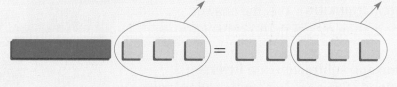

- Taking 3 ones from each side of the equation leaves the variable alone. The equation is now solved.

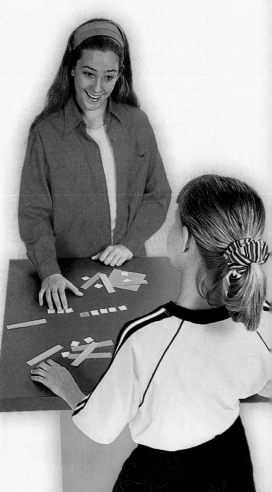

$\leftarrow$ *Rectangle is alone.*

$$x \quad = \quad 2$$

The value of the variable that makes the equation true is called the **solution** of the equation.

Think and Discuss

- What is the solution of the equation $x + 3 = 5$?

- How can you tell this from the model?

- What operation did you model in the second step?

- $\boxed{\text{CRITICAL THINKING}}$ What would you do differently to solve the equation $x - 3 = 5$?

Try This

- Solve each equation by using algebra tiles.
  $x + 4 = 5$ $6 = x + 1$ $x + 7 = 10$ $x + 5 = 7$

Solving Addition and Subtraction Equations

What You'll Learn
How to solve addition and subtraction equations

Why Learn This?
To solve problems involving amounts, distances, and quantities

VOCABULARY
inverse operations

CULTURAL LINK

Agriculture is a big part of Brazil's landscape. In Brazil the chief crops are coffee, cotton, soybeans, sugar, cocoa, rice, corn, and fruits. It was reported in the *Guinness Book of Records* that a 28-pound 11-ounce pineapple was grown in Tarauacá, Brazil. An average pineapple weighs about 4 pounds. How many times greater in pounds is the pineapple reported in the *Guinness Book of Records* than an average-sized pineapple?

Tamika helps her parents in their family restaurant. Tamika got 12 cans of peaches from the pantry. There were 17 cans left. How many were originally in the pantry?

One way to solve this problem is by representing it with an equation. You may choose any variable to represent what you are trying to find out. If c represents the original number of cans, then you can use the equation $c - 12 = 17$ to solve this problem.

Talk About It

• What does the variable c represent?

• Is $c - 12 = 17$ an addition equation or a subtraction equation?

• **CRITICAL THINKING**   Will the answer be more than or less than 17? Explain.

Addition and subtraction are **inverse operations**. When solving any equation, you must use inverse operations to get the variable alone on one side of the equation. Since $c - 12 = 17$ is a subtraction equation, use the inverse operation, addition, to solve it.

EXAMPLE 1 Use the equation $c - 12 = 17$ to find the original number of cans in the pantry.

Let c = original number of cans.   *Choose a variable.*
Then $c - 12$ = cans left.

$$c - 12 = 17$$ *Write the equation.*
Solve the equation.
$$c - 12 + 12 = 17 + 12$$ *Add 12 to each side.*

$$c = 29$$

$$c - 12 = 17$$ *Check your solution.*

$$29 - 12 = 17$$ *Replace c with 29.*

$$17 = 17 ✓$$ *The solution checks.*

So, the original number of cans in the pantry was 29.

• Solve the equation. $y - 3 = 11$

When solving an addition equation, use the inverse operation, subtraction.

EXAMPLE 2 A carpenter cut a 57-in. board into two pieces. One board is 33 in. long. How long is the other board?

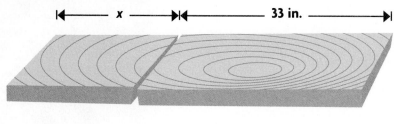

Let x = length of other piece.   *Choose a variable.*
Then $x + 33$ = the total length.

$x + 33 = 57$ *Write the equation.*

$x + 33 - 33 = 57 - 33$ *Solve the equation.*
                                   Subtract 33 from each side.
$x = 24$

$x + 33 = 57$ *Check your solution.*

$24 + 33 = 57$ *Replace x with 24.*

$57 = 57$ ✓   *The solution checks*

So, the length of the second board is 24 in.

• Solve the equation. $x + 6 = 25$

GUIDED PRACTICE

Tell the inverse of the operation in the equation.

1. $x - 5 = 7$ **2.** $r + 4 = 81$ **3.** $7 + w = 23$

4. $p - 8 = 22$ **5.** $85 = k - 12$ **6.** $27 = y + 5$

Tell how to solve the equation.

7. $x - 52 = 71$ **8.** $r + 47 = 92$ **9.** $17 + w = 36$

10. $3 = w - 7$ **11.** $8 + p = 19$ **12.** $23 = x + 1$

Use inverse operations to solve. Check your solution.

13. $b + 32 = 92$ **14.** $x - 10 = 17$ **15.** $52 = c + 25$

16. $y + 12 = 18$ **17.** $21 = m - 8$ **18.** $7 + d = 42$

Sometimes equations use decimals or fractions.

EXAMPLE 3 Solve and check. $b - 6.5 = 13$

$$b - 6.5 = 13 \qquad \textit{Solve.}$$

$$b - 6.5 + 6.5 = 13 + 6.5 \qquad \textit{Add 6.5 to each side.}$$

$$b = 19.5$$

$$b - 6.5 = 13 \qquad \textit{Check your solution.}$$

$$19.5 - 6.5 = 13 \qquad \textit{Replace b with 19.5.}$$

$$13 = 13 \checkmark \qquad \textit{The solution checks.}$$

So, for the equation $b - 6.5 = 13$, $b = 19.5$.

- Solve and check. $4.8 + p = 19.1$

EXAMPLE 4 Solve and check. $k + 5\frac{1}{2} = 12$

$$k + 5\frac{1}{2} = 12 \qquad \textit{Solve.}$$

$$k + 5\frac{1}{2} - 5\frac{1}{2} = 12 - 5\frac{1}{2} \qquad \textit{Subtract } 5\frac{1}{2} \textit{ from each side.}$$

$$k = 6\frac{1}{2}$$

$$k + 5\frac{1}{2} = 12 \qquad \textit{Check your solution.}$$

$$6\frac{1}{2} + 5\frac{1}{2} = 12 \qquad \textit{Replace k with } 6\frac{1}{2}.$$

$$12 = 12 \checkmark \qquad \textit{The solution checks.}$$

So, for the equation $k + 5\frac{1}{2} = 12$, $k = 6\frac{1}{2}$.

- Would you solve an addition or subtraction equation differently if the variable were on the right side of the equation?

- Solve and check. $10\frac{1}{2} = x + 8\frac{1}{3}$

INDEPENDENT PRACTICE

Determine whether the given value is a solution of the equation. Write *yes* or *no*.

1. $x + 7 = 10$; $x = 3$

2. $10 + y = 15$; $y = 5$

3. $d - 3 = 8$; $d = 5$

4. $j - 4 = 3$; $j = 7$

5. $18 + p = 21$; $p = 4$

6. $10 = m - 15$; $m = 25$

7. $l - 8.5 = 12$; $l = 21.5$

8. $17.4 + d = 42.9$; $d = 25.5$

9. $8 = k - 1\frac{1}{4}$; $k = 9\frac{1}{4}$

Use inverse operations to solve. Check your solution.

10. $x + 4 = 12$ **11.** $a + 7 = 9$ **12.** $y - 5 = 13$ **13.** $r - 8 = 27$

14. $9 + k = 32$ **15.** $42 = b - 13$ **16.** $93 = q - 81$ **17.** $17 + c = 71$

18. $942 = b - 1{,}367$ **19.** $m + 345 = 1{,}247$ **20.** $a + 768 = 1{,}940$ **21.** $y - 1{,}435 = 139$

22. $15 = k - 4.9$ **23.** $x + 17.8 = 32.4$ **24.** $y + 9\frac{1}{4} = 18$ **25.** $64 = s - 39\frac{2}{5}$

For Problems 26–29, write an equation.

26. Seven more than a number, x, is twenty.

27. Nine less than a number, y, is seventeen.

28. The sum of Bill's and June's monthly allowances is \$29. Bill has a monthly allowance of \$15.

29. The difference between Mary's and Jeff's heights is 6 in. Jeff's height is 60 in., and he is shorter than Mary.

Problem-Solving Applications

For Problems 30–31, choose a variable, write an equation, and solve it.

30. **SPORTS** Before gym class Angel got 21 tennis balls out of the bin. There were 32 tennis balls left in the bin. How many tennis balls were there originally?

31. Jon had \$324 from mowing lawns. He put \$125 in a checking account and the rest in a savings account. How much did he put in the savings account?

32. ✏️ **WRITE ABOUT IT** Write one addition and one subtraction problem that can be solved by using equations. Solve both problems.

Mixed Review and Test Prep

Evaluate the expression for the given value of the variable.

33. $6 \times a$, for $a = 7$ **34.** $y \times 9$, for $y = 12$ **35.** $k \div 4$, for $k = 28$ **36.** $8 \div m$, for $m = 5$

A bag contains 6 yellow, 4 blue, 2 red, and 2 orange marbles. You choose one without looking. Find each probability.

37. P(yellow)   **38.** P(orange)   **39.** P(blue)   **40.** P(yellow or red)   **41.** P(blue, red, or yellow)

42. **LOGICAL REASONING** Rick runs 2 miles each weekday and 3 miles each weekend. His friend Joan runs 1 mile each weekday and 5 miles each weekend. In a week, how much farther does Rick run?

     A 2 mi   **B** 3 mi   **C** 4 mi   **D** 5 mi

43. **TIME** Suppose it is 4:25 A.M. on Tuesday. If you have an appointment at 12:50 P.M. on Wednesday, how many hours and minutes are there until your appointment?

     F 8 hr 25 min   **G** 22 hr 25 min

     H 28 hr 25 min   **J** 32 hr 25 min

1. **VOCABULARY** A mathematical phrase that includes only numbers and operation symbols is a(n) __?__. (page 292)

Tell whether the expression is numerical or algebraic. (pages 292–293)

2. $5 + 9$ **3.** $y \div 4$ **4.** $17 \times k$ **5.** $8 - 2 \times 3$

Evaluate the algebraic expression for y = 5. (pages 294–295)

6. $y + 12$ **7.** $7 \times y$ **8.** $50 \div y$

9. $y^2 - 6$ **10.** $y^3 + 12$ **11.** $4 + y \times 2$

12. $y \div 5$ **13.** $y \times 3 + 5$ **14.** $100 - y^2$

15. $2 \times y - 1$ **16.** $11 - y$ **17.** $10 + 3 \times y$

Copy and complete the input-output tables. (pages 296–297)

	INPUT	ALGEBRAIC EXPRESSION	OUTPUT
	y	$y \div 3$	
18.	12	?	?
19.	15	?	?
20.	18	?	?

	INPUT	ALGEBRAIC EXPRESSION	OUTPUT
	m	$3 \times m$	
21.	7	?	?
22.	8	?	?
23.	9	?	?

Make an input-output table for the algebraic expression.
Use 7, 7.5, and 8 as input.

24. $4.5 \times a$ **25.** $a - 6.25$ **26.** $12 + a$ **27.** $a \div 0.5$

28. VOCABULARY Addition and subtraction are __?__. (page 300)

Tell whether the value given is a solution. (pages 300–303)

29. $r + 7 = 42; r = 35$ **30.** $d - 9 = 15; d = 6$ **31.** $t + 14 = 42; t = 32$

32. $w - 15 = 10; w = 25$ **33.** $x + 4 = 12; x = 8$ **34.** $y - 5 = 10; y = 25$

Solve and check.

35. $x + 7 = 10$ **36.** $9 + y = 19$ **37.** $x - 6 = 3$

38. $19 = k - 12$ **39.** $21 + n = 52$ **40.** $x + 3 = 4$

1. Justin has a soccer game every fifth day in September. His first game is on September 5. How many games will he play in September?

   A 3
   B 4
   C 5
   D 6
   E Not Here

2. Gordon walked $1\frac{1}{4}$ miles on Tuesday, $1\frac{7}{8}$ miles on Thursday, and $2\frac{1}{3}$ miles on Saturday. What is a reasonable estimate of the total distance Gordon walked?

   F $3\frac{1}{2}$ mi

   G $4\frac{1}{2}$ mi

   H $5\frac{1}{2}$ mi

   J $6\frac{1}{2}$ mi

3. Which statement is always true about the four angles in a rectangle?

   A Some angles may be acute.
   B All are obtuse.
   C Some angles may be obtuse.
   D All are right angles.
   E Not Here

4. If a solid figure is a cone, then it has —

   F one flat, circular base
   G a vertex
   H a curved lateral surface
   J All of the above

5. Ten joggers ran the following number of miles: 5, 4, 3, 6, 4, 7, 5, 3, 4, 6. What is the mode of the miles jogged?

   A 3
   B 4
   C 5
   D 6
   E 7

6. Carla was successful in 17 of 50 foul shots. Based on this information, what is a reasonable estimate of the number of shots she will make in 150 attempts?

   F 20   **G** 34
   H 51   **J** Not Here

7. Which of the following word expressions is represented by the algebraic expression $b + 6$?

   A six more than b
   B b decreased by six
   C the product of b and six
   D b divided by six

8. What is the output when the input is 7?

INPUT	ALGEBRAIC EXPRESSION	OUTPUT
s	$s - 3$	
4	$4 - 3$	1
5	$5 - 3$	2
6	$6 - 3$	3
7	$7 - 3$	?

   F 4
   G 5
   H 6
   J Not Here

ALGEBRA: REAL-LIFE RELATIONSHIPS

LOOK AHEAD

In this chapter you will solve problems that involve

- solving multiplication and division equations

- money and temperature relationships

- length, capacity, and mass/weight relationships

- distance, rate, and time relationships

SCIENCE LINK

The average speed and fuel consumption for several large jets are given in the table below.

	AIRBORNE SPEED (mph)	FUEL (gallons per hour)
B747-400	538	3,420
DC-10-40	506	2,644
L-1011-100/200	490	2,356

- Estimate the flight time of a 2,500-mile flight by jet.

- How many gallons of fuel does an L-1011 burn during a 1,225-mile flight? Round to the nearest hundred gallons.

Recognizing Rates

There are many examples of relationships that tell how many or how much per unit. Some examples are numbers of miles per hour or pesos per dollar. These relationships can be expressed as rates. Find other examples of relationships that are rates. Write each relationship as a rate. Then, write a question that has that rate as its answer.

Giant anteaters eat about 30,000 ants a day.

My heart beats 68 times a minute.

I pay $30.11 for cable TV each month.

PROJECT CHECKLIST

✓ Did you find examples of rates?

✓ Did you write each rate?

✓ Did you write a question that can be answered by each rate?

LAB
ACTIVITY

What You'll Explore
How to model the solutions of multiplication and division equations

What You'll Need
algebra tiles or paper rectangles and squares

Exploring Equations

Multiplication equations are algebraic statements involving equal values or expressions. Here are some examples of multiplication equations.

$$5 \times a = 35 \quad 3 \times y = 27 \quad 110 = 11 \times w$$

You can use algebra tiles to model multiplication equations.

ACTIVITY 1

Explore

- A green rectangle represents a variable, and a yellow square represents the value 1. Here is a model of the multiplication equation $3 \times a = 9$. To represent $3 \times a$, use 3 green rectangles.

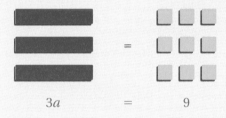

$$3a \qquad = \qquad 9$$

- Model $4 \times y = 12$.

Think and Discuss

- How is modeling multiplication equations different from modeling addition or subtraction equations?

- How would modeling $9 = 3 \times y$ be different from modeling $3 \times y = 9$?

- How do you think you will model the solving of a multiplication equation?

Try This

- Model each equation by using algebra tiles.

$$2 \times b = 10 \quad 4 \times y = 8 \quad 15 = 3 \times a$$

- Write four multiplication equations. Solve the equations. Then check your work.

You can use algebra tiles to model the solving of multiplication equations.

ACTIVITY 2

Explore

Solve the equation $3 \times y = 6$.

- Model $3 \times y = 6$.

- Divide each side of the model into 3 equal groups.

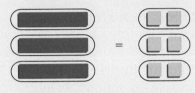

- Look at one group on each side. What is in each pair? What equation does this model?

Think and Discuss

- What is the solution to the equation $3 \times y = 6$?

- What operation did you model in the second step? How does this operation relate to multiplication?

- If the equation were $2 \times y = 6$, what would you do differently in the second step?

- How do you think you would solve an algebraic equation involving division?

Try This

- Use algebra tiles to model and solve each equation. Record each step.

$2 \times c = 6$ $4 \times a = 12$ $3 \times b = 6$ $15 = 5 \times y$

Solving Multiplication and Division Equations

What You'll Learn
How to solve multiplication and division equations

Why Learn This?
To solve for the unknown in equations such as how many hours you need to work to earn a certain amount

REMEMBER:
. . . .
Division can be written as a fraction. **See page 146.**

$2 \div 3 = \frac{2}{3}$

$x \div 3 = \frac{x}{3}$

The table below shows another method for writing algebraic expressions that involve multiplication.

In an algebraic expression with multiplication, you may omit the multiplication sign. The expression $7b$ means "7 multiplied by b."

Expression	New Method
$7 \times b$	$7b$
$3 \times y$	$3y$
$c \times 6$	$6c$
$d \times 4$	$4d$

- Rewrite the algebraic expressions $12 \times a$, $19 \times k$, and $r \times 11$ without using a multiplication symbol.

Remember that addition and subtraction are inverse operations. Multiplication and division are also inverse operations. As in equations with addition and subtraction, you can use inverse operations to solve multiplication and division equations.

EXAMPLE 1 Angela earns $4 per hour when she helps her mother with chores. She wants to save $160. How many hours does she have to work to earn $160?

Let h = hours worked.   *Choose a variable.*
Then $4h$ = money earned.

$4h = 160$ *Write the equation.*

$\frac{4h}{4} = \frac{160}{4}$ *Solve the equation. Since $4h$ involves multiplication, divide each side by 4.*

$h = 40$

$4h = 160$ *Check your solution.*

$4 \times 40 = 160$ *Replace h with 40.*

$160 = 160$ ✓ ← The solution checks.

So, Angela must work 40 hr to earn $160.

- Explain the steps you would use to solve $5m = 60$.

Talk About It

- How did you choose the variable to solve the problem in Example 1?

- How is solving a multiplication equation similar to solving an addition or subtraction equation?

- How is solving a multiplication equation different from solving an addition or subtraction equation?

- If Angela earned $5 per hour, how would the equation in Example 1 be different? Would she have had to work more or fewer hours? Explain.

To solve a division equation, use the inverse operation, multiplication.

Many states now require bicycle riders under the age of 16 to wear helmets. Bicycle helmets cost about $10 to $25, a very small price to pay for something that could save your life. And, you could use the same helmet for in-line skating too.

EXAMPLE 2 Janice bought a new bicycle. She agreed to pay $\frac{1}{5}$ of the original price every month until the bicycle was paid for. Her payment was $37. Write and solve an equation to determine the price of the bicycle.

Let p = price of the bicycle.   *Choose a variable.*

Then $\frac{p}{5}$ = amount of payment.

$\frac{p}{5} = 37$ *Write the equation.*

$5 \times \frac{p}{5} = 37 \times 5$ *Solve the equation.*
Multiply each side by 5.

$p = 185$

$\frac{p}{5} = 37$ *Check your solution.*

$\frac{185}{5} = 37$ *Replace p with 185.*

$37 = 37$ ✓   ← The solution checks.

So, the price of the bicycle is $185.

- Explain the steps you would take to solve $\frac{x}{4} = 17$.

ANOTHER METHOD You can also use the equation $\frac{1}{5}p = 37$ to find the price of the bicycle.

$\frac{1}{5}p = 37$ *Write the equation.*

$\frac{1}{5}p \div \frac{1}{5} = 37 \div \frac{1}{5}$ *Solve the equation. Divide each side by $\frac{1}{5}$.*

$p = 37 \times \frac{5}{1}$ *Multiply by the reciprocal.*

$p = 185$

REMEMBER:

To divide by a fraction, use the **reciprocal** of the fraction to write a multiplication problem. **See page 146.**

$6 \div \frac{2}{3} = 6 \times \frac{3}{2}$

↑ reciprocal

Sometimes equations use decimals.

EXAMPLE 3 Chris got a paycheck for $64.80 for 9 hours of work. Write and solve a multiplication equation to find the amount, p, that he earned each hour.

$9p = 64.80$ *Write an equation.*

$\dfrac{9p}{9} = \dfrac{64.80}{9}$ *Divide each side by 9.*

$p = 7.20$

$9p = 64.80$ *Check your solution.*

$9 \times 7.20 = 64.80$ *Replace p with 7.20.*

$64.80 = 64.80$ ✓ ← The solution checks.

So, Chris earned $7.20 each hour.

• Solve and check. $8.2x = 98.4$

EXAMPLE 4 Solve and check. $\dfrac{y}{18.6} = 2.3$

$\dfrac{y}{18.6} = 2.3$ *Write the equation.*

$18.6 \times \dfrac{y}{18.6} = 2.3 \times 18.6$ *Multiply each side by 18.6.*

$y = 42.78$

$\dfrac{y}{18.6} = 2.3$ *Check your solution.*

$42.78 \div 18.6 = 2.3$ *Replace y with 42.78.*

$2.3 = 2.3$ ✓ ← The solution checks.

So, for the equation $\dfrac{y}{18.6} = 2.3$, $y = 42.78$.

• Solve and check. $\dfrac{x}{3.1} = 4.2$

GUIDED PRACTICE

Rewrite the expression without using a multiplication symbol.

1. $12 \times k$ **2.** $r \times 7$ **3.** $s \times 3$ **4.** $8 \times k$

Tell the inverse of the operation in the equation.

5. $2x = 10$ **6.** $\dfrac{x}{7} = 8$ **7.** $24 = \dfrac{y}{9}$ **8.** $27 = 3x$

Tell how to solve the equation.

9. $\dfrac{x}{3} = 22$ **10.** $4x = 28$ **11.** $45 = 5x$ **12.** $17 = \dfrac{y}{4}$

INDEPENDENT PRACTICE

Determine whether the given value is a solution of the equation.
Write *yes* or *no.*

1. $2x = 10$; $x = 5$ **2.** $7k = 42$; $k = 7$ **3.** $\frac{y}{3} = 5$; $y = 18$ **4.** $\frac{a}{2} = 12$; $a = 24$

5. $6 = \frac{m}{7}$; $m = 42$ **6.** $40 = 5p$; $p = 8$ **7.** $72 = 12n$; $n = 5$ **8.** $9 = \frac{s}{11}$; $s = 90$

Use inverse operations to solve. Check your solution.

9. $3x = 6$ **10.** $8k = 48$ **11.** $\frac{y}{5} = 7$ **12.** $\frac{a}{6} = 3$

13. $4 = \frac{m}{3}$ **14.** $48 = 3p$ **15.** $15 = \frac{s}{5}$ **16.** $60 = 12n$

17. $160 = 16d$ **18.** $412 = \frac{a}{4}$ **19.** $1,920 = 24b$ **20.** $21 = \frac{w}{13}$

Write an equation for the word sentence. For Exercise 22,
choose a variable.

21. Eleven multiplied by a number, *x*, is sixty-six.

22. The number of cookies sold times the price, $2 each, equals $112.

Problem-Solving Applications

For Problems 23–24, choose a variable, write an equation, and solve.

23. Every time Eddie wins a game at the arcade, he gets 5 coupons. It takes 120 coupons to win a watch. How many games must he win to get the watch?

24. HOBBIES Mike gave each of 3 friends $\frac{1}{3}$ of his marbles. Each friend got 14 marbles. How many marbles did Mike have?

25. ✏ **WRITE ABOUT IT** Describe how to solve a multiplication or division equation.

Mixed Review and Test Prep

Write the numbers of quarters, dimes, nickels, and pennies in the given dollar amount.

26. $5.00   **27.** $15.00   **28.** $9.75   **29.** $42.50

Solve and check.

30. $x + 5 = 12$ **31.** $y - 5 = 17$ **32.** $24 = k + 9$ **33.** $15 = d - 5$

34. MONEY Jon has 50 coins whose total value is $1.00. What are the coins?
  A 50 pennies, 2 quarters
  B 40 pennies, 2 nickels, 1 dime
  C 35 pennies, 11 nickels, 1 dime
  D 45 pennies, 2 nickels, 2 dimes, 1 quarter

35. ALGEBRA Barry has $100 in his bank account. He withdraws *n* dollars, leaving a balance of $63. Which equation describes this situation?
  F $n - 100 = 63$ **G** $100 - n = 63$
  H $n - 63 = 100$ **J** $63 - n = 100$

Money Relationships

At Arnie's Arcade, you can play some video games for a quarter and others for a dime, nickel, or penny. Emily has $15.75. How many games can she play if she has all quarters?

Emily knows that, since there are 4 quarters in a dollar, dividing the number of quarters by 4 equals the number of dollars. She uses the equation $\frac{q}{4} = D$ to calculate how many games she can play by using only quarters.

$$\frac{q}{4} = D \qquad \textit{Let q = number of quarters.}$$
$$\textit{Let D = number of dollars.}$$
$$\frac{q}{4} = 15.75 \qquad \textit{Replace D with 15.75.}$$
$$4 \times \frac{q}{4} = 15.75 \times 4 \qquad \textit{Solve the equation.}$$
$$\textit{Multiply each side by 4.}$$
$$q = 63 \qquad \leftarrow \quad \text{There are 63 quarters.}$$

So, if Emily has $15.75 all in quarters, she can play 63 games.

Talk About It

• How many dimes, nickels, and pennies are in one dollar?

• CRITICAL THINKING What formula could you use to find out how many games Emily can play by using only dimes, only nickels, or only pennies?

Pong, the first commercial video game, was so popular when it was introduced in 1972 that its coin container filled with quarters every 6 hr. If the coin container held 250 quarters, and the arcade was open 12 hr a day, how much money did the game make in 1 week?

In the United States the basic unit of money is the dollar. In England it is the pound, and in Germany it is the mark. Arnie calls the bank to find out the current exchange rates. The formulas below show how to convert pounds and marks to dollars.
P = English pounds, M = German marks, and D = U.S. dollars.

$$1.6261 \times P = D \qquad 0.6200 \times M = D$$

EXAMPLE Find how many dollars Arnie gave a customer in exchange for 12 pounds and for 15 marks. Round your answer to the nearest cent.

Pounds		Marks	
$1.6261 \times P = D$	*Replace P with 12.*	$0.6200 \times M = D$	*Replace M with 15.*
$1.6261 \times 12 = D$		$0.6200 \times 15 = D$	
$19.5132 = D$		$9.3 = D$	

So, Arnie gave $19.51 for 12 pounds and $9.30 for 15 marks.

• How many dollars are 8 pounds? 24 marks?

Calculator Activities, page H32

GUIDED PRACTICE

Write the equations you would use to find the number of quarters
and the number of nickels in the given dollar amount.

1. $5.00   **2.** $8.00   **3.** $12.75   **4.** $3.25

Write the equations you would use to find the number of dimes
and the number of pennies in the given dollar amount.

5. $3.00   **6.** $9.00   **7.** $24.80   **8.** $2.40

INDEPENDENT PRACTICE

Find the numbers of quarters, dimes, nickels, and pennies in the
given dollar amount.

1. $8.00   **2.** $15.00   **3.** $27.00   **4.** $3.00   **5.** $6.00

6. $9.00   **7.** $19.50   **8.** $43.00   **9.** $2.50   **10.** $21.00

Use the formulas on page 314 to convert the English pounds and
German marks to U.S. dollars. Round to the nearest penny.

11. 25 pounds   **12.** 124 pounds   **13.** 60 marks   **14.** 257 marks   **15.** 82 pounds

Problem-Solving Applications

16. MONEY One roll of quarters is worth $10.
One roll of nickels is worth $2. How many
quarters and nickels are there in one roll
of each?

17. A roll of pennies is worth $0.50. A roll of
dimes is worth $5. How many pennies
and dimes are there in one roll of each?

18. COMPARE Reggie has $26.75 in quarters
and Jenny has $7.85 in nickels. Who has
more coins? Explain.

19. ✏ **WRITE ABOUT IT** Would you rather
have 10 English pounds or 10 German
marks? Explain.

Mixed Review and Test Prep

Multiply. Round to the nearest whole number.

20. $\frac{5}{9} \times 27$ **21.** $\frac{5}{9} \times 12$ **22.** $\frac{9}{5} \times 40$ **23.** $\frac{9}{5} \times 13$ **24.** $\frac{9}{5} \times 42$

Evaluate the expression for $y = 7$.

25. $8y$ **26.** $63 \div y$ **27.** $y + 15$ **28.** $y - 4$ **29.** $y + 78$

30. ALGEBRA Dana thinks of a number, n.
When Dana says "5," Bev says "15." When
Dana says "3," Bev says "9." What formula
might Bev be using?

  A $n + 10$ **B** $n \times 3$
  C $n \div 3$ **D** $n + 6$

31. COOKING Your bag of sugar has $4\frac{1}{2}$ cups.
The recipe calls for $5\frac{3}{4}$ cups of sugar. How
much more sugar do you need?

  F $5\frac{1}{3}$ c   **G** $1\frac{3}{4}$ c
  H $1\frac{1}{4}$ c   **J** $1\frac{3}{20}$ c

Temperature Relationships

REMEMBER:
....
The symbol ° represents degrees. **See page H90.**

One reason some scientists believe life may have existed or may currently exist on Mars is that the temperature on Mars's surface can be as high as 17° Celsius. You may be more familiar with the Fahrenheit temperature scale. To convert from degrees Celsius to degrees Fahrenheit, use the formula below.

$$F = \left(\frac{9}{5} \times C\right) + 32$$

EXAMPLE 1 Write the highest temperature on Mars in degrees Fahrenheit (°F). Round to the nearest degree.

$F = \left(\frac{9}{5} \times C\right) + 32$	*Write the formula.*
$F = \left(\frac{9}{5} \times 17\right) + 32$	*Replace C with 17.*
$F = 30.6 + 32$	*Operate in parentheses.*
$F = 62.6 \approx 63$	*Add. Round.*

So, the highest temperature on Mars is about 63°F.

To convert from degrees Fahrenheit to degrees Celsius, use the formula below.

$$C = \frac{5}{9} \times (F - 32)$$

EXAMPLE 2 If the temperature in Marvin's house reaches 25°C, the air conditioner comes on. The temperature is 77°F. Is the air conditioner on?

$C = \frac{5}{9} \times (F - 32)$	*Write the formula.*
$C = \frac{5}{9} \times (77 - 32)$	*Replace F with 77.*
$C = \frac{5}{9} \times 45$	*Operate in parentheses.*
$C = 25$	*Multiply.*

So, the air conditioner is on, since the temperature is 25°C.

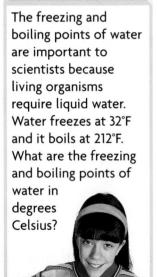

GUIDED PRACTICE

Tell whether to use $F = \left(\frac{9}{5} \times C\right) + 32$ or $C = \frac{5}{9} \times (F - 32)$ to make the conversion.

1. $20°C = \underline{\ ?\ }°F$ **2.** $3°C = \underline{\ ?\ }°F$ **3.** $18°F = \underline{\ ?\ }°C$ **4.** $85°F = \underline{\ ?\ }°C$

5. $38°C = \underline{\ ?\ }°F$ **6.** $92°F = \underline{\ ?\ }°C$ **7.** $12°C = \underline{\ ?\ }°F$ **8.** $52°F = \underline{\ ?\ }°C$

9. What formula do you use to convert temperatures from degrees Celsius to degrees Fahrenheit?

10. What formula do you use to convert temperatures from degrees Fahrenheit to degrees Celsius?

INDEPENDENT PRACTICE

Convert the temperatures from degrees Celsius to degrees Fahrenheit. Round the answer to the nearest degree.

1. $20°C$ **2.** $35°C$ **3.** $60°C$ **4.** $17°C$ **5.** $2°C$ **6.** $75°C$

7. $15°C$ **8.** $45°C$ **9.** $100°C$ **10.** $0°C$ **11.** $90°C$ **12.** $65°C$

13. $88°C$ **14.** $5°C$ **15.** $98°C$ **16.** $21°C$ **17.** $79°C$ **18.** $42°C$

Convert the temperatures from degrees Fahrenheit to degrees Celsius. Round the answer to the nearest degree.

19. $86°F$ **20.** $59°F$ **21.** $158°F$ **22.** $100°F$ **23.** $88°F$ **24.** $42°F$

25. $34°F$ **26.** $110°F$ **27.** $48°F$ **28.** $37°F$ **29.** $91°F$ **30.** $62°F$

31. $102°F$ **32.** $33°F$ **33.** $90°F$ **34.** $45°F$ **35.** $87°F$ **36.** $54°F$

Problem-Solving Applications

37. SPORTS Sidney Celsius announced that the baseball game would not be played unless the temperature reached $10°C$. Fred Fahrenheit said the temperature was $55°F$. Was the game played? Explain.

38. SCIENCE Mercury is the planet closest to the sun in our solar system. The temperature on Mercury can reach $400°C$. What is this temperature in degrees Fahrenheit?

39. WEATHER The highest temperature recorded so far in the United States was $134°F$ on July 10, 1913, in Death Valley, California. What was the temperature in degrees Celsius? Round the answer to the nearest degree.

40. ✏️ **WRITE ABOUT IT** Think about two temperatures in degrees Fahrenheit that you consider very hot and very cold. Write two problems in which those temperatures must be converted to degrees Celsius.

Technology Link

💿 In *Mighty Math Astro Algebra*, play the mission *Burning Out II* to practice solving one-step multiplication and division equations. Use Grow Slide Level Green P.

LESSON 16.4

Time and Distance Relationships

What You'll Learn
How to calculate distance, rate, or time by solving equations

Why Learn This?
To determine how far, fast, or long you will be traveling

GEOGRAPHY LINK

Saint George, Utah, near Zion National Park, and Moab, Utah, near Canyonlands National Park, are about 340 mi apart. The American Automobile Association (AAA) says the trip should take about 8.5 hours. At what rate of speed would you need to drive to make the trip in 8.5 hours?

Have you ever wondered how long a trip will take? If you know two of the three parts of the formula $d = r \times t$, or distance = rate × time, you can solve for the third part.

EXAMPLE 1 Mindy and her family are on vacation. They are 135 mi from their destination. If they are traveling 60 mi per hr, how much longer will it take them to get there?

$d = r \times t$ *Write the formula.*

$135 = 60t$ *Replace d with 135 and r with 60. Solve the equation.*

$\dfrac{135}{60} = \dfrac{60t}{60}$ *Divide each side by 60.*

$2\dfrac{1}{4} = t$

So, it will take $2\dfrac{1}{4}$ hr to get to their destination.

- **CRITICAL THINKING** How did you know the unit of time was hours?

- What would you need to know to calculate the distance?

EXAMPLE 2 How far would Mindy and her family go if they traveled for 5 hr at 55 mi per hr?

$d = r \times t$ *Write the formula.*

$d = 55 \times 5$ *Replace r with 55 and t with 5.*

$d = 275$ *Multiply.*

So, they would go 275 mi.

EXAMPLE 3 Suppose Mindy and her family take 3 hr to travel 129 mi. What is their rate of speed?

$d = r \times t$ *Write the formula.*

$129 = r \times 3$ *Replace d with 129 and t with 3.*

$\dfrac{129}{3} = \dfrac{3r}{3}$ *Divide each side by 3.*

$43 = r$

So, their rate of speed is 43 mi per hr.

GUIDED PRACTICE

Use the formula $d = r \times t$ to find the distance for Exercises 1–6.

1. $r = 50$ mi per hr
   $t = 4$ hr

2. $r = 32.5$ mi per hr
   $t = 8$ hr

3. $r = 32$ ft per sec
   $t = 15$ sec

4. $r = 18.6$ ft per sec
   $t = 42.5$ sec

5. $r = 88$ km per hr
   $t = 3$ hr

6. $r = 125.4$ km per hr
   $t = 7.3$ hr

INDEPENDENT PRACTICE

Solve each equation for distance (in feet), rate (in feet per second), or time (in seconds).

1. $d = (6.5 \times 8)$

2. $1,000 = 4r$

3. $200 = 50t$

Use the formula $d = r \times t$ to complete.

4. $d = \underline{\ ?\ }$
   $r = 25$ mi per hr
   $t = 3$ hr

5. $d = \underline{\ ?\ }$
   $r = 19$ ft per sec
   $t = 45$ sec

6. $d = 88$ km
   $r = \underline{\ ?\ }$
   $t = 11$ min

7. $d = 100$ mi
   $r = \underline{\ ?\ }$
   $t = 4$ hr

8. $d = 1,250$ km
   $r = \underline{\ ?\ }$
   $t = 250$ min

9. $d = 90$ ft
   $r = \underline{\ ?\ }$
   $t = 15$ sec

10. $d = 600$ mi
    $r = 50$ mi per hr
    $t = \underline{\ ?\ }$

11. $d = 384$ ft
    $r = 32$ ft per sec
    $t = \underline{\ ?\ }$

12. $d = 600$ ft
    $r = 30$ ft per min
    $t = \underline{\ ?\ }$

13. $d = \underline{\ ?\ }$
    $r = 47$ mi per hr
    $t = 8$ hr

14. $d = \underline{\ ?\ }$
    $r = 12.7$ km per hr
    $t = 4.2$ hr

15. $d = 2,250$ ft
    $r = 18$ ft per sec
    $t = \underline{\ ?\ }$

Problem-Solving Applications

16. **SCIENCE** The vehicle moving the space shuttle from the Vehicle Assembly Building to the launchpad travels at about 2 mi per hr. If the launchpad is 8 mi from the Vehicle Assembly Building, about how long will it take the shuttle to get there?

17. **RATES** The Sears Tower in Chicago is one of the tallest buildings in the world. It is 1,454 ft tall. If it takes a person 60 min to climb the stairs to the top of the building, how fast is that person traveling? Round the answer to the nearest whole number.

18. **HISTORY** In December 1903 Wilbur Wright flew the *Kitty Hawk* 826 ft at 14 ft per sec in North Carolina. How long did the flight last?

19. ✏️ **WRITE ABOUT IT** What would happen to the distance if the rate of speed remained the same and the time increased? if the time decreased? Explain.

What You'll Learn
How to *make a table* to see relationships

Why Learn This?
To convert measurements from one system to another

Making a Table to Relate Measurement

Sometimes you can use tables instead of formulas to look at relationships.

Lisa has a pen pal, Enrique, who lives in Colombia, South America. Enrique wrote that he has a dog that weighs 18 kg. Lisa's dog weighs 35 lb. Lisa knows that 1 kg is about 2.2 lb. Whose dog is heavier?

PROBLEM SOLVING

• **Understand**
• **Plan**
• **Solve**
• **Look Back**

UNDERSTAND What are you asked to find?

What facts are given?

PLAN What strategy will you use?

You can *make a table* to show the relationship between kilograms and pounds.

SOLVE How can you compare the weights?

You can double the number of kilograms and pounds as shown at the right. Then use the table to change the weight of Enrique's dog, 18 kg, to pounds.

Kilograms	Pounds
1	2.2
2	4.4
4	8.8
8	17.6

Use a combination of the weights in the table.

$$8 \text{ kg} + 8 \text{ kg} + 2 \text{ kg} \quad = 18 \text{ kg}$$
$$\downarrow \qquad \downarrow \qquad \downarrow \qquad \downarrow$$
$$17.6 \text{ lb} + 17.6 \text{ lb} + 4.4 \text{ lb} = 39.6 \text{ lb}$$

Compare the weights: 39.6 lb > 35 lb.

So, Enrique's dog weighs more than Lisa's dog.

LOOK BACK How can you check your answer?

What if . . . Enrique's dog weighed 14 kg? Whose dog would weigh more?

Technology Link

In *Data ToolKit*, you can make a table to help you solve problems.

PRACTICE

Solve by *making a table*.

1. Scott has a friend, Charles, who lives in France. During a phone conversation, Charles mentioned his weight is 42 kg. Scott knows his own weight is 103 lb. How many pounds more does Scott weigh than Charles?

2. Brenda and her friend Lakisha enjoy running in long races. Last week Brenda ran a 15-mi race and Lakisha ran a 25-km race. Which of them ran farther? (HINT: 1 mi ≈ 1.609 km)

3. Renato loves to collect baseball cards. His mom tells him she will buy him 4 packs of cards for every 5 times he cleans up the kitchen. Renato wants to get a case of cards, which contains 60 packs. How many times will he have to clean up the kitchen?

4. Rudy's parents are having a party this weekend. Rudy is in charge of making the punch. Each batch of punch uses 3 cans of fruit juice and 2 bottles of ginger ale. How many bottles of ginger ale does Rudy need if he uses 18 cans of fruit juice?

MIXED APPLICATIONS

Solve.

CHOOSE a strategy and a tool.
- **Make a Table**
- **Write an Equation**
- **Find a Pattern**
- **Use Estimation**
- **Guess and Check**
- **Work Backward**

Paper/Pencil   Calculator   Hands-On   Mental Math

5. Lisa has 24 tomatoes. A recipe for spaghetti sauce calls for 3 onions for every 6 tomatoes. How many onions will Lisa need if she uses 24 tomatoes?

6. On a picnic, the Evanses always take 14 pieces of fruit. If they take 2 more apples than oranges, how many of each do they take?

7. Jon is on the ticket committee for the school dance. On the first four days he sells 12, 17, 27, and 42 tickets. If the pattern continues, how many tickets will he sell on the fifth, sixth, and seventh days?

8. Donna is on the decoration committee for the school dance. She spent $15.90 on streamers, $12.15 on balloons, $6.84 on tape, $19.98 on banner paper, and $13.22 on banner paint. What is a reasonable estimate of the amount she spent?

9. Over a period of 5 months, Lori read 23 books. She read 6 books the fifth month, 2 books the fourth month, and 5 books the third month. How many books did she read during the first two months?

10. Frank earns $6 per hour helping his father around the family store. Frank wants to earn $144 to buy a new bicycle. How many hours must he work?

11. ✏️ **WRITE ABOUT IT** Write a problem that can be solved by *making a table*. Then solve.

Review/Test

Solve and check. (pages 310–313)

1. $6x = 24$ **2.** $9y = 72$ **3.** $28 = 4a$ **4.** $11n = 55$

5. $\frac{x}{5} = 3$ **6.** $\frac{y}{7} = 11$ **7.** $19 = \frac{a}{12}$ **8.** $6 = \frac{x}{3}$

9. Bill gets $0.28 for each pound of aluminum he recycles. He got $14.84 recycling aluminum. How many pounds did Bill recycle?

10. Arlene walks at a rate of 4 mi per hr. If she walks for $1\frac{1}{2}$ hr, how far does she walk?

Write the number of quarters and the number of nickels in the amount. (pages 314–315)

11. $6.00 **12.** $25.00 **13.** $17.50 **14.** $3.50

15. $8.00 **16.** $1.25 **17.** $4.75 **18.** $5.00

Convert to U.S. dollars. Round the answer to the nearest cent.
Use the formulas on page 314.

19. 8 marks **20.** 3 pounds **21.** 15 marks **22.** 2 pounds

Convert to degrees Celsius. Round to the nearest degree. (pages 316–317)

23. 59°F **24.** 89°F **25.** 72°F **26.** 38°F

Convert to degrees Fahrenheit. Round to the nearest degree. (pages 316–317)

27. 20°C **28.** 42°C **29.** 93°C **30.** 55°C

Use the formula $d = r \times t$ to complete. (pages 318–319)

31. $d = \underline{\,?\,}$
$r = 65$ mi per hr
$t = 4$ hr

32. $d = 1{,}146$ ft
$r = 3$ ft per sec
$t = \underline{\,?\,}$

33. $d = 925$ km
$r = \underline{\,?\,}$
$t = 25$ min

34. $d = 576$ mi
$r = \underline{\,?\,}$
$t = 9$ hr

35. $d = 22$ mi
$r = 55$ mi per hr
$t = \underline{\,?\,}$

36. $d = 220$ mi
$r = \underline{\,?\,}$
$t = \frac{1}{2}$ hr

37. $d = \underline{\,?\,}$
$r = 10$ ft per sec
$t = 36$ sec

38. $d = 8{,}000$ ft
$r = 40$ mi per hr
$t = \underline{\,?\,}$

Solve. (pages 320–321)

39. Suppose one British pound is equal to 1.5 U.S. dollars. Julia sees that a computer game sells for $60 in a U.S. catalog. The same computer game sells for 44 pounds in an English catalog. Which is cheaper?

40. Jeremy knows that 1 in. ≈ 2.54 cm. One model airplane is 22 in. long, and another is 60 cm long. Jeremy wants to buy the longer one. Which one should he buy?

1. One pencil is $4\frac{5}{8}$ inches long and a second is $5\frac{7}{12}$ inches long. Estimate how much longer the second pencil is.

A $\frac{1}{2}$ in.

B 1 in.

C $1\frac{1}{2}$ in.

D 2 in.

2. Which solid figure has triangles in all of its views?

F Triangular pyramid
G Cone
H Rectangular prism
J Triangular prism

3. A random sample of 50 middle school students showed that 23 students chose mysteries as their favorite type of book. The survey was expanded to include 200 students. About how many students would you expect chose mysteries?

A 46
B 69
C 73
D 92
E 112

4. The air temperature at Lake Pleasure was 86°F. What was the temperature in degrees Celsius?

F 12°C
G 30°C
H 41°C
J 57°C

5. Which fraction is equivalent to $\frac{32}{40}$?

A $\frac{4}{8}$

B $\frac{3}{4}$

C $\frac{4}{5}$

D $\frac{16}{18}$

6. Which algebraic expression represents the word expression *r less than* $\frac{2}{3}$?

F $\frac{2}{3} + r$

G $\frac{2}{3}r$

H $r - \frac{2}{3}$

J $r \div \frac{2}{3}$

K Not Here

7. The Chen family is driving a distance of 320 miles. They expect to drive at an average speed of 57 miles per hour. What is a reasonable estimate of how long the trip will take?

A Less than 3 hr
B Between 3 and 4 hr
C Between 5 and 6 hr
D Between 7 and 8 hr

8. Rahul is making punch for a student council meeting. The recipe calls for 4 cans of juice and 3 bottles of ginger ale per batch. How many bottles of ginger ale are needed if Rahul uses 24 cans of juice?

F 12
G 15
H 18
J 21
K Not Here

MATH FUN!

PAY ATTENTION TO THE EXPRESSION

PURPOSE To practice writing word, numerical, and algebraic expressions (pages 292–293)

YOU WILL NEED set of word, numerical, and algebraic expression cards

In turn, each player turns two cards face up. If the cards make a pair, the player removes the pair and chooses two more cards. If the player cannot make a pair, it is the next player's turn. The player with the most pairs wins.

$y \times 5$

the product of a number, y, and five

HOME NOTE Take your cards home and play with a family member. Make up a new set of cards and play for practice.

The Last One Wins!

PURPOSE To practice evaluating algebraic expressions (pages 294–295)

YOU WILL NEED a copy of the equations, 8-section spinner

You will replace the variable x in the equation with the number you spin on the spinner.

Take turns. Spin the spinner. If the number can solve one of the equations, cross out the equation and keep spinning. If it is not a solution, it's the next player's turn. The player to solve the *last* equation wins!

The Last One Wins!

$43 + X = 44$ $X - 3 = 2$

$23 - X = 16$ $X + 68 = 76$

$9 \div X = 3$ $X \div 3 = 2$

$6 \times X = 24$ $X \times 22 = 44$

TEMPERATURE EVENTS

PURPOSE To practice converting between Celsius and Fahrenheit temperature scales (pages 316–317)

YOU WILL NEED poster board and markers

Make a poster, starting with the figure at the right. Mark a dozen or so temperatures between 0°C and 100°C. Then, label each with something that happens at that temperature. "Water boils," "nice day for a ball game," and "blizzard" are possibilities. Be sure to label the temperature in both scales.

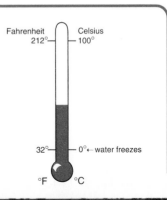

Fahrenheit 212°   Celsius 100°

32°   0° ← water freezes

°F   °C

Input-Output Tables in Spreadsheets

In this activity you will see how easy it is to make input-output tables using a spreadsheet. For the expression $a \div 2$, enter the values for a in the Input column. Then, enter the expressions in the Algebraic Expression column.

All	A	B	C
1	Input	Algebraic Expression	Output
2	a	$a \div 2$	
3	5	$5 \div 2$	2.5
4	6	$6 \div 2$	3
5	7	$7 \div 2$	3.5
6			

To find the output for 5, enter the following formula into cell C3.

All	C
1	Output
2	
3	=A3/2

To find the output for 6, enter the following formula into cell C4.

All	C
1	Output
2	
3	2.5
4	=A4/2

To find the output for 7, enter the following formula into cell C5.

All	C
1	Output
2	
3	2.5
4	3
5	=A5/2

1. Suppose you add 8 as an input in cell A6. What formula would you use for the output in cell C6?

2. What would the output be for an input of 8?

USING A SPREADSHEET

Make an input-output table in a spreadsheet. Find the output, using 6, 8, and 10 as inputs.

3. $x - 3$ **4.** $b \times 6$ **5.** $h \div 4$ **6.** $y + 5$

7. $w - 6$ **8.** $p \times 3$ **9.** $k \div 5$ **10.** $j + 9$

11. $m + 0.5$ **12.** $t \times \frac{1}{2}$ **13.** $c - 1.35$ **14.** $\frac{3}{8} + s$

Study Guide and Review

Vocabulary Check

1. A mathematical phrase that includes only numbers and operation symbols is called a(n) ___?___. **(page 292)**

2. An algebraic or numerical sentence that shows two quantities are equal is called a(n) ___?___. **(page 298)**

EXAMPLES

EXERCISES

- **Write numerical and algebraic expressions.**
  (pages 292–293)

  Translate the phrase into an algebraic expression.

  five times a number, y
  $5 \times y$

Tell whether the expression is numerical or algebraic.

3. $9 - 4$ **4.** $50 \div y$ **5.** $x + 10$

Write a numerical or algebraic expression.

6. two less than ten   **7.** seven more than a number, x

- **Evaluate numerical and algebraic expressions.**
  (pages 294–295)

  Evaluate $y + 12$ for $y = 10$.
  $\quad 10 + 12$ *Replace y with 10.*
  $\quad 22$

Evaluate the expression for $a = 15$.

8. $6 \times a$ **9.** $20 - a$

10. $a^2 - 5$ **11.** $45 \div a$

- **Use input-output tables.** **(pages 296–297)**

  Find the output of $x - 4$ for inputs of 5 and 7.

  $x - 4$
  $5 - 4 = 1$ *Replace x with 5.*
  $7 - 4 = 3$ *Replace x with 7.*

Copy and complete the input-output table.

	Input	Algebraic Expression	Output
	a	$a + 7$	
12.	5	?	?
13.	10	?	?

Make an input-output table for each algebraic expression. Use 3, 3.5, and 4.1 as input.

14. $2.5 \times y$ **15.** $y - 9.2$

- **Solve addition and subtraction equations.**
  (pages 300–303)

  Solve. $x + 6 = 12$
  $\quad x + 6 = 12$
  $\quad x + 6 - 6 = 12 - 6$ *Subtract 6.*
  $\quad x = 6$

Tell whether the given value is a solution.

16. $x - 9 = 17$; $x = 8$ **17.** $x + 8 = 36$; $x = 28$

Solve and check.

18. $b + 14 = 42$ **19.** $a - 18 = 28$

- **Solve multiplication and division equations.**
  (pages 310–313)

Solve and check. $6x = 54$

Solve. $6x = 54$ Check. $6x = 54$

$$\frac{6x}{6} = \frac{54}{6}$$ $6 \times 9 = 54$

$$x = 9$$ $54 = 54 ✓$

Solve and check.

20. $8x = 64$ **21.** $10x = 90$ **22.** $33x = 132$

23. $\frac{x}{5} = 4$ **24.** $10 = \frac{y}{15}$ **25.** $12 = \frac{y}{20}$

26. $80x = 320$ **27.** $7y = 84$ **28.** $5 = 60x$

- **Use equations to show money relationships.**
  (pages 314–315)

Tell how many dimes are in $66.80.

$$\frac{d}{10} = 66.80$$

$$10 \times \frac{d}{10} = 66.80 \times 10$$

$$d = 668$$

So, there are 668 dimes in $66.80.

Write the number of quarters in the amount.
Write the number of nickels in the amount.

29. $7.00   **30.** $12.50   **31.** $35.00

Write the number of dimes and the number of pennies in the amount.

32. $7.00   **33.** $12.50   **34.** $35.00

- **Convert between the Fahrenheit and Celsius temperature scales.** (pages 316–317)

Convert 20°C to degrees Fahrenheit.

$$F = \left(\frac{9}{5} \times C\right) + 32$$ *Use the formula.*

$$F = \left(\frac{9}{5} \times 20\right) + 32$$ *Replace C with 20.*

$$F = 36 + 32 = 68$$ So, $20°C = 68°F$.

Convert to degrees Fahrenheit. Round to the nearest degree.

35. 40°C   **36.** 100°C   **37.** 12°C

Convert to degrees Celsius. Round to the nearest degree. Use the formula
$$C = \frac{5}{9} \times (F - 32).$$

38. 70°F   **39.** 32°F   **40.** 45°F

- **Calculate distance, rate, or time by solving an equation.** (pages 318–319)

Find the distance if the rate = 60 mi per hr and the time = 4 hr.

$d = r \times t$ *Use the formula.*

$d = 60 \times 4$ *Replace r with 60. Replace t with 4.*

$d = 240$

So, the distance is 240 mi.

Use the formula $d = r \times t$ to complete.

41. $d = \underline{\ ?\ }$ **42.** $d = 416$ km
     $r = 70$ mi per hr $r = \underline{\ ?\ }$
     $t = 3$ hr $t = 8$ hr

43. $d = 2{,}145$ ft   **44.** $d = 150$ km
     $r = 5$ ft per sec $r = \underline{\ ?\ }$
     $t = \underline{\ ?\ }$ $t = 30$ min

Problem-Solving Applications

Solve. Explain your method.

45. Karen swam 2 mi. Rick swam 3 km. Who swam farther? (1 mi ≈ 1.6 km)
(pages 320–321)

46. Etienne's bag weighs 48 kg. Becky's bag weighs 100 lb. Which bag is heavier?
(1 kg ≈ 2.2 lb) (pages 320–321)

Performance Assessment

Tasks: Show What You Know

1. Explain how you would use the inverse operation to solve $n - 8.5 = 24$. Then solve and check. **(pages 300–303)**

2. Explain how to find the rate if the distance = 25 mi and the time = 10 min. $d = r \times t$ **(pages 318–319)**

Problem Solving

Solve. Explain your method.

CHOOSE a strategy and a tool.

- Find a Pattern
- Make a Table
- Act It Out
- Write an Equation
- Make a Model
- Use a Graph

Paper/Pencil   Calculator   Hands-On   Mental Math

3. Morgan wants skates that cost $125.95. He has saved $31.50. He earns $4.50 per hour. Write an algebraic expression to represent his earnings. Find how many hours he must work to earn enough money.
(pages 296–297)

4. The Waltons went on a car trip. They averaged 60 miles per hour. How far did they travel in a day if they drove 4 hours? 6 hours? 10 hours? How long would it take them to drive 810 miles? **(pages 318–319)**

Cumulative Review

Solve the problem. Then write the letter of the correct answer.

1. Estimate. 621×38 (pages 50–53)

   A. 18,000   **B.** 21,000
   C. 24,000   **D.** 28,000

2. Write $\frac{40}{80}$ in simplest form. (pages 94–95)

   A. $\frac{10}{20}$ **B.** $\frac{5}{10}$

   C. $\frac{4}{8}$ **D.** $\frac{1}{2}$

3. What is the relationship of the second figure to the first? (pages 184–187)

   A. reflection   **B.** rotation
   C. translation   **D.** no relationship

4. How many vertices does the figure have? (pages 200–201)

   A. 5 vertices   **B.** 6 vertices
   C. 7 vertices   **D.** 9 vertices

5. Jan surveys 1 out of every 10 people about their favorite school subject. This is a __?__ sample. (pages 222–223)

   A. biased   **B.** misleading
   C. nonrandom   **D.** random

6. The youngest person at the concert was 12, and the oldest was 58. What was the range of ages? (pages 228–231)

   A. 12 years   **B.** 46 years
   C. 58 years   **D.** 70 years

7. What kind of graph is shown? (pages 240–241)

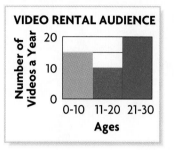

   A. circle graph   **B.** histogram
   C. line graph   **D.** stem-and-leaf plot

8. Find the median. 30, 25, 40, 35, 25 (pages 258–261)

   A. 25   **B.** 30
   C. 35   **D.** 40

9. Evaluate $x^2 + 5$ for $x = 5$. (pages 294–295)

   A. 5   **B.** 15
   C. 20   **D.** 30

10. Solve for y. $y - 6 = 3$ (pages 300–303)

   A. $y = 2$ **B.** $y = 3$
   C. $y = 9$ **D.** $y = 18$

11. How many quarters are in $6.00? (pages 314–315)

   A. 24 quarters   **B.** 60 quarters
   C. 120 quarters   **D.** 600 quarters

12. Use the formula $C = \frac{5}{9} \times (F - 32)$. Convert 60°F to degrees Celsius. Round to the nearest degree. (pages 316–317)

   A. 14°C   **B.** 16°C
   C. 50°C   **D.** 51°C

LOOK AHEAD

In this chapter you will solve problems that involve

- finding ratios and rates

- relating percents to fractions and decimals

- writing proportions

ART **LINK**

Many artists use mathematics such as ratios and proportions in their works. The table below names some artists and the types of art they are known for.

Andy Warhol (1928–1987)	pop art
Gutzon Borglum (1867–1941)	sculpture
Salvador Dalí (1904–1989)	surrealism
Claude Monet (1840–1926)	impressionism
Tom Lochray (1959–)	graphic art

- Make a time line showing the years of the artists' lives.

What Do You See?

The painting on these two pages is one of Jane Wooster Scott's works, titled *Memories Past*. Jane Wooster Scott is known for her American folk art. What is the ratio of kites to buildings in this painting?

Find a painting or picture of a group of people. Write statements containing fractions, decimals, percents, or ratios that compare members of the group. Use your statements to write a description of the group in the painting or picture.

PROJECT CHECKLIST

- ✓ Did you find a painting or picture of a group?
- ✓ Did you write statements comparing members of the group?
- ✓ Did you write a description of the group?

Ratios

Scale models are often described with ratios. For this model car, the ratio 1:64 was printed on the package. What do you think the ratio 1:64 means?

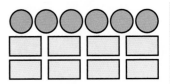

A **ratio** is a comparison of two numbers. You can write a ratio in three ways.

Write: 1 to 64 or 1:64 or $\frac{1}{64}$ ← first term
← second term

Read: one to sixty-four

You can write a ratio to compare two amounts—a part to a part, a part to the whole, or the whole to a part.

> **EXAMPLE 1** A recipe for 3 quarts of punch calls for 2 quarts of juice and 1 quart of ginger ale. Write the following ratios.
>
> **A.** quarts of juice to quarts of ginger ale → $\frac{2}{1}$ *part to part*
>
> **B.** quarts of juice to total quarts → $\frac{2}{3}$ *part to whole*
>
> **C.** total quarts to quarts of ginger ale → $\frac{3}{1}$ *whole to part*

Equivalent ratios are ratios that name the same comparisons. You can find equivalent ratios by multiplying or dividing both terms of a ratio by the same number.

> **EXAMPLE 2** Write three equivalent ratios to compare the number of circles with the number of rectangles.
>
> $\dfrac{\text{number of circles}}{\text{number of rectangles}}$ → $\dfrac{6}{8}$
>
> $\dfrac{6}{8} \to \dfrac{6 \div 2}{8 \div 2} = \dfrac{3}{4}$ *Divide first and second terms by a common factor.*
>
> $\dfrac{6}{8} \to \dfrac{6 \times 3}{8 \times 3} = \dfrac{18}{24}$ *Multiply terms by same number.*
>
> So, $\frac{6}{8}$, $\frac{3}{4}$, and $\frac{18}{24}$ are equivalent ratios that compare the number of circles to the number of rectangles.

GUIDED PRACTICE

Write the ratio in three ways.

1. one to seven　　**2.** two to three　　**3.** four to three　　**4.** eight to five

5. nine to ten　　**6.** eleven to fifteen　　**7.** twelve to one　　**8.** eighteen to seventeen

Write three equivalent ratios.

9. $\frac{1}{4}$　　**10.** 2:3　　**11.** 4:5　　**12.** $\frac{5}{3}$　　**13.** 9 to 12　　**14.** 18 to 6

INDEPENDENT PRACTICE

Write the ratio in three ways.

1. five to nine　　**2.** eight to fifteen　　**3.** ten to one　　**4.** seven to five

5. nine to twenty　　**6.** twelve to thirty-five　　**7.** twenty-one to ten　　**8.** forty to three

For Exercises 9–10, use the figure at the right.

9. Find the ratio of red sections to blue sections. Then write three equivalent ratios.

10. Find the ratio of blue sections to all the sections. Then write three equivalent ratios.

Find the missing term that makes the ratios equivalent.

11. $\frac{3}{4}, \frac{\blacksquare}{8}$　　**12.** 5 to 4, $\blacksquare$ to 12　　**13.** 20:$\blacksquare$, 8:10　　**14.** $\frac{9}{12}, \frac{45}{\blacksquare}$

15. $\frac{3}{8}, \frac{\blacksquare}{16}$　　**16.** $\frac{4}{5}, \frac{8}{\blacksquare}$　　**17.** 3 to 2, $\blacksquare$ to 8　　**18.** $\frac{\blacksquare}{2}, \frac{6}{12}$

19. 5 to $\blacksquare$, 15 to 21　　**20.** 6:2, 3:$\blacksquare$　　**21.** 15:4, $\blacksquare$: 8　　**22.** 5:$\blacksquare$, 20:24

Problem-Solving Applications

23. COMPARE Look at the photo below. Write as many ratios as you can.

25. DATA Out of 16 customers, 12 ordered juice. What is the ratio of juice drinkers to others?

26. DATA Thirty-two people wore white socks. The ratio of those who wore white socks to those who wore black socks is 1:2. How many people wore black socks?

27. ✏️ **WRITE ABOUT IT** In your own words, explain the meaning of ratio.

24. SPORTS Of the 10 members on a team, 2 are girls. What is the ratio of girls to all the members? to boys?

MORE PRACTICE Lesson 17.1, page H66

Rates

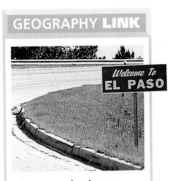
In some grocery stores, stickers on the shelves tell the prices per pound or per ounce. Consumers can use this information to see which size of a product is the better buy.

Pops Cereal $**3.28** 20 oz $0.16 per oz

Pops Cereal $**2.50** 14 oz $0.18 per oz

• Which box of cereal is the better buy? How can you tell?

A **rate** is a ratio that compares two quantities having different units of measure.

rate: $\dfrac{\text{price}}{\text{number of ounces}} \rightarrow \dfrac{\$3.28}{20\text{ oz}}$ $3.28 for 20 oz.

When the second term of a rate is 1, the rate is called a **unit rate**.

unit rate: $\dfrac{\$3.28}{20\text{ oz}} = \dfrac{\$3.28 \div 20}{20\text{ oz} \div 20} = \dfrac{\$0.16}{1\text{ oz}}$ $0.16 for 1 oz.

EXAMPLE The Howards are on a 300-mi road trip. They have traveled 180 mi in 3 hr. Find the unit rate of speed. At this rate, how many hours will it take them to go 300 miles?

$\dfrac{\text{miles} \rightarrow}{\text{hours} \rightarrow} \dfrac{180}{3}$ *Write a ratio to compare miles to hours.*

$\dfrac{180}{3} = \dfrac{180 \div 3}{3 \div 3} = \dfrac{60}{1} \leftarrow \text{unit rate}$ *Divide both terms by 3 to make the second term 1.*

So, the unit rate of speed is 60 mi per hour.

$\dfrac{60}{1} = \dfrac{60 \times 5}{1 \times 5} = \dfrac{300}{5} \dfrac{\leftarrow \text{miles}}{\leftarrow \text{hours}}$ *Write an equivalent ratio with 300 as the first term.*
Think: $60 \times ? = 300$

So, it will take the Howards 5 hr to travel 300 mi.

Talk About It

• **CRITICAL THINKING** What operation do you use to find a unit rate?

• What are the two quantities you compared in the Example?

• What if the Howards travel 150 mi in 3 hr? What would be their unit rate of speed?

GUIDED PRACTICE

Write a ratio in fraction form for each rate.

1. 150 points in 10 games
2. $64 in 16 hr
3. 30 teachers for 900 students
4. $2.59 for 4 pencils
5. $10 for 3 lb
6. 120 mi in 2 hr

Find the unit rate.

7. $5.00 for 2 T-shirts
8. 200 words in 4 min
9. $1.99 for 10 lb of potatoes
10. 32 lb in 16 weeks
11. 126 people in 3 buses
12. 150 mi on 10 gal of gasoline

INDEPENDENT PRACTICE

Write a ratio in fraction form for each rate.

1. 12 eggs for $1.10
2. $0.05 per page
3. 90 words in 2 min
4. 8 pages in 2 hr
5. 60 mi on 3 gal
6. $0.25 for 6
7. $1.89 for 3 pens
8. $1.20 for 6 goldfish
9. $15 for 5 tapes

Write the unit rate in fraction form.

10. $3.00 for 12
11. $1.80 for 12
12. 240 people per 2 mi^2
13. 100 mi in 2 hr
14. $1.50 for 10
15. 300 mi per 15 gal
16. 200 words per 5 min
17. $208 for 4 tires
18. $32 for 8 tickets

Problem-Solving Applications

19. A 10-min telephone call from Los Angeles to Miami costs $3.50. At this rate how much does it cost to talk for 1 min?

20. **SPORTS** It takes Deana 45 min to run 5 mi. How long does it take her to run 1 mi?

21. **CONSUMER MATH** A roll of film has 24 pictures and costs $8.88 to develop. How much does it cost to develop 1 picture?

22. Film costs $4.50 a roll. How much do 4 rolls cost?

23. **CONSUMER MATH** The post office sells 100 stamps for $32.00 a roll. The machine at the drugstore sells 5 stamps for $1.75. What is the unit rate at each place?

24. **CONSUMER MATH** Manuel bought 2 lb of peaches for $3.59. Jeff bought 4 lb of peaches for $6.15. Which was the better buy? Explain.

25. ⟐ **WRITE ABOUT IT** Explain what a unit rate is.

Percents

Did you know that about 25 out of 100 alpine skiers are between the ages of 7 and 17? Another way you can express this ratio is as a percent.

A **percent** is the ratio of a number to 100. *Percent* means "per hundred." The symbol for percent is %.

• The grid at the right shows the ratio of 25 shaded squares to 100 squares. What percent do you think 25 out of 100 represents?

You can express a ratio of a number to 100 as a percent. For example, the ratio 25:100 is 25%.

EXAMPLE 1 What percent of the squares in the figure below are shaded?

$$\frac{\text{shaded} \to 55}{\text{total} \to 100}$$ *Write the ratio of shaded squares to total squares.*

$$\frac{55}{100} = 55\%$$ *Write the ratio as a percent.*

So, 55% of the squares are shaded.

You can change a decimal to a percent by using place value. The decimal 0.55 is 55 hundredths. It can be expressed as $\frac{55}{100}$, which is 55%.

EXAMPLE 2 Write 0.8 as a percent.

$$0.8 = \frac{8}{10}$$ *Use place value to express the decimal as a ratio in fraction form.*

$$= \frac{8 \times 10}{10 \times 10} = \frac{80}{100}$$ *Write an equivalent ratio with a denominator of 100.*

$$= 80\%$$ *Write the ratio as a percent.*

ANOTHER METHOD You can also multiply any decimal by 100 to change it to a percent.

$$0.8 \text{ of } 100 = 0.8 \times 100$$ *Multiply by 100.*

$$= 80$$

$$0.8 = 80\%$$

• CRITICAL THINKING What pattern do you see when you write a decimal as a percent?

To change a ratio that does not have 100 as its second term to a percent, think about equivalent ratios. The ratio of shaded squares to total squares in Figure 1 below is 1:4. The ratio for Figure 2 is 25:100. The shaded areas in both figures are equivalent, and the ratios $\frac{1}{4}$ and $\frac{25}{100}$ are equivalent.

Figure 1

Figure 2

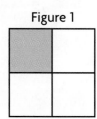

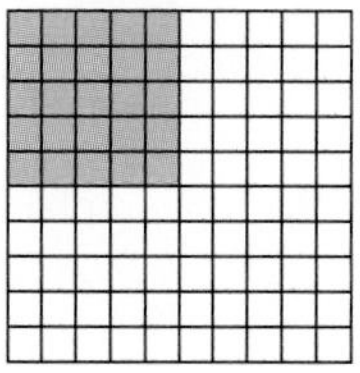

- What percent of Figure 1 is shaded? What percent of Figure 2 is shaded?

- Suppose the ratio of shaded to total squares in Figure 1 were $\frac{2}{4}$. What percent would this be?

You can use equivalent ratios to write a ratio as a percent.

EXAMPLE 3 Sammy has taken his dog to 3 out of 5 obedience classes. What percent of the obedience classes have Sammy and his dog attended?

$$\frac{3}{5} = \frac{3 \times 20}{5 \times 20} = \frac{60}{100} \qquad \textit{Write an equivalent ratio with a denominator of 100.}$$

$$\frac{60}{100} = 60\% \qquad \textit{Write the ratio as a percent.}$$

So, Sammy has taken his dog to 60% of the classes.

ANOTHER METHOD You can also use division.

$\frac{1}{3} \rightarrow 3\overline{)1.00}$ *Divide first term by second term.*

$0.33\frac{1}{3} = 33\frac{1}{3}\%$ *Change decimal to percent. Multiply by 100 by moving decimal point two places to the right.*

REMEMBER:
....
Multiplying a number by 10 is the same as moving the decimal point one place to the right.

$12 \times 10 = 120$
$120 \times 10 = 1,200$
$1,200 \times 10 = 12,000$

See page H12.

GUIDED PRACTICE

Write a percent, a ratio in fraction form, and a decimal to describe the shaded part.

1.

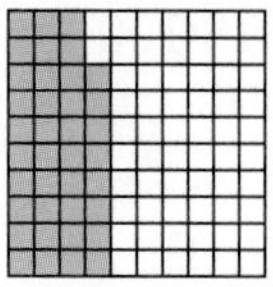

2.

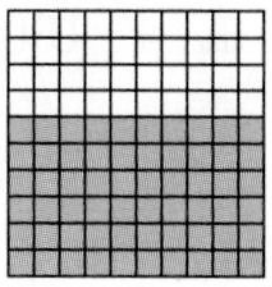

3.

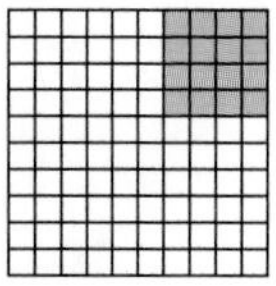

Percents as Ratios and Decimals

You can write a percent as a ratio.

> **EXAMPLE 4** Write 60% as a ratio in fraction form.
>
> $60\% = \dfrac{60}{100}$ *Write percent as ratio with second term of 100.*
>
> $\dfrac{60}{100} = \dfrac{6}{10} = \dfrac{3}{5}$ *Write ratio in simplest form.*
>
> So, 60% written as a ratio is $\dfrac{3}{5}$.
>
> • Write an equivalent ratio for 12%.

You can write a percent as an equivalent decimal.

> **EXAMPLE 5** Suppose a shoe store marks up its prices 333%. What decimal can you write for the percent of markup?
>
> $333\% = \dfrac{333}{100}$ *Write percent as ratio with second term of 100.*
>
> $= 3.33$ *Write ratio as decimal.*
>
> So, the markup written as a decimal is 3.33.

ANOTHER METHOD You can also use division. What decimal can you write for 70%?

$70\% = 0.70$ *Divide by 100 by moving decimal point two places to left.*

So, 70% written as a decimal is 0.70.

The key sequence below shows how to change a percent to a decimal or a ratio by using the *TI-Explorer Plus*.

70   2nd   %   `0.7`   F↔D   `7/10`

• What key would you press to change the ratio back to a decimal?

REMEMBER:

Dividing a number by 10 is the same as moving the decimal point one place to the left.

$120 \div 10 = 12$

$12 \div 10 = 1.2$

$1.2 \div 10 = 0.12$

See page H12.

INDEPENDENT PRACTICE

Write the ratio as a percent.

1. $\dfrac{35}{100}$ **2.** $\dfrac{28}{100}$ **3.** 84:100   **4.** $\dfrac{32}{100}$ **5.** 40:100   **6.** 15:100

Write the decimal as a percent.

7. 0.1   **8.** 0.75   **9.** 0.38   **10.** 0.15   **11.** 0.8   **12.** 0.91

Calculator Activities, page H35

Write the ratio as a percent.

13. $\frac{4}{5}$ **14.** $\frac{11}{100}$ **15.** $\frac{6}{25}$ **16.** $\frac{4}{50}$ **17.** $\frac{2}{3}$ **18.** $\frac{1}{8}$

Write the percent as a decimal and as a ratio in simplest form.

19. 20% **20.** 45% **21.** 90% **22.** 12% **23.** 48% **24.** 64%

25. 17% **26.** 22% **27.** 9% **28.** 20.5% **29.** 12.5% **30.** 7.5%

Complete.

31. $75\% = \frac{75}{\blacksquare} = \frac{3}{4}$

32. $40\% = \frac{40}{100} = \frac{2}{\blacksquare}$

33. $0.15 = \frac{3}{20} = \blacksquare\%$

34. $\frac{3}{5} = \frac{\blacksquare}{100} = \blacksquare\%$

35. $\blacksquare\% = \frac{29}{100} = 0.29$

36. $\frac{623}{100} = 6.23 = \blacksquare\%$

Complete by using $<$, $>$, or $=$. If $<$ or $>$, tell why.

37. $\frac{2}{50}$ ● 5% **38.** 45% ● $\frac{9}{20}$ **39.** 0.0018 ● 1.8% **40.** $\frac{10}{16}$ ● 0.625

Problem-Solving Applications

41. Out of 100 students, 22 want cheese pizza, 42 want pepperoni pizza, and the rest want a supreme pizza. What percent of the students want a supreme pizza?

42. **DATA** A survey of customers at an arcade indicates that $\frac{3}{5}$ are in the age group of 13–17. What percent of the customers are in this age group?

43. **CRITICAL THINKING** There are 35 students in Ms. Rivera's class and $\frac{1}{5}$ of them are in the band. What percent are in the band? How can you check your answer?

44. **WRITE ABOUT IT** How can you write a percent as a decimal? as a fraction?

Mixed Review and Test Prep

Name a fraction that tells what part of the figure is shaded.

45.   **46.**   **47.**   **48.**

Solve the equation.

49. $4n = 20$ **50.** $2y = 60$ **51.** $\frac{81}{t} = 9$ **52.** $\frac{a}{12} = 4$

53. **CHOOSE A STRATEGY** This week Matt made 5 more than twice the number of cookies he made last week. Last week he made 20 cookies. How many did he make this week?

A 25 **B** 40 **C** 45 **D** 100

54. **TIME** The talent show has 20 students. Each student takes 3 minutes to set up and 6 minutes to perform. How long does the show last?

F 3 hr **G** 2 hr **H** $1\frac{1}{2}$ hr **J** 1 hr

MORE PRACTICE Lesson 17.3, page H67

339

Technology Link

In *Mighty Math Astro Algebra,* sort equivalent ratios, decimals, and percents in the mission *Fuel for the Glop.* Use Grow Slide Level Red D.

Percents and the Whole

What You'll Learn
How to express parts of a whole area as percents

Why Learn This?
To describe parts of pizzas, parts of a gameboard, and other parts

Many home computer programs can be used to make greeting cards. Even though a program allows you to design the outside and the inside of the card separately, the entire card is printed on one side of a sheet of paper. Then you fold the paper as in the Activity here.

Stacey makes her own greeting cards for birthdays. On the back of each card, she writes her trademark, *Personal Designs*.

ACTIVITY

• Fold a piece of notebook paper in half two times.

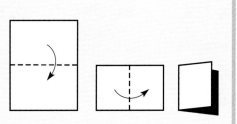

• Write *Happy Birthday* or another greeting on the front of the card.

• Write a message on the inside of the card, at the right.

• Mark the back of the card with your own design or logo.

• Unfold the card. What percent does each of the four parts of the card represent? What percent of the whole are the parts you used to write on or create a design on?

It's easy to see ratios or percents on 10 × 10 grids and on figures with rectangular parts. But some figures are not that simple.

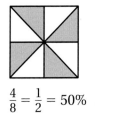

$$\frac{4}{8} = \frac{1}{2} = 50\%$$

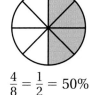

$$\frac{4}{8} = \frac{1}{2} = 50\%$$

$$\frac{4}{8} = \frac{1}{2} = 50\%$$

EXAMPLE On the gameboard at the right, the blue squares give a player an additional turn. What percent of the squares are blue?

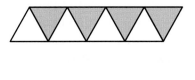

$$\frac{\text{number of blue squares}}{\text{total number of squares}} = \frac{8}{40} \qquad \textit{Write a ratio in fraction form.}$$

$$\frac{8}{40} \rightarrow 40\overline{)8.00}^{\,0.20} = 20\% \qquad \textit{Write the ratio as a percent.}$$

So, 20% of the squares are blue.

• What percent of the squares on the gameboard are not blue?

GUIDED PRACTICE

Tell what percent of the figure is shaded.

1.

2.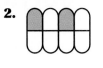

3.

4.

INDEPENDENT PRACTICE

Tell what percent of the figure is shaded.

1.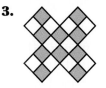

2.

3.

4.

5.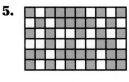

6.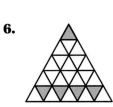

7.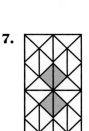

8.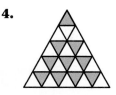

For Exercises 9–10, use the figure at the right.

9. One pizza was divided into 10 slices. Of these slices, 2 had only cheese, 4 had mushrooms, and 4 had pepperoni. What percent of the pizza had only cheese? What percent had mushrooms?

10. Gunther took 3 slices of pepperoni pizza. What percent of the pizza was left?

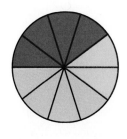

Problem-Solving Applications

11. HOBBIES Janice was playing a game that had 60 square spaces. She landed on 12 spaces and needed to land on 18 more to win. What percent of all the spaces did she need to land on to win?

12. When Janice had landed on 21 of the 60 spaces, she decided to take a break. What percent of all the spaces had she landed on?

13. PERCENT Draw a figure that has equal parts. Then shade some of the parts. Determine what percent of the figure is shaded.

14. ✏️ **WRITE ABOUT IT** Explain how to find what percent of a figure is shaded.

LAB ACTIVITY

What You'll Explore
How to find equivalent ratios to form a proportion

What You'll Need
two-color counters

Exploring Proportions

A **proportion** is a number sentence, or an equation, that states that two ratios are equivalent. For example, $\frac{1}{3} = \frac{2}{6}$ is a proportion, since $\frac{1}{3}$ and $\frac{2}{6}$ are equivalent ratios.

You can use counters to model proportions.

ACTIVITY 1

Explore

- Make the model shown at the right. Write the ratio of red counters to yellow counters.

- Separate the red counters into two equal groups. Divide the yellow counters equally between the two groups. Write the ratio of red to yellow in each group.

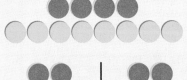

- Separate the red counters into four equal groups. Divide the yellow counters equally among the four groups. Write the ratio of red to yellow in each group.

Think and Discuss

- How are the ratios you modeled related?

- You can use the ratios $\frac{4}{8}$ and $\frac{2}{4}$ to write the proportion $\frac{4}{8} = \frac{2}{4}$. What other proportions can you write with the ratios you modeled?

Try This

- Model the ratio of 8 red counters and 12 yellow counters. Then model as many equivalent ratios as you can. What proportions can you write?

  - Double the numbers of red and yellow counters. Use this ratio and an equivalent ratio to write a proportion.

You can use models to determine whether two ratios form a proportion.

ACTIVITY 2

Explore

- Model these ratios: $\frac{2}{3}$, $\frac{8}{12}$, and $\frac{9}{16}$.

- Place the models for $\frac{2}{3}$ and $\frac{8}{12}$ side by side. Separate $\frac{8}{12}$ into groups of $\frac{2}{3}$ if you can.

- Place the models for $\frac{2}{3}$ and $\frac{9}{16}$ side by side. Separate $\frac{9}{16}$ into groups of $\frac{2}{3}$ if you can.

Think and Discuss

- Can $\frac{2}{3}$ and $\frac{8}{12}$ be used to write a proportion? Explain.
- Can $\frac{2}{3}$ and $\frac{9}{16}$ be used to write a proportion? Explain.

Modeling is not the only way to determine whether two ratios form a proportion. Another way is to find the cross products. Equivalent ratios have equal cross products.

$$\frac{2}{3} \longleftrightarrow = \longleftrightarrow \frac{8}{12}$$

$2 \times 12 \overset{?}{=} 8 \times 3 \leftarrow$ cross products
$24 = 24$

The ratios form a proportion.

$$\frac{2}{3} \longleftrightarrow = \longleftrightarrow \frac{9}{16}$$

$2 \times 16 \overset{?}{=} 9 \times 3 \leftarrow$ cross products
$32 \neq 27$

The ratios do not form a proportion.

Try This

- Use models or cross products to determine whether the ratios form a proportion.

$\frac{1}{3}$ and $\frac{4}{12}$ $\frac{3}{4}$ and $\frac{6}{9}$ $\frac{2}{4}$ and $\frac{3}{6}$

Technology Link

You can practice making equivalent ratios that form a proportion by using E-Lab, Activity 17. Available on CD-ROM and on the Internet at **www.hbschool.com/elab**

PROBLEM-SOLVING STRATEGY

Write a Proportion

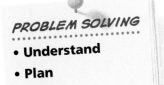

PROBLEM SOLVING
· **Understand**
· **Plan**
· **Solve**
· **Look Back**

Some problems can be solved
by finding equivalent ratios.
Remember that a proportion is
an equation that states that
two ratios are equivalent.

Renata went to a science
museum. One of the exhibits
shows that a person who
weighs 60 lb on Earth would
weigh 10 lb on the moon.
Renata weighs 75 lb. Predict
what she would weigh on
the moon.

UNDERSTAND   What are you asked to find?

What facts are given?

PLAN   What strategy will you use?

You can *write a proportion.* Write two ratios that compare the
weight on Earth and the weight on the moon. Since you don't
know Renata's weight on the moon, use *n* to represent it.

SOLVE   Write a proportion.

$$\frac{\text{weight on Earth}}{\text{weight on moon}} \rightarrow \frac{60}{10} = \frac{75}{n} \leftarrow \frac{\text{Renata's weight on Earth}}{\text{Renata's weight on moon}}$$

$$\frac{60}{10} = \frac{75}{n} \qquad \textit{Find the cross products.}$$

$$60 \times n = 75 \times 10$$

$$60n = 750 \qquad \textit{Solve the equation.}$$

$$\frac{60n}{60} = \frac{750}{60}$$

$$n = 12.5$$

So, Renata would weigh 12.5 lb on the moon.

LOOK BACK   How can you check your answer?

What if . . . Renata weighed 84 lb? What would her weight be on
the moon?

PRACTICE

Write a proportion to solve.

1. A machine fills 4 bottles every 6 sec. How long does it take to fill 12 bottles?

2. Five bags of potatoes weigh 45 lb. How much do 2 bags weigh?

3. Paul's car uses about 10 gal of gasoline on a 180-mi trip. Predict how much gasoline Paul will need to travel 450 mi.

4. A map uses a scale of 2 in. for every 15 mi. If the map shows a distance of 32 in., what is the actual distance?

MIXED APPLICATIONS

Solve.

CHOOSE a strategy and a tool.

- Find a Pattern
- Work Backward
- Write a Proportion
- Write an Equation
- Use a Formula
- Guess and Check

Paper/Pencil   Calculator   Hands-On   Mental Math

5. Twenty quarters and nickels are worth a total of $1.40. How many of each coin are there?

6. Juan has a ratio of 3 fish to 5 plants in his aquarium. He has 33 fish. How many plants does he have?

7. At 5:00 P.M. the temperature was 60°F. It started to fall at a steady rate. At 7:00 P.M. it was 45°F. If the pattern continues, what will the temperature be at 9:00 P.M.?

8. Patty is making 12 invitations. Each invitation is 6 in. long and 4 in. wide. How many square inches of paper does she need?

9. Tom has an average of 80 on four tests. What must he get on his fifth test to have an average of 82?

10. Ten tangerines cost $1.60. How much do 8 tangerines cost?

11. A dairy company paid $42,000 for 35 acres of farmland. The company wants to buy 20 acres more at the same price per acre. How much will 20 acres cost?

12. ▭ **WRITE ABOUT IT** How does writing a proportion to represent a problem help you solve it?

Mixed Review and Test Prep

Write a percent.

13. 0.09   14. $\frac{29}{100}$ 15. 2:5   16. $\frac{9}{20}$ 17. $\frac{4}{50}$ 18. 0.8

Change to degrees Fahrenheit or Celsius. Use the formulas on page 316.

19. 5°C   20. 10°C   21. 20°C   22. 32°F   23. 302°F   24. 212°F

25. **HOBBIES** Jamie and Lisa made a total of 54 pot holders to sell. Jamie made twice as many as Lisa. How many did Lisa make?

   A 18   **B** 24   **C** 36   **D** 42

26. **SPORTS** The Tigers lost 15 out of 20 games. What percent of the games did they lose?

   F 25%   **G** 50%   **H** 75%   **J** 90%

1. **VOCABULARY** A comparison of two numbers is a(n) __?__ .
   (page 332)

Write the ratio in three ways. (pages 332–333)

2. two to four
3. ten to one
4. three to ten
5. five to three

Write three equivalent ratios. (pages 332–333)

6. 5:7
7. $\frac{6}{9}$
8. 33 to 11
9. $\frac{5}{10}$

10. **VOCABULARY** When the second term of a rate is 1, the rate is called a(n) __?__ . (page 334)

Write the rate and the unit rate in fraction form. (pages 334–335)

11. $0.75 for 3
12. 210 km in 3 hr
13. $4.00 for 10
14. 20 mi in 5 hr

15. **VOCABULARY** A ratio of one number to 100 is called a(n) __?__ . (page 336)

Write the ratio or decimal as a percent. (pages 336–339)

16. 0.38
17. $\frac{2}{8}$
18. 0.99
19. $\frac{5}{25}$

Write the percent as a decimal and as a ratio. (pages 336–339)

20. 20%
21. 18%
22. 35%
23. 10%
24. 80%
25. 45%
26. 75%
27. 90%

Tell what percent is shaded. (pages 340–341)

28.
29.
30.
31.

32. **VOCABULARY** A number sentence, or an equation, that states that two ratios are equivalent is a(n) __?__ . (page 342)

Solve. (pages 344–345)

33. Beach towels are on sale at 2 for $18. Rich buys 5 of them. How much does he pay?

34. Oranges are selling for $3.00 a dozen. Find the cost of 2 oranges.

1. A regular octagon has rotational symmetry. What is the angle measure of each turn?

   A 30°
   B 45°
   C 60°
   D 90°

2. Which of the following is best suited for display in a bar graph?

   F heights of the tallest buildings in a city
   G frequency of cars stopping at a tollbooth
   H changes in a person's weight over a year
   J part of a day a person spends reading

3. The owner of a music shop made a line graph to show the number of CDs sold during a 4-week period. What trend does the graph show?

   MUSIC SHOP SALES

   A sales are increasing
   B sales are even
   C sales are decreasing
   D no trend is shown

4. Danny can eat pancakes, cereal, or eggs and drink milk or orange juice for breakfast. How many choices does Danny have?

   F 4   **G** 5
   H 6   **J** 7

5. If an insect moves at the rate of 7 feet per second, how long does it take to move 21 feet?

   A $\frac{1}{3}$ sec
   B 3 sec
   C 7 sec
   D 21 sec

6. A survey showed 1 out of every 25 students at Phillip High takes Latin. Which of the following decimals represents this ratio?

   F 0.04   **G** 0.12
   H 0.4   **J** 0.54

7. Daria correctly answered 17 out of 20 questions on a math test. What percent did she answer correctly?

   A 70%   **B** 80%
   C 85%   **D** 92%

8. If 6 notebooks cost $2.50, how much will 9 notebooks cost at the same rate?

   F $2.75
   G $2.95
   H $3.25
   J $3.75
   K $4.00

PERCENT AND CHANGE

LOOK AHEAD

In this chapter you will solve problems that involve

- finding a percent of a number

- making and reading circle graphs that use percents

- finding the amount of a discount

- finding the percent one number is of another

SCIENCE **LINK**

Scientists use a tagging procedure to estimate animal populations. The table below shows types of animals whose populations are decreasing.

ANIMAL SPECIES		
Type	Number of Endangered in U.S.	Total Number of Endangered
Mammals	55	307
Birds	76	254
Reptiles	14	79
Amphibians	7	15
Fish	65	76

- What ratio can you write for the number of endangered reptiles in the U.S. to the total number of endangered reptiles?

- What percent of the total number of endangered species are in the U.S.?

Making a Fact Sheet

FACT SHEET

Here are some facts about the black rhinoceros:

Black rhino populations are being reduced at an alarming rate. In 1980, there were fewer than 15,000 animals—a reduction of over 50,000 animals in just 10 years. Today experts estimate the wild population at less than 2,400. There are approximately 200 black rhinos in zoological institutions.

- Today's black rhino population is about _____% of the 1980 rhino population.

- The 1980 black rhino population was about _____% of the 1970 rhino population.

- About _____% of the remaining rhinos are in zoos.

Some facts about the endangered black rhinoceros are shown at the left. Select an endangered plant or animal. Find out how the population of your plant or animal has changed over the years. Make a fact sheet that includes population changes expressed as percents. Share your fact sheet with the class.

PROJECT CHECKLIST

✓ Did you look up information about an endangered plant or animal?

✓ Did you make a fact sheet using percents?

✓ Did you share the facts with the class?

PROBLEM-SOLVING STRATEGY

Acting It Out to Find a Percent of a Number

PROBLEM SOLVING
....
• **Understand**
• **Plan**
• **Solve**
• **Look Back**

In a recent study, it was found that, of the 350 million cans of chicken soup sold each year, 60% are bought in the period from October through March. How many cans of soup are bought during these months?

UNDERSTAND   What are you asked to find?

What facts are given?

PLAN   What strategy will you use?

You can *act it out*. Use index cards to model equal parts of 100%. This will help you visualize the percent of the number.

SOLVE   Let each card represent 10%.

Put down 10 cards to represent 100%, or 350 million cans. Each 10% represents 35 million cans.

←				350					→
10%	10%	10%	10%	10%	10%	10%	10%	10%	10%
35	35	35	35	35	35	35	35	35	35

Now separate the whole into 60% and 40%. Since each card represents 35 million, the 6 cards that make up 60% represent 6×35 million, or 210 million.

10%	10%	10%	10%	10%	10%
35	35	35	35	35	35

10%	10%	10%	10%

So, 210 million cans of soup are sold in the period from October through March.

LOOK BACK   What other strategy could you use to solve the problem?

What if . . . 20% of the chicken soup is sold in October? How many cans is this?

PRACTICE

Solve the problem by acting it out.

1. The Shop for Food supermarket sold 1,000 bags of potatoes in May. In June it sold 60% of that amount. How many bags of potatoes did it sell in June?

2. The Arena sold 20 million tickets for different events during one year. Of these tickets, 80% were for sporting events. How many tickets were for sporting events?

3. A late-night talk-show host has done 3,000 shows, 20% of which were live, not taped. How many were live?

4. Miss Smith paid an interest rate of 10% last year on her $90,000 home loan. How much did she pay in interest for the year?

MIXED APPLICATIONS

Solve.

CHOOSE a strategy and a tool.
- Draw a Diagram
- Guess and Check
- Write an Equation
- Work Backward
- Act It Out
- Make a Table

Paper/Pencil   Calculator   Hands-On   Mental Math

5. Ms. Seville gets a 10% bonus for every $100,000 worth of merchandise she sells. How much is her bonus for $100,000 worth of merchandise?

6. A play was performed 2,880 times on Broadway in 8 years. On average, how many shows were performed each month?

7. Tickets to an opera rehearsal cost $8 for seats in the front and $5 for seats in the back. Ms. Henderson bought 17 tickets for $109. How many of each kind did she buy?

8. Victor is in a rowboat 6 mi from shore. He rows 4 mi an hour. How long will it take Victor to reach the shore if he rests for 30 min at the end of an hour?

9. Keith went shopping at the mall. He spent $30 on a part for his lawn mower and $6 each for two tools to fix his lawn mower. He has $17 left. How much money did he take to the mall?

10. The Laysan duck lives on the Laysan Island. Of the 600 ducks left, some live in captivity. Suppose 35% of the remaining ducks live in captivity. How many live in captivity?

11. Carlos, Bob, and Shelley live in the same apartment building. Carlos lives on the floor below Bob. Shelley lives on the first floor. Bob lives on the floor above Shelley. What floor does Carlos live on?

12. ✏️ **WRITE ABOUT IT** Write a problem that can be solved by using the strategy *act it out*. Then explain the steps you would use to solve the problem.

Percent of a Number

What You'll Learn
How to find a percent of a number

Why Learn This?
To understand surveys, free-throw percents, and other percents

REMEMBER:

To change a percent to a ratio in fraction form, write the percent over 100. Then simplify. **See page H29.**

$$12\% = \frac{12}{100}$$
$$= \frac{3}{25}$$

CULTURAL LINK

Pizza originated near Naples, Italy. Today pepperoni is the favorite topping of $\frac{2}{3}$ of American pizza eaters. Out of 33 students, how many would you expect to want pepperoni?

Derek surveyed 40 of his schoolmates about the food in the cafeteria. He found that 75% of them prefer pizza. How many of the 40 students prefer pizza, or what is 75% of 40?

ACTIVITY

WHAT YOU'LL NEED: two-color counters or colored paper squares

• To find 75% of 40, use 40 yellow counters.

• Since 75% equals $\frac{3}{4}$, separate the counters into four groups. Change the color of three groups to red.

• Count the number of red counters. How many of the 40 students prefer pizza?

• Use two-color counters to find 75% of 12.

Talk About It

• Why did you separate the counters into four groups in the first part of the activity?

• How many groups would you separate the counters into if 30% of the classmates preferred pizza? Explain.

Sometimes it is difficult to use counters to find the percent of a number. There are many ways to find the percent of a number. One way is to change the percent to a fraction and multiply.

EXAMPLE 1 Jerry is the goalie for his ice hockey team. He was injured during one of the games. At the end of the season he had played in only 40% of the 75 games. In how many games did he play? Find 40% of 75.

$40\% = \frac{40}{100}$ *Write the percent as a ratio in fraction form.*

$= \frac{2}{5}$ *Write the ratio in simplest form.*

$\frac{2}{5} \times \frac{75}{1} = \frac{150}{5} = 30$ *Multiply the ratio by the number.*

So, Jerry played in 30 games.

• What is 20% of 90?

SPORTS LINK

In basketball, a player who shoots 33% is a rather poor shooter. In baseball, however, a batter who hits 0.333 (33.3%) usually is considered a star. If a batter hits 0.250, what percent of the time does he or she get a hit?

Sometimes it is easier to change the percent to a decimal to find the percent of a number.

EXAMPLE 2 During the basketball season, Regina shot 175 free throws. She made 88% of them. How many free throws did she make?

To solve, find 88% of 175.

$88\% = 0.88$ *Change the percent to a decimal.*

175×0.88 *Multiply the number by the decimal.*

$$
\begin{array}{r}
175 \\
\times\, 0.88 \\
\hline
1400 \\
+14000 \\
\hline
154.00
\end{array}
$$

So, Regina made 154 free throws.

Talk About It

• What percent can you write to describe the number of free throws Regina missed?

• What is 68% of 275?

• **CRITICAL THINKING** Without finding the percents, tell how the answer for 60% of 80 compares to the answer for 30% of 80.

Use two-color counters or paper squares to find the percent of the number.

1. 25% of 4   **2.** 30% of 20   **3.** 75% of 36   **4.** 80% of 30

Write the simplest form of the ratio you would use to find the percent of the number.

5. 20% of 8   **6.** 50% of 80   **7.** 75% of 48   **8.** 15% of 20

Write the decimal you would use to find the percent of the number.

9. 52% of 6   **10.** 40% of 90   **11.** 28% of 75   **12.** 76% of 25

Use any method to find 25% of the number.

13. 60   **14.** 88   **15.** 280   **16.** 156

_____

The percent of a number is used in many everyday situations. When you make a purchase, the amount of sales tax you pay is determined by finding the percent of a number.

> **EXAMPLE 3** Diane bought an electronic organizer that cost $25. The sales tax was 6%. How much did she pay in sales tax?
>
> 6% = 0.06   *Change percent to decimal.*
>
> $25 × 0.06   *Multiply price by the decimal.*
>
> $$\begin{array}{r} 25 \\ \times\, 0.06 \\ \hline 1.50 \end{array}$$
>
> So, she paid $1.50 in sales tax.

You can use a calculator to find the percent of a number.

> **EXAMPLE 4** Use a calculator to find 150% of 92.
>
> Use the key sequence shown below.
>
> 92 ⊗ 150 [2nd] [%] [=] ⟦ 138 ⟧
>
> • What did the percent key on the calculator do? Explain why this is correct.
>
> • Why is the answer, 138, greater than the original number, 92?

Calculator Activities, page H36

INDEPENDENT PRACTICE

Use a ratio in simplest form to find the percent of the number.

1. 10% of 6 **2.** 25% of 80 **3.** 45% of 10 **4.** 80% of 55 **5.** 90% of 120

Use a decimal to find the percent of the number.

6. 12% of 9 **7.** 28% of 20 **8.** 32% of 60 **9.** 71% of 84 **10.** 95% of 18

Use the method of your choice to find the percent of the number.

11. 15% of 20 **12.** 21% of 88 **13.** 35% of 92 **14.** 40% of 106 **15.** 2% of 12

16. 51% of 30 **17.** 99% of 99 **18.** 82% of 150 **19.** 16% of 200 **20.** 63% of 85

21. 5.5% of 70 **22.** 125% of 50 **23.** 200% of 100 **24.** 150% of 38 **25.** 300% of 75

Find the sales tax. Round to the nearest cent when necessary.

26. price: $20
tax rate: 7%

27. price: $19.99
tax rate: 6.5%

28. price: $8.19
tax rate: 9%

29. price: $125.00
tax rate: 5%

Problem-Solving Applications

30. **CONSUMER MATH** Richard wants to buy a radio that costs $32. He has $34. The sales tax rate is 7%. Does he have enough money for the radio and sales tax?

31. Avi wants to buy a bike that costs $124. The store manager will hold the bike for Avi if he puts down a 20% deposit. How much is the deposit?

32. **SCIENCE** The surface area of the Earth is about 200,000,000 mi². If 70% of the surface is water, about how many square miles are water?

33. ✏️ **WRITE ABOUT IT** Explain how you can find the percent of a number.

Mixed Review and Test Prep

Find the product.

34. $\frac{1}{4} \times 360$ **35.** $\frac{1}{10} \times 360$ **36.** $\frac{3}{4} \times 360$ **37.** $\frac{1}{8} \times 360$ **38.** $\frac{3}{10} \times 360$

Write the percent as a ratio in simplest form and as a decimal.

39. 10% **40.** 25% **41.** 45% **42.** 84% **43.** 32%

44. **COMPARE** Eileen has $\frac{5}{8}$ cup of fruit punch. June has $\frac{2}{3}$ cup. Which conclusion is reasonable?

A June has more punch than Eileen.
B Both have the same amount of punch.
C Eileen has more punch than June.
D Eileen has twice as much punch.

45. **GEOMETRY** Two perpendicular lines form an angle which is a(n) __?__.

F Straight angle
G Obtuse angle
H Right angle
J Acute angle

Circle Graphs

Using what you have learned about writing percents as decimals and ratios can help you make a circle graph showing percents. Remember that a circle graph shows the whole, or 100% of the data.

FRENCH FRIES SURVEY		
Preference	Frequency	Percent
Thin	78	65%
Thick	30	25%
Do not like	12	10%

REMEMBER:

To make a **circle graph**:
• Find the size of the angle for each category.
• Use a compass to draw a circle.
• Draw a radius.
• Use a protractor to mark each angle.
• Label each section.

See page 246.

EXAMPLE 1 Use the data above to make a circle graph.

$65\% = 0.65, 0.65 \times 360° = 234°$

$25\% = 0.25, 0.25 \times 360° = 90°$

$10\% = 0.10, 0.10 \times 360° = 36°$

Write the percents as decimals or ratios. Find the degree measure of each section.

Make the graph.

FRENCH FRIES

Thin 65%

Thick 25%

Do not like 10%

Look at the circle graph at the right. You can find the amount Dave spent on each category by finding the percent of $300.

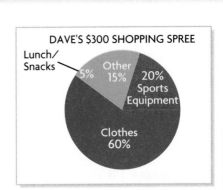

DAVE'S $300 SHOPPING SPREE

Lunch/Snacks 5%

Other 15%

20% Sports Equipment

Clothes 60%

EXAMPLE 2 How much did Dave spend on sports equipment?

$20\% = \dfrac{20}{100} = \dfrac{1}{5}$

$\dfrac{1}{5} \times 300 = 60$

Write the percent as a ratio in fraction form. Then multiply the ratio by the total amount.

So, Dave spent $60 on sports equipment.

GUIDED PRACTICE

For Exercises 1–2, use the data in the table at the right.

1. Find the angles you would use to make a circle graph.

2. Make a circle graph.

For Exercises 3–5, use the circle graph at the right.

3. If 200 people were surveyed, how many people chose tennis as their favorite sport?

4. If 200 people were surveyed, how many people chose baseball as their favorite sport?

5. What percent of the people didn't pick golf as their favorite sport?

CARLA'S VEGETABLE GARDEN	
Vegetable	**Percent**
Tomatoes	45%
Celery	20%
Carrots	35%

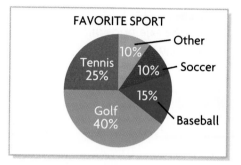

FAVORITE SPORT

INDEPENDENT PRACTICE

For Exercises 1–2, use the data in the table at the right.

1. Find the angles you would use to make a circle graph.

2. Make a circle graph.

For Exercises 3–5, use the circle graph at the right.

3. If 500 people were surveyed, how many use the stove the most?

4. If 1,000 people were surveyed, how many use the TV the most?

5. If 1,000 people were surveyed, how many use the stereo the most?

BUDGET	
Item	**Percent**
Rent	30%
Food	25%
Clothes	12.5%
Other	32.5%

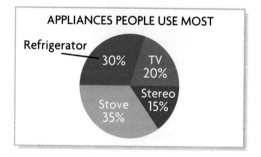

APPLIANCES PEOPLE USE MOST

Problem-Solving Applications

6. RECREATION In a survey, people were asked about their favorite type of vacation. Of 100 people, 15% chose camping, 20% chose ocean cruise, 25% chose big city, and 40% chose other. Use the results to make a circle graph.

8. DATA Make a table with data that can be used to make a circle graph.

7. Ned had $500. He spent 50% on clothes and 30% on stereo equipment. He put the rest in his stock fund. Make a circle graph to show how Ned spent his $500. How much did he put in his stock fund?

9. ✏️ **WRITE ABOUT IT** Explain how you find the angle for each percent in a circle graph.

Discount

SNEAKER SALE!

regular price
$79.70
now 30% off!

What You'll Learn

How to find the amount of discount and the sale price of an item if given the price and the discount rate

Why Learn This?

To find the sale price of an item you want to buy

VOCABULARY

discount

CONSUMER LINK

A *clearance* might mean that an additional discount will be taken from an already discounted price. If a sign says "Clearance Sale! Take another 20% off our already-low prices," how much will you pay for a jacket that was marked down 35% from an original price of $89.95?

Darrin needs new sneakers for basketball season. He noticed this advertisement in the newspaper. How much will Darrin pay for the sneakers with a discount of 30%?

To find the sale price of the sneakers, you must find the amount of the discount. To find the amount of the **discount**, multiply the regular price by the discount rate.

discount = regular price × discount rate
$$= \$79.70 \times 30\%$$
$$= \$79.70 \times 0.30 \quad \text{\textit{Change the percent}}$$
$$= \$23.91 \quad\quad\quad\quad \text{\textit{to a decimal.}}$$

So, the amount of discount on the sneakers is $23.91.

To find the sale price, subtract the discount from the regular price.

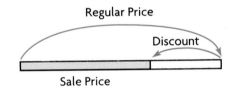

Regular Price

Discount

Sale Price

sale price = regular price − discount
$$= \$79.70 - \$23.91$$
$$= \$55.79$$

So, Darrin will pay $55.79 for the sneakers on sale.

• Look at the diagram above. How is the sale price related to the discount and the regular price?

EXAMPLE 1 A computer game regularly sells for $32. Today it is on sale at a discount of 25%. Find the amount of discount and the sale price of the computer game.

Discount = $32 × 25%   Sale Price = $32 − $8
        = $32 × 0.25   = $24
        = $8

So, the discount is $8 and the sale price is $24.

• **CRITICAL THINKING** How can you estimate to find the 25% discount on a $95 coat?

GUIDED PRACTICE

The regular price is given. The discount rate is 40%. Find the
amount of discount.

1. $99.00   **2.** $45.00   **3.** $30.00   **4.** $51.25   **5.** $109.50

The regular price is given. The discount rate is 25%. Find the
sale price.

6. $80.00   **7.** $55.00   **8.** $96.00   **9.** $110.00   **10.** $49.92

INDEPENDENT PRACTICE

Find the amount of discount.

1. regular price:
$84.00

50% OFF

2. regular price:
$60.00

SAVE 75%

3. regular price:
$24.00

15% DISCOUNT

4. regular price:
$38.00

10% OFF

Find the sale price.

5. regular price:
$12.50

DISCOUNT 30%

6. regular price:
$21.00

10% OFF

7. regular price:
$62.50

SAVE 40%

8. regular price:
$15.80

SALE 20% OFF

The discount rate is 25%. Find the amount of discount and
the sale price.

9. regular price:
$47.95

10. regular price:
$120.00

11. regular price:
$51.89

12. regular price:
$88.50

Problem-Solving Applications

13. **CONSUMER MATH** Rory bought a watch for
$75. The next day, he noticed that the
same watch was on sale for 20% off. He
took the watch back to the store and got a
refund of 20%. How much was his refund?

14. Dennis is looking for the best buy in
videotapes. Store A's price is $28, with a
discount of 15%. Store B is selling
videotapes for $35, with a discount of
25%. Which has the better buy? Explain.

15. **FITNESS** Donna is buying exercise
equipment. The salesperson suggested
she wait until it goes on sale for 25% off
the regular price of $450. How much will
she save if she waits for the sale? What will
the sale price be?

16. **WRITE ABOUT IT** Explain how to find
the sale price once you know the amount
of discount.

Simple Interest

VOCABULARY

simple interest
principal

REAL-LIFE LINK

Most banks pay *compound interest,* interest on any previous interest as well as on the principal. If a bank paid you 5% interest on a $150.00 principal, you would have $157.50 after 1 year. After 2 years you would have $165.38, 5% interest on the original $150.00 and on the $7.50 interest from the first year. How much would you have after 5 years?

Did you know that money you put in the bank can grow even if you don't add to it? When money is invested in a bank, it earns interest.

Simple interest is the amount of interest earned on the amount deposited, or the **principal**. You can find the simple interest for one year by multiplying the interest rate by the principal.

| Simple interest rate | × | Principal | = | Interest earned for 1 year |

EXAMPLE Carol invested $150 at a simple interest rate of 4%. Find the interest she will earn in 1 year.

$4\% = 0.04$ *Write interest rate as a decimal.*

$0.04 \times 150 = 6$ *Multiply interest rate by principal.*

So, the interest earned in 1 year is $6.

• If Carol keeps her $150 in the bank, how much simple interest will she have earned at the end of 2 years?

ACTIVITY

WHAT YOU'LL NEED: for each group of four players: four principal cards and four interest-rate cards, as shown below

Turn over and stack the principal and interest-rate cards.

Principal	Interest Rate
$75	5%
$100	6%
$250	8%
$198	$7\frac{1}{2}\%$

• Each player draws a principal card and an interest-rate card.

• Players find the interest earned, using the interest rate and principal on their cards.

• The player with the greatest interest wins points equal to the interest earned by all the players.

• Play for five rounds. What is the greatest interest possible for a single player in one round?

GUIDED PRACTICE

Find the interest for one year.

1. principal: $200
rate: 5% per year

2. principal: $350
rate: 6% per year

3. principal: $1,300
rate: 10% per year

4. principal: $15,000
rate: 8% per year

5. principal: $890
rate: 6.5% per year

6. principal: $480
rate: 7.5% per year

INDEPENDENT PRACTICE

Find the simple interest.

	Principal	Yearly Rate	Interest for 1 Year	Interest for 2 Years
1.	$50	2%	$ ▆	$ ▆
2.	$180	3.5%	$ ▆	$ ▆
3.	$230	5%	$ ▆	$ ▆
4.	$500	4.1%	$ ▆	$ ▆
5.	$1,200	6.9%	$ ▆	$ ▆

Problem-Solving Applications

6. **RATES** Norma put $1,600 in the bank for 1 year. She earned interest at the rate of 6% a year. How much did she earn?

7. **CONSUMER MATH** Juanita had $100 that earned 4.5% simple interest a year in a bank. After 2 years, she withdrew her money. How much did she withdraw?

8. ✏️ **WRITE ABOUT IT** Explain how to find the amount of simple interest that money will earn in a year.

Mixed Review and Test Prep

Find the product.

9. $\frac{1}{10} \times 10$ **10.** $\frac{2}{5} \times 25$ **11.** 0.12×30 **12.** 0.2×45

Write the fraction or decimal as a percent.

13. 0.25   **14.** $\frac{3}{10}$ **15.** 0.333   **16.** $\frac{3}{5}$

17. **TRAVEL** A plane leaves New York at 4:30 P.M. It stops in Atlanta for $1\frac{1}{4}$ hours. Then it flies for $1\frac{3}{4}$ hours and arrives in Miami at 9:30 P.M. How long was the trip from New York to Atlanta?

A 2 hr   **B** 3 hr   **C** 4 hr   **D** 5 hr

18. **MEASUREMENT** Lena Sharp added a square room to her house. The room is 100 ft². What is the perimeter of Lena's new room?

F 10 ft   **G** 40 ft   **H** 50 ft   **J** 100 ft

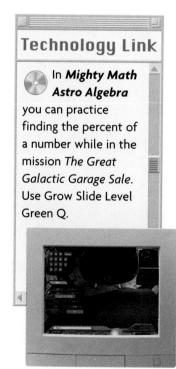

Technology Link

In *Mighty Math Astro Algebra* you can practice finding the percent of a number while in the mission *The Great Galactic Garage Sale*. Use Grow Slide Level Green Q.

What You'll Explore
How to use a model to find the percent one number is of another

What You'll Need
graph paper, scissors, and pen or pencil

REMEMBER:
....

A **percent** can be thought of as the number per hundred. So, *n*% means *n* per hundred. **See page H28.**

Finding the Percent One Number Is of Another

Julie plays shortstop in the city softball league. Last year, out of 25 attempts, she made 7 errors. You can use a model to find what percent 7 is of 25.

Explore

• Cut a 5 × 5 square out of graph paper. To represent 7 out of 25, shade 7 of the 25 squares.

• Since a percent is the number per hundred, cut out 3 more 5 × 5 squares to make a total of 100 squares. Shade 7 of the 25 squares in each 5 × 5 square.

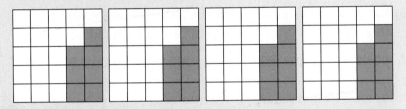

• How many of the 100 squares are shaded? What percent does this represent? Explain.

• What percent of 25 is 7? What percent of Julie's attempts were errors?

• Suppose that this year Julie made 3 errors in 20 attempts. Model the percent 3 is of 20. Use 4 × 5 rectangles. How many 4 × 5 rectangles will you need? How many squares will you shade in each 4 × 5 rectangle?

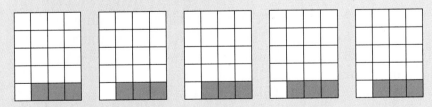

• What percent of 20 is 3? What percent of Julie's attempts were errors? What is the difference in percent of Julie's errors from last year to this year?

Think and Discuss

- Why did you use 100 squares in the models?

You can write the proportion $\frac{10}{25} = \frac{40}{100}$ for the model below.

One square shows the ratio $\frac{10}{25}$.

Four squares show the ratio $\frac{40}{100}$.

- What proportion can you write for the five 4 × 5 rectangles on page 362?

To find the percent one number is of another without a model, you can write and solve a proportion.

What percent of 20 is 3?

$$\frac{3}{20} = \frac{n}{100}$$

Think: What number, n, is to 100 as 3 is to 20?

$$20n = 300$$

$$\frac{20n}{20} = \frac{300}{20}$$

$$n = 15 \rightarrow \frac{15}{100} = 15\%$$

So, 3 is 15% of 20.

- Why do you use 100 in the proportion?

- What proportion can you write to find what percent 13 is of 25?

- CRITICAL THINKING What percent of 20 is 5? What percent of 5 is 20? How are these percents related?

Try This

- Use a model or a proportion to solve.

What percent of 20 is 17?

What percent of 50 is 39?

What percent of 25 is 2?

What percent of 40 is 18?

Technology Link

You can estimate the percent one number is of another by using E-Lab, Activity 18. Available on CD-ROM and on the Internet at **www.hbschool.com/elab**

Review/Test

Solve. (pages 350–351)

1. During lunch, 70% of 1,400 students ordered french fries. How many students ordered french fries?

2. If it rains 80% of the days of the year, how many days does it rain? Use 365 as the number of days in a year.

Find the percent of the number. (pages 352–355)

3. 40% of 300   **4.** 25% of 24   **5.** 18% of 350   **6.** 95% of 80

7. 20% of 200   **8.** 15% of 24   **9.** 11% of 80   **10.** 90% of 300

11. What is the sales tax on an item that costs $33.20 and has a sales tax rate of 7.5%?

12. What is the sales tax on an item that costs $98 and has a sales tax rate of 8%?

13. Use these data to make a circle graph of a family budget of $2,000: fun, 25%; food, 40%; rent, 30%; other, 5%. (pages 356–357)

Find how much money was spent on each part of the family budget in Exercise 13.

14. food   **15.** rent   **16.** fun   **17.** other

18. Use these data to make a circle graph of the results of 300 voters: Smith, 50%; Jones, 30%; Anderson, 20%. (pages 356–357)

Find the number of people who chose each candidate in Exercise 18.

19. Smith   **20.** Jones   **21.** Anderson

The regular price is given. The discount rate is 30%. Find the amount of discount. (pages 358–359)

22. $880   **23.** $120   **24.** $18   **25.** $50

The regular price is given. The discount rate is 50%. Find the sale price. (pages 358–359)

26. $72   **27.** $15.88   **28.** $21   **29.** $66

30. VOCABULARY The amount of interest earned on the principal is called __?__ . (page 360)

Find the interest earned. (pages 360–361)

31. principal: $200; interest rate: 6%

32. principal: $8,000; interest rate: 7%

33. principal: $500; interest rate: 4%

1. Rod wants to plant a rectangular garden. He wants it to have an area of 33 square feet. The length of the garden is 8 feet. How wide will the garden be?

   A $3\frac{3}{8}$ ft

   B $4\frac{1}{8}$ ft

   C $4\frac{1}{2}$ ft

   D $4\frac{3}{4}$ ft

   E $5\frac{1}{8}$ ft

2. At which of the following times do the hands of a clock form an acute angle?

   F 2:00
   G 3:00
   H 6:00
   J 12:00

3. Circles form the top and bottom views of which solid figure?

   A Cylinder
   B Rectangular prism
   C Triangular pyramid
   D Rectangular pyramid

4. Don is surveying members of a soccer club. Which of the following methods should he use to produce a random sample?

   F A survey of 1 out of every 8 club members
   G A survey of 2 out of every 9 goalies
   H A survey of the 25 oldest club members
   J A survey of the top 12 scoring champions
   K Not Here

5. Guy is twice as old as Beth, plus 3 years. Guy is 27 years old. Which of the following equations can be used to find b, Beth's age?

   A $b + 5 = 27$
   B $2b = 27$
   C $3 + b = 27$
   D $2b + 3 = 27$
   E $3b + 2 = 27$

6. A candidate for governor shook 135 hands in 9 minutes. Which of the following ratios shows the number of hands she can shake in one minute?

   F 14:1   **G** 15:1
   H 16:1   **J** 17:1

7. The Scottsville Baseball Club has 120 members. Boys make up 85% of the club. How many boys are in the club?

   A 18
   B 24
   C 96
   D 102
   E 108

8. Marie earns $900 a month. The circle graph below shows her monthly budget. Based on the graph, how much does Marie spend on food each month?

   F $90
   G $180
   H $225
   J $270

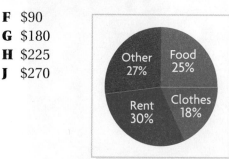

RATIO, PROPORTION, AND SIMILAR FIGURES

LOOK AHEAD

In this chapter you will solve problems that involve

- identifying similar and congruent figures

- using ratios to identify similar figures

- using proportions to find unknown lengths of sides of similar figures

- using proportions to find measures that are difficult to measure directly

SCIENCE LINK

The table below compares the diameters of the planets in our solar system.

- Which planets are nearly congruent?

- The planets are listed in order based on their distance from the sun. Look for a pattern comparing each planet's size to its distance from the sun. Which planet stands out as not fitting your pattern?

PLANET	DIAMETER (NEAREST 100 MI)
Mercury	3,000
Venus	7,500
Earth	7,900
Mars	4,200
Jupiter	86,900
Saturn	72,400
Uranus	31,500
Neptune	30,600
Pluto	1,400

Proportions in Models

Choose an everyday object and make a clay model of it. Measure parts of the object and the same parts of the model. Record your measurements in a table. Then use your table to write ratios that compare parts of your model to the same parts on the object. Decide if your model is a scale model. If it is not, make changes to it. Share your model with the class.

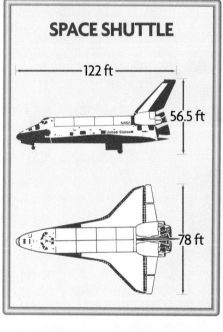

SPACE SHUTTLE

122 ft

56.5 ft

78 ft

PROJECT CHECKLIST

✓ Did you make a model?

✓ Did you measure the object and model and record your measurements?

✓ Did you make changes so that your model is a scale model?

What You'll Explore
How to make similar figures

What You'll Need
graph paper, scissors, ruler, pencil, geoboard and bands

VOCABULARY
similar figures

Making Similar Figures

You can use graph paper and what you know about ratios to make enlargements of rectangles and triangles.

ACTIVITY 1

Explore

- Draw a rectangle 2 units wide and 3 units long on the graph paper. Cut it out.

- Draw a second rectangle, with the sides twice the length of the sides in the first rectangle. Cut it out.

- Draw a third rectangle, with the sides 3 times the length of the sides in the first rectangle. Cut it out.

Think and Discuss

- Compare the three rectangles. How are they alike? How are they different?

- What kind of angles are in each rectangle?

- Geometric figures that have the same shape and angles of the same size are called **similar figures**. Are the rectangles similar? Explain.

- Are the rectangles congruent? Explain.

- When you place the rectangles on top of each other, how can you arrange them to compare their sizes?

- Lay the rectangles next to each other. If you translate, rotate, or reflect one of the rectangles, will they still be similar? Explain.

Try This

- Draw each rectangle below on graph paper. Then draw a similar rectangle for each.

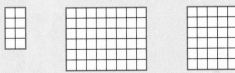

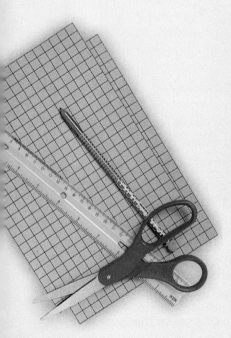

Another way to make similar figures is to use a geoboard. You can make similar triangles using the geoboard and bands.

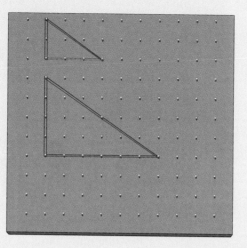

Technology Link

You can find the steepness of a line with similar trangles by using E-Lab, Activity 19. Available on CD-ROM and on the Internet at **www.hbschool.com/elab**

ACTIVITY 2

Explore

• Use the geoboard to make a right triangle with one side 2 units long and the other side 3 units long.

• Make another right triangle, with sides twice as long.

Think and Discuss

• Compare the two right triangles you made on the geoboard. How are they alike? How are they different?

• Are the two triangles similar? Explain.

Try This

• Use the geoboard to make each figure below. Then make a similar figure for each. Record your drawings on graph paper.

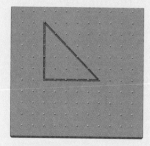

• Make a new right triangle on the geoboard. Then make a similar right triangle. Record your drawings on graph paper.

Similar and Congruent Figures

What You'll Learn
How to identify similar and congruent figures

Why Learn This?
To recognize whether pairs of figures are alike in shape or in both shape and size

Mount St. Helens looks powerful in a picture of any size. These two pictures show the same image in different sizes. Since the pictures are not the same size, they are not congruent.

The length and the width of the smaller picture have been increased in the same proportion to produce the larger picture. The shapes of the pictures are similar rectangles.

Congruent figures have the same shape and size. Similar figures have the same shape.

The Venn diagram shows the relationship between pairs of congruent figures.

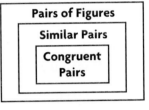

Talk About It
• Look at the Venn diagram. Are all pairs of congruent figures also similar? Explain.

• Are all pairs of similar figures congruent? Explain.

• Two figures have the same size and shape. Are they similar, congruent, or both? Where would you place them in the Venn diagram?

REMEMBER:

Congruent polygons have all sides congruent and all angles congruent. **See page H23.**

All the sides of the two hexagons are the same length.

All the angles of the two hexagons have the same measure. The two hexagons are congruent.

EXAMPLES Decide whether the figures in each pair appear to be similar, congruent, both, or neither.

A. *same shape, not same size*
The figures are similar.

B. *same shape, same size*
The figures are both similar and congruent.

C. *not same shape, not same size*
The figures are neither similar nor congruent.

D. *same shape, not same size*
The figures are similar.

GUIDED PRACTICE

Tell whether the figures in each pair appear to be *similar, congruent, both,* or *neither.*

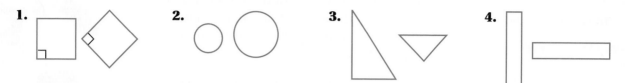

1.   **2.**   **3.**   **4.**

INDEPENDENT PRACTICE

Look at each figure. Tell whether each pair of figures appear to be *similar, congruent, both,* or *neither.*

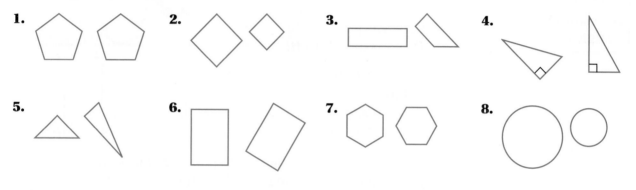

1.   **2.**   **3.**   **4.**

5.   **6.**   **7.**   **8.**

Use the diagram to answer Exercises 9–12.

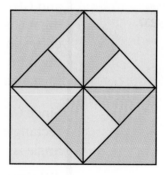

9. How many pairs of similar yellow triangles can you find?

10. How many pairs of congruent yellow triangles can you find?

11. How many pairs of similar green triangles can you find?

12. How many pairs of congruent green triangles can you find?

Problem-Solving Applications

13. LOGICAL REASONING Sharon has two cups without handles. The cups are similar but not congruent. Can Sharon store one cup inside the other cup? Explain.

14. GEOMETRY If triangle *A* is congruent to triangle *B* and triangle *B* is congruent to triangle *C*, what conclusion can you make about triangle *A* and triangle *C*?

15. On graph paper, draw two figures that are congruent. Then draw two figures that are similar but not congruent.

16. **WRITE ABOUT IT** Explain why all congruent figures are similar but not all similar figures are congruent.

Ratios and Similar Figures

What You'll Learn
How to use ratios to identify similar figures

Why Learn This?
To make an object smaller or larger while keeping its shape the same

VOCABULARY

corresponding sides

corresponding angles

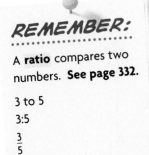

REMEMBER:

A **ratio** compares two numbers. **See page 332.**

3 to 5

3:5

$\frac{3}{5}$

The larger soccer ball and the soccer ball on the key ring have similar black pentagons. What other similar polygons can you find on them?

Similar figures have **corresponding sides** and **corresponding angles**. The corresponding sides and angles for similar rectangles *ABCD* and *EFGH* are shown below.

$\overline{AB}$ corresponds to $\overline{EF}$.
$\overline{BC}$ corresponds to $\overline{FG}$.
$\overline{CD}$ corresponds to $\overline{GH}$.
$\overline{DA}$ corresponds to $\overline{HE}$.

$\angle A$ corresponds to $\angle E$.
$\angle B$ corresponds to $\angle F$.
$\angle C$ corresponds to $\angle G$.
$\angle D$ corresponds to $\angle H$.

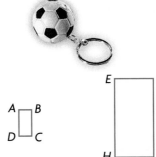

If figures are similar, their corresponding angles are congruent and their corresponding sides have the same ratio.

EXAMPLE 1 Rectangles *CDEF* and *GHIJ* are similar. Find the measures of corresponding angles and the ratio of the lengths of the corresponding sides.

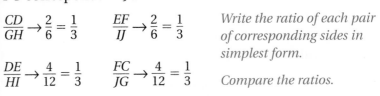

Since *CDEF* and *GHIJ* are rectangles, the measure of each angle is 90°.

Check that corresponding angles are congruent.

$\overline{CD}$ corresponds to $\overline{GH}$.
$\overline{DE}$ corresponds to $\overline{HI}$.
$\overline{EF}$ corresponds to $\overline{IJ}$.
$\overline{FC}$ corresponds to $\overline{JG}$.

List the corresponding sides.

$\dfrac{CD}{GH} \rightarrow \dfrac{2}{6} = \dfrac{1}{3}$ $\dfrac{EF}{IJ} \rightarrow \dfrac{2}{6} = \dfrac{1}{3}$

Write the ratio of each pair of corresponding sides in simplest form.

$\dfrac{DE}{HI} \rightarrow \dfrac{4}{12} = \dfrac{1}{3}$ $\dfrac{FC}{JG} \rightarrow \dfrac{4}{12} = \dfrac{1}{3}$

Compare the ratios.

So, the ratio of the lengths of corresponding sides is $\frac{1}{3}$.

GUIDED PRACTICE

For Exercises 1–8, use the similar rectangles at the right.

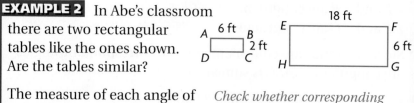

R [6 ft] S
4 ft
U T

M [3 ft] N
2 ft
P O

1. $\overline{UT}$ corresponds to __?__ .

2. $\angle O$ corresponds to __?__ .

3. __?__ corresponds to $\overline{ST}$.

4. __?__ corresponds to $\angle N$.

Write the ratio of each pair of corresponding sides in simplest form.

5. $\dfrac{RS}{MN}$ **6.** $\dfrac{RU}{MP}$ **7.** $\dfrac{UT}{PO}$ **8.** $\dfrac{ST}{NO}$

You can decide if two polygons are similar by checking that the ratios of the corresponding sides are equivalent and that the corresponding angles are congruent.

EXAMPLE 2 In Abe's classroom there are two rectangular tables like the ones shown. Are the tables similar?

A [6 ft] B
2 ft
D C

E [18 ft] F
6 ft
H G

The measure of each angle of a rectangle equals 90°.

Check whether corresponding angles are congruent.

$\overline{AB}$ corresponds to $\overline{EF}$.
$\overline{BC}$ corresponds to $\overline{FG}$.
$\overline{CD}$ corresponds to $\overline{GH}$.
$\overline{DA}$ corresponds to $\overline{HE}$.

List the corresponding sides.

$\dfrac{AB}{EF} \rightarrow \dfrac{6}{18} = \dfrac{1}{3}$ $\dfrac{CD}{GH} \rightarrow \dfrac{6}{18} = \dfrac{1}{3}$

Write the ratio of each pair of corresponding sides in simplest form.

$\dfrac{BC}{FG} \rightarrow \dfrac{2}{6} = \dfrac{1}{3}$ $\dfrac{DA}{HE} \rightarrow \dfrac{2}{6} = \dfrac{1}{3}$

Compare the ratios.

Since the corresponding angles are congruent and the ratios of the corresponding sides are equivalent, the two tables are similar.

Talk About It

• **CRITICAL THINKING** In Example 2, can the ratio for corresponding sides be $\frac{3}{1}$? Explain.

• Are congruent figures also similar figures? Why or why not?

You can also find out whether two triangles are similar by comparing the ratios of the corresponding sides and the measures of the corresponding angles.

When writing the name of a triangle, like triangle *DEF,* you can use the symbol △ for the word *triangle*. Triangle *DEF* can be written as △*DEF*.

ART LINK

Quilt patterns often use similar figures. What similar figures can you find in the Evening Star pattern?

EXAMPLE 3 Searon is using a copy machine to make an enlargement of △*ABC*. He labeled his new triangle *DEF*. Are the triangles similar?

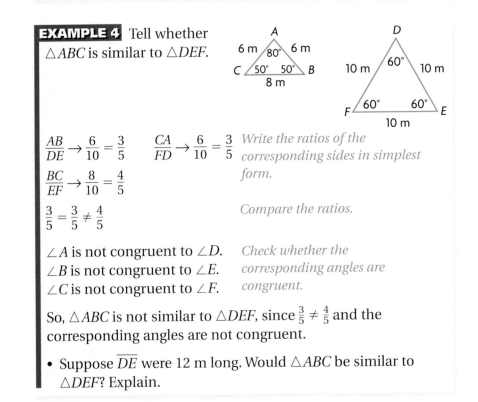

$\dfrac{DE}{AB} \rightarrow \dfrac{20}{5} = \dfrac{4}{1}$ $\dfrac{FD}{CA} \rightarrow \dfrac{12}{3} = \dfrac{4}{1}$ *Write the ratios of the corresponding sides in simplest form.*

$\dfrac{EF}{BC} \rightarrow \dfrac{16}{4} = \dfrac{4}{1}$

Compare the ratios.

∠*D* and ∠*A* are congruent.
∠*E* and ∠*B* are congruent.
∠*F* and ∠*C* are congruent.

Check whether corresponding angles are congruent.

Since the ratios are equivalent and the corresponding angles are congruent, △*ABC* is similar to △*DEF*.

• Are all right triangles similar? Explain.

EXAMPLE 4 Tell whether △*ABC* is similar to △*DEF*.

$\dfrac{AB}{DE} \rightarrow \dfrac{6}{10} = \dfrac{3}{5}$ $\dfrac{CA}{FD} \rightarrow \dfrac{6}{10} = \dfrac{3}{5}$ *Write the ratios of the corresponding sides in simplest form.*

$\dfrac{BC}{EF} \rightarrow \dfrac{8}{10} = \dfrac{4}{5}$

$\dfrac{3}{5} = \dfrac{3}{5} \neq \dfrac{4}{5}$ *Compare the ratios.*

∠*A* is not congruent to ∠*D*.
∠*B* is not congruent to ∠*E*.
∠*C* is not congruent to ∠*F*.

Check whether the corresponding angles are congruent.

So, △*ABC* is not similar to △*DEF*, since $\dfrac{3}{5} \neq \dfrac{4}{5}$ and the corresponding angles are not congruent.

• Suppose $\overline{DE}$ were 12 m long. Would △*ABC* be similar to △*DEF*? Explain.

INDEPENDENT PRACTICE

Name the corresponding sides and angles. Write the ratio of the corresponding sides in simplest form.

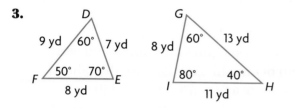

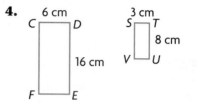

Tell whether the figures in each pair are similar. Write *yes* or *no*. If you write *no*, explain.

3.

4.

Problem-Solving Applications

5. RATIOS The rectangular front of a building is 70 ft high and 50 feet wide. The rectangular front of another building is 35 ft high and 25 ft wide. Are the fronts similar?

6. Jeri bought two postcards while on vacation. One postcard is 3 in. × 5 in. The other postcard is 4 in. × 6 in. Are the two postcards similar? Explain.

7. ✏️ **WRITE ABOUT IT** Explain how to determine if two figures are similar.

Mixed Review and Test Prep

Solve.

8. $\frac{4}{7} = \frac{n}{14}$

9. $\frac{2}{6} = \frac{8}{x}$

10. $\frac{6}{8} = \frac{a}{12}$

Find the interest for one year.

11. principal: $150
rate: 6% per year

12. principal: $400
rate: 5% per year

13. principal: $590
rate: 7.5% per year

14. PATTERNS Mike is using brown (B) and white (W) tiles to cover a floor. If the pattern Mike uses is B, B, W, B, B, W, what will be the colors of the eighteenth and nineteenth tiles?

A B, B   **B** B, W   **C** W, B   **D** W, W

15. CONSUMER MATH Rob bought a stereo on credit. He will pay $26 per month for 24 months. How much will Rob pay in all?

F $624   **G** $604   **H** $602   **J** $584

Technology Link

💿 In *Mighty Math Cosmic Geometry,* you can practice showing figures that are similar and congruent in the *Geo Academy.* Use Grow Slide Level M.

What You'll Learn
How to use similar figures to find the unknown length of a side

Why Learn This?
To make enlargements or reductions of models or other figures

ALGEBRA CONNECTION

Proportions and Similar Figures

Many artists use projections to draw or paint a picture. The triangles in the projection at the right are similar.

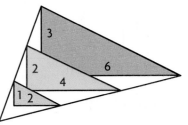

- What ratio of corresponding sides relates the blue triangle to the yellow triangle? the yellow triangle to the green triangle?

Sometimes you know that two figures are similar and you need to find the length of one side. You can use the ratio of corresponding sides to write a proportion to find the unknown length.

REMEMBER:
. . . .
A **proportion** is an equation that states that two ratios are equivalent.
See page 342.

COMPUTER LINK

Some computer software allows the user to resize photographs. If a 4-in. × 6-in. photo is resized to 6 in. × 9 in., will the enlargement be similar to the original?

EXAMPLE Margo is a designer and needs to enlarge a picture so that it is 10 in. wide. She wants the pictures to be similar. How long does her picture need to be?

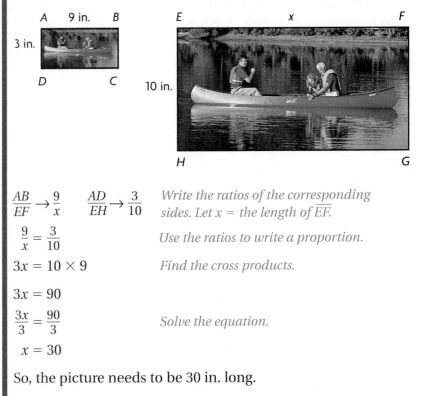

$$\frac{AB}{EF} \rightarrow \frac{9}{x} \qquad \frac{AD}{EH} \rightarrow \frac{3}{10}$$ *Write the ratios of the corresponding sides. Let x = the length of $\overline{EF}$.*

$$\frac{9}{x} = \frac{3}{10}$$ *Use the ratios to write a proportion.*

$$3x = 10 \times 9$$ *Find the cross products.*

$$3x = 90$$

$$\frac{3x}{3} = \frac{90}{3}$$ *Solve the equation.*

$$x = 30$$

So, the picture needs to be 30 in. long.

- Suppose Margo decides that she wants the image to be 9 in. wide. How long will her picture need to be?

GUIDED PRACTICE

The figures in each pair are similar. Write the proportion you would use to find *n*.

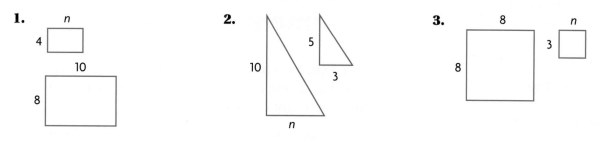

1. *n* 4 10 8

2. 5 10 3 *n*

3. 8 *n* 3 8

INDEPENDENT PRACTICE

The figures in each pair are similar. Find *n*.

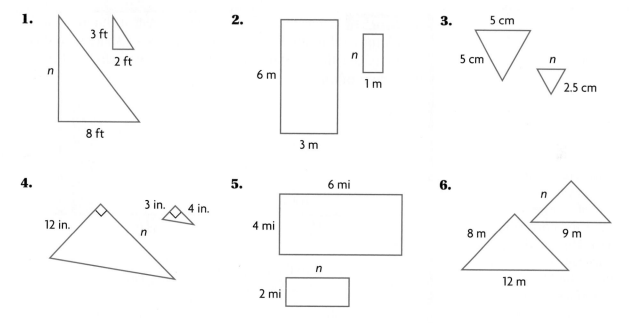

1. 3 ft 2 ft *n* 8 ft

2. 6 m *n* 1 m 3 m

3. 5 cm 5 cm *n* 2.5 cm

4. 12 in. 3 in. 4 in. *n*

5. 6 mi 4 mi *n* 2 mi *n*

6. *n* 8 m 9 m 12 m

Problem-Solving Applications

7. LOGICAL REASONING Two common envelope sizes are $3\frac{1}{2}$ in. $\times$ $6\frac{1}{2}$ in. and 4 in. $\times$ $9\frac{1}{2}$ in. Are these envelopes similar? Explain.

8. ART Kirsten has a photo that is 8 in. $\times$ 10 in. She wants to reduce the photo so that it is 5 in. long. How wide should she make the photo?

9. HOBBIES Byron is making a model of his dad's sports car. The actual length of the car is 15 ft, or 180 in. The actual width of the car is 5 ft, or 60 in. He wants to make his model 6 in. long. How wide should he make his model?

10. WRITE ABOUT IT When you know two figures are similar, how can you use a proportion to find an unknown length of a side?

ALGEBRA CONNECTION

Proportions and Indirect Measurement

Suppose you want to know how tall a saguaro cactus has grown. You can't get close enough to the cactus to measure it. How can you measure the height of the cactus?

One way to estimate the height of the cactus is to use indirect measurement. When you use similar figures and proportions to find a measure, you are using a technique called **indirect measurement**.

SCIENCE LINK

The saguaro cactus grows only in the desert of southern Arizona and northern Sonora, Mexico. Using the diagram in the example, find the height of a saguaro that casts a shadow 64 ft long.

EXAMPLE On a sunny day, the cactus casts a shadow that is 58 ft long. At the same time, a yardstick casts a shadow that is 4 ft long. Use the similar right triangles shown below to find the height of the cactus.

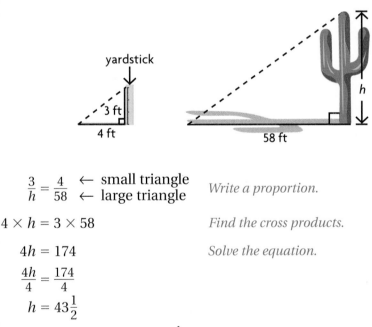

$$\frac{3}{h} = \frac{4}{58} \quad \begin{array}{l} \leftarrow \text{small triangle} \\ \leftarrow \text{large triangle} \end{array} \qquad \textit{Write a proportion.}$$

$$4 \times h = 3 \times 58 \qquad\qquad \textit{Find the cross products.}$$

$$4h = 174 \qquad\qquad\qquad \textit{Solve the equation.}$$

$$\frac{4h}{4} = \frac{174}{4}$$

$$h = 43\frac{1}{2}$$

So, the saguaro cactus is $43\frac{1}{2}$ ft tall.

Talk About It

- **CRITICAL THINKING** Why can you use similar figures for indirect measurement?

- Name three objects that would be difficult to measure directly.

GUIDED PRACTICE

Write a proportion. Then solve for *x*.

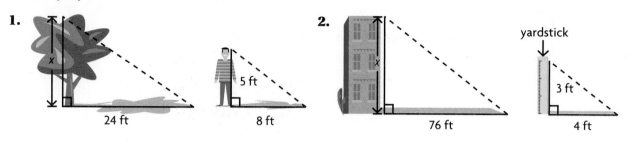

1.

24 ft   5 ft   8 ft

2.

yardstick

76 ft   3 ft   4 ft

INDEPENDENT PRACTICE

Find the unknown length.

1.

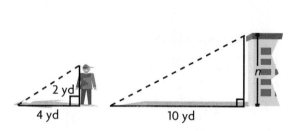

2 yd   4 yd   *n*   10 yd

2.

n   72 in.   40 in.   48 in.

Problem-Solving Applications

3. PROPORTION A flagpole casts a shadow that is 4 m long. At the same time, a nearby pole that is 2 m high casts a 1-m shadow. What is the height of the flagpole?

4. ✏️ **WRITE ABOUT IT** The shadow of a 6-ft man measures 5 ft, and his son's shadow is 2 ft. Explain how to find the son's height.

Mixed Review and Test Prep

Complete.

5. $\frac{2}{4} = \frac{\blacksquare}{12}$ **6.** $\frac{5}{7} = \frac{15}{\blacksquare}$ **7.** $\frac{3}{9} = \frac{\blacksquare}{18}$ **8.** $\frac{6}{9} = \frac{30}{\blacksquare}$

Find the sales tax. Round to the nearest cent when necessary.

9. price: $10
tax rate: 5%

10. price: $5.35
tax rate: 7.5%

11. price: $36.50
tax rate: 6%

12. price: $73
tax rate: 6.5%

13. MONEY Jay spent $17.50 at the movies. It cost $7.25 for a movie and $3.00 for a snack. How many movies did Jay see?

     A 1   **B** 2
     C 3   **D** 4

14. TIME A 1-hour 42-minute production ends at 8:15. What time did the production start?

     F 6:00   **G** 6:13
     H 6:33   **J** 6:42

Tell whether the figures in each pair appear to be *similar, congruent, both,* or *neither.* **(pages 370–371)**

1.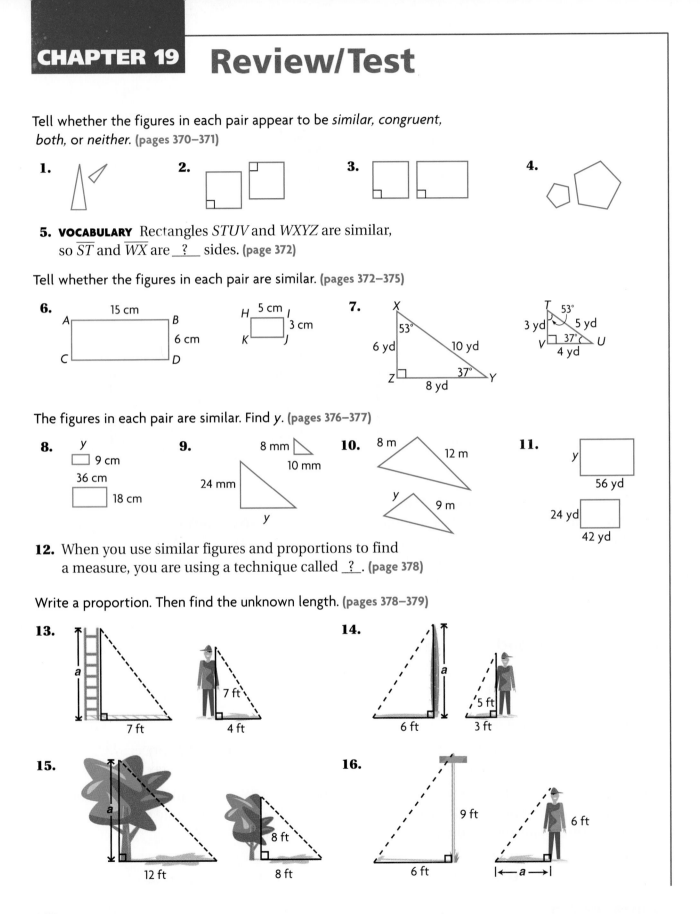

2.

3.

4.

5. **VOCABULARY** Rectangles *STUV* and *WXYZ* are similar, so $\overline{ST}$ and $\overline{WX}$ are __?__ sides. **(page 372)**

Tell whether the figures in each pair are similar. **(pages 372–375)**

6.
15 cm
A ———— B
6 cm
C ———— D

H 5 cm I
3 cm
K J

7.
X
53°
6 yd   10 yd
Z   37° Y
8 yd

T 53°
3 yd   5 yd
V 37° U
4 yd

The figures in each pair are similar. Find *y*. **(pages 376–377)**

8.
y
9 cm
36 cm
18 cm

9.
8 mm
10 mm
24 mm
y

10.
8 m
12 m
y
9 m

11.
y
56 yd
24 yd
42 yd

12. When you use similar figures and proportions to find a measure, you are using a technique called __?__ . **(page 378)**

Write a proportion. Then find the unknown length. **(pages 378–379)**

13.
a
7 ft

7 ft
4 ft

14.
a
6 ft

5 ft
3 ft

15.
a
12 ft

8 ft
8 ft

16.
9 ft
6 ft

6 ft
6 ft
|←— *a* —→|

1. Last week, a basketball team practiced $5\frac{1}{3}$ hours. The team usually practices $8\frac{3}{4}$ hours a week. How many fewer hours did the team practice last week?

A $2\frac{3}{4}$ hr

B $3\frac{1}{3}$ hr

C $3\frac{5}{12}$ hr

D $4\frac{1}{5}$ hr

2. Rico is building a model of a prism. He will use balls of clay for the vertices and straws for the edges. How many straws will he use if he uses 8 balls of clay for his model?

F 4
G 8
H 12
J 16

3. Jon has $300 in his savings account. Out of the $300, $150 was for mowing lawns, $100 was for delivering newspapers, and $50 was from his birthday. He wants to make a circle graph to show this data. What does the circle graph represent?

A Amount of money he spent
B Amount of money he received as gifts
C Amount of money he earned
D Total amount of money he has
E Amount of money in his savings account

4. A survey was done to find out how many students take a bus, ride a bicycle, or walk to Main School. A bar graph displaying the results shows that 8 out of the 50 students surveyed walk to the school. Based on this information, which is the best estimate of the number of walkers identified in a survey of 200 Main School students?

F 16
G 24
H 32
J 40

5. Paul made some birdhouses. He sold 9 at a craft fair. Now he has 9 left. Which equation can be used to find b, the number of birdhouses he made?

A $9 - b = 8$
B $8 + b = 9$
C $b - 9 = 8$
D $9b = 8$
E Not Here

6. Chloe put $1,800 in the bank for two years. The account earns simple interest at the rate of 7% a year. How much interest did Chloe earn?

F $70

G $140

H $252

J $360

7. Rectangles *ABCD* and *GHJK* are similar. Which of the following ratios is equivalent to $\frac{AB}{GH}$?

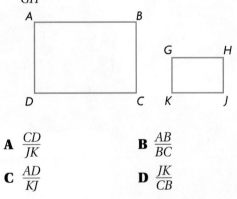

A $\frac{CD}{JK}$

B $\frac{AB}{BC}$

C $\frac{AD}{KJ}$

D $\frac{JK}{CB}$

8. A model of a building is 5 inches wide and 12 inches high. The actual building is 75 feet wide. How high is the actual building?

F 27 ft
G 60 ft
H 120 ft
J 150 ft
K 180 ft

APPLICATIONS OF RATIO AND PROPORTION

LOOK AHEAD

In this chapter you will solve problems that involve

- changing the size of scale drawings

- using map scales

- using the Golden Ratio and Golden Rectangles

Wingspan: 4 inches

Largest Butterfly

Females of the Queen Alexandra's birdwing butterfly have a wingspan of up to 11 in. and can weigh 0.9 oz.

Smallest Moth

The micro-moth *Stigmella ridiculosa*, found in the Canary Islands, has a wingspan of about 0.08 in.

- Extend your arms and measure your armspan. Compare your armspan to the wingspan of the largest Queen Alexandra's birdwing butterfly.

- Compare the wingspan of the largest butterfly to that of the smallest moth.

Estimating Scale

Find a photograph of an object and a description of its measurement. For each measurement given in the description, measure that part in the photograph. Write ratios of the photograph measurements to the actual measurements. Estimate the scale of the photograph of the object compared to its actual size.

Rajah Brooke's Birdwing
Wingspan: 17 cm

PROJECT CHECKLIST

✔ Did you find a photograph of an object and a description of its measurements?

✔ Did you measure the object in the photograph?

✔ Did you write ratios?

✔ Did you estimate the scale?

What You'll Learn
How to use scale to find the dimensions for a drawing or the actual dimensions of an object

Why Learn This?
To interpret scale drawings or to find the dimensions needed to make a scale drawing

VOCABULARY

scale drawing

scale

ALGEBRA CONNECTION

Scale Drawings: Changing the Size

A **scale drawing** is a picture or diagram of a real object but is smaller or larger than the real object. Some common scale drawings are floor plans of houses, road maps, and diagrams for bikes or other objects that need assembly.

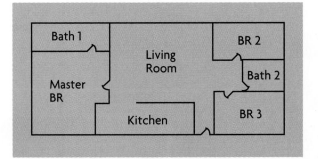

A scale drawing can be an enlargement or a reduction of an actual object. A floor plan such as the one above or other types of architectural drawings are usually called blueprints. These types of blueprints are usually a reduction of an actual object.

REAL-LIFE **LINK**

One of the easiest ways to enlarge or reduce drawings is by using a copy machine. Most office copiers can enlarge drawings by as much as 150% or reduce to 64%. Suppose you have a map with the scale 1 in. = 1 mi. If you reduce the map to 75% of its original size, what will the scale be?

ACTIVITY

WHAT YOU'LL NEED: centimeter graph paper, metric ruler

• Use a metric ruler to measure the length and width of the scale drawing above in centimeters.

• On graph paper, make a drawing that is twice the size of the scale drawing above.

• What are the length and width of your scale drawing, in centimeters?

• Write the ratio of length to width for the original drawing.

• Write the ratio of length to width for your drawing. Are the two ratios equivalent? Are the drawings similar? Explain.

• Make a drawing that is half the size of the original drawing.

• Write the ratio of length to width for your new drawing. Is the ratio equivalent to the length-to-width ratio of the original scale drawing? Are the drawings similar? Explain.

A **scale** is a ratio between two sets of measurements. The scale 1 in. = 3 ft on the drawing below means that 1 in. on the drawing represents 3 ft on the actual bike.

The scale can be expressed in three ways.

1 in.:3 ft   or   1 in. = 3 ft   or $\dfrac{1 \text{ in.}}{3 \text{ ft}}$ ← $\dfrac{\text{size of scale drawing}}{\text{actual size}}$

scale: 1 in. = 3 ft

You can use proportions to find an actual length of an object such as the bike.

EXAMPLE 1 Measure the length of the bike in the scale drawing above. Then use the scale to find the actual length of the bike.

scale drawing length: 2 in.   *Use an inch ruler to measure the length on the scale drawing.*

$\dfrac{\text{drawing (in.)}}{\text{actual (ft)}} \rightarrow \dfrac{1}{3} = \dfrac{2}{n}$ *Write a proportion. Let n represent the actual length of the bike.*

$1 \times n = 2 \times 3$ *Find the cross products. Multiply.*

$n = 6$

So, the actual length of the bike is 6 ft.

- Suppose the scale on the drawing of the bike were 1 in. = 4 ft. What would the actual length of the bike be ?

Talk About It

- **CRITICAL THINKING** How many times greater is the actual length of the bike than the length of the bike in the scale drawing?

- The scale for the bike is 1 in.:3 ft. What would the scale be if both measurements were in inches?

- What types of objects have you and your family assembled that came with a scale drawing?

GUIDED PRACTICE

Write the ratio of the length to the width.

1. $l = 3, w = 5$ **2.** $l = 9, w = 4$

3. $l = 8, w = 2$ **4.** $l = 5, w = 6$

Find the missing dimension. Use the scale 1 in. = 2 ft.

5. drawing length: 2 in.
actual length: ■ ft

6. drawing height: 4 in.
actual height: ■ ft

7. drawing length: ■ in.
actual length: 6 ft

8. drawing width: ■ in.
actual width: 16 ft

9. drawing length: ■ in.
actual length: 7 ft

10. drawing length: ■ in.
actual length: 9 ft

You can use proportions to determine how long you need to make an object in a drawing with a given scale. Sometimes you want your scale drawing to be bigger than the actual object.

EXAMPLE 2 Suppose you want to make a scale drawing of a microchip. The microchip is rectangular and has a length of 3 mm. Write a proportion to determine the length to draw the microchip, using a scale of 5 cm:1 mm.

$$\frac{\text{drawing (cm)}}{\text{actual (mm)}} \rightarrow \frac{5}{1} = \frac{n}{3}$$ *Write a proportion. Let n represent the length of the microchip in your drawing.*

$$n \times 1 = 5 \times 3$$ *Find the cross products. Multiply.*

$$n = 15$$

So, in your scale drawing, the microchip will have a length of 15 cm.

Talk About It

• If the scale were 6 cm:1 mm, what would the length of the microchip be in your drawing?

• **CRITICAL THINKING** How would the scale drawing of the microchip be different from the scale drawing of the bicycle?

• What are some other objects for which a scale drawing larger than the actual object might be useful?

Microchips are in many things, such as computers, appliances, and pets. Yes, pets! Microchips can be used to identify lost pets. A chip the size of a grain of rice is put under the pet's skin. The chip contains a number that can be read by a scanner. Find out from a vet why these chips are replacing other methods of pet identification.

INDEPENDENT PRACTICE

Find the ratio of length to width.

1. $l = 4, w = 3$

2. $l = 6, w = 11$

3. $l = 2.4, w = 1.2$

Find the missing dimension.

4. scale: 1 in.:5 ft
drawing length: 2 in.
actual length: ▦ ft

5. scale: 1 in.:5 ft
drawing length: 3 in.
actual length: ▦ ft

6. scale: 1 in.:5 ft
drawing length: ▦ in.
actual length: 30 ft

7. scale: 1 in.:9 ft
drawing length: 4 in.
actual length: ▦ ft

8. scale: 1 in.:9 ft
drawing length: ▦ in.
actual length: 81 ft

9. scale: 4 cm:1 mm
drawing length: ▦ cm
actual length: 12 mm

Problem-Solving Applications

10. PROPORTION Mr. Whitman made a scale drawing of his porch with a length of 5 in. The actual length is 20 ft. The width in the scale drawing is 3.5 in. Write a proportion, and find the actual width of the porch.

11. ALGEBRA Suppose you are putting together a desk. The scale drawing in the directions has the scale 1 in. = 2 ft. The drawing length is 3 in. Write a proportion, and find the actual length of the desk.

12. ✏ **WRITE ABOUT IT** Explain what the scale 1 cm = 2 m on a scale drawing means.

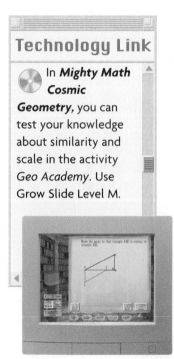

Technology Link

In *Mighty Math Cosmic Geometry,* you can test your knowledge about similarity and scale in the activity *Geo Academy.* Use Grow Slide Level M.

Mixed Review and Test Prep

Solve for *x*.

13. $\dfrac{1}{2} = \dfrac{5}{x}$

14. $\dfrac{1}{8} = \dfrac{x}{10}$

15. $\dfrac{3}{1} = \dfrac{60}{x}$

Tell whether the figures in the pair are similar. Corresponding angles are congruent.

16.

17.

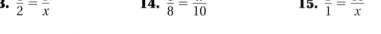

18. NUMBER SENSE Prime numbers that differ by two, such as 3 and 5 or 5 and 7, are called *twin primes.* Which is a pair of twin primes?

A 9 and 11   **B** 19 and 23
C 27 and 29   **D** 41 and 43

19. MENTAL MATH Lynn had 60,000 pesos on a trip to Mexico. She bought 2 dresses for 15,000 pesos each and 1 ring for 20,000 pesos. How much did she have left?

F 30,000 pesos   **G** 20,000 pesos
H 10,000 pesos   **J** 5,000 pesos

LAB ACTIVITY

What You'll Explore
How to stretch or shrink a figure by changing one dimension

What You'll Need
graph paper, ruler

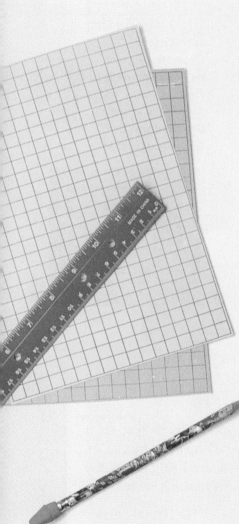

Stretching and Shrinking

Designers often change the look of a design by enlarging or reducing one dimension.

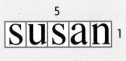

You can stretch a figure by increasing one dimension.

ACTIVITY 1

Explore

- On graph paper, copy the robot face shown at the right.

- Write the dimensions of the robot face, eyes, and mouth.

- Change the length of the robot face, including the eyes and the mouth, by multiplying each length by 2.

- Draw the new robot face.

- What are the dimensions of the robot face, eyes, and mouth after you have changed the length by a factor of 2?

Think and Discuss

- What happened to the shape of the robot face when you changed the lengths?

- Are the two figures similar? Explain.

Try This

- Determine which dimension of the original was stretched to make the new figure. Then determine what factor was used.

Original figure   New figure

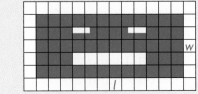

GUIDED PRACTICE

Use the scale 1 in. = 100 mi. Write the proportion you would use to find the actual miles.

1. $\frac{1}{2}$ in. **2.** 2 in. **3.** 3 in. **4.** $4\frac{1}{2}$ in. **5.** 10 in.

Use the scale 1 in. = 60 mi to find the actual miles.

6. $\frac{1}{2}$ in. **7.** 2 in. **8.** 3 in. **9.** 5 in. **10.** $7\frac{1}{2}$ in.

INDEPENDENT PRACTICE

Write and solve a proportion to find the actual miles. Use a map scale of 1 in. = 50 mi.

1. map distance: $1\frac{1}{2}$ in. **2.** map distance: 2 in. **3.** map distance: $4\frac{1}{2}$ in.

4. map distance: 10 in. **5.** map distance: 12 in. **6.** map distance: 20 in.

7. map distance: $8\frac{1}{4}$ in. **8.** map distance: 15 in. **9.** map distance: 7 in.

Problem-Solving Applications

For Problems 10–14, use the map of Colorado. The scale is $\frac{1}{2}$ in. = 46 mi.

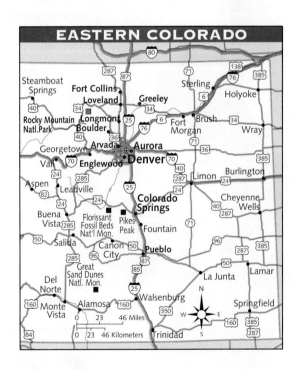

10. MAPS Find the straight-line distance from Colorado Springs to Limon in both map inches and actual miles.

11. MAPS Find the straight-line distance in miles from Pueblo to Denver and then from Denver to Steamboat Springs.

12. MAPS Find the straight-line distance from Pueblo to Steamboat Springs in map inches and actual miles. Can you travel this way? Explain.

13. What is the straight-line distance from Denver to Wray in inches and miles?

14. TIME Andy drove from Denver to Wray. How far did he drive? If he drove 55 mi per hour, about how long did it take him?

15. RATES The scale of the map to Ms. Beck's destination is 1 in. = 25 mi. Ms. Beck drives 500 mi the first day. What map distance has she covered?

16. ✏ WRITE ABOUT IT Explain how you can find an actual distance between two cities if you know the map distance and the scale.

What You'll Learn
How to draw a map from directions

Why Learn This?
To understand how scale is used to draw maps

PROBLEM SOLVING
....
• **Understand**
• **Plan**
• **Solve**
• **Look Back**

PROBLEM-SOLVING STRATEGY
Draw a Diagram

Jan and Kent want to study for a test after school. Jan asked Kent to give her directions from the school to his house. Here is what Kent said: "At the front of the school, turn east. Go 12 blocks on Main Street to Jackson Street, and turn north. Go 8 blocks, and turn east on Oak Street. Then go 5 blocks on Oak Street." Jan asked Kent to draw her a map. What did his map look like?

UNDERSTAND   What are you asked to find?

What facts are given?

PLAN   What strategy will you use?

You can *draw a diagram* according to Kent's directions.

SOLVE   How will you solve the problem?

One way to see what Kent's map might look like is to draw a map to scale. Use the scale 1 in. = 4 blocks, or the ratio $\frac{1}{4}$, to find the map distances.

Main St.: $\frac{1}{4} = \frac{x}{12}$ Jackson St.: $\frac{1}{4} = \frac{x}{8}$ Oak St.: $\frac{1}{4} = \frac{x}{5}$

$4x = 12$ $4x = 8$ $4x = 5$

$x = 3$ in. $x = 2$ in. $x = 1\frac{1}{4}$ in.

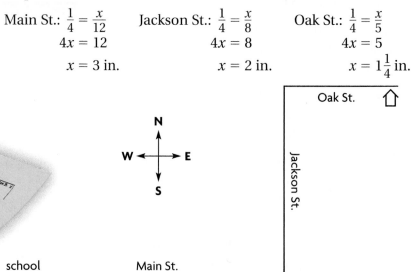

Kent's map should have looked like the one above.

LOOK BACK   How can you check each of the distances on your map?

What if . . . you used the scale 1 in. = 3 blocks? How long would you draw the line for the distance on Main Street?

PRACTICE

Draw a diagram to solve.

1. To get to Paul's house, Arnie walks 4 blocks west, 2 blocks south, and then 2 blocks east. Arnie drew a map of his walk, using the scale 2 in. = 1 block. Show what Arnie's map looks like.

2. To go home from work, Ms. Steel drives $1\frac{1}{2}$ mi east, 2 mi north, $\frac{1}{2}$ mi west, and then 1 mi south. Use a scale of $\frac{1}{2}$ in. = 1 mi to draw a map of her route.

3. Jody is on a 5-day hiking trip. The first day she hikes 6 mi north. The second day she hikes 8 mi east. The third day she hikes 10 mi south. The fourth day she hikes 8 mi west. In what direction and how many miles does she go on the fifth day to return to the starting point? Use the map scale 1 in. = 1 mi.

4. Manuel jogs 20 blocks. He starts out going 5 blocks north and then goes 4 blocks east. Next he goes north again for 1 block and west for 4 blocks. In what direction and how far must he go from that point to meet the first part of his path? How far is it then to his starting point? Use the map scale 1 in. = 1 block.

MIXED APPLICATIONS

Solve.

CHOOSE a strategy and a tool.

- Write an Equation
- Guess and Check
- Make a Table
- Draw a Diagram
- Account for All Possibilities
- Act It Out

Paper/Pencil   Calculator   Hands-On   Mental Math

5. Each swan boat at a park is carrying either 3 or 4 people. There are 5 boats carrying 17 people in all. How many four-person boats are there?

6. There are 50 boys and girls in the school band. There are 4 times as many girls as boys. How many girls are there?

7. A teacher had 330 books. She divided them evenly among classes with 30 students each. How many 30-student classes are there?

8. Joyce walks 10 blocks south, 4 blocks east, 2 blocks north, and 9 blocks west. At what point does she cross her own path?

9. Mike has a square-shaped garden. His garden has an area of 225 ft². He wants to put a border around his garden. What is the length of each side of the garden?

10. Each of 4 friends bowled one game with each of the other friends. How many total games were bowled?

11. Bill's recipe calls for 4 tomatoes, 2 onions, and 5 oz of cheese. He needs to triple the recipe, so he uses 12 tomatoes. How many onions and how much cheese should he use to triple the recipe?

12. ✏️ **WRITE ABOUT IT** Write directions for a route you take, and draw a map of it, using a scale.

LESSON 20.4

What You'll Learn
How to use ratios to recognize similar rectangles that are Golden Rectangles

Why Learn This?
To identify a common feature in architecture and art

VOCABULARY

Golden Ratio

Golden Rectangle

ART **LINK**

The ancient Greeks felt that the Golden Rectangle was more pleasing to the eye than any other rectangle. If you wanted to make a garden in the shape of a Golden Rectangle on a piece of land 25 ft × 20 ft, how big to the nearest foot could your garden be?

GEOMETRY CONNECTION

Golden Rectangles

You can find certain ratios in nature, art, and architecture. Any ratio equivalent to the value of about 1.6 is a **Golden Ratio**. When the length-to-width ratio of a rectangle is approximately 1.6 to 1, the rectangle is called a **Golden Rectangle**. The front of the United Nations Secretariat Building is a Golden Rectangle.

You can write a ratio to identify a Golden Rectangle.

EXAMPLE Are these both Golden Rectangles?

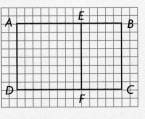

small: $\frac{l}{w} = \frac{8}{5}$ large: $\frac{l}{w} = \frac{15}{6}$ *Write the length-to-width ratio for each rectangle.*

$\frac{8}{5} = 1.6$ $\frac{15}{6} = 2.5$ *Write the decimal equivalent for each ratio.*

$1.6 = 1.6$ 2.5 is not about 1.6.   *Compare the ratios with the Golden Ratio.*

So, the small rectangle is a Golden Rectangle, but the large one is not.

If you start with a Golden Rectangle, you can divide it into a square and another Golden Rectangle.

ACTIVITY **WHAT YOU'LL NEED:** graph paper

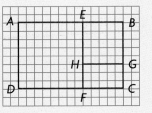

- Draw a 13 × 8 Golden Rectangle on your graph paper. Label it *ABCD*.

- Use the smaller dimension of rectangle *ABCD* to draw a line that makes an 8 × 8 square as shown. Label the 8 × 5 rectangle *EBCF*.

- Use the smaller dimension of rectangle *EBCF* to draw a line that makes a 5 × 5 square. Label the 5 × 3 rectangle *HGCF*.

GUIDED PRACTICE

Tell whether each ratio is a Golden Ratio. Write *yes* or *no* and explain.

1. $\frac{5}{3}$
  2. $\frac{6.5}{3}$
  3. $\frac{10}{6}$
  4. $\frac{24}{15}$

Write the ratio of length to width and the decimal equivalent.
Then tell whether the rectangle is a Golden Rectangle.

5. $l = 8, w = 6$
  6. $l = 15, w = 9$
  7. $l = 16, w = 10$
  8. $l = 26, w = 16$

INDEPENDENT PRACTICE

For Exercises 1–6, tell whether the rectangle is a Golden Rectangle. Explain.

1.
2.5 ft
4 ft

2.
6 m
9.5 m

3.
4.5 yd
12 yd

4.
12 in.
20 in.

5.
10 cm
12 cm

6.
18 ft
30 ft

Problem-Solving Applications

7. HISTORY The Parthenon, an ancient Greek temple in the city of Athens, was built around 450 B.C. The front of the Parthenon has a length of 101 ft and a height of 60 ft. Is the front a Golden Rectangle? Explain.

8. ART Draw and label a Golden Rectangle.

9. ✏ **WRITE ABOUT IT** Explain how to tell whether a rectangle is a Golden Rectangle.

Mixed Review and Test Prep

Complete.

10. 1 ft = __?__ in.
  11. 1 yd = __?__ ft
  12. 16 oz = __?__ lb

The figures are similar. Solve for *x*.

13.
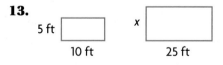
5 ft
10 ft
x
25 ft

14.
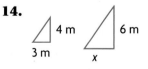
4 m
3 m
6 m
x

15. MONEY You earn $5.00 an hour. If you get a 5% raise, how much do you then earn per hour?

A $5.05
  B $5.20
C $5.25
  D $5.50

16. PROBABILITY A number cube has faces numbered 1–6. What is the probability of tossing a 3 or 5?

F $\frac{1}{6}$
  G $\frac{1}{4}$
  H $\frac{1}{3}$
  J $\frac{1}{2}$

1. **VOCABULARY** A ratio between two sets of measurements is a(n) __?__ . (page 385)

Use the scale 2 in. = 5 ft. Find the actual length. (pages 384–387)

2. scale drawing: 8 in.

3. scale drawing: $11\frac{1}{4}$ in.

4. scale drawing: 16 in.

Use the scale 2 cm = 1 mm. Find the dimension for the drawing. (pages 384–387)

5. actual width of a straw: 7 mm

6. actual length of a staple: 11 mm

7. actual length of a dollar: 156 mm

Use the scale 1 in. = 150 mi to find the actual distance for each map distance. (pages 390–391)

8. 5 in.

9. 12 in.

10. $3\frac{1}{2}$ in.

11. $6\frac{3}{4}$ in.

12. 30 in.

13. $\frac{1}{2}$ in.

14. $1\frac{1}{2}$ in.

15. 20 in.

Use the scale 1 in. = 150 mi to find the map distance. (pages 390–391)

16. 375 mi

17. 1,305 mi

18. 75 mi

19. 1,575 mi

20. 250 mi

21. 700 mi

22. 450 mi

23. 1,125 mi

Solve. (pages 392–393)

24. To get home after school, Jesse walks four blocks north on Main Street and then takes a right on Fox Run Road. His house is the second on the right. Draw a map to show how to get from his house to the school, and give directions.

25. Caryn lives between the gym and the video store. She lives $1\frac{1}{2}$ mi from the gym and $2\frac{1}{4}$ mi from the video store. Draw a map to show how far the gym is from the video store. Use the scale $\frac{1}{4}$ in. = 1 mi.

26. **VOCABULARY** A rectangle that has a length-to-width ratio of about 1.6 to 1 is a(n) __?__ . (page 394)

Write the ratio of length to width and the decimal equivalent. Tell whether the rectangle is a Golden Rectangle. (pages 394–395)

27. $l = 11$ mi
    $w = 4$ mi

28. $l = 6.4$ cm
    $w = 4$ cm

29. $l = 24$ yd
    $w = 15$ yd

30. $l = 3.68$ m
    $w = 2.3$ m

1. Lila sold 47 red or white flowers. She sold 13 more red than white flowers. How many red flowers did she sell?

   A 17
   B 30
   C 33
   D 34
   E Not Here

2. Dana gathered data about the number of hits made by each member of a baseball team. What is her first step in making a histogram of her data?

   F Title the graph.
   G Label the axes.
   H Make a frequency table.
   J Label the scale.
   K Not Here

3. A spinner is divided into 8 equal sections. Four sections are colored red, 3 are colored blue, and 1 is colored green. What is the probability that the spinner will stop on a red section?

   A $\frac{1}{8}$
   B $\frac{1}{2}$
   C $\frac{3}{5}$
   D $\frac{7}{8}$

4. When $y = 3$, what is the value of $4y - 3 \times 2$?

   F 6
   G 9
   H 11
   J 15
   K 18

5. Express 37% as a decimal.

   A 0.37%
   B 37
   C 3.7
   D 0.37

6. The diagram below shows similar figures. What proportion would you use to find x?

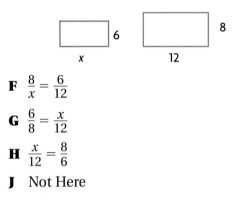

   F $\frac{8}{x} = \frac{6}{12}$
   G $\frac{6}{8} = \frac{x}{12}$
   H $\frac{x}{12} = \frac{8}{6}$
   J Not Here

7. On a map, the distance between Newsport and Hampton is 8 inches. The scale is 1 inch = 40 miles. Find the actual distance.

   A 48 mi
   B 128 mi
   C 240 mi
   D 302 mi
   E 320 mi

8. Which rectangle shown in the diagram below is approximately a Golden Rectangle?

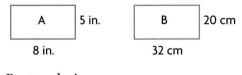

   F Rectangle A
   G Rectangle B
   H Both
   J Neither

MATH FUN!

The Square Percent Puzzle

PURPOSE To practice expressing parts of a whole area as a percent (pages 340–341)

YOU WILL NEED centimeter graph paper, colored markers or pencils

Cut out six 4 × 4 grids. Find 6 ways you can shade 50% of the 4 × 4 grid. Each shaded square must touch at least one other shaded square.

One has been done for you. Write a proportion to prove that your design matches 50%.

A Taxing Situation

PURPOSE To practice applying percents (pages 358–359)

Suppose there is a 5% sales tax on all purchases, and the item you're buying is on sale at 20% off.

Sale!!
20% Off

Decide which method gives you the lower price.

• to calculate the discount first and then figure the tax on the lower price, or

• to add the tax first and then figure the discount on the total.

Take a guess. Test your answer on a price of $100, and again on $20. What you learn may surprise you!

CHECK WHAT YOU SEE

PURPOSE To practice identifying similar and congruent figures (pages 370–371)

How many squares of various sizes are in the checkerboard shown? Are these squares similar? How many congruent squares are there of each size? Make a table to tally your count of each size. Then find the total number of squares in the checkerboard.

SIZE	NUMBER OF SQUARES
1 × 1	16

HOME NOTE Look for other gameboards that have similar and congruent figures.

Earning Interest

In this activity you will see how easy it is to find the amount of interest earned in a bank account by using a spreadsheet.
Look at the data on the spreadsheet below.

ALL	A	B	C	D
1	Principal	Yearly Rate	Interest for 1 Year	Interest for 5 Years
2	$100	0.04	$4.00	$20.00
3	$500	0.06		
4				

To find the interest for 1 year, this formula was used in cell C2.

ALL	C
1	Interest for 1 Year
2	=A2*B2

To find the interest for 5 years, this formula was used in cell D2.

ALL	D
1	Interest for 5 Years
2	=A2*B2*5

1. What formula would you enter in cell C3 to find the interest?

2. What interest would appear in cell C3?

3. What formula would you enter in cell D3 to find the interest?

4. What interest would appear in cell D3?

USING A SPREADSHEET

Copy and complete the table.

	Principal	Yearly Rate	Interest for 1 Year	Interest for 5 Years
5.	$80.00	3%	?	?
6.	$150.00	4%	?	?
7.	$400.00	5.5%	?	?
8.	$800.00	6%	?	?
9.	$1,200.00	6.4%	?	?

Study Guide and Review

Vocabulary Check

1. A comparison of two numbers, such as 1 to 3, is a(n) __?__.
(page 332)

2. The amount of interest earned on the principal is the __?__ interest. (page 360)

3. Any ratio equivalent to a value of about 1.6 is a(n) __?__.
(page 394)

EXAMPLES

• **Find rates and unit rates.** (pages 334–335)

$4.80 for 8 pens

$$\text{rate} \rightarrow \frac{\$4.80}{8}$$

$$\text{unit rate} \rightarrow \frac{\$4.80 \div 8}{8 \div 8} = \frac{\$0.60}{1}$$

• **Use percents to make and interpret circle graphs.** (pages 356–357)

favorite day: 60%, Saturday; 40%, Sunday

60% = 0.60   ← circle = 360°
0.60 × 360° = 216°   *Find the angle*
             measurements.
40% = 0.40
0.40 × 360° = 144°

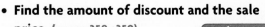

Sunday 40%
Saturday 60%

• **Find the amount of discount and the sale price.** (pages 358–359)

regular price: $88.00

Sale 40%

discount = $88.00 × 40%
         = $88.00 × 0.40
         = $35.20

sale price = $88.00 − $35.20
           = $52.80

EXERCISES

Write the rate and the unit rate in fraction form.

4. $0.90 for 5   **5.** 240 mi in 4 hr

6. $3.60 for 10   **7.** 630 words in 9 min

Use the data below to make a circle graph.

8. Out of 200 people surveyed about their favorite color car, 25% liked red, 50% blue, 15% green, and 10% white.

Use the data from Exercise 8 to find how many people liked each color.

9. red   **10.** blue   **11.** green   **12.** white

The regular price is given. The discount rate is 35%. Find the amount of discount.

13. $24.00   **14.** $62.00   **15.** $110.00

16. $28.50   **17.** $18.00   **18.** $40.75

The regular price is given. The discount rate is 20%. Find the sale price.

19. $45.20   **20.** $16.00   **21.** $250.00

22. $48.99   **23.** $110.75   **24.** $320.00

- **Use similar figures to find the unknown length of a side.** (pages 376–377)

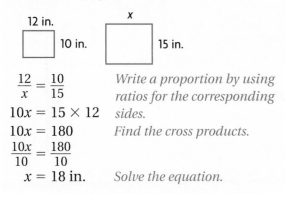

$$\frac{12}{x} = \frac{10}{15}$$ *Write a proportion by using ratios for the corresponding sides.*

$10x = 15 \times 12$ *sides.*

$10x = 180$ *Find the cross products.*

$$\frac{10x}{10} = \frac{180}{10}$$

$x = 18$ in. *Solve the equation.*

The figures in each pair are similar. Find y.

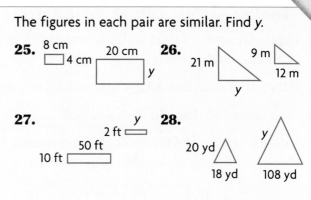

25. **26.**

27. **28.**

- **Use proportions to measure indirectly.**
  (pages 378–379)

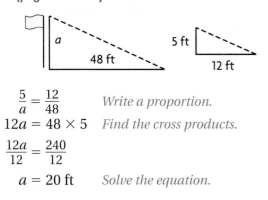

$$\frac{5}{a} = \frac{12}{48}$$ *Write a proportion.*

$12a = 48 \times 5$ *Find the cross products.*

$$\frac{12a}{12} = \frac{240}{12}$$

$a = 20$ ft *Solve the equation.*

Use the two similar right triangles to write a proportion. Then find the missing length.

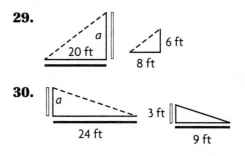

29.

30.

- **Read and use scales on a map.** (pages 390–391)

map scale: 1 in. = 20 mi
map distance: $5\frac{1}{2}$ in.

$$\frac{1}{20} = \frac{5.5}{x}$$ *Write a proportion.*

$1x = 20 \times 5.5$ *Find cross products. Multiply.*

$x = 110$ The actual distance is 110 mi.

Use the scale 1 in. = 50 mi to find the actual distance for each map distance.

31. 6 in. **32.** 9 in. **33.** $10\frac{1}{2}$ in.

Use the scale 1 in. = 40 mi to find the map distance.

34. 160 mi **35.** 320 mi **36.** 380 mi

Problem-Solving Applications

Solve. Explain your method.

37. Golf balls are 3 for $6. Thalia buys 9 of them. How much does she pay?
(pages 344–345)

38. Sonny walks to work. He walks east 2 blocks, then north 4 blocks, and east 1 block. Use a scale of 1 cm = 1 block to draw a map of his route. (pages 392–393)

Performance Assessment

Tasks: Show What You Know

1. Explain how you would set up a proportion to solve the problem. Then solve. **(pages 344–345)**

If 3 cans of soup cost $3.29, how much will 7 cans cost?

2. Find 60% of 85. Explain your method. **(pages 352–355)**

3. Explain how to find the missing dimension of the figure in these two similar rectangles. **(pages 376–377)**

6 in.   *x*   9 in.

2 in.

4. Measure the length of the desk in the scale drawing. Then use the scale to find the actual length of the desk. **(pages 384–387)**

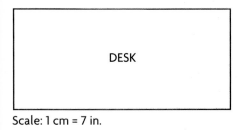

DESK

Scale: 1 cm = 7 in.

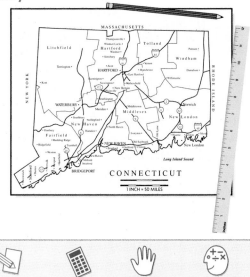

Problem Solving

Solve. Explain your method.

CHOOSE a strategy and a tool.

- **Find a Pattern**
- **Make a Model**
- **Write an Equation**
- **Act It Out**
- **Make a Table**
- **Draw a Diagram**

Paper/Pencil   Calculator   Hands-On   Mental Math

5. One sixth grade class read 720 books during the school year. In May, the class read 20% of the books. How many books did the class read in May? **(pages 352–355)**

6. Ted has a picture that is 4 in. wide × 6 in. long. He wants to reduce the size of the picture so it is 3 in. long. How wide will the new picture be? **(pages 376–377)**

7. A map uses the scale 2 in. = 25 mi. How many inches on the map would equal 375 mi? **(pages 390–391)**

8. To get from her house to school, Janice walks 6 blocks east, 8 blocks north, and 3 blocks west. Draw a map to show the route. Use the scale $\frac{1}{2}$ in. = 1 block. **(pages 392–393)**

Cumulative Review

Solve the problem. Then write the letter of the correct answer.

1. Find the quotient. $42.84 \div 14$ (pages 70–73)

    A. 3.06   **B.** 3.6
    C. 30.6   **D.** 306

2. Subtract. Write in simplest form. $\frac{7}{9} - \frac{4}{6}$
(pages 110 –111)

    A. $\frac{1}{18}$ **B.** $\frac{1}{9}$

    C. $\frac{9}{18}$ **D.** $\frac{3}{3}$

3. Identify the figure. (pages 160–161)

    ◄————•————•

    A. line   **B.** line segment
    C. point   **D.** ray

4. Find the mode. 30, 25, 40, 35, 40
(pages 258–261)

    A. 15   **B.** 25
    C. 30   **D.** 40

5. A number cube is numbered 1–6. What is the probability of rolling an even number?
(pages 272–275)

    A. $\frac{1}{6}$ **B.** $\frac{2}{6}$, or $\frac{1}{3}$

    C. $\frac{3}{6}$, or $\frac{1}{2}$ **D.** $\frac{6}{6}$, or 1

6. Solve for x. $6 + x = 18$ (pages 300–303)

    A. $x = 3$ **B.** $x = 12$
    C. $x = 24$ **D.** $x = 108$

7. Find the number of nickels in $6.00.
(pages 314–315)

    A. 12   **B.** 24
    C. 60   **D.** 120

8. Find the unit rate. 400 km in 8 hr
(pages 334–335)

    A. 5 km per hr   **B.** 50 km per hr
    C. 392 km per hr   **D.** 1,200 km per hr

9. Write as a percent. $\frac{2}{25}$ (pages 336–339)

    A. 2%   **B.** 8%
    C. 25%   **D.** 80%

10. In a circle graph, what angle measurement would represent 20%?
(pages 356–357)

    A. 20°   **B.** 36°
    C. 72°   **D.** 720°

11. The regular price is $118.00. The discount rate is 25%. Find the sale price.
(pages 358–359)

    A. $2.95   **B.** $29.50
    C. $88.50   **D.** $93.00

12. The pair of figures are similar. Find x.
(pages 376–377)

    [4 cm by x rectangle; 10 cm by 6 cm rectangle]

    A. $x = 2.4$ cm   **B.** $x = 12$ cm
    C. $x = 20$ cm   **D.** $x = 24$ cm

13. A map has a scale of 1 in. = 120 mi. Find the actual distance in miles for $2\frac{1}{2}$ in. on the map. (pages 390–391)

    A. 60 mi
    B. 180 mi
    C. 300 mi
    D. 360 mi

MEASUREMENT

LOOK AHEAD

In this chapter you will solve problems that involve

- customary and metric measurements

- perimeter

- circumference

SPORTS LINK

Believe it or not, walking is an Olympic event! In fact, competitive walking is extremely demanding. The table shows some of the winning times from the 1996 Summer Olympics.

EVENT	WINNING TIME
Men's 20-km Walk	1 hr 20 min 7 sec
Men's 50-km Walk	3 hr 43 min 30 sec
Women's 10-km Walk	41 min 49 sec

- Convert all the times to minutes. Give the times to the nearest minute.

- A distance of 20 km is about 12 mi. Estimate the average speed of the winner in the men's 20-km walk. Give your answer to the nearest mph.

Walk It Out

You can find the length of the classroom without using measuring tools. Walk 5 regular steps. Measure that distance in feet or meters. Call this amount your 5-step distance. Walk the length of the classroom counting your steps. Then, using your 5-step distance, estimate the length of the classroom.

PROJECT CHECKLIST

✓ Did you measure your 5-step distance?

✓ Did you walk the length of the classroom?

✓ Did you estimate the length of the classroom?

What You'll Learn
How to change one customary unit of measurement to another

Why Learn This?
To solve everyday problems such as finding the number of feet in 32 yd

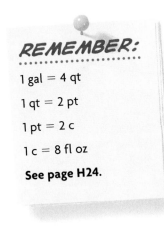

REMEMBER:

1 gal = 4 qt

1 qt = 2 pt

1 pt = 2 c

1 c = 8 fl oz

See page H24.

HISTORY LINK

The United States is the only major country that uses customary units; the rest of the world uses metric units. What are some ways you use customary units?

Customary Measurements

Suppose you have been asked to make punch for a party for 36 people. If you expect each person to drink 1 pt of punch, how many quarts of punch should you make?

To change one unit of measurement to another, you multiply or divide.

To change larger units to smaller units, multiply.

$8\frac{1}{2}$ ft = ■ in.

Think: 1 ft = 12 in.

$8\frac{1}{2} \times 12 = 102$

So, $8\frac{1}{2}$ ft = 102 in.

To change smaller units to larger units, divide.

36 pt = ■ qt

Think: 2 pt = 1 qt

$36 \div 2 = 18$

So, 36 pt = 18 qt.

Another way to change units of measurement is to write a proportion.

EXAMPLE 1 Cheryl measured the length of her classroom as 14 yd. How many feet is this?

$$\frac{1 \text{ yd}}{3 \text{ ft}} = \frac{14 \text{ yd}}{x \text{ ft}} \rightarrow \frac{1}{3} = \frac{14}{x}$$ *Use the relationship of yards to feet to write a proportion.*

$$x = 3 \times 14$$ *Find the cross products.*

$$x = 42$$ *Solve for x.*

So, the classroom is 42 ft long.

• What proportion can you write to find the number of feet in 12 yd?

EXAMPLE 2 Carrie's puppy weighs 72 oz. How many pounds does the puppy weigh?

$$\frac{1 \text{ lb}}{16 \text{ oz}} = \frac{x \text{ lb}}{72 \text{ oz}}$$ *Use the relationship of pounds to ounces to write a proportion.*

$$\frac{1}{16} = \frac{x}{72}$$

$$16x = 72$$

$$\frac{16x}{16} = \frac{72}{16}$$ *Find the cross products. Solve for x.*

$$x = 4.5$$

So, Carrie's puppy weighs 4.5 lb, or $4\frac{1}{2}$ lb.

• How many pounds are in 112 oz?

GUIDED PRACTICE

Tell what you would multiply or divide by to change the unit.

1. feet to inches

2. pounds to tons

3. cups to pints

Change to the given unit.

4. 2 qt = ■ pt

5. 84 in. = ■ ft

6. 8 lb = ■ oz

INDEPENDENT PRACTICE

Change to the given unit.

1. 20 ft = ■ in.

2. 60 in. = ■ ft

3. 30 yd = ■ ft

4. 10 c = ■ fl oz

5. 6 qt = ■ pt

6. 8 days = ■ hours

Use a proportion to change to the given unit.

7. 5 years = ■ months

8. 2 T = ■ lb

9. 8 c = ■ pt

10. 3 pt = ■ fl oz

11. 5 gal = ■ qt

12. 32 pt = ■ gal

Problem-Solving Applications

13. CONSUMER MATH Peter bought 108 in. of fabric on Wednesday and 12 yd of fabric on Saturday. How many yards of fabric did Peter buy altogether?

14. ESTIMATION Pat's 10-step distance is 40 ft. She walks 210 steps to school each day. How many feet is Pat's house from the school?

15. MEASUREMENT A sign on a bridge says, "limit 12,000 lb." How many tons are allowed on the bridge?

16. ✏ WRITE ABOUT IT Explain how you know whether to multiply or divide when changing one unit to another.

Mixed Review and Test Prep

Tell which unit of measurement is larger.

17. millimeter or centimeter

18. kilometer or meter

19. liter or milliliter

20. gram or kilogram

Use the scale 2 cm = 5 m to solve.

21. ■ cm = 15 m

22. 18 cm = ■ m

23. 17 cm = ■ m

24. ■ cm = 21.25 m

25. GEOMETRY The angles of an equilateral triangle are congruent and have a sum of 180°. What is the measure of each angle?

A 180°   **B** 90°   **C** 60°   **D** 45°

26. PROBABILITY If you roll a number cube numbered 7–12, what is the probability of it showing an odd number less than 10?

F $\frac{1}{3}$ **G** $\frac{1}{4}$ **H** $\frac{1}{6}$ **J** $\frac{1}{12}$

ALGEBRA CONNECTION

Metric Measurements

What You'll Learn
How to change one metric unit of measurement to another

Why Learn This?
To solve everyday problems such as how many milliliters there are in 4 L of water

Metric measurement uses the base 10 number system. The relationships between metric units are the same as the relationships between place-value positions. Each unit is 10 times greater than the preceding unit.

To change larger metric units to smaller ones, multiply by a power of 10.	To change smaller metric units to larger ones, divide by a power of 10.
$4.5 \text{ m} = \blacksquare \text{ cm}$	$300 \text{ mL} = \blacksquare \text{ L}$
Think: 1 m = 100 cm	**Think:** 1,000 mL = 1 L
$4.5 \times 100 = 450$	$300 \div 1,000 = 0.3$
So, 4.5 m = 450 cm.	So, 300 mL = 0.3 L.

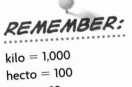

REMEMBER:
....
kilo = 1,000
hecto = 100
deka = 10
deci = 0.1
centi = 0.01
milli = 0.001

See page H24.

You can also write a proportion to change metric units of measurement.

EXAMPLE 1 Rosa is at the beach. Her sand pail holds 0.75 L of water. How many milliliters is this?

$$\frac{1 \text{ L}}{1,000 \text{ mL}} = \frac{0.75 \text{ L}}{x \text{ mL}}$$ *Use the relationship of liters to milliliters to write a proportion.*

$$\frac{1}{1,000} = \frac{0.75}{x}$$

$$x = 1,000 \times 0.75$$ *Find the cross products.*
$$x = 750$$ *Solve for x.*

So, Rosa has 750 mL of water in her pail.

SCIENCE LINK

In metric measurement there is a link between volume and mass. One milliliter of water has a mass of one gram. So, in Example 1, what is the mass of the 0.75 L of water?

EXAMPLE 2 When Rosa's pail is full of sand, it weighs 3,500 g. How many kilograms is this?

$$\frac{1 \text{ kg}}{1,000 \text{ g}} = \frac{x \text{ kg}}{3,500 \text{ g}}$$ *Use the relationship of kilograms to grams to write a proportion.*

$$\frac{1}{1,000} = \frac{x}{3,500}$$

$$1,000x = 3,500$$ *Find the cross products.*

$$\frac{1,000x}{1,000} = \frac{3,500}{1,000}$$

$$x = 3.5$$ *Solve for x.*

So, Rosa's pail weighs 3.5 kg.

• How many kilograms are in 1,800 g?

GUIDED PRACTICE

Tell what you would multiply or divide by to change the unit.

1. meters to centimeters

2. milliliters to liters

3. meters to kilometers

4. kilograms to grams

Change to the given unit.

5. 5 m = ■ km

6. 35 cm = ■ mm

7. 9 L = ■ mL

8. 2 g = ■ kg

INDEPENDENT PRACTICE

Tell what you would multiply or divide by to change the unit.

1. liters to kiloliters

2. decimeters to centimeters

3. millimeters to meters

4. kiloliters to liters

5. grams to kilograms

6. decigrams to grams

7. kilometers to meters

8. centimeters to kilometers

Complete the pattern.

9. 1 m = 100 cm
   0.1 m = ■ cm
   0.01 m = ■ cm

10. 1 g = 0.001 kg
   10 g = ■ kg
   100 g = ■ kg

11. 100 mL = ■ L
   10 mL = ■ L
   1 mL = ■ L

Change to the given unit.

12. 200 kL = ■ L

13. 40 m = ■ cm

14. 500 m = ■ km

15. 10 g = ■ kg

16. 1,000 m = ■ km

17. 1 mm = ■ m

Use a proportion to change to the given units.

18. 440 g = ■ mg

19. 9 L = ■ mL

20. 18 kL = ■ L

21. 0.34 m = ■ km

22. 425,000 mg = ■ g

23. 320 mm = ■ cm

Problem-Solving Applications

24. GEOGRAPHY The north rim of the Grand Canyon is about 2,512 m above sea level. How many kilometers is this?

25. SPORTS Jason's gym bag weighs 2,000 g. How many kilograms does it weigh?

26. ART Shawna bought 3 m of tissue paper to use for an art project. She used 200 cm of the paper. Did she have any paper left over? If so, how much?

27. COOKING Maria needs 2,500 mL of soup so each of her 5 guests will have the same amount. How many liters of soup does Maria have to make to have 2,500 mL?

28. ✎ WRITE ABOUT IT Is it easier to change customary units or metric units? Explain.

Measuring One Dimension

You can use customary or metric units to measure the length of line segments. The line segment below is 4 in. long.

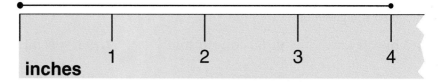

When you measure a length, the measurement you record depends on the tool you are measuring with and how precise you want the measurement to be. The **precision** of a measurement is related to the unit of measure. The smaller the unit of measure used, the more precise the measurement.

Measured to the nearest centimeter, this segment is about 5 cm.

A more precise measurement of the length is 47 mm.

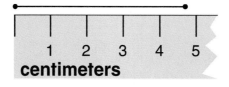

You can measure more precisely with a ruler marked in millimeters than with a ruler marked only in centimeters.

EXAMPLES Tell which measurement is more precise.

A. 3 cm or 32 mm
The millimeter is the smaller unit.
So, 32 mm is more precise.

B. 4 ft or 50 in.
The inch is the smaller unit.
So, 50 in. is more precise.

C. 1 m or 104 cm
The centimeter is the smaller unit.
So, 104 cm is more precise.

GUIDED PRACTICE

Measure each line segment to the nearest centimeter.

1. ●━━━━━━━━━●   **2.** ●━━━━●

Write the letter of the most precise measurement.

3. the length of a steel pipe
   a. 3 m   **b.** 25 dm   **c.** 254 cm

4. the length of a piece of lumber
   a. $7\frac{1}{2}$ ft   **b.** 93 in.   **c.** 8 ft

5. the length of fabric for a shirt
   a. 27 in.   **b.** 2 ft   **c.** 3 ft

6. the height of a teenager
   a. 1.5 m   **b.** 175 cm   **c.** 2 m

INDEPENDENT PRACTICE

Use the ruler to measure the line segment to the given length.

1. nearest inch; nearest $\frac{1}{2}$ inch

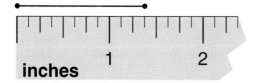

2. nearest $\frac{1}{2}$ inch; nearest $\frac{1}{4}$ in.

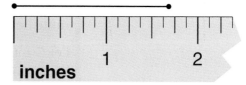

Tell which measurement is more precise.

3. 7 ft or 85 in.   **4.** 53 mm or 5 cm   **5.** 71 mm or 7 cm   **6.** 25 in. or 2 ft

7. 111 in. or 3 yd   **8.** 45 mm or 4 cm   **9.** 395 in. or 32 ft   **10.** 8.2 cm or 82.5 mm

11. 400 m or 0.4 km   **12.** 27 yd or 83 ft   **13.** 10,520 ft or 2 mi   **14.** 30 cm or 300 mm

15. 65 mm or 6 cm   **16.** 5 km or 4,902 m   **17.** 6 cm or 62 mm   **18.** 2 m or 201 cm

19. 211 mm or 21 cm   **20.** 1,508 m or 1.5 km   **21.** 4 km or 4,000 m   **22.** 6.98 m or 699 cm

Problem-Solving Applications

23. CRITICAL THINKING Caryn said the city repaired 2,650 yd of the road in front of her house. Her brother said the city repaired 1.5 mi of road. Which measurement is more precise?

24. MEASUREMENT Jason needs to replace the stick on his kite. He measures the length of the stick as $2\frac{1}{2}$ ft. His sister measures it as $28\frac{3}{4}$ in. Which measurement is more precise?

25. HOBBIES Sylvia measures the length of a piece of yarn. She says it is about 10 yd long. Katrina measures the same piece of yarn and says it is 30 ft long. Which measurement is more precise?

26. ✏ **WRITE ABOUT IT** Explain how to make a measurement more precise.

Networks

What You'll Learn
How to use networks to find the distance from one place to another

Why Learn This?
To solve everyday problems such as finding the shortest route when traveling

VOCABULARY
network

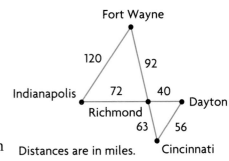

Fort Wayne

120   92

Indianapolis   72   40   Dayton

Richmond

63   56

Distances are in miles.   Cincinnati

You can show the distances between several places or things on a network. A **network** is a graph with vertices and edges. In the network at the right, the towns are the vertices, and the connecting roads between them are the edges.

You can use a network to help you find the shortest or longest route between two locations.

EXAMPLE 1 Cindy's family is planning a trip from Cincinnati to Fort Wayne, with a stop in Dayton. Use the distances shown on the network above to help Cindy's family find the shortest route between Cincinnati and Fort Wayne.

Let C = Cincinnati, D = Dayton, R = Richmond, F = Fort Wayne, and I = Indianapolis.

$CDRF = 56 + 40 + 92 = 188$ *List the different routes*
$CDRIF = 56 + 40 + 72 + 120 = 288$ *and the number of miles*
$CRDRF = 63 + 40 + 40 + 92 = 235$ *for each one.*

$188 < 235 < 288$ *Compare the distances.*

So, the shortest route is 188 mi.

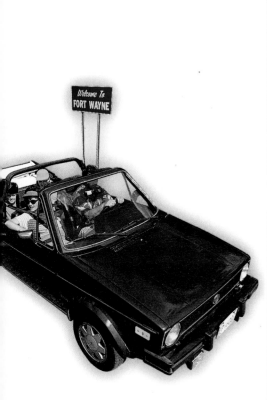

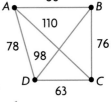

EXAMPLE 2 The network to the right shows distances in kilometers. Starting from C, find the shortest route that includes all four landmarks.

A   80   B
110
78   76
98
D   63   C

$CBAD = 76 + 80 + 78 = 234$ *List the different*
$CBDA = 76 + 98 + 78 = 252$ *routes and the total*
number of kilometers for each.
$CDAB = 63 + 78 + 80 = 221$
$CDBA = 63 + 98 + 80 = 241$ *Compare the distances.*
$CABD = 110 + 80 + 98 = 288$
$CADB = 110 + 78 + 98 = 286$

So, the shortest route is 221 km.

• What is the distance for the route *DBCA*?

GUIDED PRACTICE

Use the network to find the distance for each route.
Distances are in kilometers.

1. *ABDA*   **2.** *CBAD*   **3.** *DBAC*

Use the network to find which route is shorter.

4. *BACD* or *DCBA*   **5.** *CBAD* or *ABDA*   **6.** *CADB* or *BCAD*

INDEPENDENT PRACTICE

Use the network to find the shorter route. Give the distance.

1. *BCDE* or *BEDC*   **2.** *DEFB* or *DEBC*

3. *FEBC* or *FCBE*   **4.** *EFBC* or *CBFD*

5. *EBFD* or *EFDC*   **6.** *BEDF* or *CFDE*

7. *CBFE* or *CBEF*   **8.** *CBFD* or *BCDF*

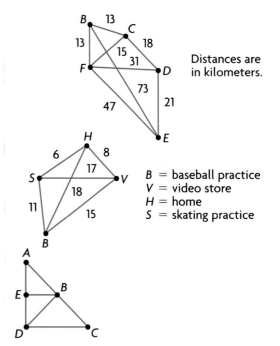

Distances are in kilometers.

Problem-Solving Applications

9. SPORTS Use the network to find the shortest route Kendra can take from home to get her brother to his baseball practice and then go to her skating practice. Distances are in miles.

B = baseball practice
V = video store
H = home
S = skating practice

10. TRANSPORTATION A bus starts at city *A* and travels to the other 4 cities. Use the network to find all the possible routes.

11. ✏️ **WRITE ABOUT IT** Explain how to use a network to find the shortest route.

Mixed Review and Test Prep

Add.

12. 36 + 104 + 27 + 19 + 63   **13.** 56.1 + 113 + 4.6 + 34.7   **14.** 19 + 3.9 + 408 + 86.2

Complete the proportions.

15. $\frac{2}{4} = \frac{\blacksquare}{12}$ **16.** $\frac{5}{7} = \frac{15}{\blacksquare}$ **17.** $\frac{3}{9} = \frac{\blacksquare}{18}$ **18.** $\frac{6}{9} = \frac{30}{\blacksquare}$

19. PERCENT A toy train was worth $8 in 1995. Its value now is 450% of that amount. How much is the train worth now?
  A $3.60   **B** $36.90
  C $44.00   **D** $360.00

20. FRACTIONS Ed rode $3\frac{1}{2}$ miles to the park. Riding home along a scenic route, he traveled $4\frac{1}{4}$ miles. What was the round-trip distance?
  F $7\frac{1}{4}$ mi   **G** $7\frac{1}{2}$ mi   **H** $7\frac{3}{4}$ mi   **J** 8 mi

Perimeter

For exercise, John walks every evening. He walks a path around the triangular park shown at the right. How can John find the total distance around the park?

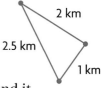
2 km
2.5 km
1 km

The **perimeter** of a polygon is the distance around it.
To find the perimeter, add the lengths of the sides of the polygon.

ACTIVITY

WHAT YOU'LL NEED: metric ruler

• Draw a square, a rectangle, a rhombus, a trapezoid, and a parallelogram. Make a table like this one for your data.

Figure	Side 1	Side 2	Side 3	Side 4	Perimeter	Formula
Square						

• Measure and record the lengths of the sides of each of your quadrilaterals. Then find and record each perimeter.

• For each quadrilateral, describe the relationship between the lengths of the sides and the perimeter. Then generate a formula for finding the perimeter and record.

• Draw and measure another square, rectangle, rhombus, trapezoid, and parallelogram. Use your formulas to find the perimeters. Record in your table. Did your formulas work for these figures?

• **CRITICAL THINKING**   Could you use your formula for the perimeter of a rectangle to find the perimeter of any rectangle? Explain.

You can compare opposites sides to find a missing length.

EXAMPLE   The figure below is a floor plan of Jason's bedroom. All the lengths are in feet. Find the missing lengths. Then find the perimeter.

$x = 7 - 5 = 2$
$y = 8 + 6 = 14$

$P = 8 + 2 + 6 + 5 + 14 + 7$
$P = 42$

8
x   6
7
5
y

So, $x = 2$ and $y = 14$. The perimeter is 42 ft.

GUIDED PRACTICE

Find the perimeter. Write the formula you used.

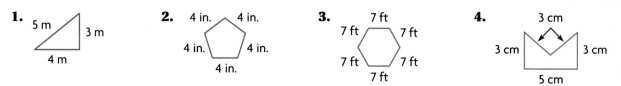

1. 5 m, 3 m, 4 m

2. 4 in., 4 in., 4 in., 4 in., 4 in.

3. 7 ft, 7 ft, 7 ft, 7 ft, 7 ft

4. 3 cm, 3 cm, 3 cm, 5 cm

INDEPENDENT PRACTICE

Find the missing lengths. Then find the perimeter.

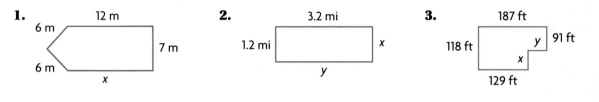

1. 12 m, 6 m, 6 m, 7 m, x

2. 3.2 mi, 1.2 mi, x, y

3. 187 ft, 118 ft, y, 91 ft, x, 129 ft

The perimeter is given. Find the missing length.

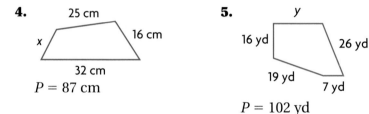

4. 25 cm, 16 cm, x, 32 cm
$P = 87$ cm

5. y, 16 yd, 26 yd, 19 yd, 7 yd
$P = 102$ yd

6. 38 mi, 24 mi, d, 53.7 mi, 45.3 mi
$P = 179.6$ mi

Problem-Solving Applications

7. TRANSPORTATION Laura's route to the doctor's office is shown at the right. She travels from home, to the store, to the doctor's office, and then straight home. She drives a total of 34 km. How many kilometers is the straight route from the doctor's office to her home?

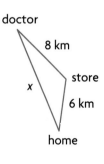

doctor, 8 km, store, x, 6 km, home

8. SPORTS A football field is 120 yd long and $53\frac{1}{3}$ yd wide. What is the perimeter of a football field?

9. GEOGRAPHY Each side of the square base of the Great Pyramid of Cheops, in Egypt, measures 115 m. What formula would you use to find the perimeter? What is the perimeter of the base?

10. ✏️ **WRITE ABOUT IT** Explain how to find the perimeter of your desk top.

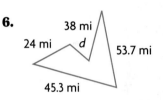

Technology Link

In *Mighty Math Cosmic Geometry*, you can practice finding the perimeter of different figures in the *Amazing Angles* activity. Use Grow Slide Level E.

LAB ACTIVITY

What You'll Explore
How to find the circumference of a circle

What You'll Need
compass, string, ruler, calculator

VOCABULARY
circumference
pi

REMEMBER:

The **radius** of a circle is the distance from the center to a point on the circle. The **diameter** is the distance along a straight line from one point on a circle, through the center, to another point on the circle. **See page H20.**

radius
diameter

ALGEBRA CONNECTION
Circumference

Perimeter is found by adding the lengths of the sides of a polygon. Circles do not have sides, so to find the distance around a circle, use a different method. The distance around a circle is called the **circumference**.

You can use a compass, a string, and a ruler to explore the circumference of a circle.

ACTIVITY 1
Explore

- Open the compass to a width of $3\frac{1}{2}$ in. Use the compass to construct a circle with a $3\frac{1}{2}$-in. radius.

$3\frac{1}{2}$ in.

- Lay the string around the circle. Mark the string where it meets itself.

- Use a ruler to measure the string from its beginning to the mark you made.

mark

Think and Discuss

- What is the diameter of the circle?

- What is the circumference of the circle?

- Use a calculator to divide the circumference by the diameter. How many times as long is the circumference?

- **CRITICAL THINKING**   Do all circles have the same circumference? Explain.

Try This

- Construct a circle with a different diameter. Use string to measure the circumference and a ruler to measure the string. How many times as long as the diameter is the circumference?

- Use the string and ruler to find the circumference of different circular objects. Find how many times longer the circumference is than the diameter. Is the relationship between the diameter and the circumference always the same? Explain.

Instead of using string and a ruler to find the circumference, you can use a formula.

You know that the circumference of a circle is always a little more than 3 times the diameter. This means that the ratio of the circumference to the diameter, $\frac{C}{d}$, is a little more than 3. This ratio, $\frac{C}{d}$, is called **pi**, or π.

When you know the diameter, you can use the formula $C = \pi d$. Since the diameter is twice the radius, use $C = 2\pi r$ when you know the radius. The value of π is an approximation, so write $\approx$ instead of $=$ in your answers.

ACTIVITY 2

Explore

- Use a calculator with a π key, or use 3.14 as an approximation of π.

- Use this formula and key sequence. To the nearest whole number, find the circumference of a circle with a diameter of 4 cm.

  4 cm

  $C = \pi d$

  | π | × | 4 | = | 12.56637061 |

  $C \approx 12.56637061$, or about 13 cm

- Use this formula and key sequence. To the nearest whole number, find the circumference of a circle with a radius of 2.25 m.

  2.25 m

  $C = 2\pi r$

  2 | × | π | × | 2.25 | = | 14.13716694 |

  $C \approx 14.13716694$, or about 14 m

Think and Discuss

- What appeared in the display when you entered π into the calculator?

- Why are there two formulas for the circumference of a circle?

Try This

- To the nearest whole number, find the circumference of a circle with a radius of 5.75 in.

Calculator Activities, page H37

Tell what you would multiply or divide by to change the units.
(pages 406–407)

1. inches to feet   **2.** tons to pounds   **3.** quarts to pints   **4.** ounces to pounds

Change to the given unit. (pages 406–407)

5. 16 qt = ■ gal   **6.** 2 mi = ■ yd   **7.** 30 in. = ■ ft   **8.** 24 oz = ■ lb

Use proportions to change to the given units. (pages 408–409)

9. 320 mm = ■ cm   **10.** 16 L = ■ mL   **11.** 18 mg = ■ g

12. 0.0086 kg = ■ g   **13.** 50 mm = ■ cm   **14.** 50 cm = ■ mm

15. 3 g = ■ mg   **16.** 5 kg = ■ g   **17.** 1.02 kg = ■ g

18. 380 L = ■ mL   **19.** 325 cm = ■ m   **20.** 0.04 km = ■ m

Tell which measurement is more precise. (pages 410–411)

21. 14 cm or 143 mm   **22.** 4 ft or 1 yd   **23.** 14 in. or 1 ft

24. 1 m or 104 cm   **25.** 20 in. or 2 ft   **26.** 2 m or 215 cm

27. 3 L or 3,100 mL   **28.** 975 g or 1 kg   **29.** 3 yd or 10 ft

30. VOCABULARY A graph with vertices and edges is called
a(n) __?__. (page 412)

Use the network to find the length of each route. (pages 412–413)

31. *EFGH*   **32.** *FHGE*

33. *HFGE*   **34.** *HGFE*

35. *EGHF*   **36.** *GEFH*

37. VOCABULARY The distance around a polygon is its __?__.
(page 414)

Find *x*. Then find the perimeter. (pages 414–415)

38.   **39.** a regular pentagon   **40.**

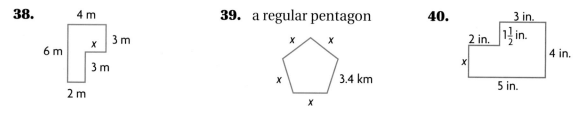

1. Seth jogged $3\frac{1}{4}$ laps around the track. His friend Susan jogged $5\frac{3}{4}$ laps. How many laps did Seth and Susan jog altogether?

   A 8
   B 9
   C $12\frac{1}{4}$
   D 15

2. The cost of a pizza was shared equally among 5 friends. If p represents the cost of the pizza, which algebraic expression shows the amount each friend paid?

   F $\frac{5}{p}$
   G $p + 5$
   H $5 - p$
   J $\frac{p}{5}$
   K $5p$

3. The area of a rectangular room is 42 square feet. The width of the room is 6 feet. What is the length of the room?

   A 7 ft
   B 12 ft
   C 14 ft
   D 36 ft
   E Not Here

4. Which of the following numbers are both equivalent to $\frac{3}{5}$?

   F 3.5; 35%
   G 66%; 6.6
   H 0.6; 60%
   J 40%; 0.4
   K Not Here

5. The regular price of a bedspread is $56. If it is discounted 30%, what is the sale price?

   A $16.80
   B $22.60
   C $34.50
   D $39.20

6. Julie wants to make a scale drawing of her bedroom. Which of the following would be a good scale for her to use?

   F 1 in.:1 cm
   G 1 in.:2 ft
   H 1 in.:25 ft
   J 1 in.:1 mi

7. Carla needs to change 8 kilometers into meters. What should she do?

   A Divide by 1,000
   B Multiply by 100
   C Multiply by 1,000
   D Divide by 10
   E Not Here

8. The diagram shows the dimensions of a lot. The perimeter of the lot is 211 feet. What is the missing length?

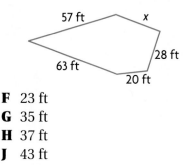

   F 23 ft
   G 35 ft
   H 37 ft
   J 43 ft

LOOK AHEAD

In this chapter you will solve problems that involve

- **estimating area**

- **the strategy** *use a formula*

- **finding area of triangles, parallelograms, and circles**

- **changing length and width**

SCIENCE **LINK**

Gardens come in all different sizes. A backyard garden is usually a fraction of an acre in size. On the other hand, the average-size farm in the United States is about 470 acres.

The acre is a unit of area. You can estimate an acre by picturing a square approximately 200 feet × 200 feet.

- About how many square feet is one acre?

- A garden is about 50 ft × 100 ft. What fraction of an acre is the garden?

- How many quarter-acre gardens can fit in the average farm in the United States?

Landscaping with Trees

In a landscaping plan, a tree is shown as a circle. The diameter of the circle is the width of the tree.

Design a landscaping plan for a garden. Decide what size garden you want. Then, make a scale drawing of the garden. Decide what types of trees to plant, and find out the width of each tree when it is fully grown. Draw the circles for your trees to scale. Estimate how much of the garden's area is covered with trees.

MEASURING AREA

Tree	Height	Width
Norway Spruce	80 ft	30 ft
Gingko	80 ft	60 ft
Sugar Maple	60 ft	40 ft
Norway Maple	50 ft	50 ft
Flowering Dogwood	15 ft	20 ft

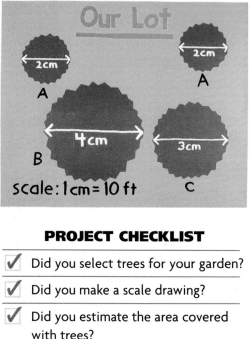

Our Lot

A — 2cm
A — 2cm
B — 4cm
C — 3cm

scale: 1cm = 10 ft

PROJECT CHECKLIST

✓ Did you select trees for your garden?

✓ Did you make a scale drawing?

✓ Did you estimate the area covered with trees?

Estimating Area

You have learned that area is the number of square units needed to cover the surface of a figure. To find area, you can count the number of square units inside the figure. How would you find the area of each figure?

 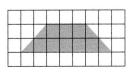

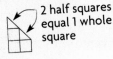

REMEMBER:
....
When you count square units, two half squares equal one whole square.
See page H25.

2 half squares equal 1 whole square

$A = 4$ units^2

Because it is difficult to find the area of an irregular figure, you can use estimation. To estimate an area, use a grid and count each square that is full, almost full, or about half full.

EXAMPLE Estimate the area of the figure on the grid above. Each square of the grid represents 1 mi^2.

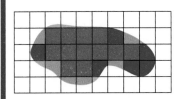

Count full squares.

Count almost-full squares.

Count half-full squares.

Do not count squares that are less than half full.

full (red): 12
almost full (blue): 6
half full (green): 6

$12 + 6 + 6(\frac{1}{2})$

$12 + 6 + 3 = 21$

Find the sum of the squares counted.

So, the area of the figure is about 21 mi^2.

• The area of the grid is 50 mi^2. Is the answer 21 mi^2 a reasonable estimate of the area of the figure on the grid?

• Estimate the shaded area of each figure below. Each square represents 1 yd^2.

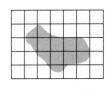

GUIDED PRACTICE

Estimate the area of the figure. Each square is 1 cm².

1.

2.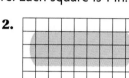

3.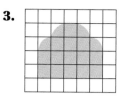

4.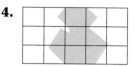

INDEPENDENT PRACTICE

Estimate the area of the figure. Each square is 1 in.²

1.

2.

3.

4.

5.

6.

7.

8.

Problem-Solving Applications

9. **LANDSCAPING** In the diagram of Sean's landscaping design, each square is 1 ft². Sean needs to find the area so that he can buy enough plants to cover the area. Estimate the area.

10. Draw an irregular figure on graph paper. Then estimate the area in square units.

11. ✏️ **WRITE ABOUT IT** Explain how to estimate the area of an irregular figure.

Mixed Review and Test Prep

Find the value.

12. $a \times b$ for $a = 4$ and $b = 5$

13. $\frac{1}{2} \times c$ for $c = 24$

Change to the given unit.

14. 2 qt = ▧ pt

15. 60 in. = ▧ ft

16. 16 oz = ▧ lb

17. 90 ft = ▧ yd

18. **ALGEBRA** If Jim is twice as tall as Cissy, and Jim's height is x, what expression represents Cissy's height?

   A $2x$ **B** x

   C $\frac{1}{2}x$ **D** $x - 2$

19. **PATTERNS** For the pattern 4, 9, 15, 22, . . . , the next number is—

   F 28   **G** 29

   H 30   **J** 31

PROBLEM-SOLVING STRATEGY

Use a Formula

PROBLEM SOLVING
• • • • • • • • • • • • • • • •
- **Understand**
- **Plan**
- **Solve**
- **Look Back**

One way to find the area of a figure is to count the number of square units. You can find the area of a figure another way.

Brad is using square tiles to cover the lid of his brother's rectangular toy box and the top of his mother's square table. Each tile is 1 in.^{2} The toy box is 36 in. long and 18 in. wide. Each side of the tabletop is 22 in. How many tiles will Brad need?

UNDERSTAND   What are you asked to find?

What information is given?

PLAN   What strategy will you use?

You can *use a formula* to find the area of the toy box lid and the area of the table top. Since each tile is 1 in.^{2}, the number of tiles equals the total area.

SOLVE   How will you solve the problem?

You need to find the areas of two figures. Use the formulas for the area of a rectangle and the area of a square. Then add the two areas.

rectangle: area = length × width
$$A = lw$$
$$A = 36 \times 18$$
$$A = 648$$

square: area = side × side
$$A = s^2$$
$$A = 22^2$$
$$A = 484$$

$$648 + 484 = 1{,}132 \leftarrow \text{sum of the areas}$$

So, Brad needs 1,132 in.^{2} of tile, or 1,132 tiles.

LOOK BACK   What other strategy could you use to solve the problem?

What if . . . the toy box lid were 24 in. long and 13 in. wide? What would the area of the lid be?

CONSUMER LINK

A baseball "diamond" is actually a square measuring 90 ft on each side. If sod costs $2.50 per square yard and grass seed costs $0.25 per square foot, which method is the cheaper way to replace the infield grass? Explain.

PRACTICE

Use a formula to solve.

1. Matilda is putting tile flooring in a dollhouse. The living room is 3 ft × 2.5 ft. How many square feet of tile does she need for the living room?

2. A tablecloth measures 81 cm on each side. What is the area of the tablecloth?

3. Abby is going to replace two sides of her nylon tent. Each rectangular side is 2.5 m × 3.2 m. How much nylon does she need? If nylon costs $2.60 per square meter, how much should she plan to spend?

4. Sonny is buying carpet for his porch, which is 15 ft × 10 ft. How much carpet does Sonny need? If the carpet costs $5 per square foot, how much will the carpet cost?

MIXED APPLICATIONS

Solve.

CHOOSE a strategy and a tool.
- Work Backward
- Draw a Diagram
- Find a Pattern
- Write an Equation
- Use a Formula
- Make a Chart

Paper/Pencil   Calculator   Hands-On   Mental Math

5. Simone's yard is 24 ft long and 16 ft wide. Fencing costs $2.50 a foot. How much will it cost to fence in her yard?

6. Dean leaves at 8:00 A.M. His destination is 118 mi away. He travels 60 mi per hour. At what time will he arrive?

7. At a craft show, Polly sold 38 wicker baskets in 3 days. She sold 10 baskets on the second day and 16 baskets on the third day. How many baskets did she sell on the first day?

8. Dolly's dining room is 25 ft × 27 ft. Will a dining room set that takes up an area of 300 ft^2 fit in her dining room? Explain.

9. Ken bounces a ball. With each bounce, the ball goes half as high as on the previous bounce. On the first bounce, it bounces 116 in. How high does it bounce on the sixth bounce?

10. Ace's Computer Company sells 4 more computers each day than the day before. The company sold 56 computers on Friday. How many had it sold on Monday?

11. Glenda spent a total of $34.71 at a zoo. She spent $18.00 for admission, $12.21 for a picture, and the rest on lunch. How much did Glenda spend on lunch?

12. Mark wants to put a $2\frac{1}{2}$-ft chest against a $12\frac{1}{4}$-ft wall. If he centers the chest, how far will it be from each end of the wall?

13. Temporary Hires employs 340 people. Women make up 75% of its workforce. How many women work at Temporary Hires?

14. ✏️ **WRITE ABOUT IT** Write a problem that you can solve with the formula for the area of a rectangle.

ALGEBRA CONNECTION

Area of Triangles and Parallelograms

In the activity below, you will explore how a rectangle and a parallelogram are related.

ACTIVITY

WHAT YOU'LL NEED: scissors, graph paper

• Copy parallelogram *ABCD* on graph paper. Draw a dashed line perpendicular to the base as shown. This is the height.

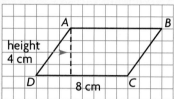

• Count the number of squares and half-squares to find the area. Record the length of the base, the height, and the area in a table like the one below.

Figure	Base (Length)	Height (Width)	Area
Parallelogram	?	?	?
Rectangle	?	?	?

• Cut out the parallelogram. Then cut along the dashed line. Slide the triangle across until it is on the right, forming a rectangle.

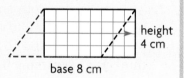

• Find the area of the rectangle. Record the length, width, and area in your table.

Talk About It

• **CRITICAL THINKING** What is the relationship between the dimensions and area of the parallelogram and the dimensions and area of the rectangle?

• What is the formula for the area of a rectangle? What formula can you write for the area of a parallelogram?

• Draw a parallelogram. Use your formula to find the area.

CULTURAL LINK

Geometric figures were first used more than 15,000 years ago in cave paintings and pottery decorations. Today, some native people of the American Southwest still decorate their pottery with the traditional geometric patterns of their ancient ancestors, the Anasazi. What geometric shapes do you recognize in the pottery shown below?

EXAMPLE 1 Find the area of the parallelogram at the right.

$A = bh$ *Write the formula.*

$A = 4\frac{3}{4} \times 3\frac{1}{2} = \frac{19}{4} \times \frac{7}{2}$ *Replace b with $4\frac{3}{4}$ and h with $3\frac{1}{2}$.*

                                                  Multiply.

$A = 16\frac{5}{8}$

So, the area of the parallelogram is $16\frac{5}{8}$ ft^2.

$3\frac{1}{2}$ ft

$4\frac{3}{4}$ ft

EXAMPLE 2 Find the area of the parallelogram with the given measurements.

A. $b = 14$ yd, $h = 8$ yd

$A = bh$ *Write the formula.*

$A = 14 \times 8$ *Replace b with 14 and h with 8.*

$A = 112$ ← 112 yd^2

B. $b = 1.5$ m, $h = 2.6$ m

$A = bh$

$A = 1.5 \times 2.6$

$A = 3.9$ ← 3.9 m^2

GUIDED PRACTICE

Use a formula to find the area of each parallelogram.

1. 4 ft   6 ft

2. 3.5 m   7.2 m

3. 5 cm   8.5 cm

4. $b = 2.5$ m
    $h = 1.5$ m

5. $b = 9$ in.
    $h = 12$ in.

6. $b = 14$ ft
    $h = 20\frac{1}{2}$ ft

Area of Triangles

You can also use a formula to find the area of a triangle.

• Draw three parallelograms on graph paper. Find the area of each. Record the base, height, and area in a table.

Figure	Base	Height	Area	Area of One Triangle
1				

• Cut each parallelogram in half to form congruent triangles. Find the area of each. Record in your table.

• What is the relationship between the area of any of your parallelograms and one of its triangles? What formula can you write for the area of a triangle?

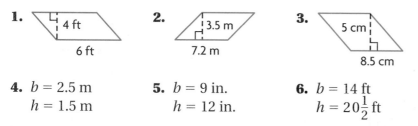

The formula for the area of a triangle is $A = \frac{1}{2}bh$.

You can use this formula to find the area of any triangle.

EXAMPLE 3 Sara needs to replace the sail on her remote-controlled model sailboat. She needs to find the area of the sail shown in the diagram at the right. What is the area of the sail?

4 ft

2 ft

$A = \frac{1}{2}bh$ *Write the formula.*

$A = \frac{1}{2} \times 2 \times 4$ *Replace b with 2 and h with 4.*

$A = \frac{1}{2} \times 8$ *Multiply.*

$A = 4$

So, the area is 4 ft².

EXAMPLE 4 Find the area of each triangle.

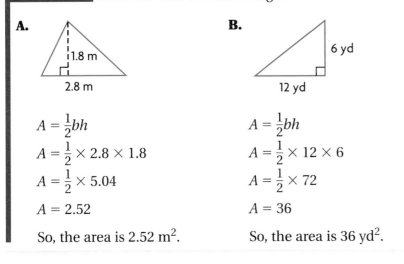

A.

1.8 m

2.8 m

B.

6 yd

12 yd

$A = \frac{1}{2}bh$

$A = \frac{1}{2} \times 2.8 \times 1.8$

$A = \frac{1}{2} \times 5.04$

$A = 2.52$

So, the area is 2.52 m².

$A = \frac{1}{2}bh$

$A = \frac{1}{2} \times 12 \times 6$

$A = \frac{1}{2} \times 72$

$A = 36$

So, the area is 36 yd².

INDEPENDENT PRACTICE

Use a formula to find the area of the parallelogram.

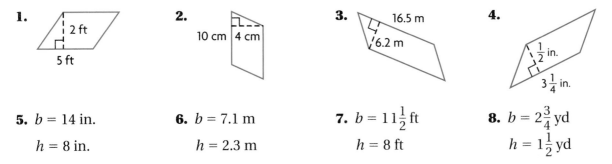

1. 2 ft   5 ft

2. 10 cm   4 cm

3. 16.5 m   6.2 m

4. $\frac{1}{2}$ in. $3\frac{1}{4}$ in.

5. $b = 14$ in.

    $h = 8$ in.

6. $b = 7.1$ m

    $h = 2.3$ m

7. $b = 11\frac{1}{2}$ ft

    $h = 8$ ft

8. $b = 2\frac{3}{4}$ yd

    $h = 1\frac{1}{2}$ yd

Use a formula to find the area of the triangle.

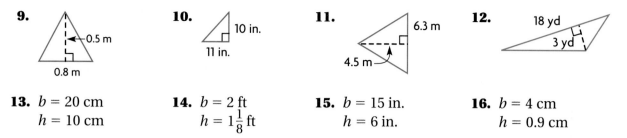

9.   **10.**   **11.**   **12.**

13. $b = 20$ cm   **14.** $b = 2$ ft   **15.** $b = 15$ in.   **16.** $b = 4$ cm
   $h = 10$ cm $h = 1\frac{1}{8}$ ft $h = 6$ in. $h = 0.9$ cm

17. Look at parallelogram $ABCD$. Count the squares. One square equals 1 cm². What is the area of parallelogram $ABCD$? of $\triangle BCD$?

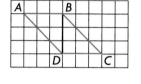

Problem-Solving Applications

For Problems 18–19, use the figure at the right.

18. CRITICAL THINKING A square table has 4-ft sides. When triangular flaps are raised on two opposite sides, the table takes the shape of a parallelogram. The base of each triangle is 2 ft. What is the area of the parallelogram?

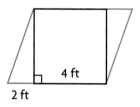

19. GEOMETRY If the sides of the square were 6 ft and the base of each triangle were 4 ft, what would the area of the parallelogram be?

20. ✏️ **WRITE ABOUT IT** Explain how the formula for the area of a triangle relates to the formula for the area of a parallelogram.

Mixed Review and Test Prep

Find the product.

21. 2×16 **22.** 14.5×2 **23.** $\frac{1}{2} \times 12$

Change to the given unit.

24. 2 cm = ▓ mm   **25.** 1 cm = ▓ m   **26.** 3 g = ▓ kg

27. ALGEBRA Use the table to decide which equation shows the relationship between x and y.

A $y = x + 6$ **B** $y = 4x$
C $x = 4y$ **D** $x = y - 6$

x	2	3	4
y	8	12	16

28. CRITICAL THINKING A sack of flour held 40 pounds. Six pounds of flour leaked out through a hole in the sack. Half of the rest was damaged by water. How much good flour remains?

F 17 lb   **G** 18 lb
H 19 lb   **J** 20 lb

Technology Link

💿 In *Mighty Math Cosmic Geometry,* you can practice finding the area of different figures in the *Amazing Angles* activity. Use Grow Slide Level H.

Changing Length and Width

SPORTS **LINK**

An American football field is 100 yd long and $53\frac{1}{3}$ yd wide. Each end zone is 10 yd deep. A Canadian football field is 110 yd long and 65 yd wide. Each end zone is 25 yd deep. How much more area does a Canadian football field require than an American field?

How do you think that the perimeter and area of a figure change when the dimensions change?

Look at the rectangle at the right. The perimeter is 16 cm. The area is 12 cm².

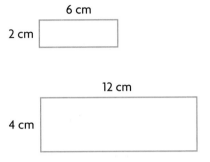

The next rectangle was made by doubling the length and width of the original rectangle. The perimeter is 32 cm, and the area is 48 cm².

Talk About It

- Compare the perimeters of the rectangles. How did the perimeter change when the dimensions were doubled?

- Compare the areas of the rectangles. How does the area of the larger rectangle compare with the area of the smaller rectangle?

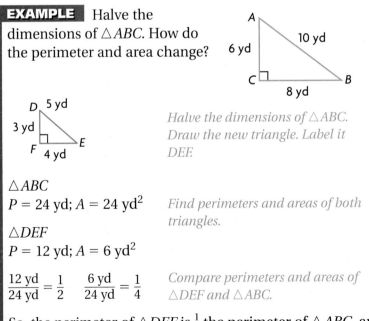

EXAMPLE Halve the dimensions of $\triangle ABC$. How do the perimeter and area change?

Halve the dimensions of $\triangle ABC$. Draw the new triangle. Label it DEF.

$\triangle ABC$
$P = 24$ yd; $A = 24$ yd²

$\triangle DEF$
$P = 12$ yd; $A = 6$ yd²

Find perimeters and areas of both triangles.

$\dfrac{12 \text{ yd}}{24 \text{ yd}} = \dfrac{1}{2}$ $\dfrac{6 \text{ yd}}{24 \text{ yd}} = \dfrac{1}{4}$

Compare perimeters and areas of $\triangle DEF$ and $\triangle ABC$.

So, the perimeter of $\triangle DEF$ is $\frac{1}{2}$ the perimeter of $\triangle ABC$, and the area of $\triangle DEF$ is $\frac{1}{4}$ the area of $\triangle ABC$.

GUIDED PRACTICE

Find the perimeter and area of each figure. Then double the dimensions, and find the new perimeter and area.

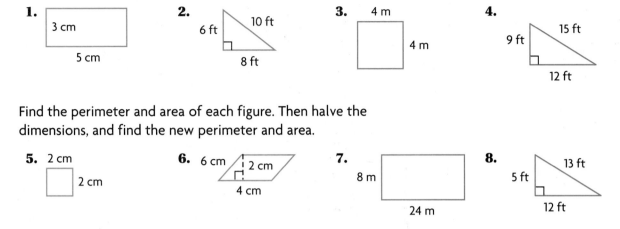

1. 3 cm, 5 cm

2. 6 ft, 10 ft, 8 ft

3. 4 m, 4 m

4. 9 ft, 15 ft, 12 ft

Find the perimeter and area of each figure. Then halve the dimensions, and find the new perimeter and area.

5. 2 cm, 2 cm

6. 6 cm, 2 cm, 4 cm

7. 8 m, 24 m

8. 5 ft, 13 ft, 12 ft

INDEPENDENT PRACTICE

Find the perimeter and area of each figure. Then double the dimensions, and find the new perimeter and area.

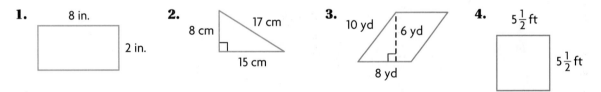

1. 8 in., 2 in.

2. 8 cm, 17 cm, 15 cm

3. 10 yd, 6 yd, 8 yd

4. $5\frac{1}{2}$ ft, $5\frac{1}{2}$ ft

Find the perimeter and area of each figure. Then halve the dimensions, and find the new perimeter and area.

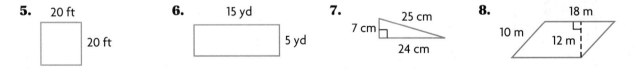

5. 20 ft, 20 ft

6. 15 yd, 5 yd

7. 7 cm, 25 cm, 24 cm

8. 18 m, 10 m, 12 m

Problem-Solving Applications

9. GARDENING Suppose Diantha decides to double the dimensions of her 10-ft × 12-ft garden. How much fencing will she need?

10. GARDENING Nancy is putting a fence around a garden that is 8 ft × 15 ft. How much fencing does she need?

11. ENTERTAINMENT Ralph put two equal-size rectangular tables together for a party. The total length doubled. Did the area become four times as great? Explain.

12. ✏️ WRITE ABOUT IT Explain what happens to the area of a polygon when the dimensions are doubled.

LAB ACTIVITY

What You'll Explore
How to find the area of a circle

What You'll Need
compass, scissors, calculator

The Area of a Circle

You solved problems about the circumference of a circle in Chapter 21. You saw that the circumference is about 3 times the length of the diameter. In this activity you will find the relationship between the area of a circle and its radius.

You can use paper and a compass to explore the area of a circle.

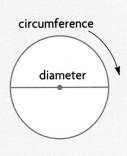

circumference

diameter

ACTIVITY 1

Explore

• Use the compass to draw a circle of any size on your paper.

• Draw a square around the circle.

• The length of each side of the square is equal to the diameter of the circle, or 2*r*. Label your drawing as shown.

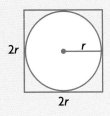

2*r*   *r*

2*r*

Think and Discuss

• What is the area of the square?

• Is the area of the circle greater than, equal to, or less than the area of the square? Explain.

Try This

• Find the area of the square to estimate the area of the circle.

A.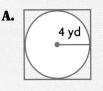
4 yd

B.
3 yd

C.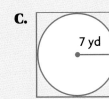
7 yd

Another way to see how the area of a circle and its radius are related is to rearrange the circle to form a parallelogram.

ACTIVITY 2

Explore

- Use the compass to construct a circle on a piece of paper.

- Cut out the circle. Fold it three times, as shown.

- Unfold it, and trace the folds. Shade one half of the circle, as shown.

- Cut along the folds. Fit the pieces together to make a figure that approximates a parallelogram.

Think and Discuss

Think of the figure as a parallelogram. How do the base and the height of the parallelogram relate to the parts of the circle?

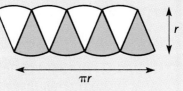

base $(b) = \frac{1}{2}$ the circumference of the circle, or πr

height $(h) =$ the radius of the circle, or r

- What is the formula for the area of a parallelogram?

- Use the formula for the area of a parallelogram to write a formula for the area of a circle. Use πr for the base and r for the height.

Try This

- Use the formula you wrote for the area of a circle to find the areas of each circle in Try This on page 432. Find the area to the nearest square yard. How do your answers compare with the estimates you made?

Technology Link

You can see how the radius or diameter of a circle relates to its area by using E-Lab, Activity 22. Available on CD-ROM and on the Internet at **www.hbschool.com/elab**

What You'll Learn
How to use a formula to find the area of a circle

Why Learn This?
To solve everyday problems such as finding how much of a lawn a sprinkler will water

ALGEBRA CONNECTION

Finding the Area of a Circle

In the Lab Activity, you found that the area of a circle is related to the radius and π. You can use the formula $A = \pi r^2$ to find the area of a circle.

The diagram at the right is of Jon's circular garden. He needs to replace his broken sprinkler. He wants to find the area of the garden so that he will know which sprinkler to buy.

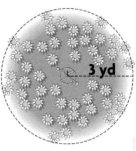

3 yd

EXAMPLE 1 Find the area of Jon's garden to the nearest square yard. Use 3.14 for π.

$A = \pi r^2$ *Write the formula.*

$A \approx 3.14 \times 3^2$ *Replace π with 3.14 and r with 3.*

$A \approx 3.14 \times 9$ *Multiply.*

$A \approx 28.26$

So, the area of Jon's garden is about 28 yd^2.

• Suppose Jon's garden had a radius of $2\frac{1}{2}$ yd. What would the area be, to the nearest square yard?

Sometimes you know only the diameter of a circle.

EXAMPLE 2 Find the area of the circle.

20 cm ÷ 2 = 10 cm   *Find the radius.*

20 cm

$A = \pi r^2$ *Write the formula.*

$A = \pi \times 10^2$ *Replace r with 10. Use a calculator.*

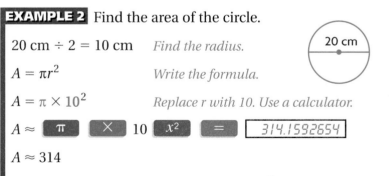

$A \approx$ π   ×   10 x^2 =   314.1592654

$A \approx 314$

So, the area of the circle is about 314 cm^2.

• Suppose a circular dartboard has a diameter of $1\frac{1}{2}$ ft. What is the area of the dartboard, to the nearest square foot?

• **CRITICAL THINKING** Which has the greater area: a circle with a radius of 5 m or a circle with a radius of 50 cm?

Some people believe that crop circles (large, circular flattened areas of wheat fields) are made by alien spacecraft. Actually, most have been done by college students with a good knowledge of algebra and geometry.

Calculator Activities, page H37

GUIDED PRACTICE

Find the area of each circle. Round to the nearest whole number.

1. 3 cm **2.** 20 m **3.** 12 ft **4.** 4.5 m

Find the area of each circle. Round to the nearest tenth.

5. $r = 4$ mm   **6.** $r = 0.9$ m   **7.** $d = 66$ in.   **8.** $d = 100$ ft

INDEPENDENT PRACTICE

Find the area of each circle. Round to the nearest whole number.

1. 8 yd **2.** 15 ft **3.** 9.1 cm **4.** 28 in.

Find the area of each circle. Round to the nearest tenth.

5. $r = 6$ yd   **6.** $d = 32$ mm   **7.** $r = 1.5$ m   **8.** $d = 48$ cm

9. $d = 10.6$ cm   **10.** $r = 2.6$ m   **11.** $r = 25$ yd   **12.** $r = 40$ ft

13. $r = 3.5$ m   **14.** $r = 24$ ft   **15.** $r = 4.4$ cm   **16.** $r = 11$ yd

17. $d = 22$ ft   **18.** $r = 5$ in.   **19.** $r = 14.3$ m   **20.** $r = 10\frac{1}{2}$ ft

21. $r = 17$ in.   **22.** $d = 22.8$ km   **23.** $d = 9.2$ cm   **24.** $r = 21.3$ m

Problem-Solving Applications

25. CONSUMER MATH Two pizzas cost the same amount. One has a radius of 9 in., and the other has an area of about 200 in.^{2} Which is the better buy? Explain.

26. COOKING Barbara has cooked a 10-in.-diameter pancake. She puts the pancake on a round plate whose area is 50 in.^{2} Does the pancake fit? Explain.

27. GEOMETRY Use what you know about the areas of squares and circles to find the area of the shaded part of the figure at the right. (HINT: The diameter of the circle is the same length as a side of the square.)

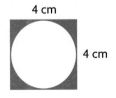

4 cm

4 cm

28. Draw a circle. Determine its radius, and find its area.

29. ✐ **WRITE ABOUT IT** Explain the difference between the area of a circle and the circumference of a circle.

Estimate the area of the figure. (pages 422–423)

1.

$\blacksquare = 1 \text{ ft}^2$

2.

$\blacksquare = 1 \text{ m}^2$

3.

$\blacksquare = 1 \text{ mi}^2$

4.

$\blacksquare = 1 \text{ cm}^2$

Solve. (pages 424–425)

5. Ms. Searl covers her bulletin board with fabric. The bulletin board is 7 yd long and 2 yd wide. How much fabric will she need to cover the board?

6. Carpeting sells for $39 per square yard, installed. A family room is 4 yd × 6 yd. What is the cost of carpeting the family room?

7. A classroom wall is 24 ft long and 9 ft high. Find the area of the wall.

8. Vanessa's room is 15 ft long and 12 ft wide. What is the area of her room?

Find the area of the figure. (pages 426–429)

9.

13 yd
15 yd

10.

8 ft
23 ft

11.

11 in.
16 in.

12. parallelogram
$b = 4.5 \text{ m}; h = 2 \text{ m}$

13. triangle
$b = 7 \text{ ft}; h = 15 \text{ ft}$

14. parallelogram
$b = 1\frac{1}{4} \text{ yd}; h = \frac{3}{5} \text{ yd}$

Double the dimensions of the figure. Then find the perimeter and area of the new figure. (pages 430–431)

15. 2.5 m / 1.25 m

16. 7 mi / 4 mi

17. 5 cm / 8 cm

18. 9 in. / 4 in.

Halve the dimensions of the figure. Then find the perimeter and area of the new figure. (pages 430–431)

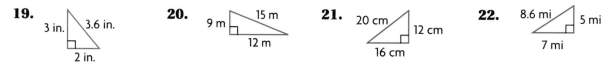

19. 3 in. 3.6 in. 2 in.

20. 9 m 15 m 12 m

21. 20 cm 12 cm 16 cm

22. 8.6 mi 5 mi 7 mi

Find the area of the circle, to the nearest whole unit. (pages 434–435)

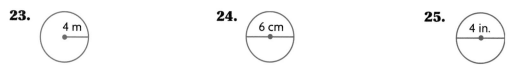

23. 4 m

24. 6 cm

25. 4 in.

1. A school band practiced for $5\frac{3}{4}$ hours during the school week and $2\frac{1}{8}$ hours on the weekend. Altogether, about how long did the band practice?

   A 5 hr   **B** 6 hr
   C 7 hr   **D** 8 hr

2. Sarah has $9\frac{3}{5}$ yards of fabric. This is 6 times the amount she needs to make one pillow. How much fabric does she need for each pillow?

   F $1\frac{1}{2}$ yd

   G $1\frac{3}{5}$ yd

   H $1\frac{3}{4}$ yd

   J $2\frac{1}{3}$ yd

   K Not Here

3. What is the area of a triangle with a base of 3.2 meters and a height of 1.5 meters?

   A 1.8 m^2
   B 2.4 m^2
   C 2.75 m^2
   D 3.2 m^2

4. During a fund-raising activity, students at Elm School sold 1,000 plants. Members of the faculty purchased 15% of the plants. How many plants were purchased by faculty members?

   F 30

   G 45

   H 105

   J 150

5. A model of Town Hall is 30 inches wide and 18 inches high. Town Hall is actually 36 feet high. What is the actual width of Town Hall?

   A 15 ft
   B 45 ft
   C 60 ft
   D 120 ft

6. A tile is in the shape of a regular pentagon. One side of the tile is 6.25 centimeters long. What is the perimeter of the tile?

   F 12.5 cm
   G 25 cm
   H 30 cm
   J 31.25 cm
   K 37.5 cm

7. Walt is making a long-distance telephone call. The operator tells him that it will cost $4.25 for the first 3 minutes. What equation can Walt use to determine the number of quarters, q, he needs to put in the phone?

   A $0.25q = 4.25$

   B $q + 0.25 = 4.25$

   C $\frac{0.25}{q} = 4.25$

   D $q - 0.25 = 4.25$

   E $\frac{q}{0.25} = 4.25$

8. The diameter of a circular garden is 20 meters. What is the area of the garden?

   F About 60 m^2
   G About 122 m^2
   H About 187 m^2
   J About 288 m^2
   K About 314 m^2

MEASURING SOLIDS

LOOK AHEAD

In this chapter you will solve problems that involve

- **estimating volume**

- **finding the volume of prisms and cylinders**

- **changing the length, width, and height of solids**

- **finding the surface area of rectangular prisms**

MUSIC **LINK**

The highest price ever paid for a saxophone at an auction was $145,545. That amount was paid in 1994 for a saxophone owned by Charlie Parker.

- Round the price paid for the saxophone to the nearest thousand dollars and to the nearest ten thousand dollars.

- The highest price ever paid for a music box was about $23,000. About how many times as much is the price of the saxophone?

Volumes of Music

The sound created by tapping a glass jar or blowing across a bottle opening can be changed by adjusting the volume of liquid in the jar or bottle. Collect at least 8 jars or bottles. Experiment with different amounts of water in the bottles and jars to hear the different sounds you can make. Then, play a musical scale, "do, re, mi, fa, sol, la, ti, do," on the bottles or jars.

PROJECT CHECKLIST

✓ Did you adjust the amount of water in the jars or bottles?

✓ Did you play a musical scale on the jars or bottles?

What You'll Learn
How to estimate or find the volume of rectangular prisms and triangular prisms

Why Learn This?
To find how much a household container will hold

VOCABULARY
volume

REMEMBER:

An arrangement of rectangles that folds to form a prism is a **net** for the prism.
See page 202.

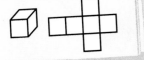

ALGEBRA CONNECTION

Estimating and Finding Volume

To measure area, find the number of square units needed to cover the given surface. When you measure **volume**, you are finding the number of cubic units needed to occupy a given space.

ACTIVITY

WHAT YOU'LL NEED: box net, scissors, tape, 30 centimeter cubes

Jerry keeps his favorite marbles in a small open box. Use centimeter cubes and the net for the box to estimate the volume of the box.

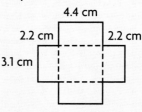

- Use scissors to cut out the net. Fold along the dashed lines, and tape the sides of the net to make a model of the open box.

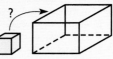

- Estimate how many centimeter cubes would fill the box. Record your estimate. Check your estimate by putting as many cubes as you can in the box. How many cubes can be put in the box?

- Was your original estimate greater or less than the actual volume of the box?

You can visualize about how many cubes will fill a prism.

EXAMPLE 1 Use the cube below to estimate the volume of the prism.

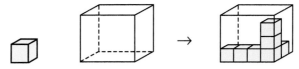

You could put about 4 cubes along the length and about 2 cubes along the width. The bottom layer would have about 8 cubes. You would have about 3 layers of 8 cubes each.

$3 \times 8 = 24$ So, the volume is about 24 cm³.

GUIDED PRACTICE

Use the cube to estimate the volume of the prism.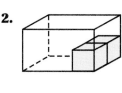

1. **2.** **3.**

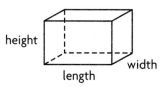

4. Estimate the order of the volumes from the least to the greatest for the cereal box, soda can, shoe box, tennis ball can, milk carton, and wastebasket at the right.

_____

The figure below is a rectangular prism, because the shape of its base is a rectangle. A rectangle has two dimensions: length and width. A rectangular prism has three dimensions: length, width, and height.

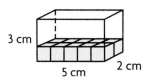

REMEMBER:
....

A **prism** is named by the shape of its base.

See page 196.

Imagine placing a layer of centimeter cubes on the base of the prism. It takes 10 centimeter cubes to make one layer in the rectangular prism shown below.

3 cm

5 cm 2 cm

You can also find the number of cubes that make one layer on the base of the prism by multiplying length and width, or 5×2.

Imagine filling the entire prism with centimeter cubes. You can stack layers of centimeter cubes until the prism is full.

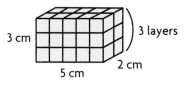

3 cm 3 layers

5 cm 2 cm

It takes 3 layers of 10 cubes to fill the prism. The number of centimeter cubes needed to fill the prism is 3×10, or 30.

Talk About It

The table shows the dimensions and the volume of the prism on page 441 and two other prisms.

- In the table, what relationship do you see between the length, width, height, and volume of the figures?

Length	Width	Height	Volume
5	2	3	30
3	3	6	54
7	4	3	▪

- What formula could you write to find the volume of these rectangular prisms?

- Use centimeter cubes to find the volume of the third rectangular prism in the table. Test your formula by using it to find the volume of the third prism. Did you get the same volume using your formula as when you used cubes?

- What formula could you write to find the volume of any rectangular prism?

EXAMPLE 2 Use the formula $V = lwh$ to find the volume of the prism.

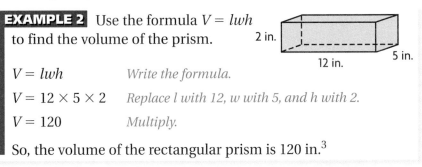

2 in.
12 in.
5 in.

$V = lwh$ *Write the formula.*

$V = 12 \times 5 \times 2$ *Replace l with 12, w with 5, and h with 2.*

$V = 120$ *Multiply.*

So, the volume of the rectangular prism is 120 in.^{3}

The volume of the rectangular prism at the right is 10 m × 6 m × 5 m, or 300 m^3. Suppose you cut the rectangular prism in half to make two triangular prisms.

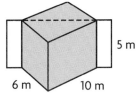

5 m
6 m
10 m

- CRITICAL THINKING What is the volume of one of the triangular prisms?

To find the volume of the triangular prism, use this formula.

$V = \frac{1}{2} \times l \times w \times h$, or $\frac{1}{2} lwh$

$V = \frac{1}{2} \times 10 \times 6 \times 5$

$V = \frac{1}{2} \times 300$

$V = 150$

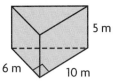

5 m
6 m
10 m

So, the volume of the triangular prism is 150 m^3.

REMEMBER:

A triangular prism has bases that are congruent triangles.

See page 197.

INDEPENDENT PRACTICE

Find the volume of each figure.

1.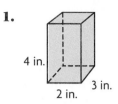
4 in. 3 in. 2 in.

2.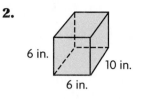
6 in. 10 in. 6 in.

3.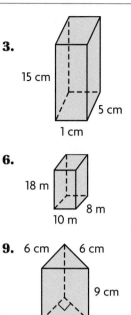
15 cm 5 cm 1 cm

4.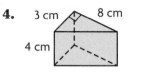
3 cm 8 cm 4 cm

5.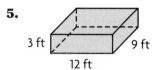
3 ft 9 ft 12 ft

6.
18 m 8 m 10 m

7.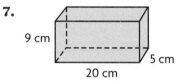
9 cm 5 cm 20 cm

8.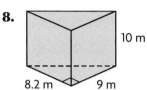
10 m 8.2 m 9 m

9.
6 cm 6 cm 9 cm

Problem-Solving Applications

10. GEOMETRY A toy has parts in the shape of cubes, 2 in. on each side. What is the volume of each part?

11. Extra parts for the toy in Problem 10 can be bought for $0.50 per cubic inch. How much will three new parts cost?

12. LOGICAL REASONING A tree house in the shape of a triangular prism has 8-ft square sides, as shown at the right. The front is a triangle with a height of 6 ft. What is the volume of the tree house?

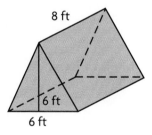
8 ft 6 ft 6 ft

13. ✏️ **WRITE ABOUT IT** Explain why volume is in cubic units.

Mixed Review and Test Prep

Double each factor, and find the product.

14. $5 \times 4 \times 2$ **15.** $10 \times 4 \times 4$ **16.** $8 \times 6 \times 3$ **17.** $2 \times 9 \times 5$ **18.** $10 \times 7 \times 4$

Find the area of each rectangle.

19. $l = 10$ ft, $w = 9$ ft   **20.** $l = 6$ in., $w = 2$ in.   **21.** $l = 4.5$ cm, $w = 3.2$ cm

22. SURVEY Marsha surveyed her math class of 35 students about their favorite song, and 40% chose the same song. How many students chose the same song?

    A 10   **B** 12
    C 14   **D** 16

23. COMPARE Which shows the fractions $\frac{3}{4}, \frac{2}{3}$, and $\frac{5}{8}$ in order from least to greatest?

    F $\frac{2}{3}, \frac{5}{8}, \frac{3}{4}$ **G** $\frac{5}{8}, \frac{3}{4}, \frac{2}{3}$

    H $\frac{5}{8}, \frac{2}{3}, \frac{3}{4}$ **J** $\frac{2}{3}, \frac{3}{4}, \frac{5}{8}$

What You'll Learn
How the volume of a rectangular prism changes when the dimensions change

Why Learn This?
To predict the volume of a solid after it is made larger or smaller

ALGEBRA CONNECTION

Changing Length, Width, and Height

You have learned that doubling or halving the length and width of a rectangle affects the perimeter and area. How do you think changing the length, width, and height of a rectangular prism affects the volume?

ACTIVITY

WHAT YOU'LL NEED: 36 cubes

- Look at prisms *A* and *B* at the right. Use your cubes to build both prisms.
- Copy the table below. Record the length, width, height, and volume of each prism.
- Find the ratios of the dimensions of prism *B* to prism *A*.

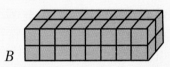

	Length (in units)	Width (in units)	Height (in units)	Volume (in units³)
Prism A	?	?	?	?
Prism B	?	?	?	?
Ratio $\frac{B}{A}$	?	?	?	?

Talk About It

- Look at the ratios for length, width, and height. Are the two prisms similar? Explain.
- How do the length, width, and height of prism *B* compare with those of prism *A*?
- How does the volume of prism *B* compare with that of prism *A*?
- Find the volume of the prism at the right. Then, double the dimensions and find the new volume.

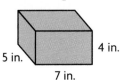

5 in.   7 in.   4 in.

A detergent company makes different-size boxes for its detergent. The dimensions of a large box are sometimes halved to make a smaller box. The volume of the box changes when the length, width, and height are halved.

EXAMPLE 1 The volume of the box shown is 576 in.^{3} Halve the dimensions of the box. How does the volume change?

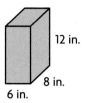

12 in.

8 in.

6 in.

length: 6 in. ÷ 2 = 3 in.   *Halve the dimensions.*
width: 8 in. ÷ 2 = 4 in.
height: 12 in. ÷ 2 = 6 in.

$V = lwh$ *Write the formula.*

$V = 3 \times 4 \times 6$ *Replace l with 3, w with 4, and h with 6.*

$V = 72$ *Multiply.*

$\frac{72}{576} = \frac{1}{8}$ *Compare the volumes.*

So, the volume of the smaller box is 72 in.^{3}, or $\frac{1}{8}$ the volume of the larger box.

- The volume of a 12 cm × 6 cm × 4 cm rectangular prism is 288 cm^3. By what number could you multiply the volume to find the volume of a rectangular prism that is 6 cm × 3 cm × 2 cm?

GUIDED PRACTICE

Double the dimensions of the prism. Find the volume of the larger prism.

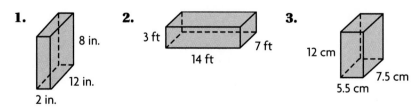

1.   **2.**   **3.**

8 in.   3 ft   7 ft   12 cm

12 in.   14 ft   7.5 cm

2 in.   5.5 cm

Halve the dimensions of the prism. Find the volume of the smaller prism.

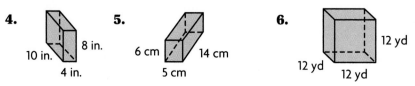

4.   **5.**   **6.**

10 in.   8 in.   6 cm   14 cm   12 yd

4 in.   5 cm   12 yd   12 yd

445

You have seen how the volume changes when the length, width, and height of a prism change. How do you think the volume will be affected if only the height changes?

ACTIVITY 2

WHAT YOU'LL NEED: 18 cubes

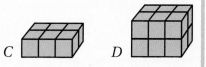

- Use your cubes to build both prisms.

C D

- Copy the table, and record the length, width, height, and volume of each prism.

	Length (in units)	Width (in units)	Height (in units)	Volume (in units³)
Prism C	?	?	?	?
Prism D	?	?	?	?
Ratio $\frac{D}{C}$	?	?	?	?

- Look at the ratios for length, width, and height. Are prisms C and D similar? Explain.

- How did the volume change when only the height doubled?

EXAMPLE 2 Some cereal products are priced according to the volume of the container. The volume of the container at the right is 240 in.³ Double the height of the container. How does the volume of the container change?

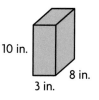

10 in.

3 in.   8 in.

height: 10 in. × 2 = 20 in.   *Double the height.*

$V = lwh$ *Write the formula.*

$V = 3 \times 8 \times 20$ *Replace l with 3, w with 8, and h with 20.*

$V = 480$ *Multiply.*

$\frac{480}{240} = \frac{2}{1}$, or 2   *Compare the volumes.*

So, the volume of the larger container is 480 in.³, or 2 times the volume of the original container.

- **CRITICAL THINKING** How do you think the volume will be affected if only the width is doubled?

INDEPENDENT PRACTICE

Find the volume. Then double the dimensions. Find the new volume.

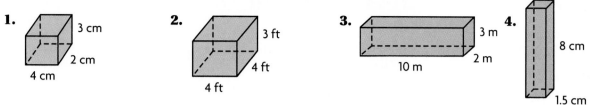

1. 3 cm, 2 cm, 4 cm

2. 3 ft, 4 ft, 4 ft

3. 3 m, 2 m, 10 m

4. 2.5 cm, 8 cm, 1.5 cm

Find the volume of each prism. Then halve the dimensions and find the new volume.

	Length	Width	Height	Volume	Volume (After the Dimensions Have Been Halved)
5.	6 m	5 m	2 m	?	?
6.	9 in.	6 in.	12 in.	?	?
7.	14 ft	10 ft	6 ft	?	?

Problem-Solving Applications

8. COMPARE Jasmine has a box of cereal that is 6 in. × 4 in. × 10 in. She wants the same volume of cereal divided into two congruent boxes. Her two new boxes have the same length and width as the old box but half the height. Does she have half the volume in each new box? Explain.

9. Jasmine used two boxes with the same height as the cereal box but half the length. What is the volume of each new box?

10. ✏ **WRITE ABOUT IT** Explain what happens to the volume of a prism when all the dimensions are doubled.

Mixed Review and Test Prep

Find the area of each circle. Round to the nearest whole number.

11. $r = 2$ ft

12. $r = 3$ cm

13. $r = 5$ m

14. $r = 16$ in.

Find the area.

15. 3 in., 2 in.

16. 5 ft, 6 ft

17. 10 m, 8 m

18. 9 cm, 11 cm

19. NUMBER SENSE Dave gives a clerk $20.00 to pay for a $5.89 notebook. How much change will he receive?

A $14.01   **B** $14.11   **C** $15.01   **D** $15.11

20. HOBBIES Larry made a model, using the scale 1 inch = 5 feet. The actual length is 200 feet. What is the model's length?

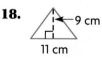

F 20 in.   **G** 30 in.   **H** 40 in.   **J** 400 in.

Exploring the Volume of a Cylinder

Sara's Soup comes in cylindrical cans. The dimensions of a can are shown below. How many cubic centimeters of soup does a can hold?

3 cm

8 cm

Explore

• Cut out the patterns for an open cylinder and an open cube. Fold and tape the models.

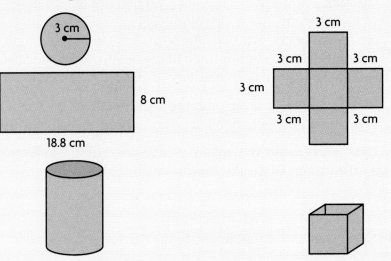

• Fill the cube with popcorn kernels. Empty the kernels into the cylinder. Continue filling the cube and emptying the kernels into the cylinder until the cylinder is full. Record the number of times you fill the cube.

• About how many cubic centimeters does a can of Sara's Soup hold?

• A volume of 1 cm^3 is equal to 1 mL. How many milliliters of soup will the cylinder hold?

Think and Discuss

Look at the rectangular prism at the right. Remember that the formula for the volume of a rectangular prism is $V = lwh$.

- If you know the area of the base of the prism, $l \times w$, how can you find the volume?

- Now look at the cylinder at the right. What figure is the base of the cylinder?

- What formula can you use to find the area of the base?

- Suppose you can find the volume of the cylinder in the same way you find the volume of the rectangular prism. By what can you multiply the area of the base to find the volume?

The volume of a cylinder can be found by multiplying the area of the base by the height. The formula is

$$V = \pi r^2 \times h \leftarrow \pi r^2 = \text{area of circle}$$

- Look back at the Sara's Soup can on page 448. Use the formula above to find the volume of the soup can. Use 3.14 for π. How does your answer compare with the answer you found with the popcorn kernels?

Try This

- Draw the following open-cylinder patterns on centimeter graph paper. Cut out each one, fold it, and tape it to make an open cylinder. Use the 27-cm^3 cube and the popcorn kernels from the Explore section to find the volume.

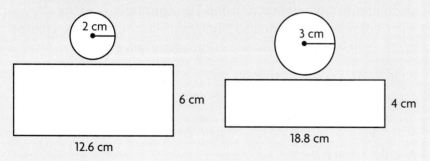

2 cm
12.6 cm
6 cm

3 cm
18.8 cm
4 cm

- Use the formula $V = \pi r^2 \times h$ to find the volumes of the two cylinders above.

Technology Link

You can see how the radius or height of a cylinder relates to its volume by using E-Lab, Activity 23. Available on CD-ROM and on the Internet at **www.hbschool.com/elab**

449

Volume of a Cylinder

What You'll Learn
How to find the volume of a cylinder

Why Learn This?
To solve everyday problems such as finding the amount of dirt needed to fill a flowerpot

In the previous Lab Activity, you discovered that you can find the volume of a cylinder by multiplying the area of the circular base by the height.

Volume = area of base × height
$$V = \pi r^2 \times h, \text{ or}$$
$$V = \pi r^2 h$$

EXAMPLE 1 Scott has a flowerpot in the shape of a cylinder. The container has a 5-in. radius and an 8-in. height. To determine the amount of potting soil to buy, he needs to know the volume of the flowerpot. What is the volume, to the nearest cubic inch?

$V = \pi r^2 h$ *Write the formula.*

$V \approx 3.14 \times 5^2 \times 8$ *Replace π with 3.14, r with 5, and h with 8.*
$V \approx 3.14 \times 25 \times 8$
$V \approx 628$ *Multiply.*

So, the volume is about 628 in.^{3}

- Suppose the radius of the flowerpot is 7 in. What would the volume be, to the nearest cubic inch?

Silos are used to store animal feed. They are cylinders, rather than rectangular prisms, because cylinders allow the feed to be stored without air pockets which could cause the feed to spoil. What volume of feed can a silo that is 60 ft tall with a 20-ft diameter hold?

EXAMPLE 2 A perfume manufacturer is considering a new container for its scented powder. The container is a cylinder with a diameter of 6 cm and a height of 15 cm. Find the volume of the cylinder, to the nearest cubic centimeter.

$6 \text{ cm} \div 2 = 3 \text{ cm}$ *Find the radius.*

$V = \pi r^2 h$ *Write the formula.*

$V \approx 3.14 \times 3^2 \times 15$ *Replace π with 3.14, r with 3, and h with 15.*
$V \approx 3.14 \times 9 \times 15$
$V \approx 423.9$ *Multiply.*

So, the volume is about 424 cm^3.

- Another container being considered has a diameter of 8 cm and a height of 14 cm. What is the volume of the container, to the nearest cubic centimeter?

GUIDED PRACTICE

Find the volume of each figure. Round to the nearest whole number.

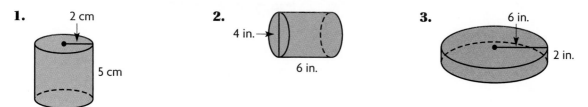

1. 2 cm / 5 cm

2. 4 in. → / 6 in.

3. 6 in. / 2 in.

INDEPENDENT PRACTICE

Find the volume of each figure. Round to the nearest whole number.

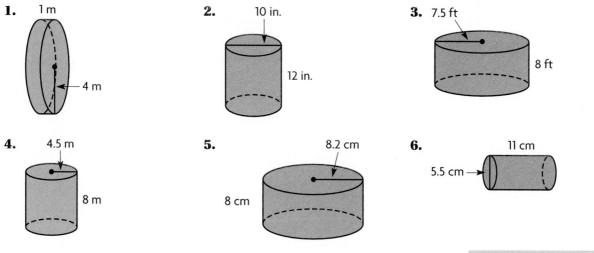

1. 1 m / 4 m

2. 10 in. / 12 in.

3. 7.5 ft / 8 ft

4. 4.5 m / 8 m

5. 8.2 cm / 8 cm

6. 11 cm / 5.5 cm →

Problem-Solving Applications

7. MEASUREMENT Cheryl has a thermos bottle with a radius of 4 cm and a height of 12 cm. She has filled it with juice. About how much juice is in the thermos?

8. SPORTS A circular swimming pool has a radius of 10 ft and a height of 6 ft. How many cubic feet of water does it take to fill the pool?

9. RATES The pool in Problem 8 was filled by a hose at the rate of 2 ft^3 per minute. About how many hours did it take to fill the pool?

10. A cylindrical pipe with a 6-ft diameter and a height of 50 ft is being put in the ground as the last attachment to a water system. What is its volume?

11. ✏ **WRITE ABOUT IT** In your own words, explain which parts of a cylinder are represented by πr^2 and by h in the formula $V = \pi r^2 h$.

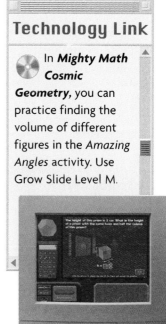

Technology Link

In *Mighty Math Cosmic Geometry,* you can practice finding the volume of different figures in the *Amazing Angles* activity. Use Grow Slide Level M.

ALGEBRA CONNECTION

Surface Area of a Prism

Andrew needs to wrap a birthday gift for his brother. The gift box is shaped like the prism at the right. How could you find the number of square inches of paper needed to wrap the gift?

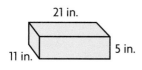
21 in.
11 in.   5 in.

You have already learned how to find the area of rectangles. You can use the formula for finding the area of a rectangle to find the surface area of a rectangular prism. The **surface area** is the sum of the areas of the faces of a solid figure.

EXAMPLE 1 A net for Andrew's gift box is shown at the right. Use it to find the surface area of the box.

$A = lw$

Area of face $A = 11 \times 5 = 55$
Area of face $B = 21 \times 11 = 231$
Area of face $C = 21 \times 5 = 105$
Area of face $D = 21 \times 11 = 231$
Area of face $E = 21 \times 5 = 105$
Area of face $F = 11 \times 5 = 55$

		A	5 in.
11 in.	5 in.	11 in.	5 in.
21 in.			
B	C	D	E
		F	5 in.

Use the formula for the area of a rectangle to find the area of each face.

$55 + 231 + 105 + 231 + 105 + 55 = 782$ *Find the sum.*

So, the surface area is 782 in.²

EXAMPLE 2 Find the surface area of the prism at the right.

Think: Opposite faces have the same area.

$A = lw$

front and back: $6 \times 8 \times 2 = 96$
top and bottom: $6 \times 4 \times 2 = 48$
left and right sides: $4 \times 8 \times 2 = 64$

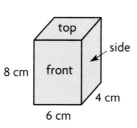
top
side
8 cm   front
4 cm
6 cm

Find the areas of the faces. Multiply by 2 to include opposite faces.

$96 + 48 + 64 = 208$ *Find the sum.*

So, the surface area is 208 cm².

• Which faces are always congruent in a rectangular prism?

GUIDED PRACTICE

Use the net to find the surface area of the prism.

1.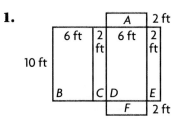

2.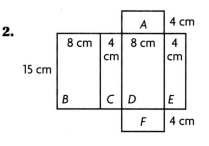

Name the dimensions of the opposite faces.

3.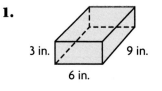

front and back: ___?___
top and bottom: ___?___
left and right side: ___?___

4.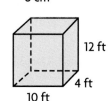

front and back: ___?___
top and bottom: ___?___
left and right sides: ___?___

INDEPENDENT PRACTICE

Find the surface area.

1. 3 in. 9 in. 6 in.

2. 11 cm 5 cm 8 cm

3. 5 m 4 m 6 m

4. 9 in. 9 in. 9 in.

5. 12 ft 4 ft 10 ft

6. 20 m 12 m 16 m

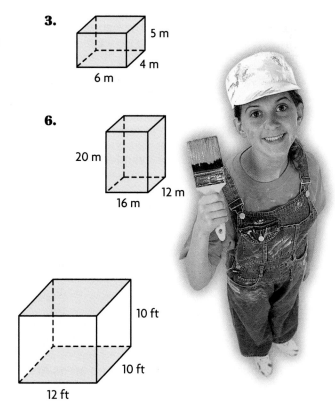

Problem-Solving Applications

Use the figure at the right for Problems 7–8.

7. MEASUREMENT Polly is painting her room, which is 12 ft × 10 ft × 10 ft. She will not paint the ceiling or the floor. How much surface area is she painting?

8. CONSUMER MATH Each can of paint covers 350 ft². How many cans of paint does Polly need?

10 ft 10 ft 12 ft

9. A rectangular prism has the following dimensions: $l = 5$ cm, $w = 3$ cm, and $h = 7$ cm. Find the surface area.

10. ✏️ **WRITE ABOUT IT** Explain what the surface area of a rectangular prism is.

MORE PRACTICE Lesson 23.4, page H76

1. **VOCABULARY** You are finding the number of cubic units needed to occupy a given space when you measure ___?___. (page 440)

Find the volume of the prism. (pages 440–443)

2.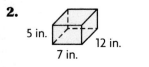
5 in. 12 in. 7 in.

3. 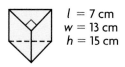 $l = 7$ cm
$w = 13$ cm
$h = 15$ cm

4.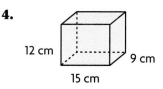
12 cm 9 cm 15 cm

Double the dimensions of the figure. Then find the volume of the larger prism. (pages 444–447)

5.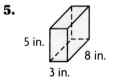
5 in. 8 in. 3 in.

6.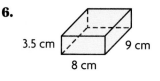
3.5 cm 9 cm 8 cm

7.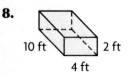
2 ft 1 ft 1.5 ft

Halve the dimensions of the figure. Then find the volume of the smaller prism. (pages 444–447)

8.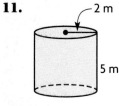
10 ft 2 ft 4 ft

9.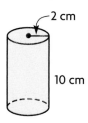
8 yd 5 yd 1 yd

10.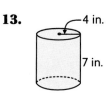
6 in. 8 in. 10 in.

Find the volume of the cylinder to the nearest whole unit. Use 3.14 for π. (pages 450–451)

11.
2 m
5 m

12.
2 cm
10 cm

13.
4 in.
7 in.

14. diameter: 5 in.
height: 4 in.

15. radius: 3 ft
height: 2 ft

16. radius: 5 m
height: 10 m

17. **VOCABULARY** The sum of the areas of all the faces of a solid figure is the ___?___. (page 452)

Find the surface area of the rectangular prism. (pages 452–453)

18.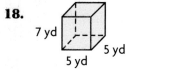
7 yd 5 yd 5 yd

19.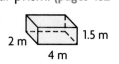
2 m 1.5 m 4 m

20.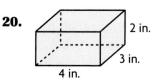
2 in. 3 in. 4 in.

1. What is $4 \times 4 \times 4 \times 4 \times 4$ written in exponent form?

A 4^2

B 4^4

C 4^5

D 5^4

2. Krista bought 8 cans of tennis balls. Each can contained 3 tennis balls and cost $1.89, including tax. Which expression can be used to find the total cost of her purchase?

F 8×1.89

G $3 \times 8 \times 1.89$

H 3×1.89

J $8 + 3 \times 1.89$

3. For three consecutive days, Zack jogged $\frac{3}{5}$ kilometer. Altogether, how far did he jog?

A $1\frac{1}{5}$ km

B $1\frac{4}{5}$ km

C 2 km

D $2\frac{1}{5}$ km

4. The table shows test scores. What was the mean (average) test score?

Test Scores					
75	85	100	91	93	87
82	76	78	100	79	98

F 25

G 75

H 82

J 87

5. Rich rolls a number cube labeled 1–6. What is the probability of him not rolling a number less than 3?

A 0

B $\frac{1}{3}$

C $\frac{1}{2}$

D $\frac{2}{3}$

6. It took Tim 4 hours to drive to college at an average speed of 52 miles per hour. Which equation can be used to find the distance Tim traveled?

F $d = 52 + 4$

G $d = 52 \times 4$

H $d = 52 - 4$

J $d = 52 \div 4$

7. Tina's computer is 18 inches wide and 24 inches long. How much space will it take up on her desk?

A 396 in.^{2}

B 412 in.^{2}

C 432 in.^{2}

D 455 in.^{2}

8. What is the volume of the figure shown below?

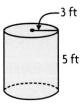

3 ft

5 ft

F about 76 ft^3

G about 84 ft^3

H about 128 ft^3

J about 141 ft^3

K Not Here

9. What is the surface area of the figure shown below?

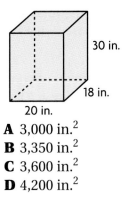

30 in.

18 in.

20 in.

A 3,000 in.^{2}

B 3,350 in.^{2}

C 3,600 in.^{2}

D 4,200 in.^{2}

MATH FUN!

NAME THAT MEASURE

PURPOSE To practice using precise measurements (pages 410–411)

YOU WILL NEED 5 index cards, ruler

Secretly measure 5 objects using two different units of measure. Write the name of an object on the front of each card, along with the two measurements.

Read the measurements to your classmates. The first person to name the object correctly receives 1 point. For an additional point, tell which measurement is more precise.

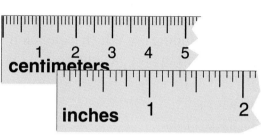

AREA OF EXPERTISE

PURPOSE To practice estimating areas of rectangles (pages 422–425)

Copy and complete the table. List four rectangles found in your classroom. Estimate their length, width, and area. Then measure them, and calculate the actual area.

How close were your estimates to the actual areas of the rectangles?

	ESTIMATE			Actual Area
	Length	Width	Area	
1.	?	?	?	?
2.	?	?	?	?
3.	?	?	?	?
4.	?	?	?	?

THOSE TASTY OATS

PURPOSE To practice finding the volume of a cylinder (pages 450–451)

YOU WILL NEED oatmeal container, light-colored construction paper, tape, crayons or markers

Measure the height and diameter of your oatmeal container. Find the volume. Design a new container that can hold the same volume but has a different shape.

Containers come in many sizes and shapes. Find five cylindrical containers at home. Estimate each cylinder's volume. Rank them from largest to smallest. Check your estimates by using the formula for volume.

Finding Perimeter and Area of Rectangles

In this activity you will see how you can enter the length and width of any rectangle and find the perimeter and area by using a spreadsheet.

Look at the spreadsheet below with data on the length and width of two rectangles.

ALL	A	B	C	D
1	Length	Width	Perimeter	Area
2	9	12	42	108
3	6	5		

To find the perimeter of the first rectangle, enter the following formula into cell C2.

ALL	C
1	Perimeter
2	=2*A2+2*B2

To find the area of the first rectangle, enter the following formula into cell D2.

ALL	D
1	Area
2	=A2*B2

1. What formula would you enter in cell C3 to find the perimeter of the second rectangle?

2. What is the perimeter of the second rectangle?

3. What formula would you enter in cell D3 to find the area of the second rectangle?

4. What is the area of the second rectangle?

USING A SPREADSHEET

Find the perimeter and area.

	Length	Width	Perimeter	Area
5.	8	5	?	?
6.	6	3	?	?
7.	12	10	?	?
8.	14	13	?	?
9.	11	9	?	?

Study Guide and Review

Vocabulary Check

1. The distance around a polygon is its __?__. (page 414)

2. The sum of the areas of the faces of a solid figure is the __?__.
(page 452)

EXAMPLES

- **Change one metric unit of measurement to another.** (pages 408–409)

$0.5 L = \blacksquare mL \quad \leftarrow 1 L = 1,000 mL$

$\dfrac{1}{1,000} = \dfrac{0.5}{x} \quad$ *Write a proportion.*

$x = 1,000 \times 0.5 \quad$ *Solve for x.*

$x = 500$

So, $0.5 L = 500 mL$.

- **Use a network to find the distance from one place to another.** (pages 412–413)

Start at B. Find the shortest route through all four points.

$BADC = 34.4$
$BACD = 30.4$
$BCDA = 32.0$
$BCAD = 30.0 \leftarrow$ shortest route

- **Find the area of triangles and parallelograms.** (pages 426–429)

$A = \dfrac{1}{2} bh$

$A = \dfrac{1}{2} \times 10 \times 6.2$

$A = \dfrac{1}{2} \times 62$

$A = 31$

So, the area is $31 m^2$.

EXERCISES

Change to the given units.

3. $24 cm = \blacksquare mm$

4. $10 mg = \blacksquare g$

5. $200 g = \blacksquare kg$

6. $5 km = \blacksquare m$

7. $100 m = \blacksquare cm$

8. $35 mL = \blacksquare L$

9. $18 km = \blacksquare m$

10. $52 cm = \blacksquare m$

Use the network to find the length of each route. Measurements are in miles.

11. *ADCBE*

12. *AEDCB*

13. *BDEAC*

14. *CDAEB*

Find the area of the figure.

15.

7 m
5.5 m

16.

8 ft
6 ft

17.

17 in.
12 in.

18.

5 cm
4.4 cm

- **Find the area of a circle.** (pages 434–435)

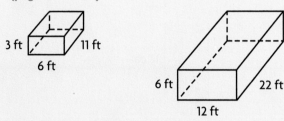

$A = \pi r^2$
$A \approx 3.14 \times 5^2$
$A \approx 3.14 \times 25$
$A \approx 78.5 \leftarrow$ area: 79 cm^2

Find the area of the circle, to the nearest whole unit. Use 3.14 for π.

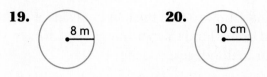

19. 8 m **20.** 10 cm

- **Change the dimensions of rectangular prisms.** (pages 444–447)

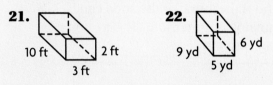

3 ft 11 ft
6 ft

6 ft 22 ft
12 ft

Find the volume. Then double the dimensions and find the volume.

$V = lwh$
$V = 6 \times 11 \times 3$
$V = 198$ cm^3

$V = lwh$
$V = 12 \times 22 \times 6$
$V = 1{,}584$ cm^3

Find the volume of the prism. Then double the dimensions of the prism and find the volume.

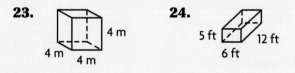

21. 10 ft 2 ft 3 ft **22.** 9 yd 6 yd 5 yd

Find the volume of the prism. Then halve the dimensions of the prism and find the volume.

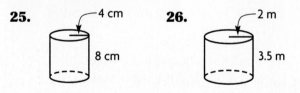

23. 4 m 4 m 4 m **24.** 5 ft 12 ft 6 ft

- **Find the volume of a cylinder.** (pages 450–451)

3 cm
10 cm

$V = \pi r^2 h$ *Use 3.14 for π.*
$V \approx 3.14 \times 3^2 \times 10$
$V \approx 3.14 \times 9 \times 10$
$V \approx 282.6$ cm^3
The volume is about 283 cm^3.

Find the volume of the cylinder, to the nearest whole unit. Use 3.14 for π.

25. 4 cm 8 cm **26.** 2 m 3.5 m

Problem-Solving Applications
Solve. Explain your method.

27. Thomas has two tables, each with a length of 10 ft and a width of 4 ft. He wants to put them side by side so that he gets the greatest area. What is the area if he arranges the tables so the length is doubled? What is the area if he arranges the tables so the width is doubled?

(pages 430–431)

28. A shed is in the shape of a triangular prism. It has 6-ft square sides, as shown in the figure at the right. The base is an equilateral triangle with a height of 5.2 ft. What is the volume of the shed?

$(V = \frac{1}{2}lwh)$ (pages 440–443)

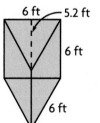

6 ft 5.2 ft
6 ft
6 ft

Performance Assessment

Tasks: Show What You Know

1. Explain how to find the length of side x and side y. Find the perimeter. Explain your method. **(pages 414–415)**

2. Find the area of the parallelogram and the area of the triangle. Explain your methods. **(pages 426–429)**

3. Find the volume of the soda can. Explain your method. ($V = \pi r^2 h$) **(pages 450–451)**

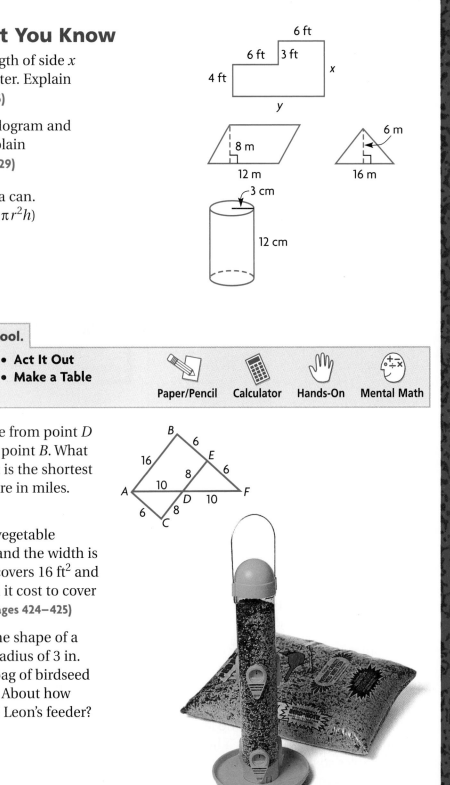

Problem Solving

Solve. Explain your method.

> ### CHOOSE a strategy and a tool.
>
> - **Find a Pattern**
> - **Make a Model**
> - **Write a Number Sentence**
> - **Act It Out**
> - **Make a Table**
>
> Paper/Pencil Calculator Hands-On Mental Math

4. Mr. Gonzalez wants to drive from point D to point F, stopping first at point B. What are 3 possible routes? What is the shortest possible route? Distances are in miles. **(pages 412–413)**

5. Mr. Wu's class is making a vegetable garden. The length is 12 ft and the width is 8 ft. A 25-lb bag of topsoil covers 16 ft² and costs $3.95. How much will it cost to cover the garden with topsoil? **(pages 424–425)**

6. Leon has a bird feeder in the shape of a cylinder. The feeder has a radius of 3 in. and is 12 in. high. A 10-lb bag of birdseed contains 1,700 in.³ of seed. About how many times will the bag fill Leon's feeder? ($V = \pi r^2 h$) **(pages 450–451)**

Cumulative Review

Solve the problem. Then write the letter of the correct answer.

1. Compare. $^-8$ ● $^-12$ (pages 30–31)

- **A.** $<$
- **B.** $>$
- **C.** $=$
- **D.** $\leq$

2. Find the product. Write it in simplest form. $3\frac{2}{5} \times 5\frac{5}{6}$ (pages 142–143)

- **A.** $\frac{102}{175}$
- **B.** $15\frac{1}{3}$
- **C.** $19\frac{25}{30}$
- **D.** $19\frac{5}{6}$

3. Name the angle. (pages 164–165)

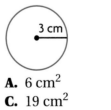

- **A.** acute
- **B.** obtuse
- **C.** right
- **D.** triangle

4. A bag holds 10 pieces of paper with names of prizes on them: 4 prizes are pencils, 3 are pens, 2 are books, and 1 is a bike. What is the probability of picking a pencil or pen as a prize? (pages 276–277)

- **A.** $\frac{1}{10}$
- **B.** $\frac{3}{10}$
- **C.** $\frac{4}{10}$, or $\frac{2}{5}$
- **D.** $\frac{7}{10}$

5. Solve for x. $\frac{x}{6} = 12$ (pages 310–313)

- **A.** $x = 2$
- **B.** $x = 6$
- **C.** $x = 18$
- **D.** $x = 72$

6. Use the formula F $= \frac{9}{5}$C $+ 32$. Convert 30°C to degrees Fahrenheit. Round to the nearest degree. (pages 316–317)

- **A.** $^-1$°F
- **B.** 22°F
- **C.** 32°F
- **D.** 86°F

7. What is the unit rate? $1.23 for 3 (pages 334–335)

- **A.** $\frac{\$0.41}{1}$
- **B.** $\frac{\$1.23}{3}$
- **C.** $3.69
- **D.** not here

8. What is 5% as a ratio in simplest form? (pages 336–339)

- **A.** $\frac{5}{100}$
- **B.** $\frac{1}{20}$
- **C.** 50:100
- **D.** 1 to 2

9. The distance around a polygon is its __?__ . (pages 414–415)

- **A.** area
- **B.** perimeter
- **C.** surface area
- **D.** volume

10. Use the formula $A = \frac{1}{2}bh$. Find the area of the triangle. (pages 426–429)

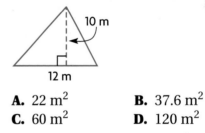

- **A.** 22 m^2
- **B.** 37.6 m^2
- **C.** 60 m^2
- **D.** 120 m^2

11. Find the area of the circle, to the nearest whole unit. Use the formula $A = \pi r^2$. Use 3.14 for π. (pages 434–435)

- **A.** 6 cm^2
- **B.** 9 cm^2
- **C.** 19 cm^2
- **D.** 28 cm^2

12. Find the volume of a rectangular prism 10 in. $\times$ 5 in. $\times$ 2.5 in. (pages 440–443)

- **A.** 17.5 in.^{3}
- **B.** 62.5 in.^{3}
- **C.** 125 in.^{3}
- **D.** 175 in.^{3}

ALGEBRA: NUMBER RELATIONSHIPS

LOOK AHEAD

In this chapter you will solve problems that involve

- **integers**
- **rational numbers**
- **terminating and repeating decimals**
- **relationships between numbers**

SCIENCE **LINK**

Many items, such as wreckage from ships, lie on the ocean floor. The greatest depth at which an item has been successfully removed is 17,251 feet. This was in the Pacific Ocean in 1991, when the wreckage of a helicopter crash was raised to the surface from that depth.

The deepest submarine dive ever recorded was the United States Navy vessel Sea Cliff's descent to a depth of 20,000 feet in 1985.

- Write integers to represent the data above.

- How do the two depths given above compare?

Hunt Between a Fraction and a Decimal

Hunt for treasures with sizes in a given range. Start by making a list of items for a treasure hunt. Then write a size range by each of the items. One number in the range should be a fraction and the other number should be a decimal. Record, sketch, and label each item you find.

TREASURE HUNT

FIND:

A toy between 0.5 and $\frac{5}{6}$ foot long.

A tool between $\frac{1}{4}$ and 0.7 meter long.

A piece of jewelry between 1.25 and $1\frac{7}{8}$ inches long.

A piece of furniture between $2\frac{2}{3}$ and 3.5 feet tall.

PROJECT CHECKLIST

✓ Did you list items for a treasure hunt?

✓ Did you write a size range for each item you found using a fraction and a decimal?

✓ Did you record, sketch, and label each item you found?

What You'll Explore
How to use decimal squares to represent decimals, fractions, and whole numbers

What You'll Need
decimal squares, index cards, large sheet of paper

Sets of Numbers

This year you have worked with whole numbers, decimals, and fractions. This activity will help you see how the three kinds of numbers are related.

ACTIVITY 1

Explore

- Write each number at the right on a card.

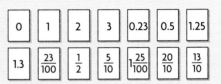

- Shade decimal squares to represent the numbers on the cards.

- Match each card with the shaded decimal square or squares for that number.

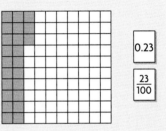

- Group your decimal squares and matching cards into three groups: fractions and mixed numbers, decimals, and whole numbers.

Think and Discuss

- What do you notice about the decimal squares for 0.5 and $\frac{5}{10}$?

- What do you notice about the decimal squares for 2 and $\frac{20}{10}$?

- Look at the decimal squares for $\frac{13}{10}$ and 1.3. What can you say about these two numbers?

Try This

- Write the fraction and decimal that are shown in each model. Write the fraction in simplest form.

ACTIVITY 2

Explore

- Divide the cards from Activity 1 into two groups: whole numbers in one group and decimals and fractions in the other.

- On a large sheet of paper, draw the diagram shown at the right.

Decimals and Fractions

> Whole Numbers

- Place your decimal and fraction cards in the correct part of the diagram. Then place your whole-number cards in the correct part of the diagram.

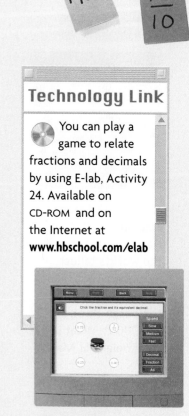

Think and Discuss

- Did you put $\frac{20}{10}$ in the Decimals and Fractions part of the diagram? If yes, why? If no, why not?

- Can any whole number be written as a fraction or a decimal? Explain.

- Why do you think the Whole Numbers part of the diagram is inside the Decimals and Fractions part?

- In which part of the diagram would you put 7? 3.5? $\frac{2}{3}$?

Try This

- Copy the diagram at the right.

- Write each of these numbers in the correct part of the diagram.

Decimals and Fractions

> Whole Numbers

$\frac{2}{5}$ $2\frac{1}{3}$ 4.04 $\frac{8}{4}$ 3.00

Technology Link

You can play a game to relate fractions and decimals by using E-lab, Activity 24. Available on CD-ROM and on the Internet at **www.hbschool.com/elab**

465

Integers

What You'll Learn
How to identify integers and find absolute value

Why Learn This?
To describe situations that involve positive or negative values

VOCABULARY

integers
absolute value

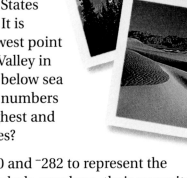

The highest point in the United States is at Mount McKinley in Alaska. It is 20,320 ft above sea level. The lowest point in the United States is in Death Valley in California. Its elevation is 282 ft below sea level. Sea level equals 0 ft. What numbers can you use to represent the highest and lowest points in the United States?

You can use the integers $^+20{,}320$ and $^-282$ to represent the elevations. **Integers** include all whole numbers, their opposites, and 0. Each integer, except 0, has an opposite that is the same distance from 0 but on the opposite side of 0. The opposite of $^+6$ is $^-6$. The opposite of $^-10$ is $^+10$.

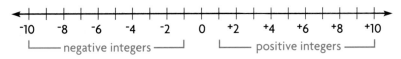

negative integers   positive integers

Negative integers are written with a $^-$ sign.

Positive integers can be written with or without a $^+$ sign.

EXAMPLE 1 Name an integer to represent the situation.

A. a loss of 5 yd in a football game
$^-5$

B. a temperature of 3° below zero
$^-3$

C. a bank deposit of $100
$^+100$

The **absolute value** of an integer is its distance from 0.

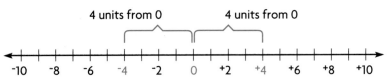

4 units from 0   4 units from 0

Write: $|^-4| = 4$ **Read:** The absolute value of negative four is four.

Write: $|^+4| = 4$ **Read:** The absolute value of positive four is four.

EXAMPLE 2 Use the number line above to find the absolute value of each integer.

A. $|^-8|$
8

B. $|^+6|$
6

C. $|^-2|$
2

D. $|^+10|$
10

At 13,796 ft, Mauna Kea, on the island of Hawaii, is not one of the world's tallest mountains. But Mauna Kea actually starts on the ocean floor, at $^-20{,}000$ ft. What is the total height of Mauna Kea?

GUIDED PRACTICE

Write the opposite integer.

1. $^-2$
2. $^+10$
3. $^+5$
4. $^-9$
5. $^+20$
6. $^-100$

7. $^-24$
8. $^+18$
9. $^+50$
10. $^-92$
11. $^+206$
12. $^-180$

Write an integer to represent each situation.

13. a $15 withdrawal from savings

14. an increase of 20 points

15. 26 degrees below zero

INDEPENDENT PRACTICE

Write an integer to represent each situation. Then describe the opposite situation, and write an integer to represent it.

1. gaining 2 pounds
2. going up 3 flights of stairs
3. losing 5 yd in a football game

Write the absolute value.

4. $|^+1|$
5. $|^-1|$
6. $|^-7|$
7. $|^+5|$
8. $|^-4|$
9. $|^-12|$

10. $|^+28|$
11. $|^-28|$
12. $|^+35|$
13. $|^-80|$
14. $|^+150|$
15. $|^-295|$

Problem-Solving Applications

16. GEOGRAPHY The elevation of the Dead Sea is about 1,310 ft below sea level. Write the elevation as an integer.

17. SCIENCE The bottom of an iceberg is 630 m below sea level. Write an integer to represent the depth of the iceberg.

18. ALGEBRA What values can n have if $|n| = 5$?

19. ✏ **WRITE ABOUT IT** Explain what the absolute value of an integer is.

Mixed Review and Test Prep

Write as a fraction in simplest form.

20. 0.25
21. 2
22. $2\frac{1}{2}$
23. 1.2
24. 0.8

Find the volume of the rectangular prism.

25. $l = 5$ ft, $w = 4$ ft, $h = 6$ ft

26. $l = 5$ m, $w = 2$ m, $h = 3$ m

27. $l = 10$ cm, $w = 8$ cm, $h = 4$ cm

28. SPORTS Six high school soccer teams compete for the regional play-offs. Each team plays each of the other teams only once. How many games are played?

A 6
B 12
C 15
D 30

29. WAGES Adrianne works for $5.50 an hour 2 hours a day, 6 days a week. How many weeks will it take her to earn more than $250.00?

F 13
G 6
H 5
J 4

Rational Numbers

In the previous lesson, you learned that integers are the whole numbers, their opposites, and 0. All integers are rational numbers. The word *rational* comes from *ratio*. A **rational number** can be written as the ratio $\frac{a}{b}$, where a and b are integers and $b \neq 0$. The numbers below are all rational numbers since they can be expressed as the ratio $\frac{a}{b}$.

$$0.5 \qquad \frac{3}{8} \qquad {}^-3 \qquad 8 \qquad 3\frac{9}{10}$$

The **Venn diagram** shows how the sets of rational numbers, integers, and whole numbers are related.

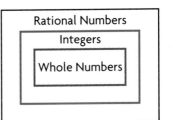

Rational Numbers
Integers
Whole Numbers

Talk About It

- Which sets of numbers would you use to express the amount of money in your savings account?

- Which sets of numbers do you use when you measure ingredients?

- The diagram shows that all whole numbers are rational numbers. Write 12 as a ratio.

- Is every integer a rational number? Explain.

EXAMPLES Use the Venn diagram above to determine in which set or sets each number belongs.

A. 21   *The number 21 belongs in the sets of whole numbers, integers, and rational numbers.*

B. ${}^-3$ *The number ${}^-3$ belongs in the sets of integers and rational numbers.*

C. $2\frac{1}{5}$ *The number $2\frac{1}{5}$ belongs in the set of rational numbers.*

D. 1.45   *The number 1.45 belongs in the set of rational numbers.*

- In which set of numbers does $\frac{{}^-3}{5}$ belong?

- **CRITICAL THINKING** Explain why a mixed number does not belong in the sets of whole numbers and integers.

GUIDED PRACTICE

Write each rational number in the form $\frac{a}{b}$.

1. 0.4 **2.** ⁻5 **3.** $1\frac{2}{5}$ **4.** 0.435 **5.** 9.71 **6.** $2\frac{1}{4}$

7. 0.87 **8.** ⁻$3\frac{1}{4}$ **9.** ⁻82 **10.** 727 **11.** 11.5 **12.** 0.7

Tell which set of rational numbers you would use for each purpose.

13. to express your height

14. to express your shoe size

15. to express money saved

16. to express temperature

INDEPENDENT PRACTICE

Write each rational number in the form $\frac{a}{b}$.

1. $5\frac{1}{4}$ **2.** 0.7 **3.** 0.32 **4.** 14.8 **5.** $1\frac{3}{7}$ **6.** 2.38

7. 100 **8.** 19 **9.** $2\frac{1}{2}$ **10.** 0.81 **11.** 260 **12.** $3\frac{2}{3}$

Use the Venn diagram at the right to determine in which set or sets the number belongs.

13. 1.2 **14.** $\frac{3}{4}$ **15.** ⁻15 **16.** 100

17. $9\frac{1}{8}$ **18.** 0.28 **19.** ⁻2.3 **20.** ⁻6

21. $1\frac{7}{8}$ **22.** 45 **23.** $\frac{1}{8}$ **24.** ⁻$4\frac{1}{2}$

25. 1,892 **26.** ⁻299 **27.** $\frac{-2}{3}$ **28.** 0

| Rational Numbers |
| Integers |
| Whole Numbers |

For Exercises 29–30, answer *true* or *false* and give an example.

29. Some rational numbers are negative integers.

30. Every whole number is an integer, and every integer is a whole number.

Problem-Solving Applications

31. NUMBER SENSE Name a rational number that is greater than one but is not an integer or whole number.

32. Name a rational number that is less than one but is not an integer.

33. CONSUMER MATH Teens were surveyed to find out how many of them belong to record clubs. The percent of teens who belong to record clubs is 20%. Write the percent of teens who belong to a record club as a rational number.

34. MUSIC In the band, 25% of the members play brass instruments. Write the percent who play brass instruments as a rational number.

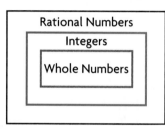

35. ALGEBRA Tamara is $61\frac{1}{2}$ in. tall. How would you write her height in the form $\frac{a}{b}$?

36. ◆ **WRITE ABOUT IT** Explain why every whole number is a rational number.

Terminating and Repeating Decimals

What You'll Learn
How to use division to write fractions as terminating or repeating decimals

Why Learn This?
To recognize decimal forms of fractions in consumer situations

VOCABULARY

terminating decimal

repeating decimal

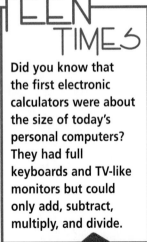

TEEN TIMES

Did you know that the first electronic calculators were about the size of today's personal computers? They had full keyboards and TV-like monitors but could only add, subtract, multiply, and divide.

Every fraction can be written as a division problem. The quotient can be written as a decimal.

$\frac{3}{4}$ can be written as $4\overline{)3}$.

$4\overline{)3.00}$ is equal to 0.75.

$\frac{2}{5}$ can be written as $5\overline{)2}$.

$5\overline{)2.0}$ is equal to 0.4.

ACTIVITY

WHAT YOU'LL NEED: calculator

Malia has a score of $\frac{3}{5}$ on her math homework. This means the answers to 3 out of 5 problems are correct. Malia wants to know the decimal form of her score so that she can determine the percent. Find Malia's decimal score for her homework.

- Use the following key sequence to enter $\frac{3}{5}$ into your calculator as a division problem.

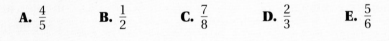

    3 ÷ 5 =

- What decimal is shown in the display on your calculator?

- Malia knows that 0.6 is equivalent to 0.60. How could you write 0.60 as a percent?

- Use the calculator to change each fraction to a decimal. Record the decimal shown in the display of your calculator.

A. $\frac{4}{5}$ **B.** $\frac{1}{2}$ **C.** $\frac{7}{8}$ **D.** $\frac{2}{3}$ **E.** $\frac{5}{6}$

Talk About It

- Look at the decimals for A–E. If you did the division with paper and pencil, which quotients would have a remainder of zero?

- Look at the decimal for $\frac{2}{3}$. Does your calculator show 0.666666666 or 0.666666667? Why would a calculator show 0.666666667?

When you divide and the remainder is zero, the quotient can be written as a **terminating decimal**. The fractions $\frac{4}{5}$, $\frac{1}{2}$, and $\frac{7}{8}$ in the activity above are examples of fractions that can be written as terminating decimals.

EXAMPLE 1 On a math quiz, Malia answered $\frac{6}{8}$ of the problems, or 6 out of 8, correctly. Use division to find the decimal. Can $\frac{6}{8}$ be expressed as a terminating decimal?

$$\frac{6}{8} \rightarrow$$

$$\begin{array}{r} 0.75 \\ 8\overline{)6.00} \\ -5\,6 \\ \hline 40 \\ -40 \\ \hline 0 \end{array}$$

Write the fraction as a division problem. Divide.

Add a zero and continue dividing.

So, since the remainder is zero, $\frac{6}{8}$ can be expressed as a terminating decimal.

- Suppose Malia answered $\frac{8}{10}$ correctly. Can $\frac{8}{10}$ be expressed as a terminating decimal? Explain.

GUIDED PRACTICE

Use a calculator to change the fractions to decimals.

1. $\frac{1}{8}$ **2.** $\frac{6}{10}$ **3.** $\frac{7}{20}$ **4.** $\frac{2}{5}$

Find the terminating decimal for each fraction.

5. $\frac{1}{4}$ **6.** $\frac{3}{20}$ **7.** $\frac{5}{8}$ **8.** $\frac{9}{40}$

Repeating Decimals

Sometimes when you divide a fraction, you continue to get a remainder. Such a quotient is a **repeating decimal**.

EXAMPLE 2 On Malia's second homework assignment, she answered $\frac{5}{6}$ of the problems correctly. What is the decimal for $\frac{5}{6}$?

$$\begin{array}{r} 0.833 \\ 6\overline{)5.000} \\ -48 \\ \hline 20 \\ -18 \\ \hline 20 \\ -18 \\ \hline 2 \end{array}$$

Write the fraction as a division problem. Divide.

If you continue to add zeros and divide, the 3 in the quotient will continue to repeat.

The decimal can be written as 0.833333333 . . . to show that it repeats endlessly.

You can draw a bar over the digit that repeats.
$\frac{5}{6}$ = 0.833333333 . . . = $0.8\overline{3}$

471

Sometimes when a fraction is written as a decimal, more than one digit repeats. Draw a bar over all the digits that repeat.

EXAMPLE 3 Find the repeating decimal for $\frac{5}{11}$.

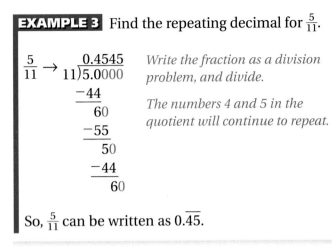

$\frac{5}{11} \rightarrow$

$$11\overline{)5.0000}$$
$$\quad\underline{-44}$$
$$\quad\quad 60$$
$$\quad\quad \underline{-55}$$
$$\quad\quad\quad 50$$
$$\quad\quad\quad \underline{-44}$$
$$\quad\quad\quad\quad 60$$

quotient: 0.4545

Write the fraction as a division problem, and divide.

The numbers 4 and 5 in the quotient will continue to repeat.

So, $\frac{5}{11}$ can be written as $0.\overline{45}$.

When you use a calculator to find the decimal that is equivalent to a fraction, look for the digits that repeat.

$\frac{1}{3} \rightarrow 1 \boxed{\div} 3 \boxed{=}$ $\boxed{0.333333333}$ $\frac{1}{3} = 0.\overline{3}$

$\frac{1}{6} \rightarrow 1 \boxed{\div} 6 \boxed{=}$ $\boxed{0.166666667}$ $\frac{1}{6} = 0.1\overline{6}$

You can use the decimal values that you know to find other decimal values.

EXAMPLE 4 Use $\frac{1}{3} = 0.\overline{3}$ to write decimals for $\frac{2}{3}$ and $\frac{4}{3}$.

$\frac{2}{3} = 2 \times \frac{1}{3} = 2 \times (0.333333\ldots) = 0.666666\ldots$, or $0.\overline{6}$

$\frac{4}{3} = \frac{3}{3} + \frac{1}{3} = 1 + 0.333333\ldots = 1.333333\ldots$, or $1.\overline{3}$

So, $\frac{2}{3} = 0.\overline{6}$ and $\frac{4}{3} = 1.\overline{3}$.

INDEPENDENT PRACTICE

Find the terminating decimal for the fraction.

1. $\frac{1}{10}$ **2.** $\frac{1}{5}$ **3.** $\frac{5}{8}$ **4.** $\frac{11}{25}$ **5.** $\frac{7}{10}$ **6.** $\frac{1}{16}$

Find the repeating decimal for the fraction.

7. $\frac{6}{11}$ **8.** $\frac{1}{15}$ **9.** $\frac{4}{3}$ **10.** $\frac{1}{30}$ **11.** $\frac{1}{9}$ **12.** $\frac{1}{45}$

The decimal for $\frac{1}{9}$ is $0.\overline{1}$, for $\frac{2}{9}$ is $0.\overline{2}$, and for $\frac{3}{9}$ is $0.\overline{3}$.
Use this information to find the decimal for the fraction.

13. $\frac{4}{9}$ **14.** $\frac{5}{9}$ **15.** $\frac{6}{9}$ **16.** $\frac{7}{9}$ **17.** $\frac{8}{9}$ **18.** $\frac{9}{9}$

Write the fraction as a decimal.

19. $\frac{3}{50}$ **20.** $\frac{11}{20}$ **21.** $\frac{7}{8}$ **22.** $\frac{11}{9}$ **23.** $\frac{3}{10}$ **24.** $\frac{7}{6}$

25. $\frac{11}{12}$ **26.** $\frac{4}{15}$ **27.** $\frac{41}{50}$ **28.** $\frac{5}{3}$ **29.** $\frac{1}{40}$ **30.** $\frac{5}{4}$

Problem-Solving Applications

For Problems 31–32, use the table at the right. The table shows the amount of rain Mr. Collins recorded and the amount of rain the local news reported.

Day	Mr. Collins	Local News
Monday	$\frac{1}{2}$ in.	0.35 in.
Tuesday	$\frac{3}{4}$ in.	0.7 in.
Wednesday	0 in.	0.1 in.
Thursday	$\frac{2}{3}$ in.	0.66 in.
Friday	$\frac{7}{10}$ in.	1.2 in.

31. WEATHER Write as a decimal the amount of rain Mr. Collins recorded for Monday. Is this amount different from the amount the local news reported? If so, what is the difference?

32. COMPARE Write as a decimal the amount Mr. Collins recorded for Friday. What is the difference between this amount and the amount the local news reported?

33. NUMBER SENSE Marsha is on a treasure hunt. Her mission is to find a jump rope that has a length of $9\frac{3}{4}$ ft. She found a jump rope that has a length of 9.75 ft. Did she complete her mission? Explain.

34. Claudia correctly solved 15 problems out of 20 on the test. Write her score as a decimal.

35. ✏️ **WRITE ABOUT IT** Explain the difference between a terminating decimal and a repeating decimal.

Technology Link

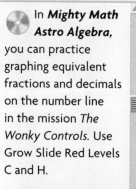

💿 In *Mighty Math Astro Algebra*, you can practice graphing equivalent fractions and decimals on the number line in the mission *The Wonky Controls.* Use Grow Slide Red Levels C and H.

Mixed Review and Test Prep

Find the LCD.

36. $\frac{1}{2}, \frac{1}{4}$ **37.** $\frac{1}{2}, \frac{2}{5}$ **38.** $\frac{1}{8}, \frac{1}{6}$

Find the volume of each cylinder. Use the formula $V = \pi r^2 h$ and 3.14 for π.

39. $r = 2$ in.,   **40.** $r = 4$ cm,
    $h = 5$ in. $h = 6$ cm

41. PERCENT Gloria's father bought a $65 camera at a discount of 30%. How much did he pay for the camera?

    A $19.50   **B** $29.50
    C $45.50   **D** $55.50

42. GEOMETRY Find the surface area of a rectangular prism that is 6 feet long, 2 feet wide, and 3 feet high.

    F 72 ft²   **G** 72 ft³
    H 36 ft²   **J** 36 ft³

Relationships on the Number Line

Ben has a pencil $4\frac{1}{4}$ in. long. Deric has a pencil $4\frac{1}{2}$ in. long. The length of Kerri's pencil is between those of Ben's and Deric's.

Think of the lengths of the pencils as rational numbers. You can use the rational numbers on a number line to find a possible length for Kerri's pencil. Notice that there is a mark between $4\frac{1}{4}$ and $4\frac{1}{2}$ that could be the length of Kerri's pencil.

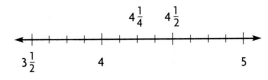

Another way to find a number between two rational numbers is to find a common denominator.

EXAMPLE 1 Find a number between $4\frac{1}{4}$ and $4\frac{1}{2}$.

$4\frac{1}{4} = 4\frac{2}{8}$ $4\frac{1}{2} = 4\frac{4}{8}$ *Use a common denominator to write equivalent fractions.*

$4\frac{3}{8}$ is one rational number   *Find a rational number between the two numbers.*
between $4\frac{2}{8}$ and $4\frac{4}{8}$.

So, $4\frac{3}{8}$ is between $4\frac{1}{4}$ and $4\frac{1}{2}$.

• **CRITICAL THINKING** Suppose you used 16 for the common denominator of $4\frac{1}{4}$ and $4\frac{1}{2}$. What rational number would be between $4\frac{4}{16}$ and $4\frac{8}{16}$?

You can also find a number between two rational numbers in decimal form.

EXAMPLE 2 Find a rational number between $^-2.7$ and $^-2.8$.

$^-2.7 = ^-2.70$ *Add a zero to each decimal.*
$^-2.8 = ^-2.80$

Think of a number line to find a number between the two decimals.

So, $^-2.73$, $^-2.75$, and $^-2.78$ are some of the numbers between $^-2.7$ and $^-2.8$.

GUIDED PRACTICE

Name a rational number between the two given numbers on the number line.

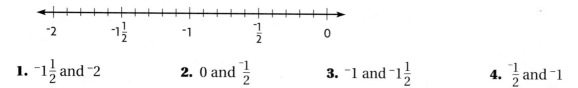

1. 1 and $1\frac{1}{2}$ **2.** 0 and $^-\frac{1}{2}$ **3.** $\frac{1}{4}$ and $\frac{3}{4}$ **4.** $^-\frac{1}{2}$ and $^-1$

Find a rational number between the two given numbers.

5. $^-1.50$ and $^-1.60$ **6.** $\frac{1}{2}$ and 1   **7.** 4 and 6   **8.** $\frac{^-2}{5}$ and $\frac{^-4}{5}$

INDEPENDENT PRACTICE

Name a rational number between the two given numbers on the number line.

1. $^-1\frac{1}{2}$ and $^-2$ **2.** 0 and $^-\frac{1}{2}$ **3.** $^-1$ and $^-1\frac{1}{2}$ **4.** $^-\frac{1}{2}$ and $^-1$

Find a rational number between the two given numbers.

5. $\frac{1}{8}$ and $\frac{3}{6}$ **6.** $\frac{1}{8}$ and $\frac{1}{2}$ **7.** $1\frac{3}{8}$ and $1\frac{1}{4}$ **8.** $^-2\frac{1}{2}$ and $^-3$

9. 2.3 and 2.4   **10.** $^-1.1$ and $^-1.2$ **11.** $^-7.4$ and $^-7.5$ **12.** 4.01 and 4.02

13. $1\frac{1}{2}$ and $1\frac{3}{4}$ **14.** $^-1.23$ and $^-1.24$ **15.** $\frac{1}{10}$ and 0.2   **16.** $\frac{1}{5}$ and 0.4

17. 1.78 and $1\frac{7}{8}$ **18.** $\frac{5}{6}$ and $\frac{8}{9}$ **19.** $^-3.25$ and $^-3.254$ **20.** $\frac{1}{8}$ and $\frac{1}{10}$

21. $\frac{^-5}{2}$ and $^-3$ **22.** 18.01 and 18.03   **23.** $\frac{3}{4}$ and $\frac{5}{6}$ **24.** 81.9 and $82\frac{1}{8}$

Problem-Solving Applications

25. CONSUMER MATH Ms. Jenkins checked the gas gauge in her car and found that the tank was between $\frac{1}{2}$ and $\frac{3}{4}$ full. Could the tank have been $\frac{5}{8}$ full? Explain.

26. FITNESS Mitch ran 10.6 km. Halfway between the 10.5-km mark and the 10.6-km mark he stopped to retie his shoe. How far had he run?

27. CRITICAL THINKING Is it easier to find a rational number between $\frac{1}{2}$ and $\frac{1}{4}$ or between 0.25 and 0.50? Explain your reasoning.

28. ✏️ **WRITE ABOUT IT** Explain how you would find a rational number between 0.3 and 0.4.

Comparing and Ordering

What You'll Learn
How to compare and order rational numbers

Why Learn This?
To compare numbers when you need to decide which product has a higher rating, or is more costly, or to organize information from a survey

TEEN TIMES

In a survey of 600 readers, 0.7 of the bike riders and about $\frac{1}{2}$ of the in-line skaters said they wore safety equipment.

You have learned that rational numbers can be located on a number line. Recall that when you compare two numbers on a number line, the number on the right is greater. On the number line below, $\frac{1}{4}$ is greater than $^-0.5$.

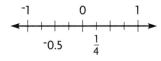

You can use the symbol $<$ or $>$ to show the comparison.

$\frac{1}{4} > ^-0.5 \rightarrow \frac{1}{4}$ is greater than $^-0.5$.

$^-0.5 < \frac{1}{4} \rightarrow ^-0.5$ is less than $\frac{1}{4}$.

It is often easier to compare rational numbers when they have the same denominator.

EXAMPLE 1 Shana correctly answered 0.75 of the questions on her math test. She correctly answered $\frac{7}{10}$ of the questions on her Spanish test. On which test did she make a higher score?

$0.75 = \frac{75}{100}$ *Write the numbers as fractions with the same denominator.*

$\frac{7}{10} = \frac{70}{100}$

$\frac{75}{100} > \frac{70}{100}$ *Use $<$ or $>$ to compare the two fractions.*

So, Shana made a higher score on her math test.

You can use $<$ or $>$ to order rational numbers.

EXAMPLE 2 Order $6\frac{3}{4}$, 6.8, and $6\frac{1}{2}$ from least to greatest.

$6\frac{3}{4} = 6.75$ $6.8 = 6.80$ *Write each rational number in decimal form. Add zeros so that all have the same number of decimal places.*

$6\frac{1}{2} = 6.50$

6.75 has seven tenths.   *Compare the decimals by looking at the place value.*
6.80 has eight tenths.
6.50 has five tenths.

$6.50 < 6.75 < 6.80$ *Order the rational numbers, using the tenths place.*
So, $6\frac{1}{2} < 6\frac{3}{4} < 6.8$.

• Use $<$ to list the numbers in order from least to greatest.
  $3.4, 3\frac{1}{9}, 3\frac{3}{8}$

Write the numbers as fractions with a common denominator.

1. 0.61 and 0.8

2. 0.4 and $\frac{1}{2}$

3. $\frac{3}{4}$ and $\frac{4}{3}$

4. 0.6 and $\frac{1}{4}$

Write each rational number in decimal form.

5. $\frac{1}{2}$

6. $2\frac{1}{4}$

7. $3\frac{1}{5}$

8. $^-6\frac{2}{5}$

9. $^-5\frac{3}{10}$

10. $9\frac{1}{8}$

Compare the rational numbers and order them from least to greatest.

11. $1\frac{1}{4}, 2, \frac{3}{4}$

12. $0.9, 0.82, 0, ^-0.9, 1.1$

13. $\frac{1}{2}, 1\frac{1}{2}, 0.6, ^-0.2, 1.3$

INDEPENDENT PRACTICE

Use the number line to tell which number is greater.

1. $1\frac{1}{4}$ or 2.5

2. $^-0.75$ or $\frac{1}{2}$

Compare. Write $<$, $>$, or $=$.

3. $0.5 \bullet 0.35$

4. $\frac{2}{8} \bullet 0.25$

5. $^-3\frac{1}{3} \bullet 3\frac{1}{3}$

6. $\frac{^-4}{5} \bullet \frac{^-1}{10}$

Compare the rational numbers and order them from least to greatest.

7. $3.8, ^-1.6, \frac{6}{3}, \frac{4}{5}$

8. $\frac{^-1}{2}, \frac{1}{6}, \frac{7}{8}, ^-0.4$

9. $^-1.25, 0.3, \frac{1}{10}, 0, \frac{^-1}{2}$

10. $9.5, ^-8.1, ^-6, 5.4, 5\frac{3}{5}, 9\frac{5}{8}$

11. $6\frac{3}{4}, 6\frac{1}{8}, 6.8, 6.02, 0.64$

Compare the rational numbers and order them from greatest to least.

12. $5.1, 8, \frac{1}{2}, 4, 6.2$

13. $1.5, ^-2, 3.3, ^-1.9, 2.4$

14. $\frac{1}{2}, \frac{3}{4}, 0.6, ^-0.9, 0.61$

Problem-Solving Applications

15. SPORTS Klaus, Hans, and John were in a skiing race. Klaus came in at 49.4 sec, Hans at 49.42 sec, and John at 49.35 sec. Who took the least amount of time?

16. SPORTS Janice's practice times for swimming laps are $1\frac{1}{2}$ min, 1.48 min, 1.51 min, and $1\frac{2}{5}$ min. What is the longest time she has taken to swim laps?

17. COMPARE Kara spent 30.5 min getting ready and Sheila spent $30\frac{5}{12}$ min. Who was ready first?

18. ▭ **WRITE ABOUT IT** Explain how you would compare 1.52 and $1\frac{2}{5}$.

1. **VOCABULARY** All whole numbers, their opposites, and 0 make up the set of ___?___. **(page 466)**

Write the opposite of each integer. **(pages 466–467)**

2. $^-32$ 3. 12 4. $^-15$ 5. $^-4$ 6. 0

Write the absolute value. **(pages 466–467)**

7. $|^-12|$ 8. $|^-4|$ 9. $|^+17|$ 10. $|^+8|$ 11. $|^-11|$

12. **VOCABULARY** A number that can be written as a ratio $\frac{a}{b}$, where a and b are integers and $b \neq 0$, is a(n) ___?___. **(page 468)**

Name the sets in which each number belongs. **(pages 468–469)**

13. $\frac{4}{2}$ 14. 0.89 15. $\frac{3}{1}$ 16. 0.75 17. 14

18. 7 19. $3\frac{3}{4}$ 20. $2\frac{1}{4}$ 21. 5 22. $^-1$

23. **VOCABULARY** When a fraction is divided and the remainder is zero, the quotient is a(n) ___?___. **(page 470)**

Write the fraction as a terminating or repeating decimal. **(pages 470–473)**

24. $\frac{1}{3}$ 25. $\frac{3}{5}$ 26. $\frac{2}{5}$ 27. $\frac{7}{10}$ 28. $\frac{1}{6}$

29. $\frac{5}{8}$ 30. $\frac{11}{15}$ 31. $\frac{2}{3}$ 32. $\frac{7}{8}$ 33. $\frac{3}{20}$

Find a rational number between the two numbers. **(pages 474–475)**

34. $\frac{1}{4}$ and $\frac{2}{3}$ 35. 1.3 and 1.32 36. $\frac{2}{5}$ and $\frac{3}{4}$ 37. $^-4.3$ and $^-4.4$

38. 3.4 and 3.52 39. $2\frac{1}{5}$ and $2\frac{1}{2}$ 40. $3\frac{5}{8}$ and $3\frac{9}{10}$ 41. 0.9 and 0.94

Use $<$ or $>$ to compare. **(pages 476–477)**

42. $2.45 \bullet 2.29$ 43. $\frac{3}{5} \bullet \frac{2}{3}$ 44. $1.3 \bullet 1\frac{2}{5}$ 45. $^-5.2 \bullet ^-6.2$

46. $3.7 \bullet 3.2$ 47. $^-8 \bullet ^-3$ 48. $^-4.8 \bullet ^-6.7$ 49. $\frac{5}{7} \bullet \frac{9}{14}$

50. Use $>$ to list the numbers in order from greatest to least.

     12.1, $12\frac{1}{8}$, 12.32

1. Last month, a soccer team practiced for $4\frac{1}{2}$ hours each week. The team usually practices a total of 15 hours a month. How many more hours did the team practice last month?

A 1 hr

B 2 hr

C $2\frac{1}{2}$ hr

D 3 hr

2. The graph shows the number of sixth-grade students at a middle school who participate in four clubs. How many more students are in the Science Club than in the Chess Club?

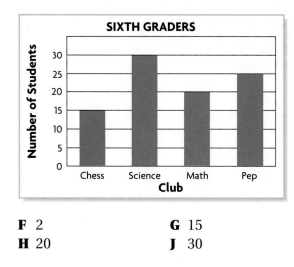

F 2   **G** 15

H 20   **J** 30

3. The ages of students in the Language Club are 13, 11, 15, 14, 12, 13, 13. What is the mode of this set of data?

A 11

B 12

C 13

D 14

E Not Here

4. A bag contains 4 blue marbles, 2 green marbles, 3 yellow marbles, and 6 red marbles. What is the probability of selecting either a blue or a red marble from the bag?

F $\frac{1}{10}$ **G** $\frac{2}{5}$

H $\frac{1}{2}$ **J** $\frac{2}{3}$

5. Sue's age is twice Mark's age plus three years. If Sue is 27, which algebraic equation can be used to find Mark's age, m?

A $2 + 3 + m = 27$

B $3m + 2 = 27$

C $27 - 3 = m$

D $2m + 3 = 27$

E Not Here

6. Vince is planning to paint a closet, which is 9 feet high, 4 feet wide, and 3 feet deep. He will paint the four walls, but not the floor or the ceiling. How much surface area will he paint?

F 105 ft^2 **G** 126 ft^2

H 150 ft^2 **J** 225 ft^2

7. Which of the following integers can be used to represent the depth of a well that goes 37 feet below ground level?

A $^-37$ **B** $^-3.7$

C $^+3.7$ **D** $^+37$

8. Which of the following rational numbers is between 8.09 and 8.3?

F 8.01

G 8.08

H 8.29

J 8.312

OPERATIONS WITH INTEGERS

LOOK AHEAD

In this chapter you will solve problems that involve

- adding integers
- subtracting integers
- multiplying integers
- dividing integers

SCIENCE **LINK**

There are several types of thermometers. The two most commonly used are liquid-in-glass and bimetallic. Both are based on the general fact that substances tend to expand when heated.

The liquid-in-glass thermometer usually contains either mercury or colored alcohol. The liquid expands as the temperature rises.

The bimetallic thermometer contains a strip consisting of two different metals which expand at different rates as they are heated. As the temperature changes, the strip bends.

- Find out whether a backyard temperature gauge and a meat thermometer are liquid-in-glass thermometers or bimetallic thermometers.

Types of Thermometers

Learn about at least three different types of thermometers. Make a table describing each thermometer, its use, and the range of temperatures it measures. Share your findings with the class.

Soil thermometer with a 5-inch stem and dial face: Measures temperatures from ⁻40°F to 120°F.

Probe thermometer: Measures temperatures of gasses and liquids from ⁻20°F to 220°F.

Maximum-minimum thermometer: Measures maximum temperatures from ⁻38°F to 130°F. Measures minimum temperatures from ⁻50°F to 120°F.

PROJECT CHECKLIST

✓ Did you investigate at least three different types of thermometers?

✓ Did you make a table showing the type, use, and range of temperatures each thermometer measures?

✓ Did you share your findings with the class?

LAB ACTIVITY

What You'll Explore
How to use two-color counters to add integers

What You'll Need
two-color counters

Modeling Addition of Integers

To keep track of points during a game, Ann and Marla are using yellow counters to represent positive points, or points gained, and red counters to represent negative points, or points lost.

ACTIVITY 1

Explore

- In the first round, Ann earned 5 points. In the second round, she earned 3 points. Use yellow counters to model Ann's total score.

First-round points   Second-round points

$$^+5 + {}^+3 = {}^+8 \leftarrow \text{total score}$$

- Marla lost 3 points in the first round. Then she lost 4 points in the second round. Use red counters to model Marla's total score.

First-round points   Second-round points

$$^-3 + {}^-4 = {}^-7 \leftarrow \text{total score}$$

Think and Discuss

- How is adding the girls' scores like adding whole numbers? How is it different?

- Would changing the order when adding Marla's and Ann's points change their scores? Why or why not?

- How would you model $^+5 + {}^+8$? $^-5 + {}^-8$?

Try This

Use counters to find the sum.

1. $^+7 + {}^+5$ **2.** $^-4 + {}^-8$

ACTIVITY 2

Explore

- Jacques gained 9 points in the first round and lost 4 points in the second round. To find his total score, Jacques paired points gained with points lost. Each point gained that is paired with a point lost equals 0.

 = 0

- Use yellow and red counters to model Jacques's total score.

$^+9 + {}^-4 = {}^+5$

- Don gained 3 points in the first round and lost 9 points in the second round. Use yellow and red counters to model Don's total score.

$^+3 + {}^-9 = {}^-6$

Think and Discuss

- Why is Jacques's score $^+5$? Why is Don's score $^-6$?

- What positive and negative integers could you combine to make a sum of 0?

Try This

Write an addition problem for the counters shown.

1.    **2.**

Use counters to find each sum.

3. $^+5 + {}^-7$ **4.** $^-4 + {}^+8$

483

ALGEBRA CONNECTION

Adding Integers

Brent and Steffan made up a game. The gameboard has a straight path, like a number line. Play starts in the middle of the board, at 0. Players use a spinner that includes positive moves and negative moves.

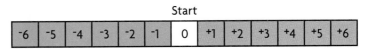

Start

| -6 | -5 | -4 | -3 | -2 | -1 | 0 | +1 | +2 | +3 | +4 | +5 | +6 |

You can show integers on a number line to represent each of Brent's and Steffan's moves. Brent's first spin was ⁻2, and his second spin was ⁻4. Where is Brent on the gameboard?

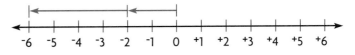

$^-2 + {}^-4 = {}^-6 \leftarrow$ Brent is at $^-6$.

Steffan's first spin was ⁺5, and his second spin was ⁻8. Where is Steffan on the gameboard?

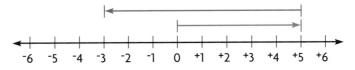

$^+5 + {}^-8 = {}^-3 \leftarrow$ Steffan is at $^-3$.

You can use a number line to find the sum of two integers.

EXAMPLE Use a number line to find the sum $^+3 + {}^-7$.

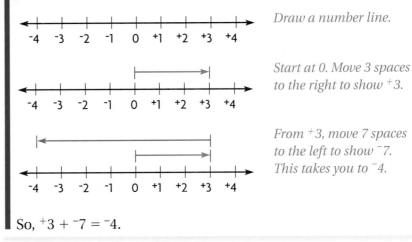

Draw a number line.

Start at 0. Move 3 spaces to the right to show $^+3$.

From $^+3$, move 7 spaces to the left to show $^-7$. This takes you to $^-4$.

So, $^+3 + {}^-7 = {}^-4$.

GUIDED PRACTICE

Write the addition equation modeled on the number line.

1.

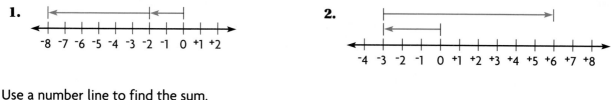

A number line from -8 to +2 with arrows.

2.

A number line from -4 to +8 with arrows.

Use a number line to find the sum.

3. $^+5 + {}^+3$ **4.** $^-4 + {}^-2$ **5.** $^-4 + {}^+5$ **6.** $^+2 + {}^-6$ **7.** $^-10 + {}^+6$

INDEPENDENT PRACTICE

Write the addition equation modeled on the number line.

1.

A number line from -2 to +6 with arrows.

2.

A number line from -2 to +6 with arrows.

Find the sum.

3. $^+9 + {}^+3$ **4.** $^-9 + {}^+3$ **5.** $^-6 + {}^+6$

6. $^+10 + {}^-1$ **7.** $^+11 + {}^-12$ **8.** $^-5 + {}^-15$

9. $^-40 + {}^-20$ **10.** $^-46 + {}^+28$ **11.** $^-100 + {}^-16$

12. $^+5 + {}^-8$ **13.** $^-5 + {}^+7$ **14.** $^+6 + {}^+20$

15. $^-35 + {}^+18$ **16.** $^+82 + {}^-93$ **17.** $^+72 + {}^-21$

18. $^-21 + {}^+21$ **19.** $^+81 + {}^-17$ **20.** $^-22 + {}^-88$

21. $^+42 + {}^+17$ **22.** $^-59 + {}^-33$ **23.** $^+91 + {}^-28$

24. $^-52 + {}^+87$ **25.** $^-77 + {}^-35$ **26.** $^+41 + {}^-16$

Problem-Solving Applications

27. TRANSPORTATION Ms. Johnson parked the car on the 5th floor of the garage. Then she took the elevator down 2 floors to the pedestrian crossover to the airport terminal. Write an addition problem to find the floor for the crossover.

28. MONEY Isabel had an account with a balance of $42. Then she made a deposit of $20. The next week she withdrew $29. Write an addition problem to find her new balance.

29. WEATHER In the morning the temperature was $^-8°F$. By noon it had risen by 12°F. What was the temperature at noon?

30. ✏ **WRITE ABOUT IT** Explain why $^-6 + {}^+2$ is not the same as $^+6 + {}^-2$.

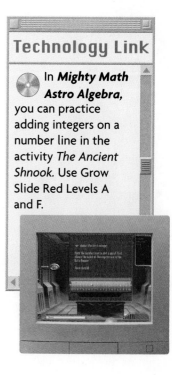

Technology Link

In *Mighty Math Astro Algebra,* you can practice adding integers on a number line in the activity *The Ancient Shnook.* Use Grow Slide Red Levels A and F.

MORE PRACTICE Lesson 25.1, page H78

LAB ACTIVITY

What You'll Explore
How to use two-color counters to subtract integers

What You'll Need
two-color counters

ALGEBRA CONNECTION

Modeling Subtraction of Integers

You can use red and yellow counters to subtract integers. Subtracting integers is similar to subtracting whole numbers.

Explore

ACTIVITY 1

- Find ⁻8 − ⁻2. First, make a row of 8 red counters.

- Then, take away 2 of them.

 ← ⁻8 − ⁻2

- How many counters are left? What is ⁻8 − ⁻2?

Using red and yellow counters, model ⁻6 − ⁺3.

- First, make a row of 6 red counters.

- Recall that a red counter paired with a yellow counter equals 0. Adding a red counter paired with a yellow counter does not change the value of ⁻6. Show another way to model ⁻6 that includes 3 yellow counters.

- Use your model to find ⁻6 − ⁺3. Take away the 3 yellow counters.

- What does your model show now? What is ⁻6 − ⁺3?

Now model ⁺5 − ⁻3.

- Model ⁺5. Put down pairs of yellow and red counters until you can take away ⁻3. What is ⁺5 − ⁻3?

Addition and subtraction of integers are related.

- Copy the two models below.

$^-8 - ^-2 = ?$ $^-8 + ^+2 = ?$

- Use the first model to find $^-8 - ^-2$. What is $^-8 - ^-2$?

- Use the second model to find $^-8 + ^+2$. What is $^-8 + ^+2$?

Think and Discuss

- The models above show that $^-8 - ^-2 = ^-8 + ^+2$. How are $^-2$ and $^+2$ related? How are subtraction and addition related?

- How are the two models different?

You can write a subtraction problem as an addition problem by adding the opposite of the number you are subtracting.

$^+5 - ^-3 = ^+5 + ^+3$ ← Add the opposite of $^-3$.

- How can you write $^-6 - ^-3$ as an addition problem?

Try This

Complete the addition problem.

1. $^-8 - ^+7 = ^-8 + $ ■   **2.** $^-11 - ^+4 = ^-11 + $ ■

3. $^-10 - ^+5 = ^-10 + $ ■   **4.** $^-12 - ^+6 = ^-12 + $ ■

Use counters to find the difference.

5. $^-6 - ^+5$ **6.** $^-5 - ^+6$

7. $^-15 - ^+9$ **8.** $^-9 - ^+4$

Technology Link

You can practice subtracting integers by using E-Lab, Activity 25. Available on CD-ROM and on the Internet at **www.hbschool.com/elab**

ALGEBRA CONNECTION

Subtracting Integers

What You'll Learn
How to use a number
line to subtract integers

Why Learn This?
To find differences
between integers
to solve problems
involving temperature

In Kansas City, Missouri, the temperature was a chilly 3°F at
9:00 A.M. A half hour later, the temperature was 7° lower. What
was the temperature in Kansas City at 9:30 A.M.?

To find the temperature in Kansas City, you need to find the
difference of ⁺3 and ⁺7. You can find the difference of two
integers by adding the opposite of the integer you are
subtracting. The opposite of ⁺7 is ⁻7. So, instead of subtracting
⁺7, add ⁻7.

$$^+3 - {}^+7 \text{ becomes } {}^+3 + {}^-7.$$

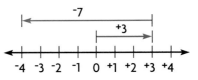

CULTURAL LINK

Several countries, such
as England, use the
Celsius temperature
scale. The people of
London, a large city in
England, enjoy mild
summers where
temperatures can
reach 21°C (70°F). In the
winter, temperatures
drop to freezing, 0°C
(32°F). What is the
temperature change
between 21°C and 0°C?

The temperature reading at 9:30 A.M. was ⁻4°F, or 4°F below zero.

EXAMPLE   In St. Louis, Missouri, the 9:00 A.M. temperature
reading was ⁻3°F. The temperature dropped 7° in a half hour.
What was the temperature reading at 9:30 A.M.?

⁻3 − ⁺7 → ⁻3 + ⁻7   *Write the subtraction problem as an
addition problem.*

*Draw a number
line.*

*Start at 0. Move 3
spaces to the left
to show ⁻3.*

*From ⁻3, move 7
spaces to the left
to add ⁻7.*

So, at 9:30 A.M. the temperature was ⁻10°F, or 10°F below zero.

Talk About It

• How is ⁻4 − ⁺2 related to ⁻4 + ⁻2?

• What is the opposite of ⁺2?

• Explain how to find ⁻5 − ⁺3.

Calculator Activities, page H40

GUIDED PRACTICE

Rewrite the subtraction problem as an addition problem.

1. $^+9 - {}^+12$ **2.** $^+5 - {}^-8$ **3.** $^-3 - {}^-10$ **4.** $^-2 - {}^+4$ **5.** $^-9 - {}^-1$

INDEPENDENT PRACTICE

Rewrite the subtraction problem as an addition problem. Use the number line to find the sum.

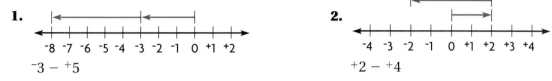

1.

$^-3 - {}^+5$

2.

$^+2 - {}^+4$

Find the difference.

3. $^+7 - {}^+12$ **4.** $^-12 - {}^-8$ **5.** $^+5 - {}^-4$ **6.** $^-6 - {}^-12$ **7.** $0 - {}^+15$

8. $^+5 - {}^+11$ **9.** $^-9 - {}^-9$ **10.** $^-7 - {}^+2$ **11.** $^+10 - {}^-10$ **12.** $^+15 - {}^+20$

13. $^+21 - {}^+37$ **14.** $^-15 - {}^+5$ **15.** $^+50 - {}^+45$ **16.** $^-35 - {}^+16$ **17.** $^-28 - {}^+26$

Problem-Solving Applications

18. The temperature outside is 15°F, but the windchill factor is ⁻2°F. What is the difference between the actual temperature and the windchill factor?

19. WEATHER The temperature at the top of the mountain is ⁻32°F. The temperature at the base is ⁻7°F. What is the difference in the temperatures?

20. CRITICAL THINKING Can you reverse the order of integers when subtracting and still get the same answer? Explain.

21. **WRITE ABOUT IT** Explain how to change a subtraction problem with integers into an addition problem.

Mixed Review and Test Prep

Find the product.

22. 12×10 **23.** 20×3 **24.** 31×5 **25.** 42×12 **26.** 50×33

Write as a decimal.

27. $\frac{1}{3}$ **28.** $\frac{2}{5}$ **29.** $\frac{2}{9}$ **30.** $\frac{2}{3}$ **31.** $\frac{3}{10}$

32. FRACTIONS Lionel gave $\frac{1}{2}$ of his baseball cards to Rick. He gave his brother $\frac{1}{2}$ of what was left. He then had 12 cards left. How many did he start with?

A 24 **B** 36
C 48 **D** 96

33. CONSUMER MATH A service call to repair a washing machine costs $35 plus $20 per hour. Raphael's machine took 2 hours to fix. How much did the repair cost?

F $50.00 **G** $55.00
H $65.00 **J** $75.00

What You'll Learn
How to multiply integers

Why Learn This?
To solve problems and notice patterns that involve multiplying integers

SCIENCE LINK

The gravitational pulls of the moon and the sun cause tidal changes on large bodies of water. When the moon and the sun are in a straight line with the Earth, very high and low tides, called spring tides, occur. Spring tides may cause tidal changes twice as great as those of average tides. If high tides at Boca Grande Pass average $^+1.75$ ft and low tides average $^-1.75$ ft, what might the spring tides be?

ALGEBRA CONNECTION

Multiplying Integers

The tide is changing at a rate of $^-2$ ft per hour. What is the change in the tide after 5 hr?

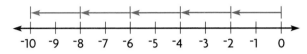

The number line shows how the tide changes $^-2$ ft per hour. You can write this as repeated addition or as multiplication.

$$^-2 + ^-2 + ^-2 + ^-2 + ^-2 = ^-10 \qquad 5 \times ^-2 = ^-10$$

So, the tide changed by $^-10$ ft.

Look at the problems below.

$$^+3 \times ^-4 = ^-12$$
$$^+2 \times ^-4 = ^-8$$
$$^+1 \times ^-4 = ^-4$$
$$0 \times ^-4 = 0$$
$$^-1 \times ^-4 = \blacksquare$$
$$^-2 \times ^-4 = \blacksquare$$

- What pattern do you see in the first factors? in the second factors? in the products?

- **CRITICAL THINKING** If this pattern continues, $^-1 \times ^-4 = ^+4$. What do you think the product is for $^-2 \times ^-4$?

> When you multiply a positive integer and a negative integer, the product is a negative integer. When you multiply two negative integers, the product is a positive integer.

EXAMPLE Find the products $^-4 \times ^-5$ and $^-4 \times ^+5$.

$^-4 \times ^-5 = ^+20$ *Multiply as with whole numbers. The product of two negative integers is a positive integer.*

$^-4 \times ^+5 = ^-20$ *Multiply as with whole numbers. The product of a negative integer and a positive integer is a negative integer.*

The calculator sequence 4 ⊠ 5 ⊞ can be used to find $^+4 \times ^-5$. What key sequence can you use to find $^-3 \times ^-4$?

GUIDED PRACTICE

Write each addition problem as a multiplication problem.

1. $^+7 + {}^+7 + {}^+7 + {}^+7$ **2.** $^-6 + {}^-6 + {}^-6$ **3.** $^-8 + {}^-8 + {}^-8 + {}^-8 + {}^-8 + {}^-8$

Write an addition problem and a multiplication problem for the model.

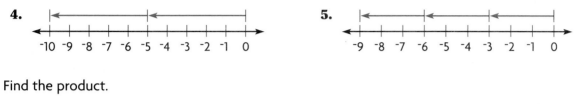

4.

5.

Find the product.

6. $^+3 \times {}^+5$ **7.** $^-3 \times {}^+5$ **8.** $^-3 \times {}^-5$ **9.** $^+3 \times {}^-5$

INDEPENDENT PRACTICE

Complete the pattern.

1.
$^+3 \times {}^-2 = {}^-6$
$^+2 \times {}^-2 = {}^-4$
$^+1 \times {}^-2 = {}^-2$
$0 \times {}^-2 = 0$
$^-1 \times {}^-2 = {}^+2$
$^-2 \times {}^-2 = \blacksquare$
$^-3 \times {}^-2 = \blacksquare$

2.
$^+2 \times {}^-6 = {}^-12$
$^+1 \times {}^-6 = {}^-6$
$0 \times {}^-6 = 0$
$^-1 \times {}^-6 = {}^+6$
$^-2 \times {}^-6 = \blacksquare$
$^-3 \times {}^-6 = \blacksquare$
$^-4 \times {}^-6 = \blacksquare$

3.
$^+3 \times {}^-3 = \blacksquare$
$^+2 \times {}^-3 = \blacksquare$
$^+1 \times {}^-3 = {}^-3$
$0 \times {}^-3 = 0$
$^-1 \times {}^-3 = {}^+3$
$^-2 \times {}^-3 = {}^+6$
$^-3 \times {}^-3 = {}^+9$

4.
$^+3 \times {}^-10 = \blacksquare$
$^+2 \times {}^-10 = \blacksquare$
$^+1 \times {}^-10 = \blacksquare$
$0 \times {}^-10 = 0$
$^-1 \times {}^-10 = {}^+10$
$^-2 \times {}^-10 = {}^+20$
$^-3 \times {}^-10 = {}^+30$

Find the product.

5. $^-6 \times {}^+8$ **6.** $^-2 \times {}^-9$ **7.** $^+5 \times {}^-4$ **8.** $^+8 \times {}^+8$ **9.** $^-7 \times {}^+3$

10. $^-1 \times {}^-11$ **11.** $^+16 \times {}^-6$ **12.** $^-20 \times {}^-3$ **13.** $^+12 \times {}^-5$ **14.** $^+10 \times {}^+9$

15. $^+6 \times {}^-5$ **16.** $^-21 \times {}^-2$ **17.** $^-25 \times {}^-3$ **18.** $^-9 \times {}^+9$ **19.** $^+18 \times {}^-10$

20. $^-40 \times {}^-6$ **21.** $^+60 \times {}^-8$ **22.** $^-100 \times {}^+4$ **23.** $^-200 \times {}^-5$ **24.** $^-15 \times {}^+3$

25. $^-15 \times {}^+7$ **26.** $^+180 \times {}^-3$ **27.** $^-22 \times {}^-7$ **28.** $^-42 \times {}^-40$ **29.** $^+32 \times {}^-6$

Problem-Solving Applications

30. SCIENCE Erosion has caused a shoreline to change by $^-2$ in. every year. Write the change in the shoreline over 4 years as a negative number.

31. At 12:00 noon the temperature was 90°F. Then the temperature fell 4° every hour until 6:00 P.M. Write the total change in temperature as a negative number.

32. CRITICAL THINKING What is the sign of the product when you multiply three negative numbers? four negative numbers?

33. ✏ **WRITE ABOUT IT** Explain how multiplying integers is different from multiplying whole numbers.

What You'll Learn
How to divide integers

Why Learn This?
To solve problems and notice patterns that involve dividing integers

REMEMBER:
....

Multiplication and division are **inverse operations**:

$3 \times 4 = 12$,
so $12 \div 3 = 4$
and $12 \div 4 = 3$.

See page H9.

ALGEBRA CONNECTION

Dividing Integers

When you want to solve a division problem, you can think about the related multiplication problem.

$36 \div 9 = $ ▨   **Think:** $9 \times 4 = 36$, so $36 \div 9 = 4$.
$48 \div 6 = $ ▨   **Think:** $6 \times 8 = 48$, so $48 \div 6 = 8$.

You can find patterns in integer division by comparing the division problems with their related multiplication problems. Look at the problems below.

Division	Multiplication
$^+32 \div {}^+8 = {}^+4$	$^+4 \times {}^+8 = {}^+32$
$^-32 \div {}^-8 = {}^+4$	$^+4 \times {}^-8 = {}^-32$
$^-32 \div {}^+8 = {}^-4$	$^-4 \times {}^+8 = {}^-32$
$^+32 \div {}^-8 = {}^-4$	$^-4 \times {}^-8 = {}^+32$

Talk About It

• When you multiply two negative integers, the product is a positive integer. What happens when you divide two negative integers?

• When you multiply a positive integer and a negative integer, the product is a negative integer. What happens when you divide a positive integer by a negative integer? a negative integer by a positive integer?

Since multiplication and division are inverse operations, the rules for multiplying integers also apply when dividing integers.

HEALTH LINK

Recreational scuba diving is limited to about ⁻100 ft because of the effects of water pressure. A diver is at ⁻50 ft and stops every 10ft before reaching the surface. At what depth will he or she make the first stop?

EXAMPLE What are $^-100 \div {}^+20$ and $^-100 \div {}^-20$?

$^-100 \div {}^+20 = {}^-5$ *Divide as with whole numbers. A negative integer divided by a positive integer is a negative integer.*

$^-100 \div {}^-20 = {}^+5$ *Divide as with whole numbers. A negative integer divided by a negative integer is a positive integer.*

• What is $^+100 \div {}^-20$?

• **CRITICAL THINKING** Is $^+100 \div {}^-5 \div {}^-4$ positive or negative? Explain.

📱 ....
Calculator Activities, page H40

GUIDED PRACTICE

Find the missing number.

1. $^+50 \div {}^+10 = {}^+5$,
$^+10 \times \blacksquare = {}^+50$

2. $^+48 \div {}^-8 = {}^-6$,
$^-8 \times {}^-6 = \blacksquare$

3. $^-2 \times {}^+15 = {}^-30$,
$^-30 \div {}^-2 = \blacksquare$

4. $^-3 \times {}^-18 = {}^+54$,
$\blacksquare \div {}^-3 = {}^-18$

Write *positive* or *negative* for each quotient.

5. $^+10 \div {}^-2$

6. $^+20 \div {}^+5$

7. $^-33 \div {}^+11$

8. $^-40 \div {}^-4$

Find the quotient.

9. $^+8 \div {}^-4$

10. $^-12 \div {}^+3$

11. $^+10 \div {}^+5$

12. $^-24 \div {}^-3$

13. $^+27 \div {}^-9$

INDEPENDENT PRACTICE

Find the quotient.

1. $^+9 \div {}^+3$

2. $^+6 \div {}^-2$

3. $^-18 \div {}^-3$

4. $^-12 \div {}^+4$

5. $^+24 \div {}^+12$

6. $^-20 \div {}^-10$

7. $^-80 \div {}^+5$

8. $^+64 \div {}^-8$

9. $^-45 \div {}^-9$

10. $^+44 \div {}^-4$

11. $^+50 \div {}^+5$

12. $^-84 \div {}^+12$

13. $^-88 \div {}^-8$

14. $^+180 \div {}^-10$

15. $^-120 \div {}^+6$

16. $^-200 \div {}^-50$

17. $^+240 \div {}^-8$

18. $^+90 \div {}^+15$

19. $^-81 \div {}^+3$

20. $^+192 \div {}^-4$

Problem-Solving Applications

21. WEATHER In 4 hr the temperature fell 16°. If it fell at the same rate every hour, how many degrees did it change every hour?

22. BUSINESS In a report, a company shows a loss for 1 yr as $^-\$42,000$. What was the average monthly loss?

23. Which problem has the greater quotient, $^-40 \div {}^+10$ or $^-40 \div {}^-10$? Explain.

24. ✐ **WRITE ABOUT IT** The product of two integers is positive. Is the quotient of those integers positive or negative?

Mixed Review and Test Prep

Find the missing number.

25. $4 + n = 12$

26. $16 - n = 8$

27. $45 + n = 50$

28. $60 - n = 40$

Compare by ordering the numbers from least to greatest.

29. $7.7, 6.2, 7.1, {}^-6.5, 8.0, {}^-5.1$

30. $2.4, 3.8, {}^-1.2, 4.6, 4.5, {}^-2.3$

31. GEOMETRY The radius of a circle is 4.2 centimeters. What is the diameter of the circle?

  A 2.1 cm   **B** 4.2 cm
  C 8.4 cm   **D** 13.2 cm

32. GEOMETRY Find the volume of a box that has a length and height of 10 inches and a width of 5 inches.

  F 25 in.^{3} **G** 400 in.^{3}
  H 500 in.^{3} **J** 1,000 in.^{3}

Write the addition problem modeled on each number line.
(pages 484–485)

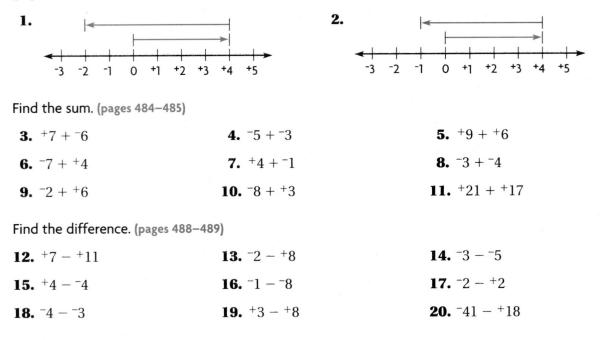

1.

2.

Find the sum. (pages 484–485)

3. $^+7 + {}^-6$ **4.** $^-5 + {}^-3$ **5.** $^+9 + {}^+6$

6. $^-7 + {}^+4$ **7.** $^+4 + {}^-1$ **8.** $^-3 + {}^-4$

9. $^-2 + {}^+6$ **10.** $^-8 + {}^+3$ **11.** $^+21 + {}^+17$

Find the difference. (pages 488–489)

12. $^+7 - {}^+11$ **13.** $^-2 - {}^+8$ **14.** $^-3 - {}^-5$

15. $^+4 - {}^-4$ **16.** $^-1 - {}^-8$ **17.** $^-2 - {}^+2$

18. $^-4 - {}^-3$ **19.** $^+3 - {}^+8$ **20.** $^-41 - {}^+18$

Solve. (pages 488–489)

21. The temperature is 8°F, but the windchill factor is ⁻5°F. What is the difference between the temperature and the windchill factor?

22. A submarine was at ⁻300 ft. Later, it was at ⁻1,250 ft. What is the difference between ⁻300 and ⁻1,250?

Write addition and multiplication problems for each number line.
(pages 490–491)

23.

24.

Find the product. (pages 490–491)

25. $^+4 \times {}^-8$ **26.** $^-9 \times {}^-5$ **27.** $^+7 \times {}^-2$ **28.** $^-19 \times {}^-5$

29. $^-5 \times {}^+4$ **30.** $^-8 \times {}^-10$ **31.** $^-12 \times {}^+5$ **32.** $^+50 \times {}^-7$

Find the quotient. (pages 492–493)

33. $^+27 \div {}^-3$ **34.** $^-18 \div {}^+2$ **35.** $^-36 \div {}^-6$ **36.** $^+180 \div {}^-9$

37. $^+144 \div {}^-12$ **38.** $^-216 \div {}^+18$ **39.** $^-220 \div {}^-10$ **40.** $^-132 \div {}^-12$

1. Jill bought a motorcycle for $2,448. She paid for it in 36 equal payments. How much was each payment?

A $42
B $57
C $68
D $72
E Not Here

2. Which of the following graphs is best suited for displaying data which are divided into parts totaling 100%?

F Line graph
G Bar graph
H Circle graph
J Histogram

3. A spinner is divided into 10 equal sections; 3 are green, 2 are blue, 4 are red, and 1 is yellow. What is the probability of the spinner stopping on either a blue or a red section?

A $\frac{1}{10}$
B $\frac{1}{8}$
C $\frac{3}{10}$
D $\frac{3}{5}$
E $\frac{9}{10}$

4. Marla can type 248 words in 4 minutes. At this rate, how long would it take her to type 558 words?

F 9 min
G 12 min
H 14 min
J Not Here

5. Zoe bought 3 pounds of Swiss cheese and 12 ounces of American cheese. Altogether, how many ounces of cheese did she buy?

A 48 oz
B 60 oz
C 68 oz
D 75 oz

6. Which of the following decimals is equal to $\frac{7}{20}$?

F 0.27
G 0.3
H 0.35
J 0.720

7. On Monday, the low temperature was $^-4°$F. The low temperature on Tuesday was 13° higher. What was the low temperature on Tuesday?

A $^-13°$F
B $^-9°$F
C 4°F
D 9°F
E 13°F

8. During a 12-hour period, the temperature fell 36°. If it fell the same number of degrees every hour, how many degrees did it fall per hour?

F 2°
G 3°
H 4°
J 5°
K Not Here

ALGEBRA: EQUATIONS AND RELATIONS

LOOK AHEAD

In this chapter you will solve problems that involve

- evaluating expressions that include integers

- solving equations that include integers

- using inequalities

- locating points and graphing relations on a coordinate plane

The table shows the 5 most popular languages of the world, based on the estimated number of people who speak that language.

LANGUAGE	TOTAL SPEAKERS
Mandarin	1,025,000,000
English	497,000,000
Hindi	476,000,000
Spanish	409,000,000
Russian	279,000,000

- About how many more English speakers are there in the world than Russian speakers?

- More people speak Mandarin than any two other languages combined. How do English and Hindi combined compare to Mandarin?

Word Equations

Many words are made up of smaller parts, such as roots, suffixes, and prefixes. Think of five different words and add or subtract prefixes or suffixes to change the words into other words. Show your new words in word equations. Then, identify the root words, prefixes, and suffixes in the equations.

succeed + ed =
SUCCEEDED

succeeded - ed + s =
SUCCEEDS

succeeds - eds + ss =
SUCCESS

success + ful =
SUCCESSFUL

un + successful =
UNSUCCESSFUL

PROJECT CHECKLIST

✓ Did you write equations to show word relationships?

✓ Did you identify root words, prefixes, and suffixes?

Evaluating Expressions

Integers can be used to describe distances above or below sea level. Death Valley, California, is 280 ft below sea level. Its position can be represented by the integer $^-280$. Sometimes numerical expressions include integers.

EXAMPLE 1 Elaine is scuba diving 25 ft below sea level. She must travel 30 ft directly up to get to the deck of her boat. Write and evaluate a numerical expression that shows how far the boat deck is above sea level.

$^-25$ ft $+ 30$ ft $= 5$ ft

So, the deck is 5 ft above sea level.

Talk About It

- In Example 1, why was Elaine's position shown as $^-25$ ft?

- Why was 30 added to $^-25$?

- Suppose Elaine traveled 13 ft deeper than 25 ft below sea level. What expression would you write?

- What is another situation in which you might have to evaluate a numerical expression involving integers?

Algebraic expressions can also involve integers. To evaluate an algebraic expression, replace the variable with the given value.

EXAMPLE 2 Evaluate the expression. Remember the order of operations.

A. $y - 8$, for $y = ^-6$ **B.** $10 + 72 \div k$, for $k = ^-9$

$\quad y - 8$ $10 + 72 \div k$

$\quad ^-6 + ^-8$ $10 + 72 \div ^-9$ *Divide first.*

$\quad ^-14$ $10 + ^-8$

                        2

GUIDED PRACTICE

Tell which operation you would perform first to evaluate the numerical expression.

1. $^-6 + 7 \times 2$ **2.** $12 \div ^-2 - 8$ **3.** $^-7 - 15 \times 4$

4. $^-14 + 10 - 4$ **5.** $17 + ^-11 \div 2$ **6.** $^-9 \times 8 \div ^-36$

INDEPENDENT PRACTICE

Evaluate the numerical expression. Remember the order of operations.

1. $7 + 3 \times 2$

2. $8 + 6 - 2$

3. $^-14 \div ^-2 \times 3$

4. $(^-18 - 17) \div 7$

5. $8^2 \times ^-3$

6. $^-9 \times (^-52 + 40)$

7. $^-3 \times 8 \div 2$

8. $6 \times ^-3 + ^-8$

Evaluate the algebraic expression for the given value of the variable.

9. $x - 8$, for $x = 15$

10. $y - 20$, for $y = 9$

11. $37 + k$, for $k = 10$

12. $15 + a$, for $a = ^-42$

13. $x^2 - ^-4$, for $x = ^-5$

14. $y^3 - 3$, for $y = ^-2$

15. $2 + k - 14$, for $k = ^-8$

16. $^-25 + a^3$, for $a = 2$

17. $9p$, for $p = ^-4.5$

18. $m \times ^-21$, for $m = ^-3$

19. $j \div 12$, for $j = ^-84$

20. $^-54 \div y$, for $y = ^-6$

21. $(w - 7) \div 4$, for $w = ^-9$

22. $a^2 \div (^-17 + 8)$, for $a = ^-6$

23. $7 \times (p + ^-9)$, for $p = ^-12$

Problem-Solving Applications

24. RECREATION The bottom of the swimming pool at the club is 10 ft below the ground. The highest diving board is 12 ft above the ground. Write and evaluate an expression to find the difference between the height of the diving board and the bottom of the pool.

25. Jared has $23 in his checking account. To find the balance in his account after he writes a check, Jared uses the expression $a - c$, where a is the amount in his checking account and c is the amount of the check. What is the balance if Jared writes a check for $32?

26. WEATHER In the late afternoon, Chrissy noticed that the thermometer read $^-3°F$. Five hours later, the temperature had dropped 8°F. What was the temperature then?

27. ▭▷ **WRITE ABOUT IT** Write a problem that involves evaluating a numerical expression that includes integers.

Mixed Review and Test Prep

Solve and check.

28. $x + 7 = 12$

29. $a - 23 = 17$

30. $9p = 45$

31. $\frac{x}{3} = 5$

Evaluate.

32. $5 - 14$

33. $^-10 - 1$

34. $^-8 - ^-6$

35. $^-8 + ^-6$

36. MONEY Jerry brings $20.00 to the ball game. He buys a hot dog for $2.50, a soda for $1.75, and a cap for $11.49. How much money does he have left?

A $4.26

B $4.46

C $5.26

D $5.46

37. CONSUMER MATH The Johnson family traveled 704 miles and used 32 gallons of gasoline. How many miles per gallon did their car get?

F 20 mpg

G 21 mpg

H 22 mpg

J 23 mpg

Solving Equations with Integers

Sometimes equations involve integers.

Early in the morning, the temperature is ⁻5°F. In the late afternoon, the temperature increases to 8°F. What is the change in temperature?

You can solve the problem by writing and solving an addition equation.

$$⁻5 + x = 8 \qquad \textit{Let x = change in temperature.}$$
$$⁻5 + 5 + x = 8 + 5 \qquad \textit{Add 5 to each side.}$$
$$x = 13$$

$$⁻5 + x = 8 \qquad \textit{Check your solution.}$$
$$⁻5 + 13 = 8 \qquad \textit{Replace x with 13.}$$
$$8 = 8 ✓ \qquad \textit{The solution checks.}$$

So, the change in temperature is 13°F.

• Solve. $y - 6 = ⁻13$

Talk About It

• Name the integers in the equation $⁻5 + x = 8$.

• Is solving equations that involve integers different from solving equations that involve whole numbers? Explain.

Integers also appear in multiplication and division equations.

EXAMPLE Solve and check. $\frac{x}{3} = ⁻5$

$$\frac{x}{3} = ⁻5$$
$$3 \times \frac{x}{3} = ⁻5 \times 3 \qquad \textit{Multiply each side by 3.}$$
$$x = ⁻15$$

$$\frac{x}{3} = ⁻5 \qquad \textit{Check your solution.}$$
$$\frac{⁻15}{3} = ⁻5 \qquad \textit{Replace x with ⁻15.}$$
$$⁻5 = ⁻5 ✓ \qquad \textit{The solution checks.}$$

So, $x = ⁻15$.

• Solve. $⁻8y = ⁻32$

GUIDED PRACTICE

Tell whether the given value is a solution of the equation. Write *yes* or *no*.

1. $b + 7 = {}^-12$;
   $b = {}^-19$

2. $k - 4 = {}^-12$;
   $k = {}^-8$

3. $c - {}^-7 = 4$;
   $c = 11$

4. $x - 10 = 43$;
   $x = {}^-53$

5. ${}^-3x = 18$; $x = {}^-6$

6. ${}^-7k = 42$; $k = 6$

7. $\frac{y}{3} = 5$; $y = {}^-15$

8. $\frac{a}{-2} = {}^-12$; $a = 24$

INDEPENDENT PRACTICE

Solve and check.

1. $x + 5 = 2$

2. $7 = y + 15$

3. $k + 4 = {}^-2$

4. ${}^-8 = n + 5$

5. $c - 7 = {}^-2$

6. ${}^-12 = w - 22$

7. $r - 4 = {}^-7$

8. ${}^-15 = u - 9$

9. $x + 135 = 22$

10. ${}^-12 = y + 145$

11. $k - 234 = {}^-179$

12. ${}^-963 = n - 548$

13. ${}^-3v = 24$

14. ${}^-8t = 56$

15. ${}^-9p = {}^-72$

16. ${}^-4m = {}^-60$

17. $5y = {}^-25$

18. $7d = {}^-63$

19. $\frac{y}{3} = {}^-8$

20. $\frac{x}{7} = {}^-10$

21. $\frac{y}{-2} = 15$

22. $\frac{a}{6} = 12$

23. $\frac{m}{-5} = {}^-3$

24. $\frac{d}{-14} = {}^-12$

25. ${}^-720 = 10p$

26. ${}^-851 = {}^-23r$

27. ${}^-235 = \frac{x}{82}$

28. ${}^-742 = \frac{a}{10}$

For Exercises 29–32, write an equation for the word sentence.

29. Negative seven multiplied by a number, x, is sixty-three.

30. A number, y, divided by negative five is negative thirteen.

31. Seven more than a number, x, is negative thirty-four.

32. Negative nine less than a number, y, is negative twelve.

Problem-Solving Applications

33. RECREATION Brenda is scuba diving. She decides to swim 30 ft deeper. She ends up 67 ft below the surface of the ocean. Write and solve an equation to find her original depth. (HINT: Let d represent her original depth.)

34. WEATHER On a cold day, Pedro noticed that the temperature had increased 7°F to the current temperature of ${}^-19$°F. Write and solve an equation to find the original temperature. (HINT: Let t represent the original temperature.)

35. ✏️ **WRITE ABOUT IT** Describe the steps necessary to solve and check the equation $\frac{x}{-7} = {}^-13$.

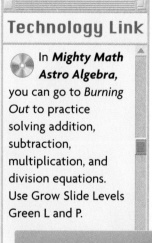

Technology Link

In *Mighty Math Astro Algebra,* you can go to *Burning Out* to practice solving addition, subtraction, multiplication, and division equations. Use Grow Slide Levels Green L and P.

What You'll Explore
How to model the solving of two-step equations

What You'll Need
algebra tiles or paper rectangles and squares

Exploring Two-Step Equations

Sometimes equations involve more than one step. In this lab you will model the solving of two-step equations.

Explore

- Use algebra tiles or paper rectangles and squares to model $2y + 1 = 7$. Use a green rectangle to represent the variable y, and a yellow square to represent 1.

$$2y \quad\quad + \quad 1 \quad\quad = \quad\quad\quad 7$$

- To solve an equation, you must arrange the model so that a rectangle is alone on one side. First, remove one square from each side.

- The model now shows $2y = 6$.

- Separate the model into two equal groups on each side.

- What does each pair of groups model?

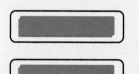

- What is the solution of the equation?

Think and Discuss

- How can you use the original model to check your solution?

- What did you do to the model so that only rectangles were alone on one side? What operation does this model?

- What operation did you model when you separated the model into two equal groups?

- How is solving a two-step equation different from solving a one-step equation?

- Look at this model. What equation does it represent? Describe how to solve it.

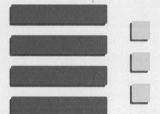 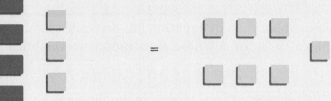

Try This

- Copy the model below. What equation does it represent? Use the model to solve the equation.

- Use a model to solve each equation.

  $2k + 3 = 11$ $\qquad$ $3a + 4 = 10$ $\qquad$ $4p + 2 = 14$

Inequalities

What You'll Learn
How to use inequality symbols and solve algebraic inequalities

Why Learn This?
To be able to visualize the solutions that make an inequality true

VOCABULARY

inequality

REMEMBER:

When comparing numbers, you can use these symbols:

=	is equal to
≠	is not equal to
<	is less than
>	is greater than
≤	is less than or equal to
≥	is greater than or equal to

See pages 18 and H90.

You must weigh less than 105 lb to play youth football. Bert, Todd, and Eduardo want to play this season. Their weights are shown in the table. Which of them will be able to play?

Name	Bert	Todd	Eduardo
Weight	101 lb	104 lb	109 lb

One way to answer the question is by using inequality symbols.

Bert	Todd	Eduardo
101 lb < 105 lb	104 lb < 105 lb	109 lb > 105 lb

So, Bert and Todd will be able to play football this season.

Remember that an equation has an equals sign. An **inequality** has $<$, $>$, $\leq$, or $\geq$. These are algebraic inequalities:

$$x < 8 \qquad y > 3 \qquad a \leq 7 \qquad b \geq 12$$

The solution of an algebraic inequality may have many possible solutions.

EXAMPLE 1 Find the whole-number solutions of $k < 3$.

$0 < 3 \qquad 1 < 3 \qquad 2 < 3$ *Replace k with whole numbers that make the inequality true.*

So, the whole-number solutions are 0, 1, and 2.

You can use a number line to visualize all the solutions of algebraic inequalities.

EXAMPLE 2 Graph all the integer solutions of $p > 2$ and of $m \leq 5$.

$p > 2$

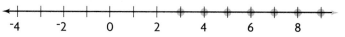

The blue arrow means the solutions go on and on.

$m \leq 5$

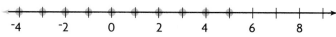

• Explain why you wouldn't graph a point at 5 on a number line for the inequality $y > 5$.

GUIDED PRACTICE

Tell what numbers the algebraic inequality represents.

1. $x < 2$ **2.** $b \geq 5$ **3.** $a \leq 6$ **4.** $y > 4$ **5.** $b \geq 0$

Give two possible values for the variable that make the inequality true.

6. $x < 8$ **7.** $x \geq 7$ **8.** $a \leq 12$ **9.** $y > 100$ **10.** $b \geq 17$

INDEPENDENT PRACTICE

Find all whole-number solutions of the inequality.

1. $x < 2$ **2.** $x < 5$ **3.** $x < 8$ **4.** $a \leq 2$ **5.** $a \leq 5$

Sketch a graph on a number line to show all integer solutions of the inequality.

6. $x < 5$ **7.** $x > 2$ **8.** $x \leq 8$ **9.** $x \geq 4$ **10.** $x < 15$

11. $x > 24$ **12.** $x \leq 0$ **13.** $x \geq 0$ **14.** $x \leq 43$ **15.** $x < 56$

Write an algebraic inequality represented by the integers graphed on the number line.

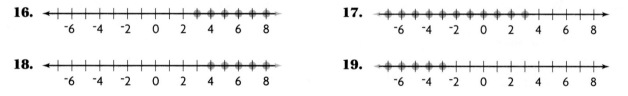

For Exercises 20–23, write an algebraic inequality for the word sentence.

20. All numbers x are less than or equal to seven.

21. All numbers y are greater than or equal to twelve.

22. All numbers x are less than eighteen.

23. All numbers y are greater than nine.

Problem-Solving Applications

24. **MEASUREMENT** A fisherman must make sure that certain fish are longer than 10 in. before he can keep them. Write an algebraic inequality representing the lengths of fish the fisherman can keep.

25. **ALGEBRA** Greg weighs more than his brother Jim. If g represents Greg's weight, and j represents Jim's weight, write an algebraic inequality that shows the relationship between their weights.

26. **WRITE ABOUT IT** What is the difference between $x \leq 8$ and $x < 8$?

Graphing on the Coordinate Plane

TEEN TIMES

Many teens like to play Battleship®. In this game competitors call out coordinates where they think their opponents have "hidden" their ships. When they guess correctly, the ships are "sunk."

Sidney lives in the city of Anytown. Some parts of Anytown are shown on the map below. Sidney's house is at the start. Each square on the map represents one city block.

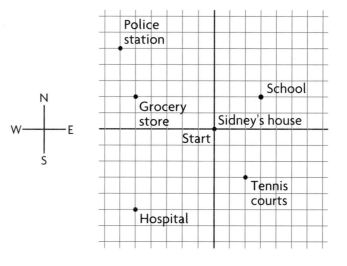

To get somewhere from Sidney's house, an **ordered pair** of numbers, such as (3,2), can be used. Since Sidney's house is the starting point, its location is (0,0). The first number tells the number of blocks to travel east or west from the start. The second number tells the number of blocks to travel north or south from the start. On this map, moving north or east is considered positive. Moving south or west is considered negative.

EXAMPLE 1 Use an ordered pair to describe how to get to the tennis courts from Sidney's home.

To get to the tennis courts, Sidney can walk 2 blocks east and 3 blocks south. This is represented by the ordered pair (2,⁻3).

2 blocks east: 2
3 blocks south:⁻3

So, the tennis courts are located at (2,⁻3).

Talk About It

• Describe how to get from Sidney's house to school.

• If the tennis courts were located 6 blocks east of Sidney's house, what ordered pair would describe their location?

Points on a coordinate plane can be located by ordered pairs just as locations on a map are. A coordinate plane is shown below.

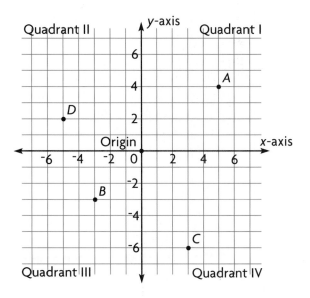

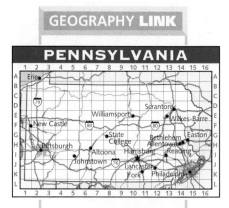

PENNSYLVANIA

Many road maps use a modified coordinate system to help locate cities. Unlike numbered coordinate planes, most maps use letters and numbers, starting with A at the top of the map and 1 at the far left or far right of the map. Suppose that a map of Pennsylvania shows Erie, which is near the top-left corner of the state, at A,2. The same map shows Philadelphia, which is near the bottom-right corner of the state, at K,15. What might the coordinates be for State College, which is near the center of the state?

Two perpendicular lines intersect to form the **coordinate plane**. These lines are called **axes**. The axes divide the coordinate plane into 4 **quadrants**. The horizontal axis is the **x-axis**. The vertical axis is the **y-axis**. The point where the x-axis and y-axis intersect, (0,0), is called the **origin**.

The first number of an ordered pair tells how far to move right or left from the origin. The second number of an ordered pair tells how far to move up or down from the origin. Moving up or to the right on the coordinate plane is positive. Moving down or to the left on the coordinate plane is negative.

To get to point A from the origin, move 5 units to the right and 4 units up. The ordered pair (5,4) gives the location of point A. In the same way, the ordered pair (3,⁻6) gives the location of point C.

• What ordered pairs give the locations of points B and D?

EXAMPLE 2 Describe how to locate the point represented by the ordered pair (6,⁻8) on the coordinate plane.

From the origin, move 6 units to the right, since the 6 is positive. Then move 8 units down, since the 8 is negative.

• **CRITICAL THINKING** Describe how the points (5,4) and (⁻5,⁻4) are different.

507

EXAMPLE 3 What ordered pair gives the location of each point on the coordinate plane below?

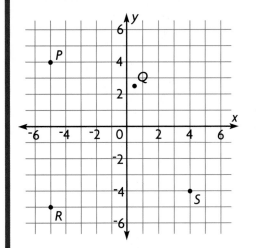

To find the ordered pair, first move right or left from the origin. Then move up or down.

Point *P* is located at (⁻5,4).　Point *Q* is located at about ($\frac{1}{2}$,2$\frac{1}{2}$).

Point *R* is located at (⁻5,⁻5).　Point *S* is located at (4,⁻4).

EXAMPLE 4 Sketch a coordinate plane, and locate the points *A* (4.5,2) and *B* (⁻3,3).

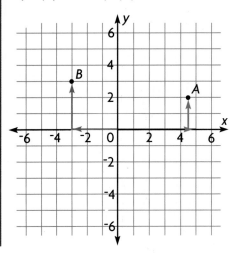

To locate point A, move 4.5 units to the right and 2 units up.

To locate point B, move 3 units to the left and 3 units up.

GUIDED PRACTICE

Describe how to locate on the coordinate plane the point for the ordered pair.

1. (2,4)　　**2.** (3,⁻4)　　**3.** (⁻1,⁻7)　　**4.** (3$\frac{1}{2}$,0)

5. (1.5,3.5)　**6.** (⁻8,6)　　**7.** (0,0)　　**8.** (6,⁻1)

9. (⁻5,⁻2)　**10.** (4,⁻3)　　**11.** (⁻4,0)　　**12.** (0,⁻6)

INDEPENDENT PRACTICE

Write the ordered pair for the point on the coordinate plane.

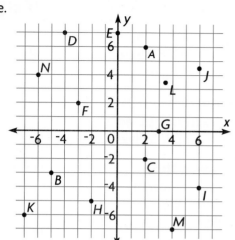

1. point A
2. point B
3. point C
4. point D
5. point E
6. point F
7. point G
8. point H
9. point I
10. point J
11. point K
12. point L
13. point M
14. point N

Sketch a coordinate plane. Locate the point for each ordered pair.

15. N $(2\frac{1}{2}, 4\frac{1}{2})$
16. P $(^-3, 7)$
17. Q $(0, 2.5)$
18. R $(^-5, ^-4)$
19. S $(6, ^-7)$
20. T $(1, ^-3)$
21. U $(^-3, ^-6)$
22. V $(^-8, 0)$
23. A $(0, ^-4)$
24. M $(^-6, ^-3)$
25. D $(4, ^-4)$
26. F $(^-3, ^-7)$

Problem-Solving Applications

27. **GEOMETRY** Locate the points (5,5), (5,⁻3), and (⁻2,6) on a coordinate plane. Connect the points in the listed order. What kind of geometric figure do you have?

28. Locate the points (2,3), (2,⁻3), and (⁻4,3) on a coordinate plane. What ordered pair tells the location of the fourth point that would make the figure a square?

29. ✏ **WRITE ABOUT IT** What do you notice about points on a coordinate plane that have both numbers positive? both numbers negative?

Mixed Review and Test Prep

Use the equation $y = 2x + 1$ to find the output, y, for the given input, x.

30. $x = 3$
31. $x = 7$
32. $x = 0$
33. $x = ^-4$

Find the product.

34. $6 \times ^-3$
35. $^-8 \times 9$
36. $^-12 \times ^-7$
37. $^-11 \times 0$

38. **LOGICAL REASONING** Jeff is half as old as Lee. In 7 years, he will be two-thirds as old as Lee. How old is Lee now?

   A 7 yr old
   B 10 yr old
   C 14 yr old
   D 20 yr old

39. **ALGEBRA** Which expression represents 3 more than twice x?

   F $2 + 3x$
   G $3x - 2$
   H $2x - 3$
   J $2x + 3$

Graphing Relations

Jaime earns $5 per hr doing yard work. The money she could earn if she worked between 1 and 6 hours is shown in the table.

Money Earned	$5	$10	$15	$20	$25	$30
Hours Worked	1	2	3	4	5	6

This relation could also be shown using ordered pairs.

(5,1)   (10,2)   (15,3)   (20,4)   (25,5)   (30,6)

The set of ordered pairs can be called a **relation**. You can use the ordered pairs to show the relation on a graph.

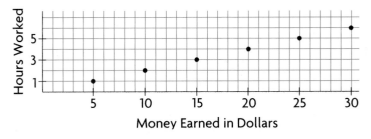

• **CRITICAL THINKING** What relation is shown by the number of hours worked and money earned?

EXAMPLE Graph the results from an input-output table on a coordinate plane. Write the expression used to get the output.

Input (x)	⁻3	⁻2	⁻1	0	1	2	3
Output (y)	⁻2	⁻1	0	1	2	3	4

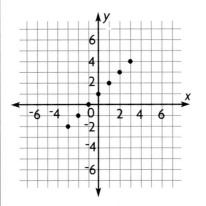

The input, x, and the output, y, can be written as ordered pairs.

(⁻3,⁻2), (⁻2,⁻1), (⁻1,0), (0,1), (1,2), (2,3), (3,4)

Since each *y*-value is one more than the *x*-value, the expression to get the output, *y*, is *x* + 1.

GUIDED PRACTICE

For Exercises 1–2, write the relation as ordered pairs.

1.

Input (x)	2	3	4	5
Output (y)	4	6	8	10

2.

Input (x)	6	7	8	9
Output (y)	3	4	5	6

3. Sketch a coordinate plane, and locate the points for the ordered pairs of the relation from Exercise 1.

4. Sketch a coordinate plane, and locate the points for the ordered pairs of the relation from Exercise 2.

5. What expression can you write, using x, to get the value of y in Exercise 1?

6. What expression can you write, using x, to get the value of y in Exercise 2?

INDEPENDENT PRACTICE

Use the first three values of x and y to complete the table.

1.

Input (x)	1	2	3	4	5
Output (y)	2	3	4	?	?

2.

Input (x)	1	2	3	4	5
Output (y)	4	8	12	?	?

3.

Input (x)	5	6	7	8	9
Output (y)	3	4	5	?	?

4.

Input (x)	2	1	0	⁻1	⁻2
Output (y)	6	3	0	?	?

5. Sketch a coordinate plane, and locate the points for the ordered pairs of the relation from Exercise 1.

6. Sketch a coordinate plane, and locate the points for the ordered pairs of the relation from Exercise 2.

7. What expression can you write, using x, to get the value of y in Exercise 1?

8. What expression can you write, using x, to get the value of y in Exercise 2?

Problem-Solving Applications

9. Use the expression $x + 5$ to make an input-output table. Use the whole numbers from 0 to 4 as the input, x. Find the output, y. Locate the points for the ordered pairs on a coordinate plane.

10. **GEOMETRY** Use the expression $5x$ to make an input-output table. Use the integers from ⁻2 to 2 as the input, x. Find the output, y. Locate the points for the ordered pairs on a coordinate plane.

11. **RECREATION** In the arcade, 5 coupons are earned for each win. This is represented by the expression $5w$, where w is the number of wins. Use ordered pairs to show the number of coupons earned, c, for 7, 8, 9, and 10 wins.

12. **WRITE ABOUT IT** Write a problem involving a relation. Make an input-output table for five inputs. Write the ordered pairs, and locate them on a coordinate plane.

Evaluate for $x = {}^-3$. (pages 498–499)

1. $x + 7$
2. $x - 12$
3. $14 - x$
4. $5x$
5. ^-12x

6. $\dfrac{18}{x}$
7. $x + 8$
8. $2 - x$
9. $7x$
10. $\dfrac{x}{-3}$

Solve and check. (pages 500–501)

11. $x + 7 = 4$
12. $^-1 = y + 8$
13. $8k = {}^-40$
14. $^-3a = 18$
15. $\dfrac{p}{7} = {}^-10$

16. $\dfrac{c}{-2} = {}^-9$
17. $5a = {}^-20$
18. $\dfrac{x}{-4} = 2$
19. $^-3m = 42$
20. $r - 21 = {}^-5$

Sketch a graph on a number line to show all the integer solutions of the inequality. (pages 504–505)

21. $x < 8$
22. $x \le 7$
23. $x \ge {}^-2$
24. $x > 3$
25. $x \le 0$

26. $x > {}^-17$
27. $x < 1$
28. $x \ge 2$
29. $x \le {}^-4$
30. $x > {}^-1$

31. VOCABULARY The two perpendicular lines that intersect to form the coordinate plane are ___?___. (page 507)

32. VOCABULARY The point where the x-axis and y-axis intersect is called the ___?___. (page 507)

Describe how to locate on the coordinate plane the point for the ordered pair. (pages 506–509)

33. $(2,4)$ **34.** $(3,^-2)$ **35.** $(^-1,^-6)$ **36.** $(1,4)$ **37.** $(^-3,2)$ **38.** $(^-2,^-1)$

For Exercises 39–44, use the tables at the right. (pages 510–511)

39. List the ordered pairs from the first table.

40. Locate the points for the ordered pairs on a coordinate plane.

x	1	2	3	4
y	4	5	6	7

41. Write an expression, using x, that gives the value of y.

42. List the ordered pairs from the second table.

a	1	2	3	4
b	3	6	9	12

43. Which point would be farthest to the right on the coordinate plane?

44. Write an expression, using a, that gives the value of b.

1. Which expression represents 2^5?

A 2×5
B $2 \times 2 \times 5 \times 5$
C 5×5
D $2 \times 2 \times 2 \times 2 \times 2$
E $2 \times 2 \times 2 \times 2 \times 2 \times 5 \times 5$

2. Ryan wants to buy an ice cream cone. He can have vanilla, chocolate, blueberry, swirl, or strawberry ice cream. He can get sprinkles, candy pieces, or peanuts for the topping. How many different cones can he choose from?

F 3
G 4
H 7
J 12
K Not Here

3. How much fencing is needed to enclose the lot shown in the drawing below?

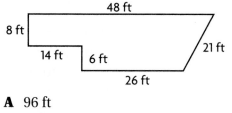

A 96 ft
B 123 ft
C 132 ft
D 155 ft

4. Jeff left $2,500 in a bank account that paid 8% simple interest per year for 2 years. How much total interest did he earn?

F $200
G $320
H $400
J Not Here

5. Jon drove a distance of 476 miles in $8\frac{1}{2}$ hours. What was his average rate of speed?

A 44 mph
B 51 mph
C 56 mph
D 58 mph
E 62 mph

6. At 7:00 A.M., the temperature was ⁻4°F. By 7:00 P.M., the temperature had risen 17°. What was the temperature at 7:00 P.M.?

F ⁻17°F
G ⁻13°F
H 3°F
J 7°F
K 13°F

7. What is the value of b in the equation $b - 8 = 37$?

A 25   **B** 29
C 37   **D** 45

8. Which point is located by the ordered pair (1,2) on the coordinate plane below?

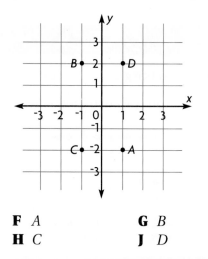

F A   **G** B
H C   **J** D

MATH FUN!

THE TERMINATOR

PURPOSE To practice writing fractions as decimals (pages 470–473)

YOU WILL NEED number cube

You can play this game with two or more players. In turn, each player rolls the number cube to form a fraction. The first number rolled is the numerator; the second, the denominator. So, rolling a 2, then a 3, gives you $\frac{2}{3}$. Then the player writes the fraction as a decimal and states whether it is terminating or repeating.

If the answer is correct, the player gets one point. The player earns a bonus point if the fraction rolled can be written as a whole number.

 Play this game at home. Explain the rules of writing fractions as decimals.

PING PONG

PURPOSE To practice adding integers (pages 484–485)

YOU WILL NEED number cube, number line, two-colored counters

$$\overset{\longleftarrow}{\underset{{}^{-10}\ {}^{-9}\ {}^{-8}\ {}^{-7}\ {}^{-6}\ {}^{-5}\ {}^{-4}\ {}^{-3}\ {}^{-2}\ {}^{-1}\ \ 0\ \ {}^{+1}\ {}^{+2}\ {}^{+3}\ {}^{+4}\ {}^{+5}\ {}^{+6}\ {}^{+7}\ {}^{+8}\ {}^{+9}\ {}^{+10}}{\rule{0pt}{0pt}}}$$

Label the number line from ⁻10 to ⁺10. Both players start at zero. In turn, roll the number cube. Even numbers are positive. Odd numbers are negative.

Each player moves his or her counter the number he or she rolls on the number line. After six turns, the player closest to zero (positive or negative) is the winner.

MATCH BATCH

PURPOSE To practice graphing integer solutions of inequalities (pages 504–505)

Match each graph with two inequalities from the list.

$$\overset{\longleftarrow}{\underset{{}^{-6}\ {}^{-5}\ {}^{-4}\ {}^{-3}\ {}^{-2}\ {}^{-1}\ 0\ 1\ 2\ 3\ 4\ 5\ 6}{\rule{0pt}{0pt}}}$$

$$\overset{\longleftarrow}{\underset{{}^{-6}\ {}^{-5}\ {}^{-4}\ {}^{-3}\ {}^{-2}\ {}^{-1}\ 0\ 1\ 2\ 3\ 4\ 5\ 6}{\rule{0pt}{0pt}}}$$

$$\overset{\longleftarrow}{\underset{{}^{-6}\ {}^{-5}\ {}^{-4}\ {}^{-3}\ {}^{-2}\ {}^{-1}\ 0\ 1\ 2\ 3\ 4\ 5\ 6}{\rule{0pt}{0pt}}}$$

a. $x < {}^-2$ **g.** $x \geq 2$

b. $x \leq 2$ **h.** $x \geq 1$

c. $x \geq {}^-4$ **i.** $x > {}^-3$

d. $x > 0$ **j.** $x < 3$

e. $x < 0$ **k.** $x \geq {}^-3$

f. $x \geq {}^-2$ **l.** $x < 4$

Graphing Ordered Pairs

In this activity you will see how you can enter ordered pairs into a spreadsheet and graph the data on a coordinate plane.

Enter the data shown below into a spreadsheet.

All	A	B	C
1		x-coordinate	y-coordinate
2		⁻1	3
3		2	0
4		3	4
5		5	2
6		2	2

Select your data by dragging from cell B2 across and down to cell C6.

Then, from the tool bar, choose the scatterplot or another type of *xy*-graph.

In some spreadsheet programs a window will open and ask whether the first column of data represents the *x*-coordinates. If this window appears, click OK.

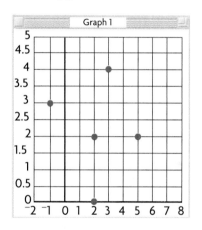

The darker horizontal and vertical lines on the graph are the *x*- and *y*-axes.

1. Write the *x*- and *y*-coordinates from the data above as ordered pairs.

2. Which ordered pair lies on the *x*-axis?

3. Where would the ordered pair (⁻2,3) appear on the graph if you added it to your data?

4. How would the graph change if you added the ordered pair (⁻3,⁻4)?

USING A SPREADSHEET
Graph the ordered pairs.

5. (8,2), (3,7), (4,5), (0,1), (7,4), (10,0), (9,9), (3,2), (6,1)

6. (5,1), (4,9), (4,12), (8,0), (9,2), (8,3), (4,7), (9,2), (0,1)

7. (⁻4,2), (1,⁻5), (3,7), (0,6), (⁻3,4), (11,1), (⁻8,8), (5,⁻5), (6,5)

8. (⁻5,7), (9,8), (⁻1,2), (4,⁻3), (⁻6,⁻3), (0,⁻7), (⁻2,⁻2)

Study Guide and Review

Vocabulary Check

1. All whole numbers, their opposites, and 0 make up the set of __?__. (page 466)

2. The two perpendicular lines that intersect to form the coordinate plane are called __?__. (page 507)

3. The point where the *x*-axis and *y*-axis intersect is called the __?__. (page 507)

EXAMPLES

EXERCISES

- **Compare sets of numbers and classify them as whole numbers, integers, and rational numbers.** (pages 468–469)

$\frac{3}{4}$ belongs in the set of rational numbers.

Name the sets in which each number belongs.

4. $^-14$ **5.** 0.51 **6.** $2\frac{1}{3}$

7. $\frac{4}{4}$ **8.** $^-10.1$ **9.** $\frac{1}{12}$

- **Write fractions as either terminating or repeating decimals.** (pages 470–473)

$\frac{2}{5} = 0.4 \leftarrow$ terminating decimal

$\frac{4}{9} = 0.\overline{4} \leftarrow$ repeating decimal

Write the fraction as a terminating or repeating decimal.

10. $\frac{1}{4}$ **11.** $\frac{2}{3}$ **12.** $\frac{5}{6}$

13. $\frac{11}{20}$ **14.** $\frac{2}{9}$ **15.** $\frac{4}{5}$

- **Compare and order rational numbers.** (pages 476–477)

$^-4.3 \bullet ^-4\frac{2}{5}$

Write equivalent fractions.

$^-4.3 = ^-4\frac{3}{10},\ ^-4\frac{2}{5} = ^-4\frac{4}{10}$

Use < or > to compare the fractions.

$^-4\frac{3}{10} > ^-4\frac{4}{10},$ so $^-4.3 > ^-4\frac{2}{5}$

Use $<$, $>$, or $=$ to compare.

16. $\frac{3}{8} \bullet 0.125$ **17.** $2\frac{3}{4} \bullet 2\frac{4}{5}$

18. $^-0.5 \bullet ^-\frac{1}{2}$ **19.** $^-2.1 \bullet ^-2.8$

- **Add and subtract integers.** (pages 484–489)

Write as an addition problem.

$^+4 - ^+6 = ^+4 + ^-6$

Start at 0. Move 4 spaces to the right. From $^+4$, move 6 spaces to the left.

```
      ◄─────────────┤
          ┌────────►
─┼──┼──┼──┼──┼──┼──┼──┼──┼──┼──┼──┼──┼─
 ‾6 ‾5 ‾4 ‾3 ‾2 ‾1  0 ⁺1 ⁺2 ⁺3 ⁺4 ⁺5 ⁺6
```

So, $^+4 - ^+6 = ^-2$.

Find the sum.

20. $^+8 + ^+4$ **21.** $^-6 + ^-11$

22. $^+7 + ^-5$ **23.** $^-3 + ^+91$

Find the difference.

24. $^+6 - ^+8$ **25.** $^-3 - ^+7$

26. $^-5 - ^-5$ **27.** $^+2 - ^-2$

- **Multiply and divide integers.** (pages 490–493)

$^-5 \times {}^-8 = {}^+40$ *The product or quotient of two negative integers is a positive integer.*

$^+12 \div {}^-2 = {}^-6$ *The product or quotient of a positive integer and a negative integer is a negative integer.*

Find the product.

28. $^+3 \times {}^-7$ **29.** $^-6 \times {}^-12$

Find the quotient.

30. $^+28 \div {}^-7$ **31.** $^-64 \div {}^-8$

32. $^-21 \div {}^+3$ **33.** $^+130 \div {}^-10$

- **Use inequality symbols and solve algebraic inequalities.** (pages 504–505)

Find the whole number solutions of $x \le 4$.

Replace x with whole numbers that make the inequality true.

So, 4, 3, 2, 1, and 0 are solutions.

Graph the integer solutions on a number line.

34. $x > {}^-5$ **35.** $x \le 9$

36. $x < {}^-1$ **37.** $x > 6$

38. $x \le {}^-8$ **39.** $x \ge {}^-10$

- **Show relations in graphs and on a coordinate plane.** (pages 510–511)

Write an expression, using x, that gets the value of y.

$(1,5), (2,10), (3,15), (4,20), (5,25)$

Since each y value is 5 times as great as each x value, the expression is $5x$.

For Exercises 40–41, use the table.

Input (x)	1	2	3	4
Output (y)	$^-1$	$^-2$	$^-3$	$^-4$

40. Graph the points for the ordered pairs on a coordinate plane.

41. Write an expression, using x, that gives the value of y.

Problem-Solving Applications

Solve. Explain your method.

42. Tom was scuba diving 50 ft below the surface of the water. Write his position as an integer. (pages 466–467)

44. The temperature is $^-10°$F. It will rise 5° during the day. What will be the high temperature? (pages 484–485)

43. Ms. Pole bought more than $\frac{1}{2}$ lb of oranges, but less than $\frac{3}{4}$ lb. Could the oranges have weighed $\frac{3}{5}$ lb? (pages 474–475)

45. Graph the points (0,0), (0,2), (2,2), and (2,0) on a coordinate plane. Connect the points. What geometric figure did you make? (pages 506–509)

Performance Assessment

Tasks: Show What You Know

1. Show each step as you find the decimal for the fraction $\frac{2}{9}$.
(pages 470–473)

2. Use a number line to show the difference $^+4 - {}^+9 = n$.
Explain. (pages 488–489)

3. Use graph paper to graph this relation on a coordinate plane.
Explain your method.

Jonathan earns $8 for each lawn he mows. (pages 510–511)

Lawns Mowed	1	2	3	4
Money Earned	$8	$16	$24	$32

Problem Solving

Solve. Explain your method.

CHOOSE a strategy and a tool.

- **Find a Pattern**
- **Make a Graph**
- **Draw a Diagram**
- **Write an Equation**
- **Make a Model**
- **Make a Table**

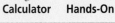 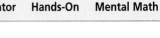

Paper/Pencil Calculator Hands-On Mental Math

4. Sid's bike weighs 25.25 lb. Chen's bike weighs $25\frac{3}{8}$ lb, Lana's bike weighs $25\frac{2}{5}$ lb, and Nova's weighs 25.2 pounds. Put the bikes in order from lightest to heaviest.
(pages 476–477)

5. Mr. Franklin weighed 173 lb. In May he lost 6 lb. In June he lost 3 more lb. In July he gained 4 lb. How much did he weigh at the end of July? (pages 484–487)

6. For every book Bess reads, her sister reads 2 more. Use the expression $x + 2$ to make an input-output table for the even numbers from 2 to 10. Then graph the line formed by the points of the ordered pairs to show this relationship. (pages 510–511)

Cumulative Review

Solve the problem. Then write the letter of the correct answer.

1. Find the difference. $18 - 3.98$ (pages 62–63)

   A. 1.402   **B.** 14.02
   C. 14.12   **D.** 15.98

2. Write the sum in simplest form. $4\frac{5}{6} + 5\frac{2}{3}$
   (pages 126–129)

   A. $9\frac{7}{9}$ **B.** $9\frac{9}{6}$
   C. $10\frac{1}{2}$ **D.** $10\frac{3}{6}$

3. Name the polygon. (pages 172–173)

   A. hexagon
   B. octagon
   C. pentagon
   D. quadrilateral

4. A spinner with numbers 1–5 is spun 100 times. The results are in the table below. What is the experimental probability of spinning an odd number? (pages 280–281)

Number	1	2	3	4	5
Times Spun	20	18	23	19	20

   A. $\frac{3}{100}$ **B.** $\frac{20}{100}$, or $\frac{1}{5}$
   C. $\frac{37}{100}$ **D.** $\frac{63}{100}$

5. Use the formula $d = r \times t$ to find the time when the distance is 50 ft and the rate is 10 ft per sec. (pages 318–319)

   A. 5 sec   **B.** 40 sec
   C. 50 sec   **D.** 500 sec

6. Find 40% of 200. (pages 352–355)

   A. 5   **B.** 8
   C. 80   **D.** 8,000

7. Find the simple interest for one year. principal: $600; interest rate: 6%
   (pages 360–361)

   A. $36   **B.** $360
   C. $636   **D.** $3,600

8. Find the perimeter. (pages 414–415)

10 ft

10 ft

   A. 10 ft   **B.** 20 ft
   C. 40 ft   **D.** 100 ft

9. Find the volume of the cylinder. Use the formula $\pi r^2 h$. Use 3.14 for π.
   (pages 450–451)

radius = 5 in.

10 in.

   A. 50 in.^{3} **B.** 314 in.^{3}
   C. 785 in.^{3} **D.** 3,140 in.^{3}

10. $^-7 + {}^+5$ (pages 484–485)

   A. $^-12$ **B.** $^-2$
   C. $^+2$ **D.** $^+12$

11. Evaluate $x - 4$ for $x = {}^-5$. (pages 498–499)

   A. $^-9$ **B.** $^-1$
   C. 1   **D.** 9

12. Solve for x. $3x = {}^-18$ (pages 500–501)

   A. $x = {}^-54$ **B.** $x = {}^-6$
   C. $x = 6$ **D.** $x = 54$

GEOMETRIC PATTERNS

LOOK AHEAD

In this chapter you will solve problems that involve

- **using transformations on the coordinate plane**

- **identifying sequences of geometric figures**

- **making tessellation shapes**

- **making fractals**

MUSIC **LINK**

In June of 1993, a marching band consisting of 6,017 people marched for a distance of 3,084 feet in Hamar, Norway. It was thought to be the largest marching band ever.

- If the band members marched 12 across, about how many rows were there? What if they marched 30 across?

- The longest musical march was by the North Allegheny Marching Tiger Band in Pennsylvania in 1996. The band members marched about 50 miles in about 14.7 hours. What was their average speed?

Marching Orders

Marching bands practice their routines for many hours. But before they can practice a routine, it must be planned.

Plan a routine. Start by deciding how you will enter the football field, the design you will form, how you will move, and how you will exit. Make a drawing to show your design and moves. Write directions for your routine. Then, practice your routine.

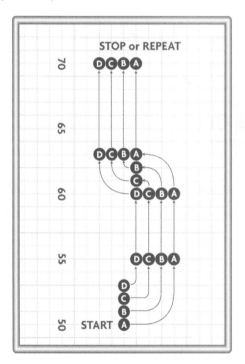

PROJECT CHECKLIST

✓ Did you plan a routine?

✓ Did you make a drawing?

✓ Did you write directions?

✓ Did you practice your routine?

ALGEBRA CONNECTION

Transformations on a Coordinate Plane

REMEMBER:
. . . .
A movement that doesn't change the size or shape of a figure is called a **transformation**. Three types of transformations are translation, reflection, and rotation.

See page H20.

Have you ever seen jets or birds fly in a V formation?

When jets move in a straight path in a V formation, the movement demonstrates a translation.

When a shape moves around a point, it demonstrates a rotation.

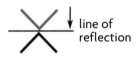

A shape can be reflected as well. Remember that when a shape is flipped over a line, it demonstrates a reflection.

Translations, rotations, and reflections are three types of transformations. When you look at a transformation on a coordinate plane, you can readily see and describe how the position of the figure changes.

ACTIVITY

WHAT YOU'LL NEED: graph paper, scissors

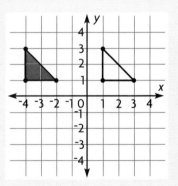

- On graph paper, draw the coordinate plane shown at the right. Graph the points (1,1), (1,3), and (3,1) and connect them to form a triangle, as shown.

- Trace the triangle on a sheet of paper. Cut it out.

- Place your cutout triangle on the triangle in the coordinate plane. Then translate it 5 units to the left along the x-axis.

- Name the coordinates for the vertices of the new triangle.

The transformation in the Activity shows a decrease of 5 for the x-coordinates and no change for the y-coordinates.

Talk About It

• Why do you think there was a decrease of 5 in the x-coordinates and no change in the y-coordinates?

• CRITICAL THINKING Which coordinates would have changed if the translation had been 5 units up or down?

You can translate a figure along both the x- and y-axes.

EXAMPLE 1 Translate square ABCD 6 units right and 5 units down. What are the coordinates of the new square? Name the coordinates of the new square as A'B'C'D'.

original

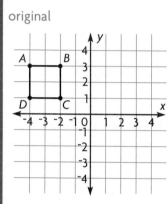

translation

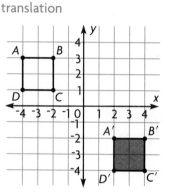

The coordinates of the new square are A'(2,⁻2), B'(4,⁻2), C'(4,⁻4), and D'(2,⁻4).

You can draw a reflection on a coordinate plane.

EXAMPLE 2 Reflect △ABC across the x-axis. What are the coordinates of the new figure, △A'B'C'?

original

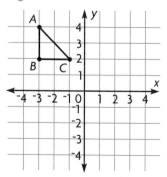

reflection

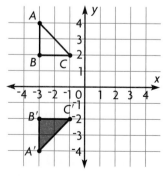

Original	New
A(⁻3,4) →	A'(⁻3,⁻4)
B(⁻3,2) →	B'(⁻3,⁻2)
C(⁻1,2) →	C'(⁻1,⁻2)

The coordinates of the new triangle are A'(⁻3,⁻4), B'(⁻3,⁻2), and C'(⁻1,⁻2).

Copy the figure onto a coordinate grid. Translate the figure 2 units left. Name the new coordinates.

1.

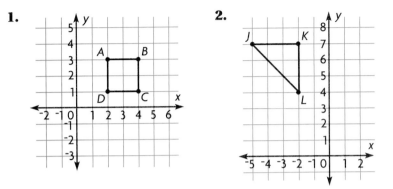

2.

Rotations on a Coordinate Plane

You can draw a rotation of a figure around the origin on a coordinate plane.

EXAMPLE 3 Rotate trapezoid *ABCD* clockwise 90° around the origin. What are the coordinates of the new figure?

original   rotation

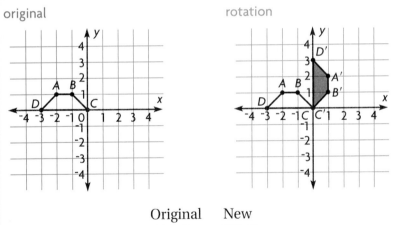

Original   New
$A(-2,1) \rightarrow A'(1,2)$
$B(-1,1) \rightarrow B'(1,1)$
$C(0,0) \rightarrow C'(0,0)$
$D(-3,0) \rightarrow D'(0,3)$

• Why didn't the coordinates of vertex *C* change from the original figure to the new figure?

• CRITICAL THINKING Would the coordinates of the new trapezoid be different if the rotation was 90° counterclockwise? Explain.

HISTORY **LINK**

The Navy's Blue Angels, established in 1946, is probably the most famous precision flying team in the world. The current team has six pilots who fly identical F/A-18 jets. In one of the team's most famous maneuvers, a vertical loop, all six planes fly in a large circle. What kind of transformation does this maneuver represent?

INDEPENDENT PRACTICE

Copy each figure onto a coordinate grid. Transform the figure according to the directions given. Give the new coordinates.

1.

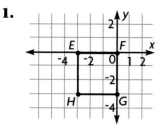

translate 3 units right and 1 unit down

2.

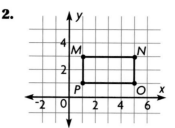

reflect across the *y*-axis

3.

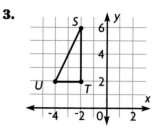

reflect across the *y*-axis

4.

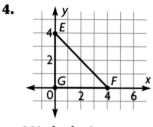

90° clockwise

5.

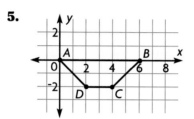

180° clockwise

6.

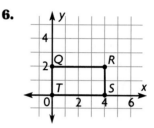

90° counterclockwise

Problem-Solving Applications

7. PATTERNS Graph rectangle *EFGH* with coordinates (1,⁻2), (4,⁻2), (4,⁻4), and (1,⁻4). Translate it 5 units up. What are the new coordinates?

8. Use the new rectangle in Problem 7, and reflect it across the *x*-axis. Graph the results. What are the new coordinates?

9. CRITICAL THINKING Triangle *ABC* has coordinates (0,0), (0,2), and (3,0). Name the coordinates after it is rotated 180° clockwise around the origin.

10. WRITE ABOUT IT Describe how the coordinates of a figure change when you translate that figure.

Mixed Review and Test Prep

Write the next term in the pattern.

11. 2, 4, 6, 8, . . .   **12.** 2, 5, 8, 11, 14, . . .   **13.** 15, 10, 5, 0, . . .   **14.** 18, 24, 30, 36, . . .

Solve for *x*.

15. $x - 7 = {}^-4$ **16.** $x + 8 = {}^-8$ **17.** ${}^-5x = {}^-30$ **18.** $\frac{x}{9} = 3$

19. TEMPERATURE It is 70°F in Miami on the same day that it is ⁻10°F in Alaska. What is the difference in temperature?

    A 60°F   **B** 70°F
    C 80°F   **D** 90°F

20. ALGEBRA To move from the point (0,3) to the point (2,5) on a coordinate plane go—

    F up 2, right 3   **G** up 3, right 2
    H up 2, right 2   **J** up 3, right 3

PROBLEM–SOLVING STRATEGY

What You'll Learn
How to find a pattern
to solve problems
that involve
transformations on
the coordinate plane

Why Learn This?
To describe the pattern
of transformations so
you'll know how
marching band
members move on
a field

Finding Patterns on the Coordinate Plane

During halftime at the next football game, the Lakeside Middle School band wants to march out and perform in the shape of an L. If the band director follows the same pattern shown in the diagram, what transformations will be used to move to positions 6 and 7?

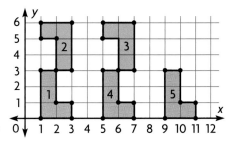

The figure labeled 1 is the first position of the band.

PROBLEM SOLVING
....
• **Understand**
• **Plan**
• **Solve**
• **Look Back**

UNDERSTAND   What are you asked to find?

What facts are given?

PLAN   What strategy will you use?

You can *find a pattern* of how the L moves from one position to the next. Then use the pattern to find the next two positions.

SOLVE   Use tracing paper to compare each position of the L with the one before.

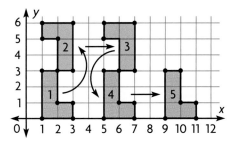

1. original position   **2.** rotation   **3.** translation
4. rotation   **5.** translation

So, to move to positions 6 and 7, the transformations will be first rotation and then translation.

LOOK BACK   What other strategy could you use?

What if . . . the marching band followed this pattern to position 8? What transformation would be illustrated?

PRACTICE

Find a pattern and solve.

1. Paul moves his vegetable garden every year so that he can plant with new soil. Use the figure at the right to tell what pattern of transformations he has used to move his garden.

2. If Paul continues the pattern from Problem 1, what transformation will he use in year 5?

3. What transformation would Paul use for moving his vegetable garden in year 8?

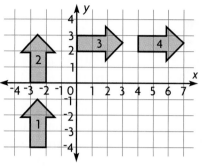

MIXED APPLICATIONS

Solve.

CHOOSE a strategy and a tool.
- **Write an Equation**
- **Use a Formula**
- **Find a Pattern**
- **Draw a Diagram**
- **Make a Chart**
- **Guess and Check**

Paper/Pencil   Calculator   Hands-On   Mental Math

4. Life insurance costs $1 a month for $1,000 of insurance coverage. Tim Jones has $10,000 in insurance, and he pays once a year. How much is his annual bill?

5. Matthew is $1\frac{1}{2}$ times as tall as his baby sister, who is half as tall as her older sister, who is 4 ft tall. How tall is Matthew?

6. Carla deposited $200 in the bank for one year at an interest rate of 3%. How much interest did she earn?

7. A stairway has 8 steps. Each step is 9 in. from the step below it. How many feet from the floor is the top step?

8. Peter and his sister solved 14 puzzles. Peter solved 4 more than his sister. How many did he solve?

9. Fran bought a $30 dress on sale at a discount of 40%. How much did she pay for the dress?

10. What transformation would be used to move the figure below to position 5? to position 6?

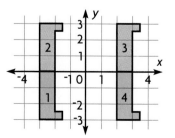

11. ✏️ **WRITE ABOUT IT** Draw the figure at the right on a coordinate plane. Move it through a pattern of transformations. Then exchange with a classmate, and try to find each other's patterns.

P

Patterns of Geometric Figures

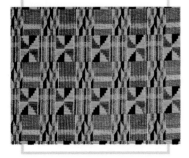
Many events that you see every day occur in a pattern and are predictable. What pattern does a traffic light show?

Patterns of geometric figures can be based on such characteristics as shape, color, size, position, or number. Recognizing the characteristics can help you determine the pattern.

EXAMPLE 1 Draw the next three figures in the geometric pattern below.

Figure 1 Figure 2 Figure 3

The next three figures are shown below.

Look for a pattern. The position of the blue triangle changes from one figure to the next, moving clockwise around the figure.

Figure 4 Figure 5 Figure 6

You can also find patterns in three-dimensional figures.

EXAMPLE 2 Draw the next two solids in the pattern below.

Figure 1 Figure 2 Figure 3

Look for a pattern. The length and height change from one solid to the next. Two cubes are added to each shape to make the next solid.

The next two solids are shown below.

Figure 4   Figure 5

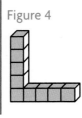 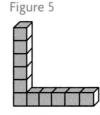

GUIDED PRACTICE

Draw the next three figures in the geometric pattern.

1.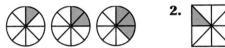

2.

3.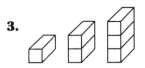

INDEPENDENT PRACTICE

Draw the next three figures in each geometric pattern.

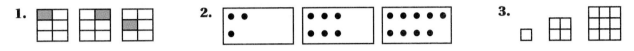

Describe the next two solid figures you would draw for the pattern.

4.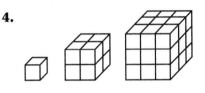

5.

Problem-Solving Applications

6. A store display has 11 cans in the bottom row. The number of cans in each other row is 2 cans less than in the row below. The top row has 3 cans. How many cans are in the display?

7. PATTERNS The figure at the right was made from rectangular prisms. The top level has 1 prism, the next level has 4 prisms, the next has 9 prisms, and the bottom level has 16 prisms. How many prisms would be in the next two levels?

8. ✏️ **WRITE ABOUT IT** Find a geometric pattern in your home. Describe it in words and with a drawing.

Mixed Review and Test Prep

Tell whether the polygon can be used to form a tessellation. Write yes or no.

9. △   **10.** ⬡   **11.** ⬡   **12.** ◯

Tell in which quadrant the ordered pair is found.

13. (2,4)   **14.** (⁻2,4)   **15.** (2,⁻4)   **16.** (⁻2,⁻4)   **17.** (9,1)   **18.** (3,9)

19. SPORTS There are 3 members on a golf team. In how many different orders can they start?

    A 3   **B** 5
    C 6   **D** 9

20. RATIOS Use the scale 1 cm = 25 cm. What is the actual length of an object when the scale length is 5 centimeters?

    F 5 cm   **G** 25 cm
    H 100 cm   **J** 125 cm

Making Figures for Tessellations

What You'll Learn
How to make figures for tessellations

Why Learn This?
To use tessellations to design tile patterns

REMEMBER:

A **tessellation** is an arrangement of shapes in a repeating pattern that completely covers a plane without overlapping or leaving gaps. **See page 188.**

SCIENCE LINK

Although most tessellations are produced by humans, a few occur in nature. A cross section of a honeycomb reveals tessellated hexagons. What other natural tessellation can you think of?

Many simple shapes are used to form tessellations, such as the patterns seen in tiled floors. You can make other tessellation shapes from regular polygons that can be tessellated.

ACTIVITY **WHAT YOU'LL NEED:** paper, scissors, and tape

Make a tessellation shape.

• Cut out a square that is 2 in. × 2 in.

• Use scissors to cut out a part of the square on one side.

• Translate the cutout part of the square to the opposite side of the square. Then tape it.

• Trace the new shape to form at least two rows of a tessellation. You will need to rotate, translate, or reflect the shape.

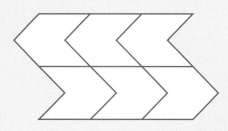

• Suppose, after you cut off part of a side of a square, you didn't tape it to the opposite side. Could the shape be tessellated? Explain.

GUIDED PRACTICE

Make the tessellation shape described by each pattern.

1.   **2.**   **3.**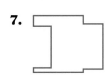

Trace the shape. Then form two rows of a tessellation.

4.   **5.**   **6.**   **7.**

INDEPENDENT PRACTICE

Make the tessellation shape described by each pattern. Then form two rows of a tessellation.

1. 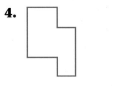   **2.** 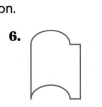   **3.**   **4.**

Write *yes* or *no* to tell whether the figure forms a tessellation.

5.   **6.**   **7.**

id="12"

Problem-Solving Applications

8. PATTERNS Cut out a square. Then change its shape by cutting out part of a side. Translate the cutout shape to the opposite side of the square. Trace repeatedly until you have two rows of a tessellation.

9. ART Make an artistic design by adding detail and color to the tessellation you made in Problem 8. Expand the tessellation, repeating the same detail and color so that the tessellated pattern stands out.

10. ✏ **WRITE ABOUT IT** Explain why you translate the cutout shape of a square to the opposite side when you make a tessellation.

Technology Link

💿 In *Mighty Math Cosmic Geometry,* you can practice identifying patterns and tessellations in the activity *Amazing Angles.* Use Grow Slide Level P.

id="5"

MORE PRACTICE Lesson 27.4, page H82

531

LAB ACTIVITY

What You'll Explore
How to use a repeating process to make a fractal

What You'll Need
ruler, isometric dot paper

REMEMBER:
....
An **isosceles triangle** has at least two congruent sides.

See page H21.

Exploring Fractal Patterns

You can make a pattern of isosceles triangles out of one triangle by repeated folding. With each fold, you make a new isosceles triangle.

ACTIVITY 1

Explore

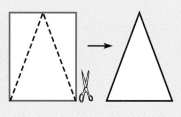

• Cut out an isosceles triangle from $8\frac{1}{2}$-in. × 11-in. paper.

• Fold the top vertex of the triangle to the middle of the opposite side. Unfold the paper, and outline the new triangle.

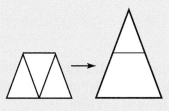

• Do the last step two more times. Fold the top vertex of the triangle to the middle of the opposite side of each new triangle.

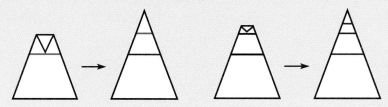

Think and Discuss

• Look at your final figure. How many triangles are there?

• How are these triangles related? How are these triangles different?

A figure has **self-similarity** if it contains a repeating pattern of smaller and smaller parts that are like the whole, but different in size.

• If the folding process above were to continue, would the resulting triangular figure appear to have self-similarity around its top vertex?

Try This

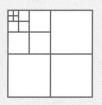

Look at the square pattern at the right. Does it appear to have self-similarity? Explain.

Imagine a tree that starts as a single trunk. It first grows 2 branches, then 4, then 8, then 16, and so on. Each time the new branches are half as long as the preceding ones, but always growing in the same pattern. Such a tree would be self-similar.

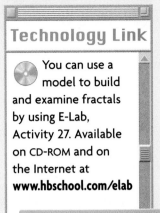

ACTIVITY 2

Explore

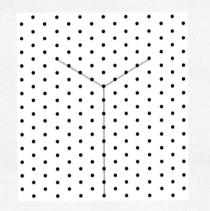

- On isometric dot paper, draw a line segment 8 units long.

- Draw two segments from the endpoint as shown. Make them 4 units long, half the length of the first.

- Turning at the same angles, draw four more branches, each 2 units long. Repeat again with eight branches, each 1 unit long.

Think and Discuss

- How many branches did you draw (not counting the "trunk")?

- How many branches would you have if you repeated this process one more time?

- How does the number of branches increase from step to step?

- How do the branches change in length?

Fractals are very special geometric figures. The mathematical tree that you drew is **fractal-like** because it appears to have self-similarity.

Try This

- Make another fractal-like tree. Start by drawing a line segment that is 12 units long.

  Add two segments that are half as long. Following the same pattern, continue by adding four segments one-quarter as long and then eight, one-eighth as long.

Solve. (pages 522–525)

1. Graph trapezoid *DEFG* with coordinates (3,3), (4,3), (5,1), and (2,1), and translate it 2 units left. What are the new coordinates?

2. Use the new trapezoid in Exercise 1 and reflect it across the *x*-axis. What are the new coordinates?

3. Triangle *ABC* has coordinates (1,5), (3,2), and (1,2). What are its coordinates after it is rotated 90° clockwise around the origin?

4. Triangle *XYZ* has coordinates (4,1), (7,1), and (7,6). It is translated 1 unit down. What are the new coordinates?

Solve. (pages 526–527)

5. What is the pattern of the shape shown on the grid for positions 1, 2, 3, and 4?

6. What transformation would be next if this pattern continued?

7. What translation pattern is shown on the grid for positions 1, 2, and 3?

8. What would be the coordinates for the vertices of the position 4 figure, following the same translation pattern?

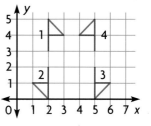

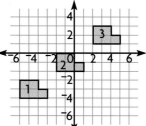

Draw the next three figures in the geometric pattern. (pages 528–529)

9. **10.** **11.** **12.**

Draw the next two solids for the pattern.

13. **14.** **15.** **16.**

Make a tessellation shape from each pattern. (pages 530–531)

17. **18.** **19.** **20.**

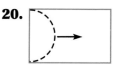

21. Trace the shape you made in Exercise 17. Then form two rows of a tessellation.

22. Trace the shape you made in Exercise 18. Then form two rows of a tessellation.

TAAS Prep

1. In a basketball game, Tara made 3 shots worth 3 points each, 5 shots worth 2 points each, and 4 free throws worth 1 point each. Which number sentence can be used to find p, the total number of points Tara scored?

A $p = (2 \times 3) + (5 \times 3) + (4 \times 2)$
B $p = (7 \times 2) + 4$
C $p = (3 \times 5) + (4 \times 2)$
D $p = (3 \times 3) + (5 \times 2) + (4 \times 1)$

2. Deb bought a shirt for $15.95, a sweater for $22.50, and socks for $1.75. All of the amounts included tax. What else do you need to know to find how much change she should receive?

F The amount of money she has saved
G The total amount she spent on the socks and the shirt
H The amount of money she gave the clerk
J Where she did her shopping

3. Chris needs to solve $\frac{r}{9} = 36$. What should he do first?

A Divide both sides by 9.
B Multiply both sides by 9.
C Add 36 to both sides.
D Subtract 36 from both sides.

4. The distance from Sweetwater to Lakeville is $6\frac{1}{2}$ inches on a map. The scale of the map is 1 inch = 30 miles. What is the actual distance between the two locations?

F 36 mi
G 75 mi
H 180 mi
J 195 mi

5. Adam bought 4 gallons of vanilla ice cream and 3 quarts of strawberry ice cream. Altogether, how many quarts of ice cream did he buy?

A 7 qt   **B** 11 qt
C 16 qt   **D** 19 qt

6. On the coordinate plane below, what point is located by ($^-2$,$^-2$)?

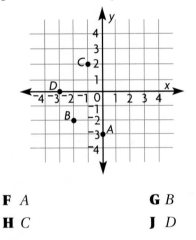

F A **G** B
H C **J** D

7. Rectangle $ABCD$ has coordinates $A(0,0)$, $B(2,0)$, $C(2,1)$, and $D(0,1)$. The rectangle is rotated 90° clockwise around the origin. What are the coordinates of point B of the new figure?

A $(0,^-2)$ **B** $(0,0)$
C $(1,0)$ **D** $(2,1)$

8. The first three figures of a geometric pattern are shown below. If the pattern is continued, what is the sixth figure?

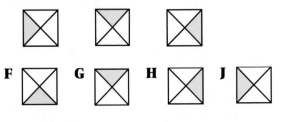

LOOK AHEAD

In this chapter you will solve problems that involve

- whole numbers and decimals

- fraction addition and subtraction

- fraction multiplication

- integers

While Samuel Morse was inventing the Morse Code in the 1830's, he made a list of what he thought to be the most common letters in the English language. A more recent list is based on a survey of 1,000,000 words in a sample of books and newspapers.

MOST-USED LETTERS IN WRITTEN ENGLISH		
	Morse's List	**Survey Results**
1.	e	e
2.	t	t
3.	a	a
4.	i	o
5.	n	i
6.	o	n
7.	s	s
8.	h	r
9.	r	h
10.	d	l

- What percent of the letters from Morse's list match the survey results ranking?

- The paragraph above the table contains 46 words, not counting the numbers. Write a fraction to compare the number of words that contain the letter "e" to the total number of words.

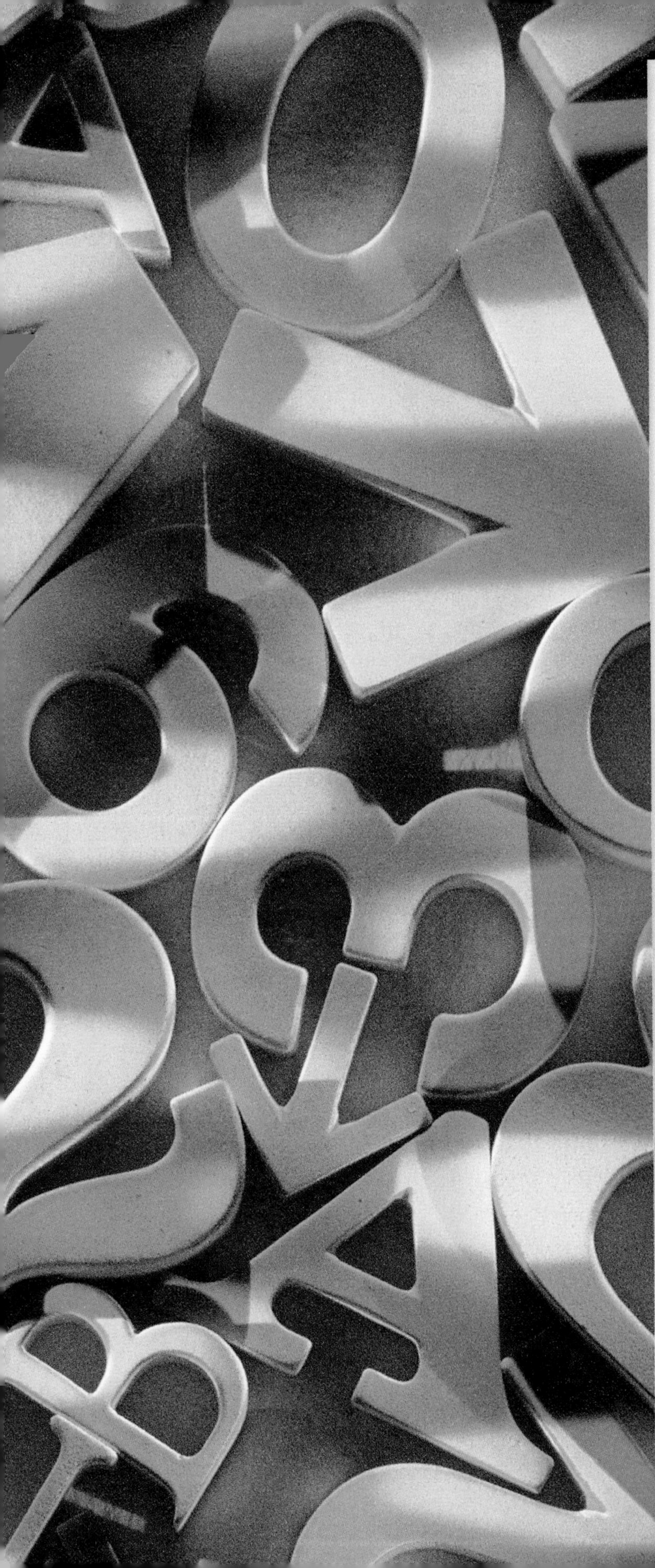

Problem-Solving Project

Pattern Codes

A code is a good way to send secret messages. Code each letter of the alphabet by using a mathematical pattern. Then, write a message using your code. Provide a key so the message can be decoded. Practice decoding other students' messages.

63 57 15

39 63 36 60 27 48 36 15 57

45 18

60 24 54 15 15

60 45

12 15 9 45 12 15

60 24 27 57

39 15 57 57 3 21 15

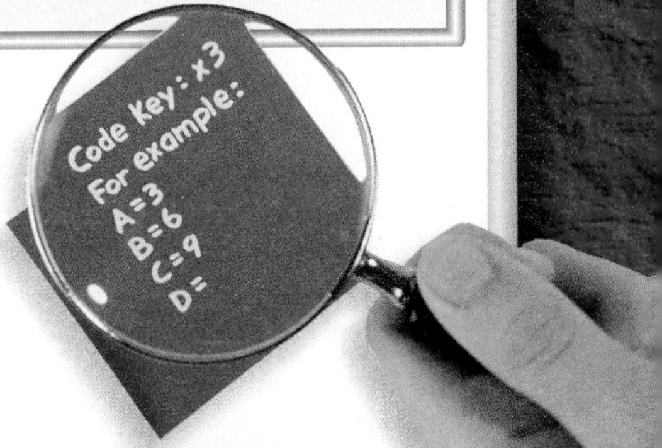
Code Key: x3
For example:
A=3
B=6
C=9
D=

PROJECT CHECKLIST

- ✓ Did you make a code?
- ✓ Did you write a message and provide a key?
- ✓ Did you decode a message?

What You'll Learn
How to identify, extend, and make number patterns with whole numbers and decimals

Why Learn This?
To relate geometric patterns to number patterns and constant patterns of operations

VOCABULARY

triangular number

sequence

term

Number Patterns

In the previous chapter, you looked at geometric patterns. Some geometric patterns show special number patterns.

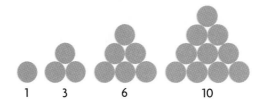

1   3   6   10

The geometric pattern above is a pattern of triangular arrays. The number of circles used to make each new figure is a **triangular number**.

You can write a pattern of numbers as a sequence. A **sequence** is an ordered set of numbers. You would write 1, 3, 6, 10, . . . to show the sequence of triangular numbers. The three dots indicate that the pattern of numbers continues in a similar fashion. Each number in the sequence is called a **term**.

BUSINESS LINK

Not all sequences follow patterns. Some phone "numbers" for businesses are clever words or phrases whose letters match the numbers on telephone keypads. The phone number for Amtrak® rail services, for example, is 1-800-USA-RAIL. What is the number sequence?

ACTIVITY

WHAT YOU'LL NEED: two-color counters

Extend the pattern of triangular arrays to find other triangular numbers. Make a table like the one below to record your data.

	First	Second	Third	Fourth	
Number	1	3	6	10	
Counters Added		+2	+3	+4	

• Make the fifth triangular array. How many counters did you use? How many counters did you add to the fourth array?

• Make the sixth and seventh triangular arrays.

• How many counters did you add to the fifth array to make the sixth? to the sixth to make the seventh? How many would you add to the seventh array to make the eighth array?

• Look at your table. Describe the relationship between each triangular number and the next triangular number.

• Use the relationship to write the ninth and tenth triangular numbers.

Many number sequences have a pattern, such as multiplying each new term by 3 to get the next term.

Term	1	3	9	27	81	243
Pattern		1×3	3×3	9×3	27×3	81×3

A sequence can have a repeating pattern of addition, subtraction, multiplication, or division that uses a constant value.

EXAMPLE 1 Look at the sequence below. Identify the pattern and the rule used to make the sequence. Then find the next three terms in the sequence.

1, 5, 9, 13, 17, . . .   *Look for a pattern.*

1   5   9   13   17   *Compare each term with the next.*

  +4   +4   +4   +4

The rule is add 4 to each term.

$17 + 4 = 21$ *Start with 17. Add 4 to each term to*
$21 + 4 = 25$ *find the next term.*
$25 + 4 = 29$

So, the next three terms in the sequence are 21, 25, and 29.

- Identify the pattern and the rule used to make the following sequence. Then find the next three terms in the sequence.
  8, 32, 128, 512, . . .

EXAMPLE 2 Brenda makes $2.00 an hour baby-sitting for the Rodriquez family. Ms. Rodriquez gives Brenda a $0.25 raise every 6 months. Write a sequence to show Brenda's rate of pay for the next 2 years. What will Brenda's rate of pay be after 2 years?

Start: $2.00
6 months: $2.00 + $0.25 = $2.25   *Add $0.25 to each term to*
12 months: $2.25 + $0.25 = $2.50   *find the next term.*
18 months: $2.50 + $0.25 = $2.75
24 months: $2.75 + $0.25 = $3.00

$2.00, $2.25, $2.50, $2.75, $3.00, . . .   *Write the terms as a sequence.*

So, after 2 years, Brenda will be earning $3.00 an hour.

- Tell how you would find Brenda's rate of pay at the end of the next year.

539

Identify the operation used to build the pattern in the sequence.

1. 3, 6, 9, 12, . . .

2. 5, 20, 35, 50, . . .

3. 2, 6, 18, 54, . . .

4. 0.001, 0.01, 0.1, 1, . . .

Make a table to show the next three terms. Describe the pattern for each sequence.

5. 1, 6, 11, 16, . . .

6. 0, 10, 21, 33, . . .

7. 0.2, 2, 20, 200, . . .

8. 1, 7, 56, 504, . . .

REMEMBER:
. . . .

Division can be expressed as a ratio in fraction form.

$3{,}645 \div 1{,}215 = \frac{3{,}645}{1{,}215}$

See page H19.

Sequences of whole numbers and decimals that show a decreasing pattern involve subtraction or division.

EXAMPLE 3 Find the next three terms in the sequence: 3,645; 1,215; 405; 135; . . .

3,645; 1,215; 405; 135; . . . *Look for a pattern.*

$\frac{3{,}645}{1{,}215} = 3$, $\frac{1{,}215}{405} = 3$, $\frac{405}{135} = 3$ *Write a ratio to compare each term with the next. The ratios are all 3. So divide each new term by 3 to get the next term.*

$135 \div 3 = 45$
$45 \div 3 = 15$
$15 \div 3 = 5$

Start with 135. Divide each term by 3 to find the next term. Repeatedly divide by 3.

So, the next three terms in the sequence are 45, 15, and 5.

• **CRITICAL THINKING** How do you know that a whole-number sequence such as 1,200; 240; 48; . . . has a rule of subtraction or division?

EXAMPLE 4 Rick has $20.00. Every day he spends $2.20 on lunch. Write a sequence to show how Rick will spend his $20.00. How much money will Rick have after 3 days?

Start: $20.00
Day 1: $20.00 − $2.20 = $17.80   *Subtract $2.20 from each*
Day 2: $17.80 − $2.20 = $15.60   *term to find the next term.*
Day 3: $15.60 − $2.20 = $13.40

$20.00, $17.80, $15.60, $13.40, . . .   *Write the terms as a sequence.*

So, after 3 days, Rick will have $13.40.

• How much money will Rick have after 5 days?

INDEPENDENT PRACTICE

Identify the operation used to build the pattern in the sequence.

1. 7, 11, 15, 19, . . .

2. 405, 135, 45, 15, . . .

3. 82, 70, 58, 46, . . .

4. 12, 30, 75, 187.5, . . .

5. 7, 7.89, 8.78, 9.67, . . .

6. 17.5, 16.3, 15.1, 13.9, . . .

Make a table to show the next three terms. Describe the pattern for each sequence.

7. 12, 16, 20, 24, . . .

8. 4, 40, 400, 4,000, . . .

9. 10, 12, 15, 19, . . .

10. 1, 4, 16, 64, . . .

11. 35, 65, 125, 215, . . .

12. 91, 90, 88, 85, . . .

13. 48, 40, 32, 24, . . .

14. 0.82, 0.75, 0.68, 0.61, . . .

15. 300, 30, 3, 0.3, . . .

16. 400, 200, 100, 50, . . .

17. 4.2, 4.95, 5.7, 6.45, . . .

18. 17.9, 17.7, 17.5, 17.3, . . .

Problem-Solving Applications

For Problems 19–21, write a sequence. Then solve.

19. BUSINESS Miguel Lopez just started a new business. His profit was $200 the first month, $220 the second month, and $240 the third month. If this pattern continues, what will his profit be in the sixth month?

20. The Treadwell Tire Company started to reduce its workforce four months ago, when it had 1,700 employees. Three months ago it had 1,590, and two months ago it had 1,480. How many employees were there one month ago?

21. Ed saves $55 of the $450 he gets paid every week. If he increases the amount he saves by $4 each week, in how many weeks will he be saving $75 a week?

22. ✏ **WRITE ABOUT IT** Think of a number sequence. Then describe the steps you would use to find the next term.

Mixed Review and Test Prep

Find the least common denominator.

23. $\frac{1}{2}, \frac{1}{3}$

24. $\frac{3}{4}, \frac{1}{6}$

25. $\frac{2}{5}, \frac{2}{3}$

26. $\frac{1}{8}, \frac{5}{6}$

27. $\frac{2}{3}, \frac{4}{9}$

Draw the next figure in each geometric pattern.

28.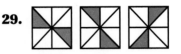

29.

30. HOBBIES Raul Sanchez has $1,000 to landscape his lawn. He plans to spend $350 of it on new trees. Each tree costs $55. How many trees can he buy?

A 6

B 7

C 8

D 9

31. MONEY Margaret trades her half dollar for 8 coins. What coins does she have now?

F 7 nickels, 1 dime

G 6 nickels, 1 dime, 1 quarter

H 5 pennies, 1 nickel, 1 dime, 1 quarter

J 5 pennies, 2 dimes, 1 quarter

ALGEBRA CONNECTION
Patterns Made with a Calculator

Some calculators have a constant function. You can use this function to find other terms in a sequence.

ACTIVITY 1

Explore

Find the next three terms in the sequence. 23, 46, 92, 184, . . .

- What operation is repeated in the sequence? What value is used with the operation?

- Follow the calculator key sequence below to find the next three terms in the sequence.

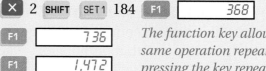

[✕] 2 [SHIFT] [SET 1] 184 [F1] | 368 |

[F1] | 736 |

[F1] | 1,472 |

The function key allows you to use the same operation repeatedly by simply pressing the key repeatedly.

- What are the next three terms in the sequence?

Think and Discuss

- Why do you enter 184 in the calculator sequence shown above?

- Which calculator key did you press to repeat a calculation?

- What is an advantage of using the constant function on a calculator to extend a sequence?

Try This

- Follow the calculator sequence to find the tenth term in this sequence: 220, 217.5, 215, 212.5, . . .

[−] 2.5 [SHIFT] [SET 1] 212.5 [F1]   ← *Press this key six times to find the tenth term.*

- What is the tenth term in the sequence?

You can use the constant function on a calculator to make your own sequences.

ACTIVITY 2

Explore

Make four sequences, one with each of the four operations.

- For each sequence, start with one of the operations and a whole number or decimal to use as a constant.

  [+] 4

- Then set the constant function by pressing the following keys.

  [SHIFT] [SET 1]

- Enter the first value of your sequence.

  180

- Repeatedly press [F1] to find at least four terms in the sequence.

- Then write the sequence.

  180, 184, 188, 192, . . .

Think and Discuss

- What do you notice about the terms of the sequences in which you used addition and multiplication as the constant operations?

- What do you notice about the terms of the sequences in which you used subtraction and division as the constant operations?

- Suppose you keep pressing [F1] for a sequence. Will the sequence continue without end? Explain.

Try This

- Find the next three terms in each of your sequences. Write the operation and the constant value that were used in each sequence.

Technology Link

You can find number patterns in a simulation by using E-Lab, Activity 28. Available on CD-ROM and on the Internet at **www.hbschool.com/elab**

ALGEBRA CONNECTION

Patterns with Fractions

What You'll Learn
How to identify, extend, and make patterns with fractions by using addition and subtraction

Why Learn This?
To be prepared to describe and extend sequences as on standardized tests

Sometimes the terms in a sequence are fractions. To identify the pattern of a sequence with fractions, it is helpful to change the fractions so that they have a common denominator. Look at the sequence below.

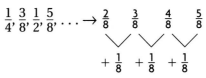

Once the fractions have the same denominator, it is easier to find a pattern in the sequence.

REMEMBER:

To add unlike fractions, you can use the **least common denominator**, or LCD. The LCD is the LCM of the denominators.

2: 2 4 6 8

4: 4 8 12 16

8: 8 16 24 32

The LCD is 8.

See page 108.

EXAMPLE 1 During the first week of training, Yoshi runs $\frac{1}{4}$ mi each day. During the second week, Yoshi runs $\frac{1}{2}$ mi each day. During the third week, she runs $\frac{3}{4}$ mi each day. During what week will Yoshi run 2 mi each day?

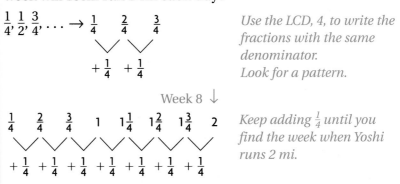

Use the LCD, 4, to write the fractions with the same denominator.
Look for a pattern.

Week 8 ↓

Keep adding $\frac{1}{4}$ until you find the week when Yoshi runs 2 mi.

So, Yoshi will run 2 mi a day during the eighth week.

EXAMPLE 2 Find the next three terms in the sequence.
$20, 18\frac{2}{3}, 17\frac{1}{3}, 16, \ldots$

20    $18\frac{2}{3}$ $17\frac{1}{3}$ 16

$-1\frac{1}{3}$ $-1\frac{1}{3}$ $-1\frac{1}{3}$

Look for a pattern. Compare each term with the next.

$16 - 1\frac{1}{3} = 14\frac{2}{3}$

$14\frac{2}{3} - 1\frac{1}{3} = 13\frac{1}{3}$

$13\frac{1}{3} - 1\frac{1}{3} = 12$

Start with 16. Subtract $1\frac{1}{3}$ from each term to find the next term.

So, the next three terms in the sequence are $14\frac{2}{3}$, $13\frac{1}{3}$, and 12.

GUIDED PRACTICE

Find a common denominator for the terms in the sequence.

1. $\frac{1}{6}, \frac{1}{3}, \frac{1}{2}, \ldots$
2. $\frac{1}{6}, \frac{1}{4}, \frac{1}{2}, \ldots$
3. $\frac{1}{9}, \frac{1}{3}, \frac{1}{1}, \ldots$
4. $\frac{1}{12}, \frac{1}{8}, \frac{1}{6}, \ldots$

Rewrite with a common denominator.

5. $\frac{1}{10}, \frac{1}{5}, \frac{3}{10}, \frac{2}{5}, \ldots$
6. $\frac{1}{12}, \frac{1}{6}, \frac{1}{4}, \frac{1}{3}, \ldots$
7. $\frac{1}{16}, \frac{1}{8}, \frac{3}{16}, \frac{1}{4}, \ldots$
8. $\frac{1}{20}, 1\frac{1}{10}, 2\frac{3}{20}, 3\frac{1}{5}, \ldots$

Find the next term in the sequence.

9. $\frac{1}{4}, \frac{1}{2}, \frac{3}{4}, \ldots$
10. $\frac{1}{8}, \frac{1}{2}, \frac{7}{8}, \ldots$
11. $\frac{3}{10}, \frac{2}{5}, \frac{1}{2}, \frac{3}{5}, \ldots$
12. $\frac{1}{2}, \frac{5}{3}, \frac{17}{6}, \ldots$

INDEPENDENT PRACTICE

Identify the pattern used to make the sequence.

1. $\frac{1}{16}, \frac{1}{8}, \frac{3}{16}, \frac{1}{4}, \ldots$
2. $\frac{1}{3}, \frac{5}{18}, \frac{4}{18}, \frac{1}{6}, \ldots$
3. $\frac{1}{9}, \frac{7}{27}, \frac{11}{27}, \frac{15}{27}, \ldots$

4. $\frac{5}{6}, \frac{7}{12}, \frac{1}{3}, \frac{1}{12}, \ldots$
5. $\frac{4}{5}, \frac{3}{5}, \frac{2}{5}, \frac{1}{5}, \ldots$
6. $\frac{1}{5}, \frac{1}{3}, \frac{7}{15}, \frac{3}{5}, \ldots$

7. $\frac{7}{6}, \frac{13}{12}, 1, \frac{11}{12}, \ldots$
8. $\frac{9}{8}, \frac{15}{16}, \frac{3}{4}, \frac{9}{16}, \ldots$
9. $3, 4\frac{1}{5}, 5\frac{2}{5}, 6\frac{3}{5}, \ldots$

Find the next three terms in the sequence.

10. $\frac{1}{3}, \frac{2}{3}, 1, 1\frac{1}{3}, \ldots$
11. $\frac{1}{16}, \frac{1}{8}, \frac{3}{16}, \ldots$
12. $2\frac{1}{2}, 2\frac{3}{4}, 3, \ldots$
13. $12\frac{1}{2}, 14, 15\frac{1}{2}, \ldots$

14. $40\frac{1}{5}, 41, 41\frac{4}{5}, \ldots$
15. $4\frac{4}{9}, 5\frac{2}{3}, 6\frac{8}{9}, \ldots$
16. $\frac{11}{12}, \frac{5}{6}, \frac{3}{4}, \ldots$
17. $21\frac{1}{4}, 20, 18\frac{3}{4}, \ldots$

18. $3, 2\frac{3}{5}, 2\frac{1}{5}, \ldots$
19. $10\frac{5}{6}, 9\frac{2}{3}, 8\frac{1}{2}, \ldots$
20. $12\frac{7}{8}, 10\frac{3}{4}, 8\frac{5}{8}, \ldots$
21. $10\frac{2}{3}, 9\frac{11}{18}, 8\frac{5}{9}, \ldots$

Problem-Solving Applications

For Problems 22–24, write a sequence. Then solve.

22. PATTERNS Stephan's favorite food is pizza. The first time he had pizza, he ate $\frac{3}{16}$ of a pizza. The second time, he ate $\frac{3}{8}$ of a pizza. If he continues this pattern, when will he be able to eat more than a whole pizza?

23. HOBBIES George spent $\frac{1}{2}$ hr on a jigsaw puzzle the first day, $1\frac{1}{4}$ hr the second day, and 2 hr the third day. If he continues this pattern, how much time will he spend on a jigsaw puzzle on the fourth day?

24. Al shipped packages in order of weight. The first package weighed 35 lb, the second weighed $33\frac{1}{2}$ lb, and the third weighed 32 lb. If this pattern continued, what package weighed 20 lb?

25. ✏️ **WRITE ABOUT IT** Explain how to determine the pattern of a sequence of fractions with different denominators.

What You'll Learn
How to identify, extend, and make sequences with fractions by using multiplication

Why Learn This?
To see the effect of a constant enlargement or reduction on fractions

ALGEBRA CONNECTION

Patterns with Fraction Multiplication

Some sequences involve adding or subtracting fractions. Sequences can also involve multiplication of fractions.

REAL-LIFE LINK

Most office copiers can reduce or enlarge images as they copy. For example, if an $8\frac{1}{2}$-in. × 11-in. page is reduced by $\frac{1}{2}$, the new image will be $4\frac{1}{4}$ in. × $5\frac{1}{2}$ in. What size will the new image be if the page is reduced by $\frac{1}{2}$ again?

EXAMPLE 1 Find the next three terms in the sequence.

$$\frac{1}{4}, \frac{3}{4}, \frac{9}{4}, \cdots$$

$$\frac{1}{4} \quad \frac{3}{4} \quad \frac{9}{4}$$

Look for a pattern.
Compare each term with the next.

$$\times 3 \quad \times 3$$

$$\frac{9}{4} \times 3 = \frac{27}{4}$$ *Multiply each term by 3 to find the next term.*

$$\frac{27}{4} \times 3 = \frac{81}{4}$$

$$\frac{81}{4} \times 3 = \frac{243}{4}$$

So, the next three terms are $\frac{27}{4}$, $\frac{81}{4}$, and $\frac{243}{4}$.

• Are the numbers increasing or decreasing? Explain.

You can make a sequence if you know the first term and the pattern of the sequence.

EXAMPLE 2 The first term of a sequence is $1\frac{1}{2}$, and the rule is to multiply by $\frac{1}{4}$. Write the first four terms of the sequence.

$$1\frac{1}{2} \rightarrow \frac{3}{2}$$ *Write the first term as a fraction.*

$$\frac{3}{2} \times \frac{1}{4} = \frac{3}{8}$$ *Multiply by $\frac{1}{4}$ three times.*

$$\frac{3}{8} \times \frac{1}{4} = \frac{3}{32}$$

$$\frac{3}{32} \times \frac{1}{4} = \frac{3}{128}$$

$$1\frac{1}{2}, \frac{3}{8}, \frac{3}{32}, \frac{3}{128}, \cdots$$ *Write the sequence.*

• Are the numbers increasing or decreasing? Explain.

• **CRITICAL THINKING** Write an increasing and decreasing sequence with fractions, using multiplication as the rule.

GUIDED PRACTICE

Identify the pattern used to make the sequence.

1. $\frac{1}{9}, \frac{2}{9}, \frac{4}{9}, \cdots$

2. $\frac{1}{11}, \frac{3}{11}, \frac{9}{11}, \cdots$

3. $\frac{5}{2}, \frac{5}{4}, \frac{5}{8}, \cdots$

4. $\frac{2}{5}, \frac{2}{15}, \frac{2}{45}, \cdots$

Find the next term in the sequence.

5. $\frac{4}{9}, \frac{8}{9}, \frac{16}{9}, \cdots$

6. $\frac{1}{2}, \frac{1}{4}, \frac{1}{8}, \cdots$

7. $\frac{1}{2}, 1, 2, \cdots$

8. $1\frac{2}{3}, \frac{5}{9}, \frac{5}{27}, \cdots$

INDEPENDENT PRACTICE

Find the next three terms in the sequence.

1. $\frac{2}{3}, \frac{4}{3}, \frac{8}{3}, \cdots$

2. $\frac{1}{8}, \frac{1}{24}, \frac{1}{72}, \cdots$

3. $\frac{1}{4}, \frac{1}{2}, 1, \cdots$

4. $\frac{13}{6}, \frac{13}{2}, \frac{39}{2}, \cdots$

Write the first four terms of the sequence.

5. pattern: multiply by 5
first term: $\frac{1}{8}$

6. pattern: multiply by $\frac{1}{2}$
first term: $1\frac{2}{3}$

7. pattern: multiply by $\frac{3}{10}$
first term: $3\frac{1}{2}$

Problem-Solving Applications

8. Dave's tree has grown over the past three years. The first year his tree was 4 ft high. The second year his tree was 5 ft high. The third year his tree was $6\frac{1}{4}$ ft high. If this pattern continues, what will the height of the tree be after the fourth year?

9. **ART** Ken is folding a piece of paper. On his first fold, each rectangle is $\frac{1}{2}$ the original. On the second fold, each rectangle is $\frac{1}{4}$ the original. On his third fold, each rectangle is $\frac{1}{8}$ the original. What fraction will each rectangle be on his fourth fold?

10. **✎ WRITE ABOUT IT** Write a multiplication sequence with fractions.

Mixed Review and Test Prep

Solve.

11. $^-2 + {^-3}$

12. $^+6 - {^-4}$

13. $^-7 - {^-9}$

14. $^-10 - {^+11}$

15. $^+5 \times {^-1}$

16. $^-6 \times {^-8}$

Write *yes* or *no* to tell whether the figure will form a tessellation.

17.

18.

19.

20.

21. **VOLUME** A truck can carry 60 cubic feet of dirt. How many trips will it take to haul away the dirt from a hole 8 feet × 6 feet × 4 feet?

A 1
B 2
C 3
D 4

22. **PROPORTIONS** Out of every 50 students, 20 belong to a club. How many students out of 300 belong to a club?

F 100
G 120
H 140
J 160

SCIENCE LINK

You probably know that the higher up on a mountain you go, the colder the air becomes. What you may not know is that the temperature drops about 5°F every 1,000 ft. So, if the temperature is 65°F in Colorado Springs (elevation 6,000 ft), about what would it be at the top of nearby Pikes Peak (elevation 14,000 ft)?

ALGEBRA CONNECTION

Patterns with Integers

You can use patterns of operations with integers to make a sequence.

ACTIVITY

Make four sequences of integers, one with each of the four operations.

- For each sequence, start with one of the operations and an integer to use as a constant.

  $+\ ^+3$

- Choose an integer, such as $^-5$, to use as the first term in each of your sequences.

- Perform the operation on each new term until you have at least four terms.

  $^-5 + {}^+3 = {}^-2$
  $^-2 + {}^+3 = {}^+1$
  $1 + {}^+3 = {}^+4$

- Write your sequence.

  $^-5,\ ^-2,\ ^+1,\ ^+4,\ \ldots$

You can extend sequences with integers in the same way as you extend sequences with whole numbers, decimals, and fractions.

EXAMPLE Find the next three terms in the sequence.

$^+2,\ ^-6,\ ^+18,\ ^-54,\ \ldots$

$^+2 \quad ^-6 \quad ^+18 \quad ^-54$
$\quad \times {}^-3 \quad \times {}^-3 \quad \times {}^-3$

Look for a pattern. Compare each term with the next.

$^-54 \times {}^-3 = {}^+162$
$^+162 \times {}^-3 = {}^-486$
$^-486 \times {}^-3 = {}^+1{,}458$

Start with $^-54$. Multiply each term by $^-3$ to find the next term.

So, the next three terms in the sequence are $^+162,\ ^-486,$ and $^+1{,}458.$

- What is the rule used in this sequence?
  $^-128,\ ^-32,\ ^-8,\ ^-2,\ \ldots$

GUIDED PRACTICE

Identify the pattern.

1. ⁻6, ⁻4, ⁻2, 0, . . .　　**2.** ⁺10, ⁺6, ⁺2, ⁻2, . . .　　**3.** ⁺1, ⁻2, ⁺4, ⁻8, . . .　　**4.** ⁻100, ⁻10, ⁻1, . . .

Find the next term in the sequence.

5. ⁻20, ⁻15, ⁻10, . . .　　**6.** ⁺3, ⁺1, ⁻1, ⁻3, . . .　　**7.** ⁺5, ⁻5, ⁺5, ⁻5, . . .　　**8.** ⁺24, ⁻12, ⁺6, ⁻3, . . .

INDEPENDENT PRACTICE

Use the given first term and pattern to write the first four terms of a sequence.

1. first term: ⁺5
pattern: add ⁻12

2. first term: ⁻1
pattern: multiply by ⁻1

3. first term: ⁺3
pattern: subtract ⁺9

Identify the pattern.

4. ⁻128, ⁺64, ⁻32, ⁺16, . . .　　**5.** ⁺141, ⁺120, ⁺99, ⁺78, . . .　　**6.** ⁺18, ⁻54, ⁺162, ⁻486, . . .

7. ⁺8, ⁺6, ⁺4, ⁺2, . . .　　**8.** ⁻271, ⁻276, ⁻281, ⁻286, . . .　　**9.** ⁺23, ⁻92, ⁺368; ⁻1,472, . . .

Find the next three terms in the sequence.

10. ⁻16, ⁻12, ⁻8, . . .　　**11.** ⁺1, ⁻2, ⁻5, ⁻8, . . .　　**12.** ⁻2, ⁺8, ⁻32, . . .　　**13.** ⁺20, ⁻2, ⁺0.2, . . .

14. ⁺10, 0, ⁻10, . . .　　**15.** ⁻12, ⁻6, 0, ⁺6, . . .　　**16.** ⁺9, ⁻18, ⁺36, . . .　　**17.** ⁺96, ⁻48, ⁺24, . . .

18. ⁻20, ⁻15, ⁻10, . . .　　**19.** ⁻180, ⁻159, ⁻138, . . .　　**20.** ⁺3, ⁻15, ⁺75, . . .　　**21.** ⁻4,096, ⁻1,024, ⁻256, . . .

22. ⁺0.2, ⁻0.1, ⁺0.05, . . .　　**23.** ⁻9, ⁺0.9, ⁻0.09, . . .　　**24.** ⁺64, ⁻32, ⁺16, . . .

Problem-Solving Applications

25. Today the temperature is 80°F. The forecast calls for a decrease of 4°F for each of the next 3 days. What is the predicted temperature 3 days from today? Write a sequence to solve.

26. CRITICAL THINKING Manuel wrote this sequence: 1, 1, 2, 3, 5, 8, . . . What is the next term in the sequence?

27. CRITICAL THINKING Amy is using a pattern of dividing by ⁻7. Her sixth number is 23. What number did she start with? Explain.

28. ✏️ **WRITE ABOUT IT** How is extending a sequence of integers the same as extending a sequence of whole numbers, decimals, or fractions?

Technology Link

In *Mighty Math Astro Algebra,* you can practice multiplying integers in the activity *The Expired Warranty.* Use Grow Slide Level Red G.

1. **VOCABULARY** An ordered set of numbers is called a(n) __?__.
(page 538)

2. **VOCABULARY** The number of circles used to make each triangular array is a(n) __?__. (page 538)

3. **VOCABULARY** Each number in a sequence is called a(n) __?__. (page 538)

Find the next three terms in the sequence. (pages 538–541)

4. 220, 22, 2.2, . . .   **5.** 5, 17, 29, . . .   **6.** 100, 99.1, 98.2, . . .   **7.** 32, 16, 8, . . .

8. 9, 15, 21, . . .   **9.** 243, 81, 27, . . .   **10.** 90, 79, 68, . . .   **11.** 8.1, 8.3, 8.5, . . .

Solve. (pages 538–541)

12. Ms. Nunn received 110 orders the first month, 125 the second month, and 140 the third month. If this pattern continues, how many orders will she receive in the fourth month?

13. The temperature was 21°F at 6 P.M. At 7 P.M. it was 18°F, and at 8 P.M. it was 15°F. If the pattern continues, what will the temperature be at midnight?

Find the next three terms in the sequence. (pages 544–545)

14. $\frac{1}{8}, \frac{5}{8}, \frac{9}{8}, \ldots$

15. $\frac{1}{3}, \frac{11}{24}, \frac{7}{12}, \ldots$

16. $7, 5\frac{3}{4}, 4\frac{1}{2}, \ldots$

17. $\frac{7}{8}, 3\frac{1}{8}, 5\frac{3}{8}, \ldots$

18. $4, 3\frac{2}{3}, 3\frac{1}{3}, \ldots$

19. $\frac{5}{6}, \frac{13}{18}, \frac{11}{18}, \ldots$

20. $\frac{2}{7}, \frac{5}{7}, \frac{8}{7}, \ldots$

21. $1\frac{1}{8}, 3\frac{1}{8}, 5\frac{1}{8}, \ldots$

22. $2, 3\frac{1}{4}, 4\frac{1}{2}, \ldots$

Find the next three terms in the sequence. (pages 546–547)

23. $\frac{3}{4}, \frac{3}{2}, 3, \ldots$

24. $\frac{1}{4}, \frac{1}{16}, \frac{1}{64}, \ldots$

25. $\frac{7}{6}, \frac{7}{3}, \frac{14}{3}, \ldots$

26. $\frac{1}{2}, \frac{1}{3}, \frac{2}{9}, \ldots$

27. $2\frac{1}{2}, 1\frac{1}{4}, \frac{5}{8}, \ldots$

28. $\frac{1}{4}, \frac{1}{40}, \frac{1}{400}, \ldots$

29. $\frac{1}{3}, \frac{1}{9}, \frac{1}{27}, \ldots$

30. $\frac{5}{4}, \frac{5}{8}, \frac{5}{16}, \ldots$

31. $2, 1\frac{1}{2}, 1\frac{1}{8}, \ldots$

Find the next three terms in the sequence. (pages 548–549)

32. ⁻1, ⁻2, ⁻3, . . .   **33.** ⁻9, ⁻12, ⁻15, . . .   **34.** ⁻3, ⁺9, ⁻27, . . .

35. ⁺32, ⁻16, ⁺8, . . .   **36.** ⁻4, ⁺8, ⁻16, . . .   **37.** ⁺7, ⁻9, ⁻25, . . .

38. ⁻4, ⁻2, 0, . . .   **39.** ⁺12, ⁻24, ⁺48, . . .   **40.** ⁻8, ⁻11, ⁻14, . . .

1. Chet wants to make a circle graph showing how much time he spends each week on homework. He spends 4 hours on math, 2 hours on reading, $3\frac{1}{2}$ hours on social studies, and $2\frac{1}{2}$ hours on language arts. What angle measure should he use for the reading section of his graph?

   A 15°
   B 30°
   C 45°
   D 60°
   E 90°

2. The mean of 5 numbers is 13.2. What is the sum of the numbers?

   F 18.2
   G 26.4
   H 66
   J Not Here

3. If $a = 7$, what is the value of $a^2 - 3a$?

   A 4
   B 18
   C 21
   D 28
   E 70

4. The temperature at Oakdale is 20°C. What is the temperature expressed in degrees Fahrenheit? Use the formula $F = (\frac{9}{5} \times C) + 32$.

   F 26°F
   G 32°F
   H 54°F
   J 68°F
   K 74°F

5. In which are the numbers ordered from least to greatest?

   A $0.3, \frac{3}{8}, \frac{2}{5}$
   B $\frac{2}{5}, 0.3, \frac{3}{8}$
   C $\frac{3}{8}, \frac{2}{5}, 0.3$
   D Not Here

6. Triangle ABC has coordinates $A(1,2)$, $B(1,0)$, and $C(3,0)$. The triangle is translated 4 units left. What are the new coordinates of point C?

   F $(^-1,0)$
   G $(^-3,^-1)$
   H $(^-2,^-3)$
   J $(3,^-4)$
   K $(7,^-4)$

7. What is the next term in the sequence?
   89, 83.4, 77.8, 72.2

   A 70.6
   B 68.3
   C 66.6
   D 65.4

8. What is the next term in the sequence?
   $3\frac{1}{4}, 3\frac{7}{8}, 4\frac{1}{2}, 5\frac{1}{8}$

   F $5\frac{1}{4}$
   G $5\frac{3}{4}$
   H 6
   J $6\frac{1}{4}$
   K $6\frac{3}{5}$

MATH FUN!

SLIDE iT

PURPOSE To practice using transformations (pages 522–524)
YOU WILL NEED 6 coins or 6 two-colored counters, 1-in. graph paper

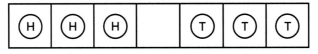

Copy the gameboard on graph paper. Try to switch the places of the heads and the tails. You must slide the coins to an empty spot either directly or around the outside. You may travel past only one coin on an outside slide, and you may not pick up a coin.

DESIGN A MOSAIC

PURPOSE To practice making shapes that tessellate (pages 530–531)
YOU WILL NEED colored construction paper, scissors, paste

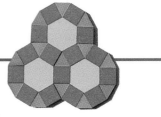

An artist fits together tiny pieces of stone or glass to create a mosaic. Cut out your own tessellating shapes from construction paper. Start with squares, rectangles, triangles, or hexagons.

Try to put the shapes together so they fit with no leftover space. Combine different colors. Then paste them in place on a sheet of construction paper to make your own mosaic. Share your design with the class.

NO FIBBING!

PURPOSE To practice finding number patterns (pages 538–549)

You may be familiar with the Fibonacci sequence. It starts with two 1's; then each number is the sum of the two previous numbers.

Find the next five terms of each sequence.
a. 1, 1, 2, 3, 5, ___?___
b. 1, 5, 6, 11, 17, ___?___
c. 3, 2, 5, 7, 12, ___?___

You can write other Fibonacci sequences. Start with any two numbers. For 2 and 4, the sequence is 2, 4, 6, 10, ...

Write your own Fibonacci sequence. Then see if your family can find the pattern.

Making Tessellations

In this activity you will see how you can make your own tessellation by using a paint program.

From the toolbar, select the rectangle. Then draw a rectangle.

Then select a circle and draw a circle. Using the selection tool and the menu bar, copy and paste the circle.

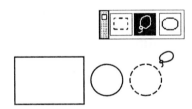

Move one of the circles to the left side of the rectangle. Then move the second circle to the right side. You may want to use a grid to help you align the circles horizontally on the rectangle.

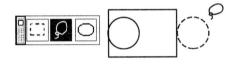

Use the eraser to erase the left and right edges of the rectangle.

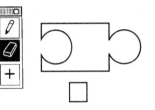

Then select, copy, and paste your new shape repeatedly to make a tessellation. Use a fill tool to color your tessellation.

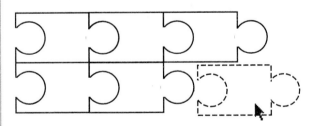 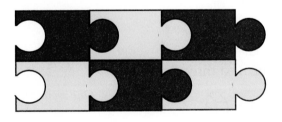

1. Why did you need to have two circles to make the tessellation shape?

2. Which shapes were selected from the toolbar to make this shape?

USING A PAINT PROGRAM

3. Use a square and a smaller square to make a tessellation shape.

4. Use a triangle and a square to make a tessellation shape.

Study Guide and Review

Vocabulary Check

1. An ordered set of numbers that follow a pattern is called a __?__ . (page 538)

2. Each number in a sequence is called a __?__ . (page 538)

EXAMPLES

EXERCISES

- **Transform figures on a coordinate plane.**
  (pages 522–525)

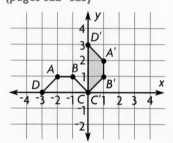

Trapezoid ABCD is rotated clockwise 90° around the origin.

ABCD coordinates: (⁻2,1), (⁻1,1), (0,0), (⁻3,0)
A'B'C'D' coordinates: (1,2), (1,1), (0,0), (0,3)

3. Graph triangle *ABC* with coordinates (1,2), (1,0), and (3,0), and translate it 4 units left. What are the new coordinates?

4. Use the new triangle in Exercise 3 and reflect it across the *x*-axis. What are the new coordinates?

5. Trapezoid *DEFG* has coordinates (1,2), (1,1), (0,0), and (0,3). What are its coordinates after it is rotated 180° clockwise around the origin?

- **Identify the next figure in a geometric pattern.** (pages 528–529)

The first three figures are:

Look for a pattern.

So, the next three figures are:

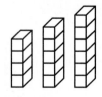

One rectangular prism is added to each figure to make the next figure.

Draw the next three figures in the geometric pattern.

6. 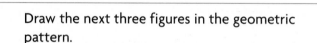   7.

Draw the next two solids in the pattern.

8. 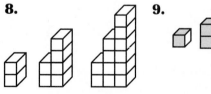   9.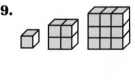

- **Make shapes for tessellations.** (pages 530–531)

Make a tessellation shape from each pattern.

10.   11.

- **Identify, extend, and make patterns with fractions by using addition and subtraction.**
(pages 544–545)

$\frac{1}{10}, \frac{1}{5}, \frac{3}{10}, \frac{2}{5}, \ldots$ *Look for a pattern.*

$\frac{1}{10} \quad \frac{2}{10} \quad \frac{3}{10} \quad \frac{4}{10}$ *Use the LCD, 10.*
Compare each term with the next term.
$+\frac{1}{10} +\frac{1}{10} +\frac{1}{10}$

$\frac{2}{5} + \frac{1}{10} = \frac{5}{10}, \text{ or } \frac{1}{2}$ *Add $\frac{1}{10}$ to $\frac{2}{5}$ to find the next term.*

Find the next three terms in the sequence.

12. $\frac{1}{6}, \frac{2}{3}, 1\frac{1}{6}, \ldots$ **13.** $\frac{11}{12}, \frac{3}{4}, \frac{7}{12}, \ldots$

14. $3, 2\frac{3}{4}, 2\frac{1}{2}, \ldots$ **15.** $\frac{1}{4}, \frac{3}{8}, \frac{1}{2}, \ldots$

16. $5\frac{5}{8}, 7\frac{7}{8}, 10\frac{1}{8}, \ldots$ **17.** $8, 6\frac{2}{3}, 5\frac{1}{3}, \ldots$

- **Identify, extend, and make patterns with fractions by using multiplication.**
(pages 546–547)

$\frac{1}{6}, \frac{1}{3}, \frac{2}{3}, \ldots$ *Look for a pattern.*
$\frac{1}{6} \quad \frac{2}{6} \quad \frac{4}{6}$ *Use the LCD, 6.*
$\times 2 \times 2$ *Compare each term with the next term.*

$\frac{2}{3} \times 2 = \frac{4}{3}, \text{ or } 1\frac{1}{3}$ *Multiply $\frac{2}{3}$ by 2 to find the next term.*

Find the next three terms in the sequence.

18. $\frac{1}{3}, \frac{2}{3}, \frac{4}{3}, \ldots$ **19.** $\frac{2}{5}, \frac{2}{15}, \frac{2}{45}, \ldots$

20. $\frac{1}{5}, \frac{2}{5}, \frac{4}{5}, \ldots$ **21.** $2\frac{1}{6}, 1\frac{1}{12}, \frac{13}{24}, \ldots$

22. $\frac{1}{2}, 1, 2, \ldots$ **23.** $\frac{1}{8}, \frac{1}{2}, 2, \ldots$

- **Identify, make, and extend sequences of integers.** (pages 548–549)

$^-100, {^+}50, {^-}25, \ldots$ *Look for a pattern.*
$\frac{^-100}{^+50} = \frac{^-2}{^+1} = {^-}2$ *Write a ratio to compare each term with the next term.*
$\frac{^+50}{^-25} = \frac{^+2}{^-1} = {^-}2$

$^-25 \div {^-}2 = {^+}12\frac{1}{2}$ *Divide by $^-2$ to find the next term.*

Find the next three terms in the sequence.

24. $^+1, {^-}2, {^+}4, {^-}8, \ldots$ **25.** $^-200, {^-}20, {^-}2, \ldots$

26. $^-6, {^-}4, {^-}2, \ldots$ **27.** $^+1, {^-}2, {^-}5, {^-}8, \ldots$

28. $^+3, {^-}3, {^-}9, \ldots$ **29.** $^+0.009, {^-}0.09, {^+}0.9, \ldots$

30. $^-90, {^-}75, {^-}60, \ldots$ **31.** $^+64, {^-}16, {^+}4, \ldots$

Problem-Solving Applications

Solve. Explain your method.

32. Draw the letter L on a coordinate plane. Then move it through a pattern of transformations. (pages 526–527)

33. Ian sold 90 books in January, 115 books in February, and 140 books in March. If this pattern continues, how many books will he sell in April? (pages 538–541)

Performance Assessment

Tasks: Show What You Know

1. Explain how to reflect $\triangle ABC$ across the x-axis. Write the coordinates of the new figure. **(pages 522–525)**

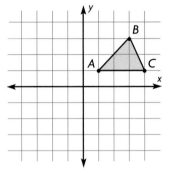

2. Find the next term in the sequence. $\frac{2}{3}, \frac{6}{3}, \frac{18}{3}, \ldots$ **(pages 546–547)**

Problem Solving

Choose a strategy and solve. Explain your method.

CHOOSE a strategy and a tool.

- **Find a Pattern**
- **Make a Model**
- **Draw a Diagram**
- **Write an Equation**
- **Make a Table**
- **Use a Graph**

Paper/Pencil   Calculator   Hands-On   Mental Math

3. Study the figures on the coordinate plane. What transformations were used to move from position 1 to positions 2, 3, and 4? Redraw the pattern and add positions 5 and 6. **(pages 522–525)**

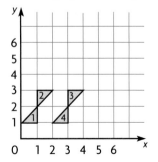

4. A company makes 400 pairs of in-line skates each week. It plans to increase production by 15 pairs each week. In how many weeks will production reach 500 pairs per week? **(pages 538–541)**

Cumulative Review

Solve the problem. Then write the letter of the correct answer.

1. Find the product. 0.075×0.9 (pages 64–67)

   A. 0.0675   **B.** 0.675
   C. 67.5   **D.** 675

2. Find the quotient. Write it in simplest form. $\frac{7}{8} \div \frac{5}{16}$ (pages 146–147)

   A. $\frac{5}{14}$ **B.** $\frac{35}{128}$
   C. $2\frac{1}{5}$ **D.** $2\frac{4}{5}$

3. Name the figure. (pages 196–199)

   A. rectangular prism
   B. rectangular pyramid
   C. triangular prism
   D. triangular pyramid

4. Find the mean. 30, 25, 40, 35, 25 (pages 258–261)

   A. 15   **B.** 25
   C. 30   **D.** 31

5. Solve for x. $4x = 32$ (pages 310–313)

   A. $x = 8$ **B.** $x = 28$
   C. $x = 36$ **D.** $x = 128$

6. Use the scale 1 in. = 200 mi. Find the actual distance. map distance: $2\frac{1}{2}$ in. (pages 384–385)

   A. 2.5 mi   **B.** 80 mi
   C. 400 mi   **D.** 500 mi

7. Which rectangle is a Golden Rectangle? (pages 394–395)

   A. 8 ft $\times$ 8 ft   **B.** 5 ft $\times$ 8 ft
   C. 3 ft $\times$ 9 ft   **D.** 4 ft $\times$ 1 ft

8. Find the surface area of the rectangular prism. (pages 452–453)

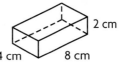

   A. 14 cm^2 **B.** 56 cm^2
   C. 64 cm^2 **D.** 112 cm^2

9. Find the difference. $^-5 - {}^-3$ (pages 488–489)

   A. $^-8$ **B.** $^-2$
   C. $^+2$ **D.** $^+8$

10. Find the quotient. $^-54 \div {}^-9$ (pages 492–493)

   A. $^-486$ **B.** $^-6$
   C. $^+6$ **D.** $^+486$

11. Rectangle *ABCD* has the coordinates $(^-5,3), (^-2,3), (^-5,1), (^-2,1)$. Reflect it across the *x*-axis. What are the new coordinates? (pages 522–525)

   A. $(^-3,^-5), (^-3,^-2), (^-1,^-2), (^-1,^-5)$
   B. $(^-5,^-3), (^-2,^-3), (^-5,^-1), (^-2,^-1)$
   C. $(2,^-2), (5,^-2), (2,^-4), (5,^-4)$
   D. $(3,4), (3,1), (1,1), (1,4)$

For Exercises 12–13, find the next term in the sequence.

12. 1.15, 1.08, 1.01, 0.94, . . . (pages 538–541)

   A. $^-0.87$
   B. 0.7
   C. 0.87
   D. 1.1

13. $^-1, ^-2, ^-4, . . .$ (pages 548–549)

   A. $^-8$
   B. $^-6$
   C. $^-1$
   D. $^+8$

STUDENT HANDBOOK

1 PRIME NUMBERS

Numbers which have exactly two different factors, one and the number itself, are **prime numbers**.

Prime	Reason	Not Prime	Reason
2	2 and 1 are the only factors of 2.	6	6 has other factors like 2 and 3.
7	7 and 1 are the only factors of 7.	8	8 has other factors like 2 and 4.
23	23 and 1 are the only factors of 23.	24	24 has other factors like 4 and 6.
47	47 and 1 are the only factors of 47.	36	36 has other factors like 3 and 12.

Examples   Determine whether the number is prime or not prime.

A. 15

$$3 \times 5 = 15$$

15 *is not prime since it has factors other than* 1 *and* 15.

B. 5

$$1 \times 5 = 5$$

5 *is prime since it has only two factors,* 1 *and* 5 *itself.*

C. 29

$$1 \times 29 = 29$$

29 *is prime since it has only two factors,* 1 *and* 29 *itself.*

PRACTICE   Determine if the numbers are prime or not prime. Write *yes* or *no.*

1. 3   **2.** 11   **3.** 12   **4.** 19   **5.** 14   **6.** 17

2 COMPOSITE NUMBERS

Numbers which have more than two factors are **composite numbers.**

Composite	Reason	Not Composite	Reason
4	1, 2, and 4 are factors of 4.	5	1 and 5 are only factors of 5.
9	1, 3, and 9 are factors of 9.	7	1 and 7 are only factors of 7.
10	1, 2, 5, and 10 are factors of 10.	17	1 and 17 are only factors of 17.
27	1, 3, 9, and 27 are factors of 27.	29	1 and 29 are only factors of 29.

Examples   Determine whether the number is composite or not composite.

A. 15

factors: 1, 3, 5, 15

15 *is composite since it has more than* 2 *factors.*

B. 13

factors: 1, 13

13 *is not composite since it has only two factors.*

C. 36

factors: 1, 2, 3, 4, 6, 9, 12, 18, 36

36 *is composite since it has more than* 2 *factors.*

PRACTICE   Determine if the numbers are composite or not. Write *yes* or *no.*

1. 6   **2.** 21   **3.** 2   **4.** 18   **5.** 11   **6.** 26

3   MULTIPLES

Multiples of a number can be found by multiplying the number by 1, 2, 3, 4, and so on.

Example   Find the first five multiples of 3.

$3 \times 1 = 3$
$3 \times 2 = 6$
$3 \times 3 = 9$ → *The numbers 3, 6, 9, 12, and 15*
$3 \times 4 = 12$ *are the first five multiples of 3.*
$3 \times 5 = 15$

PRACTICE   Write the first five multiples of each number.

1. 4	**2.** 2	**3.** 5	**4.** 7	**5.** 8	**6.** 12
7. 9	**8.** 10	**9.** 20	**10.** 15	**11.** 50	**12.** 18

4   FACTORS

When two numbers are multiplied to form a third, the two numbers are said to be **factors** of the third number.

$4 \times 8 = 32$
$\uparrow \quad \uparrow$
factors

Example   List all the factors of 32.

Think: The only possibilities for factors of 32 are whole numbers from 1 to 32. Begin with 1, since every number has 1 and itself as factors.

$1 \times 32 = 32$ ← *The numbers 1 and 32 are factors of 32.*
$2 \times 16 = 32$ ← *The number 2 and 16 are factors of 32.*
$3 \times \ ? = 32$ ← *No whole number times 3 equals 32, so 3 is not a*
                           factor of 32.
$4 \times \ 8 = 32$ ← *The numbers 4 and 8 are factors of 32.*

Think: The only remaining possibilities must be between 4 and 8. Since no whole number multiplied by 5, 6, or 7 equals 32, they are not factors of 32.

The factors of 32 are 1, 2, 4, 8, 16, and 32.

PRACTICE   List all the factors of each number.

1. 8	**2.** 20	**3.** 9	**4.** 51	**5.** 16	**6.** 27
7. 18	**8.** 63	**9.** 50	**10.** 17	**11.** 76	**12.** 23
13. 21	**14.** 48	**15.** 47	**16.** 40	**17.** 39	**18.** 52

5 DIVISIBILITY RULES

A number is divisible by another number when the division results in a remainder of 0. You can determine divisibility by some numbers with divisibility rules.

A number is divisible by	Divisible	Not Divisible
2 if the last digit is an even number.	11,994	2,175
3 if the sum of the digits is divisible by 3.	216	79
4 if the last two digits form a number divisible by 4.	1,024	621
5 if the last digit is 0 or 5.	15,195	10,007
6 if the number is divisible by 2 and 3.	1,332	44
9 if the sum of the digits is divisible by 9.	144	33
10 if the last digit is 0.	2,790	9,325

PRACTICE Determine whether each number is divisible by 2, 3, 4, 5, 6, 9, or 10.

1. 56 **2.** 200 **3.** 75 **4.** 324 **5.** 42 **6.** 812

7. 784 **8.** 501 **9.** 2,345 **10.** 555,555 **11.** 3,009 **12.** 2,001

6 PRIME FACTORIZATION

A composite number can be expressed as a product of prime numbers. This is the **prime factorization** of the number. To find the prime factorization of a number, you can use a factor tree.

Example Find the prime factorization of 24 by using a factor tree.

Think: Use 2×12, 3×8, or 4×6.

24	24	24
2×12	3×8	4×6
$2 \times 3 \times 4$	$3 \times 2 \times 4$	$2 \times 2 \times 2 \times 3$
$2 \times 3 \times 2 \times 2$	$3 \times 2 \times 2 \times 2$	
all primes	all primes	all primes

The prime factorization of 24 is $2 \times 2 \times 2 \times 3$, or $2^3 \times 3$.

PRACTICE Find the prime factorization by using a factor tree.

1. 25 **2.** 16 **3.** 56 **4.** 18 **5.** 72 **6.** 40

7. 52 **8.** 30 **9.** 22 **10.** 64 **11.** 31 **12.** 68

7 GREATEST COMMON FACTOR

The **greatest common factor** (*GCF*) of two whole numbers is the greatest factor the numbers have in common.

Example Find the *GCF* of 24 and 32.

Method 1

List all the factors for both numbers. Find all the common factors.

24: 1, 2, 3, 4, 6, 8, 12, 24
32: 1, 2, 4, 8, 16, 32

The common factors are 1, 2, 4, and 8.

So, the *GCF* is 8.

Method 2

Find the prime factorizations. Then find the common prime factors.

24: $2 \times 2 \times 2 \times 3$
32: $2 \times 2 \times 2 \times 2 \times 2$

The common prime factors are 2, 2, and 2. The product of these is the *GCF*:

So, the *GCF* is $2 \times 2 \times 2 = 8$.

PRACTICE Find the *GCF* of each pair of numbers by either method.

1. 9, 15 **2.** 25, 75 **3.** 18, 30 **4.** 4, 10 **5.** 12, 17 **6.** 30, 96

7. 54, 72 **8.** 15, 20 **9.** 40, 60 **10.** 40, 50 **11.** 14, 21 **12.** 14, 28

8 LEAST COMMON MULTIPLE

The **least common multiple** (*LCM*) of two numbers is the smallest common multiple the numbers share.

Example Find the least common multiple of 8 and 10.

Method 1

List multiples of both numbers.

 8: 8, 16, 24, 32, 40, 48, 56, 64, 72, 80
10: 10, 20, 30, 40, 50, 60, 70, 80, 90

The smallest common multiple is 40.

So, the *LCM* is 40.

Method 2

Find the prime factorizations.

 8: $2 \times 2 \times 2$
10: 2×5

The *LCM* is found by finding a product of factors.

$2 \times 2 \times 2 \times 5 = 40$.   So, the *LCM* is 40.

PRACTICE Find the *LCM* of each pair of numbers by either method.

1. 2, 4 **2.** 3, 15 **3.** 10, 25 **4.** 10, 15 **5.** 3, 7 **6.** 18, 27

7. 7, 22 **8.** 5, 40 **9.** 3, 14 **10.** 9, 12 **11.** 12, 15 **12.** 14, 16

9 PROPERTIES

The following are basic properties of addition and multiplication.

ADDITION	MULTIPLICATION

Commutative: $a + b = b + a$

Associative: $(a + b) + c = a + (b + c)$

Identity Property of Zero:
$$a + 0 = a \text{ and } 0 + a = a$$

Commutative: $a \times b = b \times a$

Associative: $(a \times b) \times c = a \times (b \times c)$

Identity Property of One:
$$a \times 1 = a \text{ and } 1 \times a = a$$

Property of Zero: $a \times 0 = 0 \text{ and } 0 \times a = 0$

Distributive: $a \times (b + c) = a \times b + a \times c$

PRACTICE Name the property illustrated.

1. $4 + 0 = 4$ **2.** $(6 + 3) + 1 = 6 + (3 + 1)$ **3.** $7 \times 51 = 51 \times 7$

4. $5 \times 456 = 456 \times 5$ **5.** $17 \times (1 + 3) = 17 \times 1 + 17 \times 3$ **6.** $1 \times 5 = 5$

7. $(8 \times 2) \times 5 = 8 \times (2 \times 5)$ **8.** $72 + 1,234 = 1,234 + 72$ **9.** $0 \times 12 = 0$

10 WHOLE-NUMBER PLACE VALUE

The place value chart below can help you read and write whole numbers. Each section is called a period. A period is made up of the hundreds, tens, and ones place. The number 527,019,346,912 is read as "five hundred twenty-seven billion, nineteen million, three hundred forty-six thousand nine hundred twelve.

BILLIONS			MILLIONS			THOUSANDS			ONES		
Hundreds	Tens	Ones	Hundreds	Tens	Ones	Hundreds	Tens	Ones	Hundreds	Tens	Ones
5	2	7,	0	1	9,	3	4	6,	9	1	2

Examples Name the place value of the digit.

A. 5 in the billions period
   5 → hundred billions place

B. 4 in the thousands period
   4 → ten thousands place

C. 9 in the ones period
   9 → hundreds place

PRACTICE Name the place value of the underlined digit.

1. 234,765,890,17<u>3</u>   **2.** 234,<u>7</u>65,890,173   **3.** 2<u>3</u>4,765,890,173   **4.** 234,765,89<u>0</u>,173

5. 234,765,8<u>9</u>0,173   **6.** 23<u>4</u>,765,890,173   **7.** 234,765,890,1<u>7</u>3   **8.** 234,7<u>6</u>5,890,173

11 COMPARING AND ORDERING WHOLE NUMBERS

The symbols $<$, $\leq$, $=$, $\geq$, $>$, $\neq$ can be used to compare whole numbers.
The symbol $<$ means "is less than," and $>$ means "is greater than."

Example

A. Use $<$, $>$, or $=$ to compare.

A. 345 ● 350 → 345 $<$ 350

B. 345 ● 340 → 345 $>$ 340

C. 345 ● 345 → 345 $=$ 345

D. 829 ● 828 → 829 $>$ 828

B. Write the numbers in order from least to greatest. Use $<$.

A. 73; 68; 84
68 $<$ 73 $<$ 84

B. 4,687; 4,874; 4,784
4,687 $<$ 4,784 $<$ 4,874

C. 123; 119; 175
119 $<$ 123 $<$ 175

PRACTICE Use $<$, $>$, or $=$ to compare.

1. 899 ● 895   **2.** 35 ● 30   **3.** 9,876 ● 9,880   **4.** 237 ● 237

Write the numbers in order from least to greatest. Use $<$.

5. 12; 8; 17   **6.** 713; 698; 624   **7.** 1,273; 1,256; 1,114   **8.** 46; 44; 49

12 ADDING AND SUBTRACTING WHOLE NUMBERS

When adding or subtracting whole numbers, align corresponding digits.

Examples

Find the sum. 247 + 1,496 + 89

Carry extra digits to the next column.

$$\begin{array}{r} \overset{22}{} \\ 247 \\ 1,496 \\ +\ \ 89 \\ \hline 1,832 \end{array}$$

Think: 7 + 6 + 9 = 22

So, 247 + 1,496 + 89 is 1,832.

Find the difference. 3,000 − 1,650

Borrow from the next column when necessary.

$$\begin{array}{r} \overset{9}{} \\ 2\ \overset{10}{\cancel{10}} \\ \cancel{3},\ \cancel{0}\ \cancel{0}\ 0 \\ -1,\ 6\ 5\ 0 \\ \hline 1,\ 3\ 5\ 0 \end{array}$$

So, 3,000 − 1,650 is 1,350.

PRACTICE Find the sum or difference.

1. 84 + 132 + 1,512   **2.** 79 + 56 + 99   **3.** 3,492 − 2,270   **4.** 895 − 689

5. 8,906 + 476 + 98   **6.** 340 + 199 + 637   **7.** 9,003 − 5,674   **8.** 1,148 − 739

9. 1,230 + 384 + 20   **10.** 88 + 821 + 9   **11.** 2,721 − 834   **12.** 5,069 − 947

13 MULTIPLYING WHOLE NUMBERS

When multiplying whole numbers, align corresponding digits.

Example Find the product. 27 × 86

STEP 1 *Multiply by the 6 ones.*

$$\begin{array}{r} 27 \\ \times\ 86 \\ \hline 162 \\ +\ 2160 \\ \hline 2{,}322 \end{array}$$

STEP 2 *Multiply by the 8 tens.*

$$\begin{array}{r} 27 \\ \times\ 86 \\ \hline 162 \\ +\ 2160 \\ \hline 2{,}322 \end{array}$$

STEP 3 *Add the partial products.*

$$\begin{array}{r} 27 \\ \times\ 86 \\ \hline 162 \\ +\ 2160 \\ \hline 2{,}322 \end{array}$$

PRACTICE Find the product.

1. 25 × 15   **2.** 84 × 37   **3.** 45 × 23   **4.** 67 × 76

5. 417 × 12   **6.** 895 × 84   **7.** 365 × 30   **8.** 642 × 205

9. 21 × 805   **10.** 128 × 12   **11.** 718 × 9   **12.** 104 × 201

14 DIVIDING WHOLE NUMBERS

In the division problem 146 ÷ 9, the **dividend** is 146 and the **divisor** is 9. In long-division the divisor is outside the long-division symbol and the dividend is inside the long-division symbol. The answer to a division problem is the **quotient**.

$$\text{divisor} \\ \downarrow \\ 9\overline{)146} \\ \uparrow \\ \text{dividend}$$

Example Find the quotient. 146 ÷ 9

STEP 1
Divide the hundreds.

$$9\overline{)146}$$

There are not enough hundreds.

STEP 2
Divide the 14 tens.

$$\begin{array}{r} 1 \\ 9\overline{)146} \\ -\ 9\downarrow \\ \hline 56 \end{array}$$

Multiply 1 and 9 and subtract. Bring down the 6.

STEP 3
Divide the 56 ones.

$$\begin{array}{r} 16 \\ 9\overline{)146} \\ -\ 9 \\ \hline 56 \\ -\ 54 \\ \hline 2 \end{array}$$

Multiply 6 and 9 and subtract.

STEP 4
Write the remainder.

$$\begin{array}{r} 16\quad \text{r2} \\ 9\overline{)146} \\ -\ 9 \\ \hline 56 \\ -\ 54 \\ \hline 2 \end{array}$$

The quotient is 16. The remainder is 2.

PRACTICE Find the quotient.

1. 167 ÷ 7   **2.** 296 ÷ 3   **3.** 510 ÷ 8   **4.** 909 ÷ 9

5. 375 ÷ 15   **6.** 780 ÷ 30   **7.** 250 ÷ 20   **8.** 896 ÷ 45

9. 249 ÷ 3   **10.** 513 ÷ 6   **11.** 382 ÷ 5   **12.** 733 ÷ 9

15 INVERSE OPERATIONS

Addition and subtraction are inverse operations. Multiplication and division are inverse operations. When performing an operation, you can use its inverse to check your answer.

Examples Use the inverse operation to check the answer.

A.
$$\begin{array}{r} 3,465 \\ -\ 2,197 \\ \hline 1,268 \end{array}$$
Check:
$$\begin{array}{r} 1,268 \\ +\ 2,197 \\ \hline 3,465 \end{array}$$

The inverse operation is addition. Since $1,268 + 2,197 = 3,465$, the answer checks.

B.
$$\begin{array}{r} 72 \\ \times\ 9 \\ \hline 648 \end{array}$$
Check:
$$9\overline{)648}$$

The inverse operation is division. Since $648 \div 9 = 72$, the answer checks.

C.
$$\begin{array}{r} 155 \\ +\ 268 \\ \hline 423 \end{array}$$
Check:
$$\begin{array}{r} 423 \\ -\ 268 \\ \hline 155 \end{array}$$

The inverse operation is subtraction. Since $423 - 268 = 155$, the answer checks.

D.
$$8\overline{)584}\ \ ^{73}$$
Check:
$$\begin{array}{r} 73 \\ \times\ 8 \\ \hline 584 \end{array}$$

The inverse operation is multiplication. Since $73 \times 8 = 584$, the answer checks.

PRACTICE Perform the operation. Check the answer.

1. 36×9 **2.** $425 - 364$ **3.** $984 + 87$ **4.** $256 \div 8$ **5.** 872×20 **6.** $967 - 92$

16 RULES FOR ROUNDING

To round to a certain place, follow these steps.
Step 1: Locate the digit in that place, and consider the next digit to the right.
Step 2: If the digit to the right is 5 or greater, round up. If the digit to the right is 4 or less, round down.
Step 3: Change each digit to the right of the rounding place to zero.

Examples

A. Round 125,439.378 to the nearest thousand.

Locate digit.
↓
12**5**,439.378
↑

The digit to the right is less than 5, so the digit in the rounding location stays the same.
↓
125,000 ← *Each digit to the right becomes zero.*

B. Round 125,439.378 to the nearest tenth.

Locate digit.
↓
125,439.**3**78
↑

The digit to the right is greater than 5, so the digit in the rounding location increases by 1.
↓
125,439.4 ← *Each digit to the right is dropped.*

PRACTICE Round 259,345.278 to the place indicated.

1. hundred thousand **2.** ten thousand **3.** thousand **4.** hundred

17 COMPATIBLE NUMBERS

Compatible numbers divide without a remainder, are close to the actual numbers, and are easy to compute mentally. Use compatible numbers to estimate quotients.

Examples

A. Use compatible numbers to estimate the quotient. 6,134 ÷ 35.

6,134 ÷ 35
6,000 ÷ 30 = 200 ← *estimate*
↑   ↑
compatible numbers

B. Use compatible numbers to estimate the quotient. 647 ÷ 7.

647 ÷ 7
630 ÷ 7 = 90 ← *estimate*
↑   ↑
compatible numbers

PRACTICE Estimate the quotient by using compatible numbers.

1. 345 ÷ 5   **2.** 5,474 ÷ 23   **3.** 46,170 ÷ 18   **4.** 749 ÷ 7   **5.** 861 ÷ 41

6. 1,225 ÷ 2   **7.** 968 ÷ 47   **8.** 3,456 ÷ 432   **9.** 5,765 ÷ 26   **10.** 25,012 ÷ 64

18 DECIMALS AND PLACE VALUE

A place-value chart can help you read and write decimal numbers. The number 912.5784 (nine hundred twelve and five thousand seven hundred eighty-four ten-thousandths) is shown. Remember that the decimal point is read as "and."

Hundreds	Tens	Ones	Tenths	Hundredths	Thousandths	Ten-Thousandths
9	1	2	5	7	8	4

Examples Name the place value of the given digit from 912.5784.

A. 7 → hundredths

B. 8 → thousandths

C. 5 → tenths

D. 4 → ten-thousandths

PRACTICE Name the place value of the underlined digit.

1. 3.0<u>5</u>94   **2.** 3.059<u>4</u>   **3.** 3.05<u>9</u>4   **4.** 3.<u>0</u>594

Write each number in words.

5. 2.345   **6.** 0.43   **7.** 10.2062   **8.** 105.0007

9. 0.87   **10.** 12.305   **11.** 121.4092   **12.** 96.1

19 COMPARING AND ORDERING DECIMALS

The symbols $<$, $\leq$, $=$, $\geq$, $>$, $\neq$ can be used to compare decimals.
To find how decimals compare, align the decimal points and
compare the corresponding digits.

Examples Use $<$, $>$, or $=$ to compare.

A. 4.135 ● 4.137

  4.135
  4.137

  $5 < 7$

So, $4.135 < 4.137$.

B. 19.010 ● 19.005

  19.010
  19.005

  $1 > 0$

So, $19.010 > 19.005$.

Example Write the numbers in order from least to greatest. Use $<$.

  2.046, 2.039, 2.064

  2.046
  2.039 $\rightarrow 3 < 4 < 6$
  2.064

So, $2.039 < 2.046 < 2.064$.

PRACTICE Use $<$, $>$, or $=$ to compare.

1. 7.456 ● 7.476
2. 19.072 ● 19.100
3. 0.029 ● 0.029
4. 8.103 ● 8.099

Write the numbers in order from least to greatest. Use $<$.

5. 2.15, 2.95, 2.45
6. 4.175, 4.173, 4.17
7. 0.043, 0.03, 0.072
8. 7.29, 7.07, 7.33

20 MODELING DECIMALS

Decimals can be modeled with decimal squares.

Examples

A. Model 0.7 with a decimal square.

Shade 7 of 10 columns since 0.7 is seven tenths.

B. Model 1.2 with decimal squares.

Shade one entire decimal square and 2 of 10 columns since 1.2 is one and two tenths.

C. Model 0.24 with a decimal square.

Shade 24 of 100 squares since 0.24 is twenty-four hundredths.

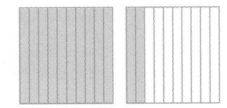

PRACTICE Model with decimal squares.

1. 0.4
2. 0.9
3. 0.1
4. 1.7

5. 1.3
6. 0.94
7. 0.47
8. 0.02

21 ADDING AND SUBTRACTING DECIMALS

When adding and subtracting decimals, you must remember to line up the decimal points vertically. You may add zeros to the right of the decimal point as place holders. Adding zeros to the right of the last digit after the decimal point doesn't change the value of the number.

Examples

A. Find the sum. $3.54 + 1.7 + 22 + 13.409$

$$
\begin{array}{r}
3.540 \\
1.700 \\
22.000 \\
+ \ 13.409 \\
\hline
40.649
\end{array}
$$

B. Find the difference. $636.2 - 28.538$

$$
\begin{array}{r}
636.200 \\
- \ 28.538 \\
\hline
607.662
\end{array}
$$

PRACTICE   Find the sum or difference.

1. $0.687 + 0.9 + 27.25$

2. $87.34 - 6.8$

3. $65 + 0.0004 + 2.57$

4. $17 - 0.095$

5. $263.7 - 102.08$

6. $27 + 3.24 + 0.256 + 0.3689$

7. $32.5 + 0.81 + 2.956$

8. $182 - 0.475$

9. $92.7 - 8.93$

22 MULTIPLYING AND DIVIDING DECIMALS BY POWERS OF 10

Notice the pattern below.

$$
\begin{array}{ll}
0.24 \times 10 = 2.4 & 10 = 10^1 \\
0.24 \times 100 = 24 & 100 = 10^2 \\
0.24 \times 1,000 = 240 & 1,000 = 10^3 \\
0.24 \times 10,000 = 2,400 & 10,000 = 10^4
\end{array}
$$

Think: *When multiplying decimals by powers of 10, move the decimal point one place to the* **right** *for each power of 10, or for each zero.*

Notice the pattern below.

$$
\begin{array}{l}
0.24 \div 10 = 0.024 \\
0.24 \div 100 = 0.0024 \\
0.24 \div 1,000 = 0.00024 \\
0.24 \div 10,000 = 0.000024
\end{array}
$$

Think: *When dividing decimals by powers of 10, move the decimal point one place to the* **left** *for each power of 10, or for each zero.*

PRACTICE   Find the product or quotient.

1. 10×9.26

2. 0.642×100

3. $10^3 \times 84.2$

4. 0.44×10^4

5. $69.7 \times 1,000$

6. $11.32 \div 10$

7. $1.276 \div 1,000$

8. $536.5 \div 10^2$

9. $5.92 \div 10^3$

10. $25 \div 10,000$

11. $2.8 \div 10^2$

12. $143.9 \div 10^3$

13. 17.4×10^2

14. $10^3 \times 0.825$

15. $10^4 \times 0.91$

23   MULTIPLYING DECIMALS

When multiplying decimals, multiply as you would with whole numbers. The sum of the decimal places in the factors equals the number of decimal places in the product.

Examples   Find the product.

A. 81.2×6.547

$$
\begin{array}{r}
6.547 \leftarrow \textit{3 decimal places} \\
\times\ 81.2 \leftarrow \textit{1 decimal place} \\
\hline
13094 \\
65470 \\
+\ 5237600 \\
\hline
531.6164 \leftarrow \textit{4 decimal places}
\end{array}
$$

B. 0.376×0.12

$$
\begin{array}{r}
0.376 \leftarrow \textit{3 decimal places} \\
\times\ 0.12 \leftarrow \textit{2 decimal places} \\
\hline
752 \\
+\ 3760 \\
\hline
0.04512 \leftarrow \textit{5 decimal places}
\end{array}
$$

PRACTICE   Find the product.

1. 6.8×3.4 2. 2.56×4.6 3. 6.787×7.6 4. 0.98×4.6 5. 0.97×0.76

6. 0.5×3.761 7. 42×17.654 8. 7.005×32.1 9. 9.76×16.254 10. 296.5×2.4

11. 32.4×7.1 12. 8.2×0.93 13. 192.1×8.5 14. 52.93×44.8 15. 0.85×0.2

24   DIVIDING DECIMALS

When dividing with decimals, set up the division as you would with whole numbers. Pay attention to the decimal places, as shown below.

Examples

A. Find the quotient. $89.6 \div 16$

Place decimal point.
$$\downarrow$$
$$
\begin{array}{r}
5.6 \\
16)\overline{89.6} \\
-\ 80 \\
\hline
96 \\
-\ 96 \\
\hline
0
\end{array}
$$

B. Find the quotient. $3.4 \div 4$

Place decimal point.
$$\downarrow$$
$$
\begin{array}{r}
0.85 \\
4)\overline{3.40} \leftarrow \textit{Insert zeros if necessary.} \\
-32 \\
\hline
20 \\
-20 \\
\hline
0
\end{array}
$$

PRACTICE   Find the quotient.

1. $242.76 \div 68$ 2. $40.5 \div 18$ 3. $121.03 \div 98$ 4. $3.6 \div 4$ 5. $1.58 \div 5$

6. $0.2835 \div 2.7$ 7. $8.1 \div 0.09$ 8. $0.42 \div 0.28$ 9. $15.12 \div 0.063$ 10. $480.48 \div 7.7$

11. $7.99 \div 0.4$ 12. $42.4 \div 0.2$ 13. $812.5 \div 5$ 14. $65.75 \div 0.25$ 15. $489.09 \div 0.05$

25 UNDERSTANDING FRACTIONS

You can use a fraction to describe part of a whole. You can also use a fraction to describe part of a group.

Examples   Write the fraction for the part that is shaded. Tell whether it represents part of a whole or part of a group.

A. 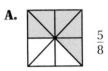 $\frac{5}{8}$

B. ◯ ◯ ◯ ◯ ◯ $\frac{7}{10}$
◯ ◯ ◯ ◯ ◯

Since 5 of 8 parts of one square are shaded, the fraction represents part of a whole.

Since 7 of 10 parts of a group of circles are shaded, the fraction represents part of a group.

PRACTICE   Write the fraction for the part that is shaded. Tell whether it represents part of a whole or part of a group.

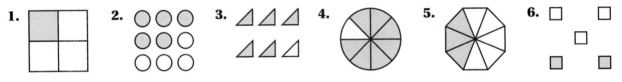

26 EQUIVALENT FRACTIONS

Equivalent fractions are fractions that name the same amount.

Examples

A. Write two equivalent fractions for $\frac{15}{30}$.

Method 1 *Multiply both the numerator and denominator by a whole number.*

$$\frac{15 \times 2}{30 \times 2} = \frac{30}{60}$$

Method 2 *Divide by a common factor of the numerator and denominator.*

$$\frac{15 \div 15}{30 \div 15} = \frac{1}{2}$$

The fractions $\frac{30}{60}$ and $\frac{1}{2}$ are equivalent to $\frac{15}{30}$.

B. Are $\frac{4}{6}$ and $\frac{12}{18}$ equivalent?

Method 1 *Write the fractions in simplest form and compare.*

$$\frac{4 \div 2}{6 \div 2} = \frac{2}{3} \qquad \frac{12 \div 6}{18 \div 6} = \frac{2}{3}$$

The fractions are equivalent.

Method 2 *Cross multiply and compare.*

$$4 \times 18 = 72 \qquad \frac{4}{6} = \frac{12}{18} \qquad 12 \times 6 = 72$$

The fractions are equivalent since the cross products both equal 72.

PRACTICE   Tell whether the fractions are equivalent. Write *yes* or *no*. Then, write two equivalent fractions for each.

1. $\frac{15}{30}, \frac{1}{2}$ **2.** $\frac{3}{18}, \frac{1}{3}$ **3.** $\frac{3}{4}, \frac{9}{12}$ **4.** $\frac{1}{2}, \frac{5}{10}$ **5.** $\frac{4}{16}, \frac{12}{18}$ **6.** $\frac{12}{21}, \frac{4}{7}$

27 SIMPLEST FORM OF FRACTIONS

A fraction is in **simplest form** when the numerator and denominator have no common factor other than 1.

Example Find the simplest form of $\frac{32}{40}$.

Method 1 *Divide the numerator and denominator by common factors until the only common factor is 1.*

$$\frac{32 \div 2}{40 \div 2} = \frac{16}{20} \qquad \frac{16 \div 4}{20 \div 4} = \frac{4}{5}$$

The simplest form of $\frac{32}{40}$ is $\frac{4}{5}$.

Method 2 *Find the GCF of 32 and 40. Divide both the numerator and the denominator by the GCF.*

$$\frac{32 \div 8}{40 \div 8} = \frac{4}{5} \leftarrow \text{GCF: 8}$$

The simplest form of $\frac{32}{40}$ is $\frac{4}{5}$.

PRACTICE Write in simplest form.

1. $\frac{20}{24}$ **2.** $\frac{4}{12}$ **3.** $\frac{14}{49}$ **4.** $\frac{60}{72}$ **5.** $\frac{40}{75}$ **6.** $\frac{12}{12}$

7. $\frac{18}{24}$ **8.** $\frac{5}{10}$ **9.** $\frac{15}{45}$ **10.** $\frac{17}{51}$ **11.** $\frac{6}{32}$ **12.** $\frac{26}{39}$

13. $\frac{12}{20}$ **14.** $\frac{32}{48}$ **15.** $\frac{8}{12}$ **16.** $\frac{34}{40}$ **17.** $\frac{14}{35}$ **18.** $\frac{27}{51}$

28 COMPARING FRACTIONS

The symbols $<, \leq, =, \geq, >, \neq$ can be used to compare fractions. To find how fractions compare, rename the fractions so they have a common denominator. Then compare the numerators.

Examples Use $<, >,$ or $=$ to compare.

A. $\frac{3}{4} \bullet \frac{4}{5}$ *Rename the fractions.*

$\frac{15}{20} \quad \frac{16}{20} \qquad 15 < 16$

So, $\frac{3}{4} < \frac{4}{5}$.

B. $\frac{2}{3} \bullet \frac{5}{8}$ *Rename the fractions.*

$\frac{16}{24} \quad \frac{15}{24} \qquad 16 > 15$

So, $\frac{2}{3} > \frac{5}{8}$.

C. $\frac{14}{21} \bullet \frac{4}{6}$ *Rename the fractions.*

$\frac{2}{3} \quad \frac{2}{3} \qquad 2 = 2$

So, $\frac{14}{21} = \frac{4}{6}$.

PRACTICE Use $<, >,$ or $=$ to compare.

1. $\frac{2}{3} \bullet \frac{3}{4}$ **2.** $\frac{7}{8} \bullet \frac{5}{6}$ **3.** $\frac{1}{4} \bullet \frac{1}{8}$ **4.** $\frac{2}{4} \bullet \frac{3}{4}$ **5.** $\frac{5}{6} \bullet \frac{3}{4}$ **6.** $\frac{1}{2} \bullet \frac{3}{6}$

7. $\frac{39}{40} \bullet \frac{7}{8}$ **8.** $\frac{1}{3} \bullet \frac{15}{45}$ **9.** $\frac{4}{5} \bullet \frac{2}{3}$ **10.** $\frac{24}{48} \bullet \frac{27}{36}$ **11.** $\frac{4}{6} \bullet \frac{7}{14}$ **12.** $\frac{9}{27} \bullet \frac{11}{33}$

13. $\frac{1}{9} \bullet \frac{2}{7}$ **14.** $\frac{4}{5} \bullet \frac{3}{4}$ **15.** $\frac{12}{17} \bullet \frac{2}{3}$ **16.** $\frac{21}{30} \bullet \frac{3}{5}$ **17.** $\frac{4}{9} \bullet \frac{36}{81}$ **18.** $\frac{4}{23} \bullet \frac{1}{4}$

29 ORDERING FRACTIONS

To order fractions, rename the fractions so they have a common denominator. Then compare the numerators.

Example Write the fractions $\frac{5}{6}$, $\frac{1}{2}$, and $\frac{3}{4}$ in order from least to greatest. Use $<$.

STEP 1 *Rename the fractions.*

$$\frac{5}{6} \qquad \frac{1}{2} \qquad \frac{3}{4}$$
$$\downarrow \qquad \downarrow \qquad \downarrow$$
$$\frac{10}{12} \qquad \frac{6}{12} \qquad \frac{9}{12}$$

STEP 2 *Compare the numerators.*

$$6 < 9 < 10$$

STEP 3 *Order the fractions.*

$$\frac{6}{12} < \frac{9}{12} < \frac{10}{12}$$
$$\text{So, } \frac{1}{2} < \frac{3}{4} < \frac{5}{6}.$$

PRACTICE Write the fractions in order from least to greatest. Use $<$.

1. $\frac{4}{5}, \frac{7}{10}, \frac{3}{5}$

2. $\frac{7}{8}, \frac{1}{2}, \frac{2}{3}$

3. $\frac{1}{2}, \frac{1}{3}, \frac{3}{4}$

4. $\frac{5}{6}, \frac{1}{2}, \frac{3}{4}$

5. $\frac{4}{6}, \frac{7}{14}, \frac{5}{7}$

6. $\frac{4}{9}, \frac{2}{3}, \frac{3}{4}$

7. $\frac{8}{9}, \frac{12}{17}, \frac{1}{3}$

8. $\frac{1}{4}, \frac{2}{3}, \frac{5}{9}$

9. $\frac{3}{7}, \frac{5}{8}, \frac{4}{5}$

10. $\frac{1}{10}, \frac{1}{7}, \frac{2}{15}$

11. $\frac{1}{6}, \frac{5}{8}, \frac{2}{11}$

12. $\frac{4}{9}, \frac{1}{3}, \frac{5}{18}$

30 MIXED NUMBERS AND FRACTIONS

Mixed numbers can be written as fractions greater than 1, and fractions greater than 1 can be written as mixed numbers.

Examples

A. Write $\frac{23}{5}$ as a mixed number.

$\frac{23}{5} \rightarrow$ *Divide the numerator by the denominator.*

$\begin{array}{r} 4 \\ 5\overline{)23} \\ -20 \\ \hline 3 \end{array} \rightarrow 4\frac{3}{5} \leftarrow$ *Write the remainder as the numerator of a fraction.*

B. Write $6\frac{2}{7}$ as a fraction.

Multiply the denominator by the whole number.   *Add the product to the numerator.*

$$6\frac{2}{7} \rightarrow 7 \times 6 = 42 \rightarrow 42 + 2 = 44$$

Write the sum over the denominator. $\rightarrow \frac{44}{7}$

PRACTICE Write each mixed number as a fraction. Write each fraction as a mixed number.

1. $\frac{22}{5}$

2. $9\frac{1}{7}$

3. $\frac{41}{8}$

4. $5\frac{7}{9}$

5. $\frac{7}{3}$

6. $4\frac{9}{11}$

7. $\frac{47}{16}$

8. $3\frac{3}{8}$

9. $\frac{31}{9}$

10. $8\frac{2}{3}$

11. $\frac{33}{5}$

12. $12\frac{1}{9}$

31 ADDING LIKE FRACTIONS

When adding like fractions, add the numerators. Simplify if necessary.

Example Add. Write the sum in simplest form. $\frac{5}{12} + \frac{3}{12}$

Since the denominators are the same, add the numerators.

$$\frac{5}{12} + \frac{3}{12} = \frac{5+3}{12} = \frac{8}{12}$$

Write in simplest form.

$$\frac{8}{12} = \frac{8 \div 4}{12 \div 4} = \frac{2}{3}$$

PRACTICE Add. Write the sum in simplest form.

1. $\frac{1}{4} + \frac{2}{4}$ 2. $\frac{3}{10} + \frac{2}{10}$ 3. $\frac{12}{21} + \frac{2}{21}$ 4. $\frac{3}{8} + \frac{1}{8}$ 5. $\frac{2}{5} + \frac{3}{5}$ 6. $\frac{2}{10} + \frac{4}{10}$

7. $\frac{4}{15} + \frac{6}{15}$ 8. $\frac{5}{9} + \frac{2}{9}$ 9. $\frac{7}{12} + \frac{2}{12}$ 10. $\frac{7}{10} + \frac{1}{10}$ 11. $\frac{15}{24} + \frac{5}{24}$ 12. $\frac{3}{48} + \frac{13}{48}$

13. $\frac{2}{7} + \frac{4}{7}$ 14. $\frac{6}{8} + \frac{2}{8}$ 15. $\frac{8}{20} + \frac{5}{20}$ 16. $\frac{5}{16} + \frac{4}{16}$ 17. $\frac{12}{14} + \frac{1}{14}$ 18. $\frac{10}{28} + \frac{10}{28}$

19. $\frac{5}{8} + \frac{1}{8}$ 20. $\frac{1}{3} + \frac{1}{3}$ 21. $\frac{2}{9} + \frac{5}{9}$ 22. $\frac{5}{16} + \frac{7}{16}$ 23. $\frac{8}{20} + \frac{4}{20}$ 24. $\frac{3}{7} + \frac{2}{7}$

32 SUBTRACTING LIKE FRACTIONS

When subtracting like fractions, subtract the numerators. Simplify if necessary.

Example Subtract. Write the sum in simplest form. $\frac{5}{12} - \frac{3}{12}$

Since the denominators are the same, subtract the numerators.

$$\frac{5}{12} - \frac{3}{12} = \frac{5-3}{12} = \frac{2}{12}$$

Write in simplest form.

$$\frac{2}{12} = \frac{2 \div 2}{12 \div 2} = \frac{1}{6}$$

PRACTICE Subtract. Write the difference in simplest form.

1. $\frac{9}{10} - \frac{8}{10}$ 2. $\frac{19}{20} - \frac{3}{20}$ 3. $\frac{11}{12} - \frac{8}{12}$ 4. $\frac{5}{8} - \frac{1}{8}$ 5. $\frac{3}{5} - \frac{1}{5}$ 6. $\frac{7}{9} - \frac{1}{9}$

7. $\frac{17}{18} - \frac{11}{18}$ 8. $\frac{7}{8} - \frac{5}{8}$ 9. $\frac{13}{16} - \frac{1}{16}$ 10. $\frac{13}{14} - \frac{3}{14}$ 11. $\frac{7}{16} - \frac{5}{16}$ 12. $\frac{19}{21} - \frac{5}{21}$

13. $\frac{8}{9} - \frac{4}{9}$ 14. $\frac{12}{15} - \frac{5}{15}$ 15. $\frac{14}{24} - \frac{7}{24}$ 16. $\frac{5}{16} - \frac{4}{16}$ 17. $\frac{2}{4} - \frac{1}{4}$ 18. $\frac{1}{12} - \frac{1}{12}$

19. $\frac{11}{12} - \frac{5}{12}$ 20. $\frac{13}{18} - \frac{7}{18}$ 21. $\frac{9}{25} - \frac{4}{25}$ 22. $\frac{11}{15} - \frac{7}{15}$ 23. $\frac{18}{30} - \frac{11}{30}$ 24. $\frac{9}{19} - \frac{3}{19}$

33 RENAMING MIXED NUMBERS

A mixed number is the sum of a whole number and a fraction. Sometimes you need to rename a mixed number in order to add or subtract.

Examples

A. Rename $5\frac{5}{4}$ so the fraction is not greater than 1.

$\frac{5}{4} = 1\frac{1}{4}$ *Rename the fraction.*

$5\frac{5}{4} = 5 + \frac{5}{4} = 5 + 1\frac{1}{4} = 6\frac{1}{4}$ *Rename the mixed number.*

So, $5\frac{5}{4} = 6\frac{1}{4}$.

B. Rename $3\frac{1}{4}$ so the whole number is 2.

$3 = 2 + \frac{4}{4}$ *Rename 3 using 2 and a fraction.*

$3\frac{1}{4} = 2 + \frac{4}{4} + \frac{1}{4} = 2\frac{5}{4}$ *Rename the mixed number.*

So, $3\frac{1}{4} = 2\frac{5}{4}$.

PRACTICE Rename the mixed number so the fraction is not greater than 1.

1. $3\frac{6}{5}$ **2.** $5\frac{7}{4}$ **3.** $8\frac{4}{3}$ **4.** $1\frac{13}{8}$ **5.** $4\frac{10}{7}$ **6.** $9\frac{3}{2}$

Rename the mixed number so the whole number is 1 less.

7. $4\frac{1}{2}$ **8.** $6\frac{2}{5}$ **9.** $3\frac{2}{7}$ **10.** $8\frac{1}{4}$ **11.** $2\frac{5}{8}$ **12.** $7\frac{6}{11}$

34 ADDING MIXED NUMBERS

To add mixed numbers, add the fractions and the whole numbers. If the fraction in the sum is greater than 1, rename it.

Example Add. $7\frac{4}{5} + 1\frac{3}{5}$

Add the fractions.

$\begin{array}{r} 7\frac{4}{5} \\ + 1\frac{3}{5} \\ \hline \frac{7}{5} \end{array}$

Add the whole numbers.

$\begin{array}{r} 7\frac{4}{5} \\ + 1\frac{3}{5} \\ \hline 8\frac{7}{5} \end{array}$

Rename so the fraction is less than 1.

$8\frac{7}{5} = 8 + \frac{7}{5} = 8 + 1\frac{2}{5} = 9\frac{2}{5}$

PRACTICE Add. Rename the sum if the fraction in the sum is greater than 1.

1. $2\frac{1}{3} + 1\frac{1}{3}$ **2.** $7\frac{3}{5} + 2\frac{1}{5}$ **3.** $3\frac{3}{7} + 5\frac{6}{7}$ **4.** $4\frac{5}{8} + 4\frac{4}{8}$ **5.** $9\frac{7}{12} + 5\frac{9}{12}$

6. $4\frac{1}{8} + 2\frac{3}{8}$ **7.** $8\frac{5}{9} + 2\frac{4}{9}$ **8.** $7\frac{2}{3} + 3\frac{2}{3}$ **9.** $4\frac{9}{10} + 1\frac{7}{10}$ **10.** $1\frac{8}{12} + 1\frac{5}{12}$

35 SUBTRACTING MIXED NUMBERS

To subtract mixed numbers, subtract the fractions first. Borrow from the ones if necessary. Then subtract the whole numbers.

Example Subtract. $6\frac{1}{5} - 2\frac{3}{5}$

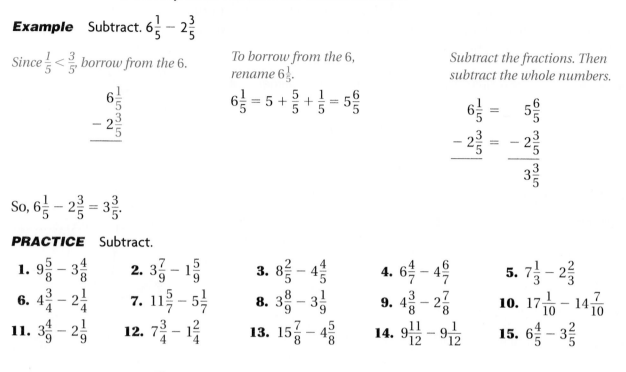

Since $\frac{1}{5} < \frac{3}{5}$, borrow from the 6.

$$6\frac{1}{5}$$
$$-\,2\frac{3}{5}$$

To borrow from the 6, rename $6\frac{1}{5}$.

$$6\frac{1}{5} = 5 + \frac{5}{5} + \frac{1}{5} = 5\frac{6}{5}$$

Subtract the fractions. Then subtract the whole numbers.

$$6\frac{1}{5} = \qquad 5\frac{6}{5}$$
$$-\,2\frac{3}{5} = \quad -\,2\frac{3}{5}$$
$$\overline{\qquad\qquad 3\frac{3}{5}}$$

So, $6\frac{1}{5} - 2\frac{3}{5} = 3\frac{3}{5}$.

PRACTICE Subtract.

1. $9\frac{5}{8} - 3\frac{4}{8}$ **2.** $3\frac{7}{9} - 1\frac{5}{9}$ **3.** $8\frac{2}{5} - 4\frac{4}{5}$ **4.** $6\frac{4}{7} - 4\frac{6}{7}$ **5.** $7\frac{1}{3} - 2\frac{2}{3}$

6. $4\frac{3}{4} - 2\frac{1}{4}$ **7.** $11\frac{5}{7} - 5\frac{1}{7}$ **8.** $3\frac{8}{9} - 3\frac{1}{9}$ **9.** $4\frac{3}{8} - 2\frac{7}{8}$ **10.** $17\frac{1}{10} - 14\frac{7}{10}$

11. $3\frac{4}{9} - 2\frac{1}{9}$ **12.** $7\frac{3}{4} - 1\frac{2}{4}$ **13.** $15\frac{7}{8} - 4\frac{5}{8}$ **14.** $9\frac{11}{12} - 9\frac{1}{12}$ **15.** $6\frac{4}{5} - 3\frac{2}{5}$

36 WRITING A FRACTION AS A DECIMAL

A fraction can be thought of as a division problem. The fraction $\frac{3}{4}$ can be read as "3 divided by 4." To write a fraction as a decimal, you can divide the numerator by the denominator.

Example Write $\frac{3}{4}$ as a decimal.

$$\begin{array}{r} 0.75 \\ 4\overline{)3.00} \\ -2\,8 \\ \hline 20 \\ -20 \\ \hline 0 \end{array}$$

Think of $\frac{3}{4}$ as a division problem. Divide the numerator, 3, by the denominator, 4.

So, $\frac{3}{4} = 0.75$.

PRACTICE Write the fraction as a decimal.

1. $\frac{1}{2}$ **2.** $\frac{1}{4}$ **3.** $\frac{3}{5}$ **4.** $\frac{9}{20}$ **5.** $\frac{1}{5}$ **6.** $\frac{7}{8}$

7. $\frac{3}{4}$ **8.** $\frac{1}{8}$ **9.** $\frac{8}{9}$ **10.** $\frac{3}{8}$ **11.** $\frac{1}{10}$ **12.** $\frac{7}{10}$

13. $\frac{9}{10}$ **14.** $\frac{4}{5}$ **15.** $\frac{1}{20}$ **16.** $\frac{2}{5}$ **17.** $\frac{3}{25}$ **18.** $\frac{15}{20}$

37 PARTS OF A CIRCLE

A circle is the set of all points a given distance from a point called the **center**. A **radius** is a line segment with one endpoint at the center of a circle and one endpoint on the circle. A **diameter** is a line segment that passes through the center of the circle and has both endpoints on the circle. A circle is named by its center.

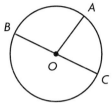

Example Name the circle, the center, a radius, and a diameter of the circle.

name: circle O radius: $\overline{OA}$, $\overline{OB}$, or $\overline{OC}$

center: O diameter: $\overline{BC}$

PRACTICE Name each.

1. the circle   **2.** the center

3. two radii   **4.** two diameters.

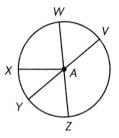

38 TRANSFORMATIONS

Three types of transformations are shown below.

 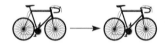

A **translation** occurs when a figure slides along a straight line.

A **reflection** occurs when a figure is flipped over a line.

A **rotation** occurs when a figure is turned around a point.

PRACTICE Identify the transformation as a *translation, reflection,* or *rotation*.

1. 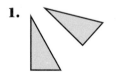   **2.** 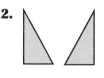   **3.**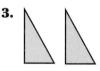

4.    **5.**   **6.**

39 TYPES OF POLYGONS

A **polygon** is a closed plane figure with at least three sides.
Polygons are classified by number of sides and angles.

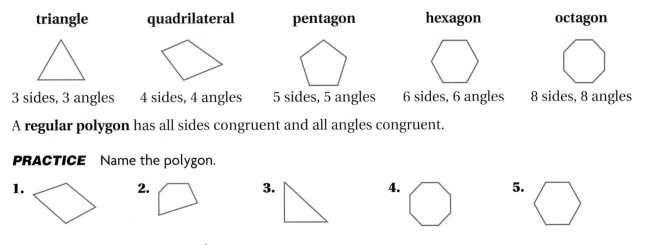

triangle	quadrilateral	pentagon	hexagon	octagon
3 sides, 3 angles	4 sides, 4 angles	5 sides, 5 angles	6 sides, 6 angles	8 sides, 8 angles

A **regular polygon** has all sides congruent and all angles congruent.

PRACTICE Name the polygon.

1.   **2.**   **3.**   **4.**   **5.**

40 TYPES OF TRIANGLES

Triangles are classified by the lengths of their sides and the
measures of their angles.

acute triangle

70°
60° 50°

all angles < 90°

obtuse triangle

20°
50°
110°

one angle > 90°

right triangle

60°
90° 30°

one angle = 90°

scalene triangle

6 m   5 m
7 m

All sides have
different lengths.

equilateral triangle

4 cm   4 cm
4 cm

All sides have the
same length.

isosceles triangle

8 ft   8 ft
5 ft

Two sides have
the same length.

PRACTICE Classify the triangles by using the given information.

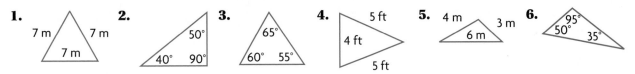

1. 7 m / 7 m / 7 m **2.** 50° / 40° 90° **3.** 65° / 60° 55° **4.** 5 ft / 4 ft / 5 ft **5.** 4 m / 3 m / 6 m **6.** 95° / 50° 35°

41 TYPES OF QUADRILATERALS

A **quadrilateral** is a polygon with 4 sides and 4 angles. There are many different kinds of quadrilaterals with special properties. Some are shown below.

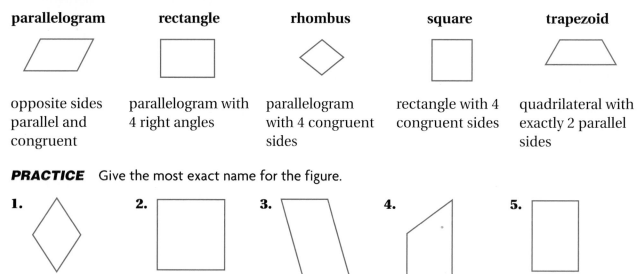

parallelogram	rectangle	rhombus	square	trapezoid
opposite sides parallel and congruent	parallelogram with 4 right angles	parallelogram with 4 congruent sides	rectangle with 4 congruent sides	quadrilateral with exactly 2 parallel sides

PRACTICE   Give the most exact name for the figure.

1.   2.   3.   4.   5.

42 SOLID FIGURES

Five basic types of solid figures are shown below.

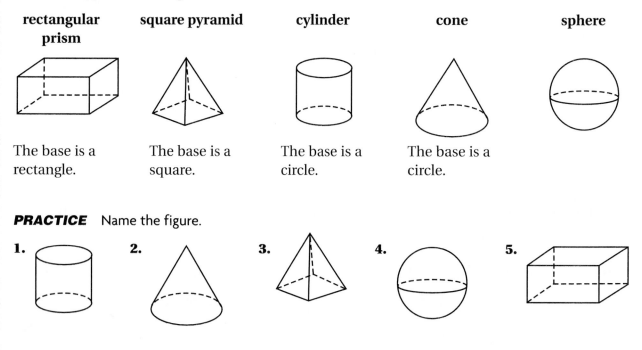

rectangular prism	square pyramid	cylinder	cone	sphere
The base is a rectangle.	The base is a square.	The base is a circle.	The base is a circle.	

PRACTICE   Name the figure.

1.   2.   3.   4.   5.

43 FACES, EDGES, AND VERTICES

Solid figures made from polygons have faces, edges, and vertices.
Each polygon is a **face**. The faces meet to form line segments
called **edges**. The edges meet to form points called **vertices**.

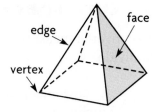

Examples Tell the number of faces, edges, and vertices for the figure.

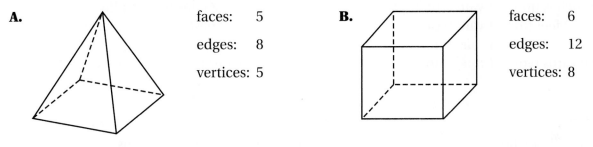

A.

faces: 5

edges: 8

vertices: 5

B.

faces: 6

edges: 12

vertices: 8

PRACTICE Tell the number of faces, edges, and vertices for the figure.

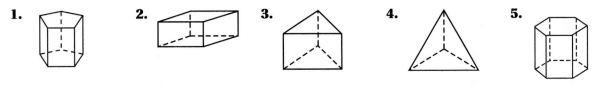

1.   **2.**   **3.**   **4.**   **5.**

44 CONGRUENT FIGURES

Congruent figures are figures that have the same size and shape.

Examples Tell if the figures appear to be congruent or not.

A.

The figures are congruent since they are
the same size and shape.

B.

The figures are not congruent since
they are not the same size.

PRACTICE Tell if the figures appear to be congruent or not. Write *yes* or *no*.

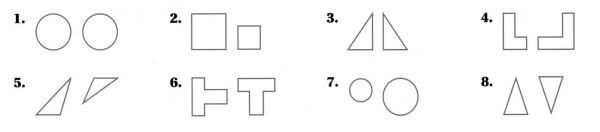

1.   **2.**   **3.**   **4.**

5.   **6.**   **7.**   **8.**

45 CUSTOMARY MEASURES

To convert a measurement from one unit to another, you multiply
or divide.

Customary Units of Length

12 inches (in.) = 1 foot (ft)
36 inches (in.) = 1 yard (yd)
3 feet (ft) = 1 yard (yd)
5,280 feet (ft) = 1 mile (mi)
1,760 yards (yd) = 1 mile (mi)

Customary Units of Capacity and Weight

8 fluid ounces (fl oz) = 1 cup (c)
2 cups (c) = 1 pint (pt)
2 pints (pt) = 1 quart (qt)
4 quarts (qt) = 1 gallon (gal)
16 ounces (oz) = 1 pound (lb)
2,000 pounds (lb) = 1 ton (T)

Examples

A. 4 feet = _?_ inches

$$4 \text{ ft} \times 12 \text{ in. per ft} = 48 \text{ in.}$$

B. 78 fluid ounces = _?_ cups

$$78 \text{ fl oz} \div 8 \text{ fl oz per c} = 9\frac{3}{4} \text{ c}$$

PRACTICE   Convert the measure to the given unit.

1. 3 ft = _?_ in.   **2.** 3 mi = _?_ ft   **3.** 72 ft = _?_ yd   **4.** 18 yd = _?_ ft

5. 55 gal = _?_ qt   **6.** 19 c = _?_ fl oz   **7.** 54 qt = _?_ gal   **8.** 64 oz = _?_ lb

46 METRIC MEASURES

The metric system is based on the decimal system. When you move
from left to right on the chart below, you multiply by powers of 10.
When you move from right to left, you divide by powers of 10.

kilometer	hectometer	dekameter	meter	decimeter	centimeter	millimeter
1 km	1 hm	1 dam	1 m	1 dm	1 cm	1 mm
=	=	=	=	=	=	=
1,000 m	100 m	10 m	1 m	0.1 m	0.01 m	0.001 m

Examples

A. 4.5 m = _?_ mm

Think: *You move from left to right 3 places, so multiply by 1,000 or move the decimal point 3 places to the right.*

$$4.5 \text{ m} = 4,500 \text{ mm}$$

B. 4.5 m = _?_ hm

Think: *You move from right to left 2 places, so divide by 100 or move the decimal point 2 places to the left.*

$$4.5 \text{ m} = 0.045 \text{ hm}$$

PRACTICE   Convert the measure to the given unit.

1. 4.5 m = _?_ cm   **2.** 7.9 m = _?_ km   **3.** 0.09 dm = _?_ m   **4.** 0.15 m = _?_ hm

5. 12 m = _?_ mm   **6.** 86 dam = _?_ m   **7.** 0.34 km = _?_ m   **8.** 480 cm = _?_ m

47 PERIMETER

Perimeter is the distance around a figure.

Examples Find the perimeter of the figure.

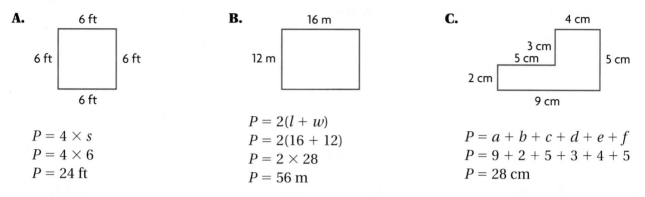

A.
6 ft
6 ft 6 ft
6 ft

$P = 4 \times s$
$P = 4 \times 6$
$P = 24$ ft

B.
16 m
12 m

$P = 2(l + w)$
$P = 2(16 + 12)$
$P = 2 \times 28$
$P = 56$ m

C.
4 cm
3 cm
5 cm
2 cm
9 cm
5 cm

$P = a + b + c + d + e + f$
$P = 9 + 2 + 5 + 3 + 4 + 5$
$P = 28$ cm

PRACTICE Find the perimeter of the figure.

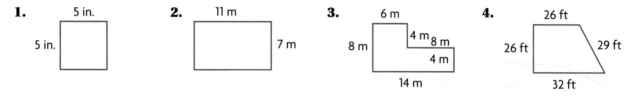

1.
5 in.
5 in.

2.
11 m
7 m

3.
6 m
4 m
8 m
8 m
4 m
14 m

4.
26 ft
26 ft 29 ft
32 ft

48 AREA

To find the area of a figure, you can count the number of square units covered by the figure.

Examples Find the area of the shaded region.

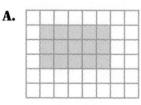

A.

15 full squares

$A = 15$ units^2

B.

12 full squares

6 half squares

$A =$ no. of full squares $+ \frac{1}{2}$(no. of half squares)
$A = 12 + \frac{1}{2}(6)$
$A = 12 + 3 = 15$ 15 units^2

PRACTICE Find the area of the shaded region.

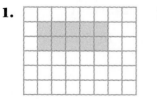

1.

2.

3.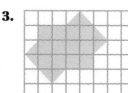

4.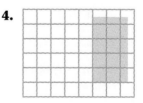

49 MEAN

The **mean** is the average of a set of numbers. To find the mean, add the numbers and divide the sum by the number of addends.

Examples   Find the mean of the set of numbers. 36, 74, 43, 36, 41

Find the sum of the numbers.   *Divide the sum by the number of addends.*

$36 + 74 + 43 + 36 + 41 = 230$ $230 \div 5 = 46$

So, the mean is 46.

PRACTICE   Find the mean of the set of numbers.

1. 32, 87, 45, 63, 73   **2.** 17, 44, 33, 10   **3.** 6, 52, 41, 21, 36, 48   **4.** 126, 99, 234.3

5. 17, 98, 24, 43, 68   **6.** 90, 88, 87, 69, 75   **7.** 138, 145, 99, 84   **8.** 22, 49, 33, 19, 24, 9

50 MEDIAN AND MODE

The **median** is the middle number of a set of numbers in numerical order. If there are two middle numbers, the median is the mean of those two numbers. The **mode** of a set of numbers is the most commonly occurring number in the set. There may be more than one mode, or there may be no mode at all.

Examples   Find the median and mode of the set of numbers.

A. 4, 25, 72, 36, 41, 2, 98

Median

Put the numbers in numerical order.

2, 4, 25, 36, 41, 72, 98

So, the median is 36.

Mode
Since no number in the set occurs more than once, there is no mode.

B. 4, 25, 72, 4, 36, 4, 2, 25, 25, 98

Median

Put the numbers in numerical order.

2, 4, 4, 4, 25, 25, 25, 36, 72, 98

$\frac{25 + 25}{2} = \frac{50}{2} = 25$ So, the median is 25.

Mode
Since both 4 and 25 occur three times, 4 and 25 are the modes.

PRACTICE   Find the median and the mode of the set of numbers.

1. 2, 1, 2, 5, 7, 9   **2.** 42, 8, 54, 192, 8, 0, 44, 16   **3.** 12, 44, 324, 17, 41

4. 0, 77, 125, 77, 2, 2, 3, 5, 3   **5.** 5, 0, 15, 25, 5, 23, 1   **6.** 297, 7, 12, 18, 5, 21, 17

7. 27, 44, 52, 27, 69, 44   **8.** 10, 82, 41, 37, 9, 16, 7   **9.** 112, 73, 222, 85, 73

51 RANGE

The **range** is the distance between extremes in a set of numbers. You can find the range by finding the difference between the greatest number and the least number. You can find the size of intervals for the data by dividing the range by the number of intervals needed.

Example Find the range of the set of numbers. Divide the range into 5 intervals for the data.

range: $44 - 2 = 42$

So, the range is 42.

2	8	12	15
15	20	25	30
35	40	42	44

5 intervals: $42 \div 5 \approx 9$

So, use 5 intervals of 9 or 10 values. Intervals of 10 are:
0–9, 10–19, 20–29, 30–39, 40–49.

PRACTICE Find the range of the set of numbers. Divide the range into 5 intervals for the data.

1. 8, 24, 32, 44, 26, 47, 17, 14, 19

2. 115, 42, 120, 99, 80, 56, 79, 84

3. 77, 72, 70, 88, 94, 99, 83, 78, 99

4. 346, 250, 225, 194, 355, 290, 323

5. 172, 141, 138, 119, 126, 193, 210

6. 49, 71, 210, 82, 185, 99, 153

7. 85, 89, 97, 51, 67, 112, 78

8. 273, 189, 141, 242, 156, 268, 263

52 RATIOS

A **ratio** is a comparison of two numbers. A ratio, like a fraction, can be written in simplest form. There are three ways to write a ratio.

Example In a group of sixth graders, there are 20 boys and 25 girls. Write the ratio of boys to girls in three different ways. Write the ratios in simplest form.

$\frac{20}{25}$, or $\frac{4}{5}$ 　　　　　　　20:25, or 4:5 　　　　　　20 to 25, or 4 to 5

The ratio in simplest form is read "4 to 5," no matter how it is written. For every 4 boys, there are 5 girls.

PRACTICE In a bag of marbles, 7 are green, 3 are red, 10 are blue, 14 are yellow, and 2 are orange. Write the ratio in simplest form in three different ways.

1. green to red 　　　**2.** blue to yellow 　　　**3.** orange to green 　　　**4.** green to orange

5. blue to green 　　　**6.** yellow to orange 　　　**7.** red to blue 　　　**8.** blue to green

53 PERCENTS

Percent means "per hundred". Percents can be modeled with 10 × 10 decimal squares.

Examples   Use the decimal square to answer the questions.

What percent of the decimal square is shaded blue?

Since 42 of 100 squares are shaded blue, 42% of the decimal square is shaded blue.

What percent of the decimal square is shaded red?

Since 39 of 100 squares are shaded red, 39% of the decimal square is shaded red.

PRACTICE   Use the decimal square at the right.

1. What percent is shaded red?

2. What percent is shaded blue?

3. What percent is shaded green?

4. What percent is shaded yellow?

5. What percent is shaded red or blue?

6. What percent is not shaded?

54 PERCENTS AND DECIMALS

You can use a 10 × 10 decimal square to show the relationship between decimals and percents.

Examples   Write a decimal and a percent for the amount shaded.

A.   0.57, 57%

B.   0.07, 7%

To change from a decimal to a percent, move the decimal point two places to the right and add a percent symbol. To change from a percent to a decimal, move the decimal point two places to the left and drop the percent symbol.

PRACTICE   Write a decimal and a percent for the amount shaded.

1.   2.   3.   4.

Change the percent to a decimal or the decimal to a percent.

5. 72%   **6.** 8%   **7.** 0.06   **8.** 0.72

55 PERCENTS AND RATIOS

A percent can be written as a ratio, and a ratio can be written as a percent.

Examples

A. Write 39% as a ratio.

Percent means "per hundred."

$39\% = \frac{39}{100}$

B. Write $\frac{47}{100}$ as a percent.

Percent means "per hundred."

$\frac{47}{100} = 47\%$

C. Write $\frac{3}{5}$ as a percent.

Change $\frac{3}{5}$ to a decimal. Then change the decimal to a percent.

$\frac{3}{5} = 3 \div 5 = 0.6 = 60\%$

So, $\frac{3}{5} = 60\%$.

PRACTICE   Write the percent as a ratio.

1. 31%   **2.** 79%   **3.** 19%   **4.** 27%   **5.** 61%

Write the ratio as a percent.

6. $\frac{53}{100}$ **7.** $\frac{7}{100}$ **8.** $\frac{91}{100}$ **9.** $\frac{4}{5}$ **10.** $\frac{13}{20}$

56 THE COORDINATE GRID

You can use two numbers to locate a point on a grid. The two numbers are called an **ordered pair**. The first number tells you how far to move horizontally from (0,0). The second number tells you how far to move vertically from (0,0).

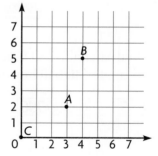

Examples

- The ordered pair for point *A* is (3,2).
- The ordered pair for point *B* is (4,5).
- The ordered pair for point *C* is (0,0).

PRACTICE   Write the the ordered pair for the point on the coordinate grid.

1. point *H*   **2.** point *J*   **3.** point *M*

4. point *K*   **5.** point *N*   **6.** point *I*

Name the point for each ordered pair.

7. (5,4)   **8.** (1,1)   **9.** (1,6)

10. (2,5)   **11.** (6,7)   **12.** (4,2)

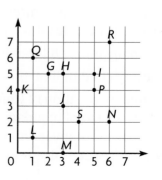

Activities

LOTS AND LOTS OF KIDS
Scientific Notation

Michali had a difficult time imagining how many young people the table represented. Linell helped her understand. Follow along as Linell explains how to use the *TI Explorer Plus* to work with the numbers.

Using the Calculator

"How many more children are there in Indonesia than in the United States?" Michali asked. Linell entered the following key sequence to answer her question.

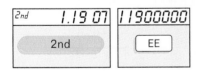

"OK," sighed Michali. "But what number is that, really?" Linell demonstrated to Michali how to show the number in decimal notation.

Total Population Under 15 Years Old	
China	3.159×10^8
India	3.044×10^8
Indonesia	6.574×10^7
United States	5.384×10^7
Brazil	5.165×10^7
Pakistan	5.163×10^7
Nigeria	5.141×10^7
Bangladesh	4.707×10^7
Mexico	3.211×10^7
Iran	2.676×10^7

PRACTICE

Use your calculator and the population data.

1. Calculate the total number of children in the North and South American countries shown.

2. How many more children live in Nigeria than in Mexico?

3. For every child in Iran, how many children live in China?

4. The total population of China is one billion, one hundred fifty-three million. The total population of the United States is two hundred fifty million. How many more people live in China than in the United States? Use scientific notation to express the answer.

> **REMEMBER:**
> Scientific notation always includes a decimal number with one digit in the ones place and the remaining digits to the right of the decimal point. The power of 10 indicates the true size of the number.

ALL AT ONCE
Division with Remainders

Hanako needs to freeze 69 dumplings she made. Each of her pans can hold 20 dumplings at a time. How many pans does she need for all of the dumplings?

$69 \div 20 = 3 \text{ r} 9$

Hanako needs 4 pans to hold all the dumplings.

Using the Calculator

There are 4 people in Hanako's family. She wants to separate the dumplings into packages of 8 dumplings so each family member can have 2 dumplings. How many packages will she be able to make? How many dumplings will be left over?

Use the following keys on the *Casio fx-65* to find the answer.

Use the following keys on the *TI Explorer Plus* to find the answer.

PRACTICE

Use your calculator to find the quotient as an integer and remainder.

1. $789 \div 12$ 2. $59 \div 5$ 3. $85 \div 8$

4. $3,258 \div 25$ 5. $359,523 \div 40$ 6. $70 \div 3$

7. $82 \div 7$ 8. $29 \div 4$ 9. $102 \div 18$

10. Tyesha waited in the line for the carousel with William, her younger brother. There were 15 horses on the carousel. Tyesha and William counted 64 people in line ahead of them. How many more times would the carousel ride begin before Tyesha and William could get on?

11. Tyesha attends Division Middle School. The principal at Division Middle School has ordered 1,190 desks. He plans to divide them equally among 45 classrooms. How many desks will be in each classroom? Will there be any extra desks? If so, how many?

> **REMEMBER:**
>
> Integers are all the positive whole numbers, all the negative whole numbers, and zero.
> These are integers:
>
> 2   ⁻56   0
> ⁻987,900   12
> 390,218,900
>
> These are not integers:
> ⁻2.3   0.009 $\frac{4}{5}$

FIRST THIS ONE, THEN THE NEXT
Solve Multistep Problems

Alison's class saved their change for several months. They had 692 pennies, 465 dimes, 274 nickels, and 178 quarters. How much money did they save?

Using the Calculator

Use the *Casio fx-65* to calculate the class savings.

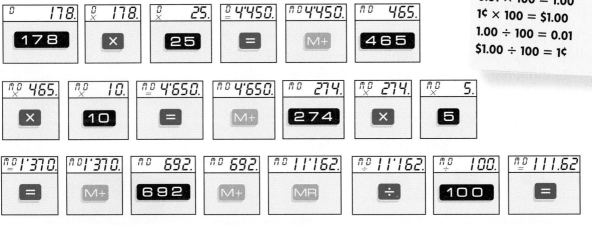

Use the *TI Explorer Plus* to calculate the class savings.

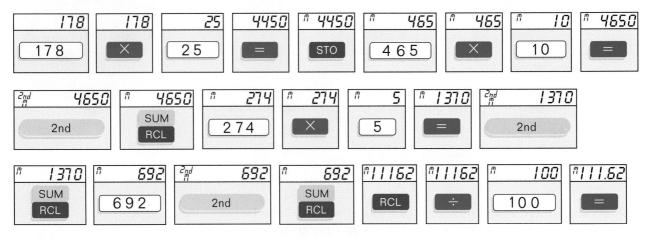

The class saved $111.62.

REMEMBER:

Money can be counted as cents or as dollars. One cent is one hundredth of one dollar; one dollar is one hundred times one cent.

$0.01 \times 100 = 1.00$
$1\text{¢} \times 100 = \$1.00$
$1.00 \div 100 = 0.01$
$\$1.00 \div 100 = 1\text{¢}$

PRACTICE

Use your calculator.

1. Mrs. Saloman counted the number of students in each school group that visited the museum. During one week the following numbers of students visited: 4 groups of 42; 5 groups of 23; 8 groups of 31; 2 groups of 57; and 3 groups of 27. How many students visited the museum as members of school groups?

FIRST THINGS FIRST
Order of Operations

Eton and Luba both found the value of $9 + 132 \div 3 - 18$.
Can you explain Eton's mistake?

Eton	Luba
$9 + 132 \div 3 - 18$	$9 + 132 \div 3 - 18$
$141 \div {}^-15$	$9 + 44 - 18$
$^-9.4$	$53 - 18$
	35

Using the Calculator

You can use the *Casio fx-65* to solve the problem.

You can use the *TI Explorer Plus* to solve the problem.

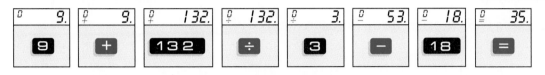

If Eton added parentheses to his work, his answer would be correct.
Here's how to include parentheses on the *Casio fx-65*.

Here's how to include parentheses on the *TI Explorer Plus*.

PRACTICE

Use your calculator.

1. $2 \times 11 + 3 \times 4$

2. $(2 \times 11 + 3) \times 4$

3. $2 \times (11 + 3 \times 4)$

4. $(270 - 140) \div 2 + 3 \times 6$

5. $270 - 140 \div (2 + 3 \times 6)$

6. $270 - 140 \div 2 + 3 \times 6$

THUMBS ACROSS THE UNITED STATES
Operations with Fractions, Simplify

Mr. Jones challenged his class to measure something in a way that it had not been measured before. Brian and Joon thought and thought. Then they decided to use the widths of their thumbs to measure across their desk map of the United States.

Brian's thumb is $\frac{7}{16}$ in. wide, and Joon's is $\frac{3}{4}$ in. wide. They worked together and each used 14 thumb widths to cover the distance from San Francisco to New York. How many inches did they cover?

REMEMBER:

When you need to simplify a fraction, you don't want to change the relationship between the numerator and denominator. To keep the relationship the same, you can divide the fraction only by a number that equals 1.
$\frac{112}{1,016} \div \frac{8}{8} = \frac{14}{127}$ $\frac{8}{8} = 1$

Using the Calculator

You can use the *Casio fx-65* to calculate the measurements with these keystrokes.

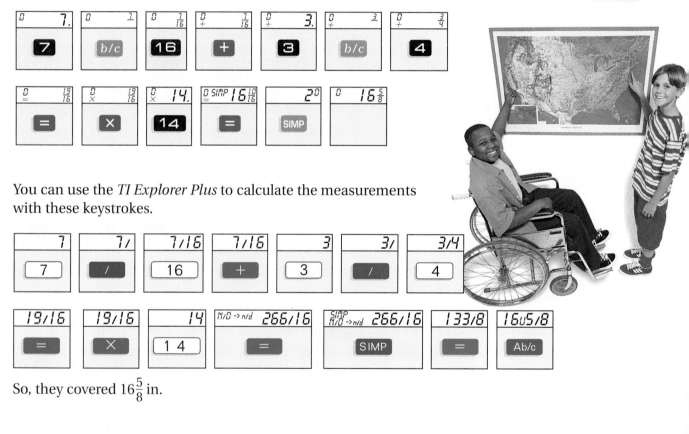

You can use the *TI Explorer Plus* to calculate the measurements with these keystrokes.

So, they covered $16\frac{5}{8}$ in.

PRACTICE
Use your calculator. Simplify your answers.

1. $\frac{6}{7} + \frac{9}{11}$

2. $\frac{67}{16} - \frac{3}{4}$

3. $\frac{484}{28} - 13$

4. $\frac{107}{12} \times \frac{19}{20}$

5. $\frac{4}{5} \times \frac{139}{8}$

IT'S THE SAME, ONLY DIFFERENT
Change Decimals to Fractions and Fractions to Decimals

Robert and Keisha are changing fractions to decimals. Keisha wrote the following:

$$\frac{2}{5} = 0.4$$

"But that doesn't make any sense!" Robert exclaimed. "How can a fraction and a decimal be the same thing? Two fifths doesn't even have a four in it!"

"Watch," Keisha explained. "I'll show you how to do the conversion with your calculator."

REMEMBER:

Fractions can be thought of as division problems. When you do the division, the quotient is in decimal form.

$$\frac{3}{4} = 4\overline{)3} = 0.75$$
$$\frac{5}{16} = 16\overline{)5} = 0.3125$$

Using the Calculator

Follow along as Keisha shows Robert how to change fractions to decimals and decimals to fractions on the *Casio fx-65*.

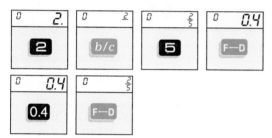

You can use the *TI Explorer Plus* to change fractions to decimals and decimals to fractions.

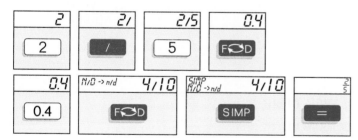

PRACTICE

Use your calculator. Change the fractions to decimals. Change the decimals to fractions. Write the fractions in simplest form.

1. 0.625

2. $\frac{4}{5}$

3. 0.55

4. $\frac{7}{4}$

5. $\frac{1}{6}$

6. 0.12

7. $\frac{16}{3}$

8. 2.125

PART OF THE WHOLE
Solve Percent Problems

The Hamilton-Martin School's 56 sixth graders plan to visit the Native American Art Museum. When Yona and Letitia telephoned to make the arrangements, they were told that the normal admission price is $4.50 per person.

However, the museum also told the girls they could have a 15% discount. What is the total admission cost for the sixth grade?

Using the Calculator

You can use the following key sequence to calculate the total cost of admission on a *Casio fx-65*.

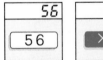

You can use the following key sequence to calculate the total cost of admission on a *TI Explorer Plus*.

So, the total admission is $214.20.

PRACTICE

Use your calculator.

1. What is 5% of 175?
2. What is 125% of 36?
3. What is 20% of 45?
4. What is 150% of 70?
5. What is 70% of 85?
6. What is 140% of 90?
7. What is 40% of 82?
8. What is 180% of 20?

9. At the museum gift shop, each of the Hamilton-Martin students bought a souvenir for $2.50. The students each paid sales tax of 4%. What was the total amount of money the students spent at the gift shop?

10. Yona bought another souvenir at the museum gift shop. She bought a small basket that had a Native American design on it. The price of the basket was $12.00, but the shop gave Yona a 20% discount. How much did Yona pay for the basket?

ROUND AND ROUND HE GOES
Area and Circumference of a Circle

Reena and Susannah tied the end of their dog's leash to a stake in their backyard. The dog's leash is 15 feet long. How large is the area that the dog can roam? Round your answer to the nearest thousandth.

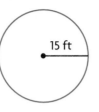

15 ft

Using the Calculator

To find the area, press these keys on a *Casio fx-65:*

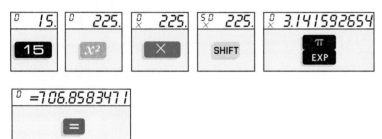

To find the area, press these keys on a *TI Explorer Plus:*

So, the area is 706.858 ft².

PRACTICE

Use your calculator to find the area and circumference of each circle. Round each answer to the nearest thousandth.

1. radius = 2.5 mi   **2.** radius = 0.32 mm   **3.** diameter = 52 yd

Solve.

4. Natalie plans to make a wreath out of six strands of bittersweet, a vine that grows near her house. She would like the wreath to have a diameter of 16 inches. About how much bittersweet does Natalie need to cut?

5. Jason is installing a sprinkler that sprays in a circle with a radius of 12 ft. How much lawn is covered by this sprinkler?

SECOND DIMENSIONS
Square and Square Root

Tiernan is saving his money to buy a Japanese fighting fish. He wants to put a small square fish tank on a shelf in his room. The shelf is only 5 in. wide. If the tank is the same width as the shelf, what is the area the tank will cover?

Using the Calculator

You can calculate the answer by using the *Casio fx-65*.

You can calculate the answer by using the *TI Explorer Plus*.

So, the area the tank will cover is 25 in.^{2}

Tiernan noticed that he had 64 in.^{2} of space on the corner of his desk. What is the greatest possible length a square fish tank could be to fit on the desk?

You can find the answer by using the *Casio fx-65*.

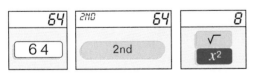

You can find the answer by using the *TI Explorer Plus*.

So, the greatest length is 8 in.

REMEMBER:

A number squared is that number multiplied by itself. You can see the multiplication pattern by making squares of blocks. The number of blocks in the square is the square of the number of blocks on each side.

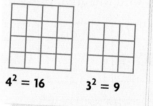

$4^2 = 16$ $3^2 = 9$

PRACTICE
Use your calculator.

1. 11^2 **2.** 20^2 **3.** 15^2 **4.** 46^2

Round to the nearest hundredth.

5. $\sqrt{1,225}$ **6.** $\sqrt{9,801}$ **7.** $\sqrt{5,869}$ **8.** $\sqrt{6,149}$

RANDOMLY CHOSEN
Random Number Generator

Ms. Salvadori, the principal, had to choose 3 of 1,000 people at Benson Corners Middle School to represent the school at the town Founders' Parade. Each deserved an equal chance. Ms. Salvadori gave each person a number between 0 and 0.999. But she didn't really know what to do next. How could she choose?

Using the Calculator

You can help Ms. Salvadori by using a *Casio fx-65*. Enter these keystrokes three times to get three different numbers.

You can help Ms. Salvadori by using a *TI Explorer Plus*. Enter these keystrokes three times to get three different numbers.

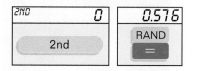

Valentin wants to survey a random sample of 10 of the 100 sixth graders in the Benson Corners Middle School. He assigns each student a number between 1 and 100. He repeats the keystrokes below ten times using his *TI Explorer Plus* to choose ten numbers at random.

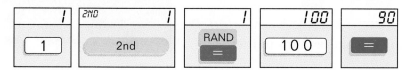

PRACTICE

Use your calculator.

1. Sarah and April played a game with the spinner at the right. If the number was odd, April got one point. If the number was even, Sarah got one point. Simulate their game by using your calculator. Assign numbers ending in 1, 3, 5, 7, and 9 to April and the rest to Sarah. Who won in your simulation?

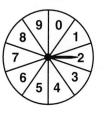

2. Pizzerias would like to survey 75 people about their favorite pizza. Write a letter to the company's president, Marshall Lee, explaining how to use a calculator to choose the 75 people randomly. Include information about random numbers.

UP WITH THE POSITIVE, DOWN WITH THE NEGATIVE

Operations with Integers

The Brooklyn Bridge was opened on May 24, 1883, and still carries pedestrian and motor vehicle traffic across the East River between Manhattan and Brooklyn. Each of the towers is sunk deep into the riverbed, 79 ft below the surface of the water, and soars 266.5 ft high above the surface of the water.

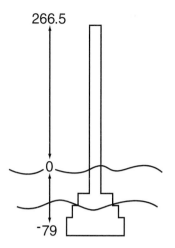

REMEMBER:

Negative can be thought of as "the opposite of." *The opposite* means "the same distance from zero on a number line."

These numbers are opposites:

3 and ⁻3
45 and ⁻45
187.4 and ⁻187.4

We can say that the height of the underwater part of the tower is ⁻79 ft in relation to the surface of the water at 0 ft. To find the total distance from the top of the tower to the bottom, we have to subtract ⁻79 from 266.5. How do you do that?

Using the Calculator

You can find the total distance from the top of the tower to the bottom by using the *Casio fx-65*.

You can find the total distance from the top of the tower to the bottom by using the *TI Explorer Plus*.

PRACTICE

Use your calculator.

1. ⁻18 ÷ 6 × 3   **2.** ⁻2 − (⁻6) + 12   **3.** 7 × (⁻4) + (⁻20)

4. 369 ÷ ⁻3 × ⁻2   **5.** 2,002 − ⁻4,100   **6.** 3 − ⁻2 + 1 − ⁻19 + ⁻9 + 6

7. 275 × 4 + ⁻5   **8.** ⁻320 ÷ 5 − 27   **9.** (16 × ⁻5 + 20) − ⁻8 + 1

CRANKY NUMBERS, MEAN DATA
Find the Mean of a Set of Data

Isabel and Felipe collected data on the number of centimeters a basketball bounced when dropped from the same height with differing amounts of air in it. Here is their data:

80 cm, 75 cm, 82 cm, 74 cm, 74 cm

What is the average bounce?

REMEMBER:

The mean of a set of data is calculated by finding the sum of the data, and then dividing by the number of items in the set of data.

Using the Calculator

You can use these keystrokes to find the sum and the mean of a data set by using the *TI Explorer Plus.*

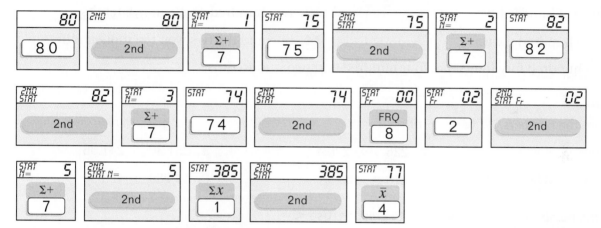

So, the sum of the data set is 385. The mean is 77.

PRACTICE

Use your calculator. Find the sum and mean of the data sets. Round the mean to the nearest hundredth.

1. Men's Long Jump: First Place, Olympics, 1960–1992
   8.67 m, 8.72 m, 8.54 m, 8.54 m, 8.35 m, 8.24 m, 8.9 m, 8.07 m, 8.12 m

2. Number of Push-ups Niko Did Each Day
   16, 18, 21, 17, 15, 15, 21, 15, 21, 14, 19

3. Number of Lawns Mowed by Kurt's Lawn Service Last Week
   3, 5, 4, 6, 5, 7, 8

4. Jared scored 80, 83, 96, 91, 78, 88, 90, 92, and 94 on his social studies tests. His goal was to have an average of 85 on his tests. Use your calculator to find Jared's mean test score. Did he meet his goal?

More Practice

CHAPTER 1

Lesson 1.1

Write the value of the digit 3.

1. 301　　　　**2.** 0.0231　　　　**3.** 8.3　　　　**4.** 0.839

Write the number in words.

5. 18　　　　**6.** 0.05　　　　**7.** 3.032

8. 2,000,500　　　　**9.** 200.05　　　　**10.** 5.16

Write the number in standard form.

11. thirteen hundredths

12. two hundred, five and five tenths

13. three hundred sixteen

14. one and thirty-seven ten thousandths

15. two hundred fifty thousand, four hundred and two hundredths

Lesson 1.2

Compare the numbers. Write $<$, $>$, or $=$.

1. 5.099 $\underset{?}{\_\_}$ 5.999　　　　**2.** 226.5 $\underset{?}{\_\_}$ 226.4　　　　**3.** 251.36 $\underset{?}{\_\_}$ 241.36

4. 18.3 $\underset{?}{\_\_}$ 18.30　　　　**5.** 418 $\underset{?}{\_\_}$ 428　　　　**6.** 49.089 $\underset{?}{\_\_}$ 49.098

Write the numbers in order from least to greatest. Use $<$.

7. 82.16, 82, 82.15　　　　**8.** 141.14, 114.41, 141.41　　　　**9.** 5.09, 5.49, 5.23

Write the numbers in order from greatest to least. Use $>$.

10. 14.63, 14.36, 13.46, 13.64　　**11.** 0.300, 3.000, 0.030, 0.303　　**12.** 12.5, 9.81, 36.3

Lesson 1.3

Use place value to change the decimal to a fraction.

1. 0.16　　　　**2.** 0.021　　　　**3.** 0.02　　　　**4.** 0.7　　　　**5.** 0.33

Write as a decimal.

6. $\frac{5}{8}$　　　　**7.** $\frac{14}{16}$　　　　**8.** $\frac{9}{100}$　　　　**9.** $\frac{2}{8}$　　　　**10.** $\frac{1}{4}$

Write $<$ or $>$.

11. $\frac{1}{7}$ $\underset{?}{\_\_}$ $\frac{2}{5}$　　**12.** $\frac{2}{7}$ $\underset{?}{\_\_}$ $\frac{1}{3}$　　**13.** $\frac{5}{9}$ $\underset{?}{\_\_}$ 0.56　　**14.** 0.6 $\underset{?}{\_\_}$ $\frac{62}{60}$　　**15.** $\frac{3}{4}$ $\underset{?}{\_\_}$ 0.7

Lesson 1.4

Write in exponent form.

1. $3 \times 3 \times 3 \times 3$ 　　　 **2.** $1 \times 1 \times 1 \times 1 \times 1 \times 1$ 　　　 **3.** 10×10

4. $12 \times 12 \times 12$ 　　　 **5.** $21 \times 21 \times 21 \times 21 \times 21$ 　　　 **6.** $1 \times 1 \times 1 \times 1$

Find the value.

7. 10^4 　　　 **8.** 5^6 　　　 **9.** 2^3 　　　 **10.** 3^2

11. 7^4 　　　 **12.** 15^1 　　　 **13.** 27^2 　　　 **14.** 1^{10}

Lesson 1.5

Compare the integers. Use $<$ or $>$.

1. $^-2 \overset{?}{=} {}^-6$ 　　　 **2.** $^-6 \overset{?}{=} {}^+6$ 　　　 **3.** $^+3 \overset{?}{=} {}^+1$ 　　　 **4.** $0 \overset{?}{=} {}^-3$

5. $^+1 \overset{?}{=} {}^-8$ 　　　 **6.** $^-5 \overset{?}{=} 0$ 　　　 **7.** $^+7 \overset{?}{=} {}^-2$ 　　　 **8.** $^-3 \overset{?}{=} {}^+3$

9. $^-2 \overset{?}{=} {}^+3$ 　　　 **10.** $0 \overset{?}{=} {}^+4$ 　　　 **11.** $^-8 \overset{?}{=} {}^-3$ 　　　 **12.** $^-7 \overset{?}{=} {}^+1$

Order the integers from least to greatest. Use $<$.

13. $^+3, {}^-2, +6, 0$ 　　　 **14.** $0, {}^-6, {}^-8, {}^-2$ 　　　 **15.** $^+2, {}^-4, {}^+6, {}^-8$

16. $^+2, {}^-2, {}^+6, {}^-6$ 　　　 **17.** $^+3, {}^-5, 0 \, {}^+2$ 　　　 **18.** $^+1, {}^-9, 0, {}^-4$

Name the opposite of the given integer.

19. $^-3$ 　　　 **20.** $^+2$ 　　　 **21.** 0 　　　 **22.** $^-17$ 　　　 **23.** $^+31$

CHAPTER 2

Lesson 2.1

Use mental math to add.

1. $15 + 17 + 5$ 　　　 **2.** $13 + 9 + 7$ 　　　 **3.** $37 + 3 + 19$

4. $51 + 7 + 29$ 　　　 **5.** $15 + 1 + 15$ 　　　 **6.** $19 + 52 + 21$

Use compensation to add.

7. $26 + 18$ 　　　 **8.** $14 + 15$ 　　　 **9.** $28 + 41$ 　　　 **10.** $32 + 36$

11. $19 + 26$ 　　　 **12.** $38 + 15$ 　　　 **13.** $19 + 31$ 　　　 **14.** $9 + 22$

Use compensation to subtract.

15. $51 - 16$ 　　　 **16.** $54 - 42$ 　　　 **17.** $47 - 29$ 　　　 **18.** $92 - 29$

19. $35 - 14$ 　　　 **20.** $52 - 21$ 　　　 **21.** $38 - 22$ 　　　 **22.** $36 - 17$

Lesson 2.2

Guess and check to solve.

1. Harry ate three meals yesterday. The total calories were 2,300. Dinner included 500 calories more than breakfast and lunch together. How many calories did dinner include?

2. Rebecca has 36 feet of fencing to create the border for her new garden. She wants the garden to be square. How long is each side?

3. Tom took his new convertible car for a drive to his friend's farm. He drove a total of 84 miles. He drove 12 miles longer to get there than he drove to get home. How long was the ride home?

4. Sean's model rocket flew for a total of 42 seconds on its first flight. The time it took to reach its highest point was 22 seconds shorter than the time it took to return to the ground. How long did it take to reach its highest point?

Lesson 2.3

Find the missing factor(s).

1. $36 \times 5 = (30 \times 5) + (\underline{?} \times 5)$

2. $51 \times 8 = (50 \times 8) + (\underline{?} \times 8)$

3. $43 \times 7 = (\underline{?} \times 7) + (3 \times 7)$

4. $26 \times 3 = (\underline{?} \times 3) + (6 \times 3)$

5. $24 \times \underline{?} = (24 \times 30) + (24 \times 1)$

6. $63 \times 17 = (63 \times \underline{?}) + (63 \times 7)$

7. $42 \times 15 = (42 \times 10) + (42 \times \underline{?})$

8. $12 \times \underline{?} = (12 \times 30) + (12 \times 7)$

9. $36 \times 12 = (36 \times \underline{?}) + (36 \times \underline{?})$

10. $3 \times 21 = (3 \times \underline{?}) + (3 \times 1)$

11. $27 \times 21 = (27 \times \underline{?}) + (27 \times \underline{?})$

12. $16 \times \underline{?} = (16 \times 20) + (16 \times 8)$

13. $36 \times 8 = (\underline{?} \times 8) + (6 \times 8)$

14. $41 \times 36 = (\underline{?} \times 36) + (1 \times 36)$

Lesson 2.4

Find the product.

1. 23×12

2. 28×18

3. 127×19

4. 156×83

5. 540×81

6. $1,206 \times 15$

7. $2,150 \times 93$

8. $4,301 \times 179$

9. $5,220 \times 860$

10. $5,825 \times 703$

11. $8,710 \times 50$

12. $7,007 \times 598$

Find the quotient.

13. $5\overline{)25}$

14. $7\overline{)630}$

15. $8\overline{)96}$

16. $11\overline{)2,541}$

17. $9\overline{)153}$

18. $13\overline{)169}$

19. $5\overline{)29}$

20. $12\overline{)53}$

21. $25\overline{)612}$

22. $21\overline{)85}$

23. $77\overline{)426}$

24. $12\overline{)4,285}$

Lesson 2.5

Estimate the sum or difference.

1. $1,200 + 4,006 + 3,210$

2. $378 + 98 + 682$

3. $319 + 277$

4. $5,403 + 4,320$

5. $482 - 71$

6. $944 + 263$

7. $340 - 208$

8. $8,710 - 4,203$

9. $5,232 - 4,828$

Estimate the product.

10. 38×4

11. 74×3

12. 69×42

13. 620×32

14. 71×29

15. 487×311

Estimate the quotient.

16. $481 \div 59$

17. $448 \div 88$

18. $2,844 \div 67$

19. $6,340 \div 71$

20. $3,480 \div 48$

21. $6,319 \div 81$

22. $4,233 \div 608$

23. $7,197 \div 810$

CHAPTER 3

Lesson 3.1

Estimate the sum.

1. $0.4 + 0.9$

2. $2.6 + 8.1$

3. $0.31 + 2.04$

4. $5.92 + 3.15$

5. $0.218 + 2.143$

6. $5.816 + 3.215$

Find the sum.

7. $2.14 + 6.08$

8. $7.04 + 0.13$

9. $0.126 + 0.408$

10. $8.360 + 5.216$

11. $17.31 + 12.06$

12. $0.08 + 7.42$

13. $\$18.56 + \$3.12$

14. $\$8.26 + \$26.15$

15. $\$31.18 + \$125.50$

Lesson 3.2

Find the difference.

1. $11.4 - 6.2$

2. $12.8 - 4.1$

3. $17.3 - 16.5$

4. $\$13.20 - \$8.17$

5. $\$21.80 - \$17.40$

6. $\$35.51 - \$23.17$

7. $8.026 - 7.317$

8. $19.408 - 1.582$

9. $5.888 - 5.261$

10. $13.601 - 10.311$

11. $5 - 3.021$

12. $628.71 - 527.21$

Find the missing number.

13. $2.56 + \underline{\,?\,} = 8.21$

14. $12.18 + \underline{\,?\,} = 17.81$

15. $503.1 - \underline{\,?\,} = 501.3$

16. $3.8 - \underline{\,?\,} = 2.6$

17. $721 - \underline{\,?\,} = 385.51$

18. $27.28 + \underline{\,?\,} = 28.27$

Lesson 3.3

Estimate the product.

1. 2.9×3.7
2. 18.8×8.3
3. 5.8×11.4
4. 32.3×0.9

5. 9.8×21.513
6. 788.3×9.2
7. 308.9×58.3
8. 11.2×8.5

Copy the problem. Place the decimal point in the product.

9. $3.12 \times 6.4 = 19968$
10. $0.48 \times 6.2 = 2976$
11. $4.54 \times 2.3 = 10442$

12. $21.3 \times 18.4 = 39192$
13. $7.03 \times 7.05 = 495615$
14. $9.2 \times 2.13 = 19596$

15. $12.31 \times 17.42 = 2144402$
16. $212.3 \times 3.26 = 692098$
17. $86.2 \times 86.3 = 743906$

18. $360.05 \times 12.6 = 453663$
19. $762 \times 3.285 = 2503170$
20. $10.11 \times 1.02 = 103122$

Find the product.

21. 3×0.4
22. 7×2.1
23. 8×1.56
24. 2×1.6

25. 8.12×5.4
26. 22.35×2.04
27. 212.021×0.24
28. 7.03×0.3

Lesson 3.4

Complete.

1. $33.6 \div 11.2 = 336 \div \underline{\ ?\ }$
2. $31.5 \div 5 = 3.15 \div \underline{\ ?\ }$
3. $20.2 \div 3 = \underline{\ ?\ } \div 0.3$

4. $16.92 \div 12 = \underline{\ ?\ } \div 1.2$
5. $661.44 \div \underline{\ ?\ } = 66{,}144 \div 312$
6. $52.2 \div 2.2 = 522 \div \underline{\ ?\ }$

Find the quotient.

7. $16.4 \div 4$
8. $36.03 \div 3$
9. $2.88 \div 0.4$
10. $0.38 \div 7.6$

11. $9.72 \div 1.2$
12. $25.0224 \div 3.12$
13. $20.801 \div 6.1$
14. $80.4 \div 4.8$

CHAPTER 4

Lesson 4.1

Name the first four multiples.

1. 19
2. 12
3. 22
4. 7

Find the missing multiple or multiples.

5. $\underline{\ ?\ }, 14, 21, \underline{\ ?\ }, 35$
6. $12, \underline{\ ?\ }, \underline{\ ?\ }, 48, 60$
7. $\underline{\ ?\ }, 12, 18, 24, \underline{\ ?\ }$

Write the factors.

8. 26
9. 32
10. 40
11. 33

Write *P* for prime or *C* for composite.

12. 32
13. 63
14. 59
15. 39

Lesson 4.2

Write the prime factorization in exponent form.

1. 18   **2.** 40   **3.** 36   **4.** 27

5. 100   **6.** 150   **7.** 280   **8.** 12

9. 28   **10.** 420   **11.** 48   **12.** 81

13. 72   **14.** 45   **15.** 147   **16.** 16

17. 98   **18.** 180   **19.** 80   **20.** 144

Solve for *n* to complete the prime factorization.

21. $2 \times 5 \times n = 50$ **22.** $2 \times n \times 5 = 30$ **23.** $2 \times 3 \times n = 18$

24. $n \times 3 \times 3 = 27$ **25.** $2 \times 2 \times 2 \times n = 16$ **26.** $n \times 2 \times 3 = 42$

Lesson 4.3

Find the LCM for each set of numbers.

1. 3, 8   **2.** 4, 15   **3.** 3, 20   **4.** 4, 16   **5.** 3, 13

6. 10, 35   **7.** 12, 20   **8.** 8, 15   **9.** 10, 16   **10.** 4, 18

Find the GCF for each set of numbers.

11. 30, 18   **12.** 20, 35   **13.** 30, 50   **14.** 36, 48   **15.** 12, 18

16. 48, 84   **17.** 52, 39   **18.** 75, 105   **19.** 45, 108   **20.** 32, 128

Lesson 4.4

Write the fraction in simplest form.

1. $\frac{15}{20}$ **2.** $\frac{16}{24}$ **3.** $\frac{9}{27}$ **4.** $\frac{21}{27}$ **5.** $\frac{8}{56}$ **6.** $\frac{10}{95}$

7. $\frac{24}{80}$ **8.** $\frac{14}{63}$ **9.** $\frac{24}{56}$ **10.** $\frac{36}{117}$ **11.** $\frac{16}{40}$ **12.** $\frac{33}{187}$

13. $\frac{7}{154}$ **14.** $\frac{27}{81}$ **15.** $\frac{9}{120}$ **16.** $\frac{15}{60}$ **17.** $\frac{32}{60}$ **18.** $\frac{50}{90}$

Write the missing number.

19. $\frac{9}{12} = \frac{\square}{4}$ **20.** $\frac{7}{42} = \frac{\square}{6}$ **21.** $\frac{25}{\square} = \frac{5}{10}$ **22.** $\frac{3}{7} = \frac{21}{\square}$

23. $\frac{1}{12} = \frac{\square}{144}$ **24.** $\frac{9}{20} = \frac{\square}{60}$ **25.** $\frac{5}{\square} = \frac{15}{27}$ **26.** $\frac{12}{\square} = \frac{3}{29}$

Lesson 4.5

Write the fraction as a mixed number or a whole number.

1. $\frac{7}{3}$ **2.** $\frac{9}{2}$ **3.** $\frac{36}{5}$ **4.** $\frac{72}{8}$

5. $\frac{10}{2}$ **6.** $\frac{13}{4}$ **7.** $\frac{36}{6}$ **8.** $\frac{12}{9}$

9. $\frac{25}{12}$ **10.** $\frac{54}{13}$ **11.** $\frac{33}{4}$ **12.** $\frac{42}{5}$

13. $\frac{52}{7}$ **14.** $\frac{14}{9}$ **15.** $\frac{44}{6}$ **16.** $\frac{9}{5}$

17. $\frac{17}{5}$ **18.** $\frac{27}{11}$ **19.** $\frac{26}{4}$ **20.** $\frac{34}{6}$

Write the mixed number as a fraction.

21. $4\frac{1}{2}$ **22.** $5\frac{2}{3}$ **23.** $2\frac{1}{8}$ **24.** $7\frac{1}{9}$

25. $7\frac{4}{7}$ **26.** $1\frac{6}{11}$ **27.** $4\frac{3}{8}$ **28.** $15\frac{3}{4}$

29. $9\frac{2}{3}$ **30.** $3\frac{4}{7}$ **31.** $8\frac{1}{3}$ **32.** $25\frac{1}{8}$

33. $6\frac{4}{9}$ **34.** $3\frac{3}{8}$ **35.** $2\frac{3}{11}$ **36.** $19\frac{5}{6}$

37. $10\frac{1}{4}$ **38.** $2\frac{5}{12}$ **39.** $8\frac{4}{5}$ **40.** $11\frac{1}{11}$

CHAPTER 5

Lesson 5.1

Find the sum or difference. Write the answer in simplest form.

1. $\frac{8}{10} + \frac{1}{10}$ **2.** $\frac{2}{7} + \frac{4}{7}$ **3.** $\frac{1}{8} + \frac{3}{8}$ **4.** $\frac{8}{11} + \frac{4}{11}$

5. $\frac{2}{5} + \frac{4}{5}$ **6.** $\frac{2}{12} + \frac{5}{12}$ **7.** $\frac{9}{11} - \frac{3}{11}$ **8.** $\frac{4}{7} - \frac{1}{7}$

9. $\frac{11}{14} - \frac{5}{14}$ **10.** $\frac{9}{12} - \frac{5}{12}$ **11.** $\frac{7}{11} - \frac{3}{11}$ **12.** $\frac{13}{20} - \frac{6}{20}$

Lesson 5.2

Draw a diagram to find each sum or difference.

1. $\frac{1}{2} + \frac{1}{3}$ **2.** $\frac{3}{4} - \frac{1}{2}$ **3.** $\frac{7}{12} + \frac{1}{3}$ **4.** $\frac{5}{8} - \frac{1}{4}$

5. $\frac{1}{8} + \frac{3}{4}$ **6.** $\frac{1}{4} + \frac{2}{3}$ **7.** $\frac{1}{8} + \frac{1}{4}$ **8.** $\frac{9}{10} - \frac{2}{5}$

9. $\frac{7}{10} - \frac{3}{5}$ **10.** $\frac{1}{4} - \frac{1}{8}$ **11.** $\frac{5}{6} - \frac{1}{2}$ **12.** $\frac{11}{12} + \frac{5}{6}$

13. $\frac{4}{6} + \frac{1}{4}$ **14.** $\frac{3}{8} + \frac{1}{4}$ **15.** $\frac{1}{2} - \frac{1}{6}$ **16.** $\frac{2}{3} + \frac{5}{12}$

Lesson 5.3

Write equivalent fractions with the LCD.

1. $\frac{3}{4} + \frac{1}{2}$ **2.** $\frac{2}{5} + \frac{8}{15}$ **3.** $\frac{1}{4} + \frac{3}{5}$ **4.** $\frac{1}{9} + \frac{5}{12}$

5. $\frac{1}{3} + \frac{5}{8}$ **6.** $\frac{1}{4} + \frac{1}{8}$ **7.** $\frac{1}{6} + \frac{2}{3}$ **8.** $\frac{1}{6} + \frac{5}{18}$

Find the sum. Write your answer in simplest form.

9. $\frac{1}{6} + \frac{3}{4}$ **10.** $\frac{5}{6} + \frac{1}{4}$ **11.** $\frac{2}{5} + \frac{1}{3}$ **12.** $\frac{5}{8} + \frac{1}{6}$

13. $\frac{1}{6} + \frac{1}{12}$ **14.** $\frac{5}{8} + \frac{1}{20}$ **15.** $\frac{5}{12} + \frac{1}{3}$ **16.** $\frac{7}{20} + \frac{1}{5}$

17. $\frac{3}{8} + \frac{1}{5}$ **18.** $\frac{4}{9} + \frac{2}{5}$ **19.** $\frac{1}{2} + \frac{1}{8}$ **20.** $\frac{2}{9} + \frac{4}{7}$

21. $\frac{1}{10} + \frac{3}{8}$ **22.** $\frac{6}{20} + \frac{3}{5}$ **23.** $\frac{4}{16} + \frac{1}{9}$ **24.** $\frac{3}{4} + \frac{1}{12}$

Lesson 5.4

Subtract. Write the answer in simplest form.

1. $\frac{5}{8} - \frac{1}{4}$ **2.** $\frac{2}{5} - \frac{1}{4}$ **3.** $\frac{5}{9} - \frac{1}{2}$ **4.** $\frac{11}{12} - \frac{1}{6}$

5. $\frac{5}{6} - \frac{3}{4}$ **6.** $\frac{4}{7} - \frac{1}{3}$ **7.** $\frac{9}{10} - \frac{2}{5}$ **8.** $\frac{3}{8} - \frac{1}{6}$

9. $\frac{7}{9} - \frac{2}{3}$ **10.** $\frac{5}{7} - \frac{2}{3}$ **11.** $\frac{2}{3} - \frac{3}{5}$ **12.** $\frac{8}{9} - \frac{1}{3}$

13. $\frac{4}{9} - \frac{1}{3}$ **14.** $\frac{5}{8} - \frac{1}{6}$ **15.** $\frac{11}{18} - \frac{1}{4}$ **16.** $\frac{5}{14} - \frac{1}{7}$

17. $\frac{3}{5} - \frac{1}{9}$ **18.** $\frac{10}{12} - \frac{1}{3}$ **19.** $\frac{7}{8} - \frac{1}{16}$ **20.** $\frac{17}{20} - \frac{7}{10}$

Lesson 5.5

Round each fraction. Write about 0, about $\frac{1}{2}$, or about 1.

1. $\frac{3}{5}$ **2.** $\frac{4}{7}$ **3.** $\frac{11}{12}$ **4.** $\frac{5}{8}$

5. $\frac{1}{5}$ **6.** $\frac{3}{11}$ **7.** $\frac{6}{7}$ **8.** $\frac{1}{9}$

Estimate the sum or difference.

9. $\frac{4}{7} - \frac{3}{8}$ **10.** $\frac{8}{9} + \frac{3}{5}$ **11.** $\frac{5}{9} - \frac{3}{7}$ **12.** $\frac{11}{12} - \frac{1}{2}$

13. $\frac{3}{7} + \frac{1}{9}$ **14.** $\frac{4}{9} + \frac{2}{3}$ **15.** $\frac{9}{10} - \frac{2}{5}$ **16.** $\frac{5}{11} - \frac{3}{8}$

17. $\frac{4}{5} - \frac{1}{3}$ **18.** $\frac{8}{9} - \frac{5}{7}$ **19.** $\frac{5}{9} + \frac{2}{6}$ **20.** $\frac{4}{7} + \frac{1}{3}$

CHAPTER 6

Lesson 6.1

Draw a diagram to find each sum. Write the answer in simplest form.

1. $1\frac{3}{8} + 1\frac{5}{8}$ **2.** $2\frac{3}{8} + 1\frac{1}{4}$ **3.** $1\frac{1}{6} + 1\frac{5}{6}$ **4.** $1\frac{1}{5} + 1\frac{7}{10}$

5. $1\frac{3}{8} + 2\frac{1}{8}$ **6.** $1\frac{2}{3} + 2\frac{1}{6}$ **7.** $2\frac{1}{2} + 1\frac{1}{8}$ **8.** $2\frac{1}{6} + 1\frac{1}{3}$

Lesson 6.2

Write each mixed number by renaming one whole.

1. $2\frac{1}{3}$ **2.** $3\frac{2}{7}$ **3.** $4\frac{1}{12}$ **4.** $5\frac{1}{8}$

5. $3\frac{4}{9}$ **6.** $2\frac{5}{7}$ **7.** $4\frac{5}{6}$ **8.** $3\frac{5}{9}$

Draw a diagram to find the difference. Write the answer in simplest form.

9. $2\frac{1}{7} - 1\frac{3}{7}$ **10.** $3\frac{3}{5} - 2\frac{1}{5}$ **11.** $2\frac{1}{3} - 1\frac{1}{6}$ **12.** $4\frac{5}{8} - 1\frac{1}{4}$

13. $3\frac{1}{3} - 1\frac{3}{4}$ **14.** $4\frac{5}{12} - 2\frac{1}{6}$ **15.** $5\frac{2}{5} - 3\frac{3}{10}$ **16.** $3\frac{1}{2} - 1\frac{1}{6}$

Lesson 6.3

Rename the fractions using the LCD. Rewrite the problem.

1. $1\frac{1}{2} + 2\frac{3}{5}$ **2.** $2\frac{5}{6} + 2\frac{2}{9}$ **3.** $3\frac{3}{4} + 1\frac{1}{5}$ **4.** $4\frac{1}{9} + 2\frac{5}{18}$

Rename the fraction as a mixed number. Write the new mixed number.

5. $2\frac{5}{4}$ **6.** $3\frac{9}{7}$ **7.** $4\frac{10}{6}$ **8.** $11\frac{12}{8}$

Tell whether you must rename to subtract. Write *yes* or *no*.

9. $3\frac{2}{7} - 1\frac{3}{5}$ **10.** $3\frac{5}{7} - 1\frac{1}{7}$ **11.** $2\frac{1}{6} - 1\frac{3}{5}$ **12.** $5\frac{1}{2} - 2\frac{1}{6}$

13. $8\frac{1}{7} - 6\frac{2}{19}$ **14.** $4\frac{1}{4} - 3\frac{2}{7}$ **15.** $3\frac{2}{3} - 2\frac{1}{6}$ **16.** $3\frac{1}{6} - 2\frac{11}{12}$

Find the sum. Write the answer in simplest form.

17. $2\frac{1}{4} + 3\frac{7}{12}$ **18.** $1\frac{1}{7} + 3\frac{2}{3}$ **19.** $3\frac{5}{6} + 4\frac{5}{9}$ **20.** $4\frac{9}{10} + 1\frac{1}{8}$

21. $1\frac{3}{4} + 5\frac{4}{5}$ **22.** $5\frac{1}{3} + 4\frac{1}{9}$ **23.** $5\frac{1}{4} + 5\frac{2}{3}$ **24.** $3\frac{7}{11} + 2\frac{1}{5}$

Find the difference. Write the answer in simplest form.

25. $4\frac{5}{9} - 3\frac{5}{6}$ **26.** $5\frac{1}{3} - 4\frac{1}{9}$ **27.** $6\frac{2}{3} - 3\frac{1}{4}$ **28.** $6\frac{1}{3} - 4\frac{7}{9}$

Lesson 6.4

Estimate the sum or difference.

1. $8\frac{7}{8} + 5\frac{1}{2}$ **2.** $3\frac{10}{13} + 4\frac{1}{9}$ **3.** $2\frac{8}{9} - 1\frac{3}{5}$ **4.** $5\frac{3}{4} - 1\frac{1}{8}$

5. $7\frac{4}{7} - 5\frac{3}{7}$ **6.** $8\frac{5}{6} - 6\frac{2}{4}$ **7.** $3\frac{3}{4} + 2\frac{7}{8}$ **8.** $7\frac{1}{9} + 2\frac{4}{5}$

9. $5\frac{4}{5} - 3\frac{1}{2}$ **10.** $3\frac{7}{9} + 1\frac{10}{11}$ **11.** $2\frac{2}{7} + 3\frac{3}{7}$ **12.** $10\frac{8}{9} - 10\frac{1}{3}$

13. $7\frac{2}{9} - 5\frac{7}{8}$ **14.** $3\frac{1}{2} + 4\frac{1}{5}$ **15.** $15\frac{1}{4} - 14\frac{1}{3}$ **16.** $4\frac{1}{7} + 1\frac{1}{4}$

17. $7\frac{4}{9} - 5\frac{1}{6}$ **18.** $8\frac{9}{11} + 2\frac{3}{8}$ **19.** $5\frac{8}{9} - 4\frac{1}{6}$ **20.** $5\frac{5}{9} - 2\frac{1}{10}$

CHAPTER 7

Lesson 7.1

Find the product. Write it in simplest form.

1. $\frac{2}{4} \times \frac{3}{4}$ **2.** $\frac{2}{5} \times \frac{1}{4}$ **3.** $\frac{3}{8} \times \frac{3}{5}$ **4.** $3 \times \frac{7}{8}$

5. $\frac{2}{7} \times 8$ **6.** $\frac{5}{6} \times \frac{3}{8}$ **7.** $\frac{5}{9} \times \frac{1}{5}$ **8.** $\frac{1}{9} \times \frac{3}{4}$

9. $\frac{1}{9} \times \frac{2}{3}$ **10.** $\frac{2}{7} \times 5$ **11.** $8 \times \frac{2}{3}$ **12.** $\frac{11}{12} \times \frac{4}{5}$

13. $\frac{1}{3} \times \frac{3}{7}$ **14.** $\frac{5}{9} \times \frac{3}{5}$ **15.** $\frac{3}{8} \times \frac{1}{6}$ **16.** $\frac{5}{6} \times \frac{2}{3}$

17. $\frac{4}{9} \times \frac{3}{8}$ **18.** $\frac{1}{4} \times \frac{8}{9}$ **19.** $\frac{4}{5} \times \frac{3}{7}$ **20.** $\frac{2}{5} \times \frac{5}{6}$

Lesson 7.2

Tell the GCF you would use to simplify the fractions.

1. $\frac{2}{5}, \frac{5}{7}$ **2.** $\frac{3}{9}, \frac{9}{27}$ **3.** $\frac{3}{4}, \frac{2}{3}$ **4.** $\frac{5}{9}, \frac{18}{21}$

5. $\frac{3}{20}, \frac{5}{6}$ **6.** $\frac{7}{27}, \frac{6}{9}$ **7.** $\frac{1}{3}, \frac{6}{10}$ **8.** $\frac{4}{7}, \frac{1}{2}$

9. $\frac{6}{8}, \frac{3}{12}$ **10.** $\frac{6}{12}, \frac{5}{18}$ **11.** $\frac{10}{25}, \frac{5}{20}$ **12.** $\frac{5}{8}, \frac{4}{10}$

13. $\frac{1}{2}, \frac{4}{9}$ **14.** $\frac{9}{10}, \frac{7}{18}$ **15.** $\frac{2}{5}, \frac{1}{4}$ **16.** $\frac{2}{3}, \frac{9}{11}$

17. $\frac{3}{26}, \frac{13}{15}$ **18.** $\frac{3}{7}, \frac{14}{27}$ **19.** $\frac{5}{12}, \frac{24}{25}$ **20.** $\frac{6}{7}, \frac{15}{18}$

Use GCF's to simplify the factors so that the answer is in simplest form.

21. $\frac{3}{8} \times \frac{10}{15}$ **22.** $\frac{9}{12} \times \frac{6}{27}$ **23.** $\frac{2}{3} \times \frac{6}{10}$ **24.** $\frac{9}{11} \times \frac{2}{3}$

25. $\frac{6}{10} \times \frac{5}{7}$ **26.** $\frac{6}{7} \times \frac{7}{9}$ **27.** $\frac{15}{20} \times \frac{20}{27}$ **28.** $\frac{5}{7} \times \frac{1}{10}$

29. $\frac{3}{4} \times 16$ **30.** $8 \times \frac{3}{4}$ **31.** $27 \times \frac{5}{6}$ **32.** $\frac{3}{4} \times \frac{12}{15}$

Lesson 7.3

Rewrite the problem by changing each mixed number to a fraction.

1. $\frac{3}{8} \times 5\frac{1}{3}$ **2.** $\frac{5}{7} \times 3\frac{2}{3}$ **3.** $2\frac{4}{9} \times 1\frac{1}{5}$ **4.** $\frac{5}{9} \times 2\frac{1}{8}$

5. $5\frac{2}{3} \times 4\frac{1}{7}$ **6.** $1\frac{5}{8} \times 3\frac{4}{5}$ **7.** $2\frac{2}{5} \times 3\frac{2}{5}$ **8.** $3\frac{3}{7} \times 2\frac{5}{6}$

9. $\frac{5}{7} \times 3\frac{1}{3}$ **10.** $6\frac{5}{6} \times 2\frac{9}{11}$ **11.** $3\frac{2}{3} \times \frac{1}{6}$ **12.** $9\frac{1}{10} \times 4\frac{5}{7}$

13. $2\frac{2}{3} \times 3\frac{2}{7}$ **14.** $2\frac{1}{5} \times 3\frac{1}{7}$ **15.** $3\frac{2}{3} \times 4\frac{3}{5}$ **16.** $5\frac{3}{4} \times 6\frac{8}{9}$

Find the product. Write it in simplest form.

17. $2\frac{2}{3} \times 2\frac{1}{4}$ **18.** $2\frac{1}{2} \times \frac{1}{5}$ **19.** $\frac{1}{5} \times 2\frac{2}{3}$ **20.** $3\frac{1}{6} \times 5\frac{1}{4}$

21. $2\frac{3}{4} \times \frac{5}{6}$ **22.** $2\frac{2}{7} \times \frac{2}{5}$ **23.** $3\frac{2}{3} \times 2\frac{3}{5}$ **24.** $6\frac{8}{9} \times 2\frac{3}{7}$

25. $3\frac{4}{9} \times 1\frac{4}{8}$ **26.** $7\frac{1}{6} \times \frac{3}{4}$ **27.** $2\frac{1}{3} \times 2\frac{3}{7}$ **28.** $7\frac{3}{4} \times 4\frac{12}{13}$

29. $8\frac{5}{8} \times 5\frac{1}{8}$ **30.** $2\frac{1}{12} \times 9\frac{3}{5}$ **31.** $6\frac{1}{10} \times 4\frac{2}{5}$ **32.** $1\frac{2}{3} \times 9\frac{9}{10}$

Lesson 7.4

Find the quotient. Write it in simplest form.

1. $\frac{2}{7} \div \frac{4}{7}$ **2.** $\frac{3}{10} \div \frac{6}{5}$ **3.** $\frac{1}{7} \div \frac{25}{49}$ **4.** $\frac{8}{9} \div \frac{2}{3}$

5. $\frac{2}{9} \div \frac{8}{18}$ **6.** $\frac{3}{8} \div \frac{9}{4}$ **7.** $2 \div \frac{8}{3}$ **8.** $\frac{11}{12} \div \frac{5}{6}$

9. $\frac{3}{4} \div \frac{1}{8}$ **10.** $\frac{5}{6} \div \frac{5}{3}$ **11.** $\frac{7}{8} \div \frac{7}{4}$ **12.** $\frac{5}{7} \div \frac{10}{11}$

13. $\frac{7}{8} \div \frac{3}{4}$ **14.** $\frac{12}{5} \div \frac{6}{10}$ **15.** $\frac{3}{7} \div \frac{18}{21}$ **16.** $\frac{3}{4} \div \frac{7}{12}$

17. $\frac{8}{3} \div \frac{24}{36}$ **18.** $\frac{4}{5} \div \frac{12}{15}$ **19.** $\frac{4}{9} \div \frac{20}{18}$ **20.** $\frac{5}{8} \div \frac{1}{6}$

Lesson 7.5

Work Backward to solve.

1. Sandy is designing a pool for her garden. It is to have an area of $24\frac{1}{3}$ square feet. The length of the pool is to be $7\frac{1}{2}$ feet. Find its width.

2. A spacecraft has left Earth for a star that is $4\frac{1}{3}$ light years away. One-fourth of the way there it will stop at a space station. How far from Earth is the space station?

3. The tallest tree in Maria's yard is $27\frac{1}{2}$ feet tall. The shortest tree is one-third the height of the tallest one. How tall is the shortest one?

4. A window in Susan's family room is $5\frac{1}{3}$ feet wide with an area of $40\frac{1}{2}$ square feet. How tall is it?

CHAPTER 8

Lesson 8.1

For Exercises 1–4, use the figure below.

1. Name four points.

2. Name three line segments.

3. Name two rays.

4. Name one line.

Name the geometric figure that is suggested in Exercises 5–8.

5. a period at the end of a sentence

6. a teacher pointing the way to the cafeteria with one arm

7. main street passing through your town

8. the line that separates the top and bottom numbers in a fraction

Lesson 8.2

Use the figure at the right for Exercises 1–2.

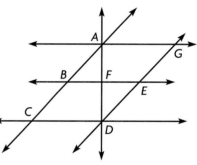

1. Name a line that is parallel to *BE*.

2. Name a line that is perpendicular to and intersects *BE*.

Write *true* or *false*. Change any false statement into a true statement.

3. Parallel lines can be perpendicular.

4. If two lines are perpendicular to the same line, then those two lines are parallel.

Lesson 8.3

Measure the angle. Write the measure and *acute, right,* or *obtuse.*

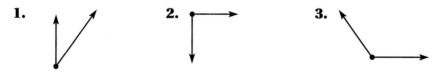

1. 2. 3.

For each time, name the angle that is formed by the hands of the clock.

4. 5:00   5. 4:15   6. 6:00

Lesson 8.4

Use a compass to determine if the line segments in each pair are congruent.
Write *yes* or *no*.

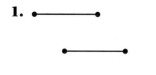

1.

2.

3.

Use a protractor to measure the angles in each pair. Tell whether they are
congruent. Write each measure, and then *yes* or *no*.

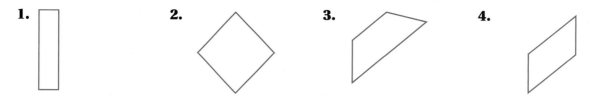

4.

5.

6.

Lesson 8.5

Name the quadrilateral. Tell whether both pairs of opposite sides are
parallel. Write *yes* or *no*.

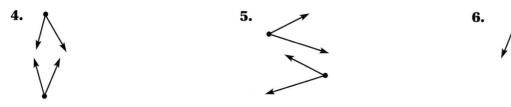

1.

2.

3.

4.

CHAPTER 9

Lesson 9.1

Trace the given figures. Draw all lines of symmetry.

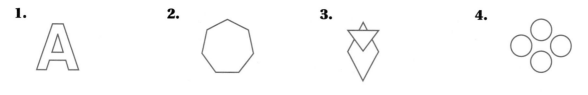

1.

2.

3.

4.

In Exercises 5–8, determine if the dashed line is a line of symmetry.
Write *yes* or *no*.

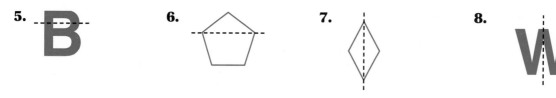

5.

6.

7.

8.

Lesson 9.2

Tell whether the second figure in each set is a *translation*, or a *rotation* of the first figure.

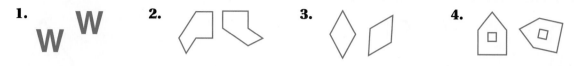

1. **2.** **3.** **4.**

Trace the figure, the line, and point. Draw a reflection about the line and then rotate that figure 90° clockwise.

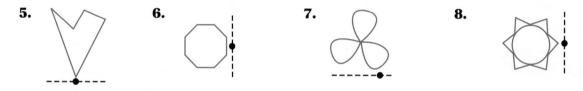

5. **6.** **7.** **8.**

Lesson 9.3

Trace each polygon several times, and cut out your tracings. Tell whether the polygon forms a tessellation. Write *yes* or *no*.

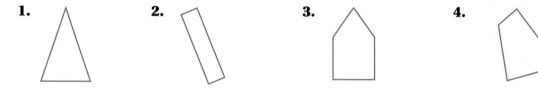

1. **2.** **3.** **4.**

Find the measure of the angles that surround the circled vertex. Then find the sum of the measures.

  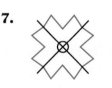

5. **6.** **7.** **8.**

Lesson 9.4

Make a model to solve.

1. Jamal is making a design from the shape shown at the right. He wants the design to tessellate a plane. Can he use this figure to tessellate a plane?

2. Draw your own design that will tessellate a plane.

3. Draw your own design that would not tessellate a plane.

CHAPTER 10

Lesson 10.1

Write *base* or *lateral face* to identify the shaded face.

1.

2.

3.

4.

For Exercises 5–8 name each figure shown in Exercises 1–4 above.

Write *true* or *false* for each statement. Change each false statement into a true statement.

9. If a cone is placed on its lateral surface, it can be rolled in a straight line.

10. In a rectangular prism, the bases are congruent.

11. A cylinder may have a polygon as its base.

Lesson 10.2

Use the figure at the right for Exercises 1–5.

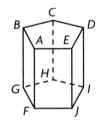

1. Name the vertices.

2. Name the edges.

3. Name the faces.

4. Name the figure.

5. Write *polyhedron* or *not a polyhedron* to describe the figure.

Lesson 10.3

Use the rectangular prism at the right for Exercises 1–3.

1. What are the dimensions of the faces?

2. Draw the faces, with the correct measurements, on an index card.

3. Cut out the faces, and arrange them to form a net for the prism. Tape the pieces together to form a prism.

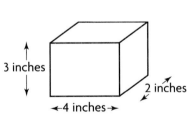

3 inches

2 inches

←4 inches→

Will the arrangement of squares fold to form a cube? Write *yes* or *no*.

4.   5.   6.

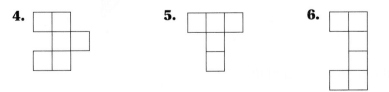

Lesson 10.4

Name the solid figure that has the given views.

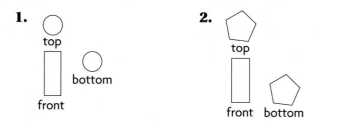

1.   2.

3.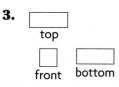

Name the solids that have the following features.

4. The figure contains circles.

5. The figure has one rectangle and several triangles.

6. The figure has six rectangles.

7. The figure has two triangles and three rectangles.

Lesson 10.5

Solve by first solving a simpler problem.

1. Jeremy wants to make a model of a prism whose base has 12 sides. He will use balls of clay for the vertices and straws for the edges. How many balls of clay will he need?

2. Look back at Exercise 1. How many straws will Jeremy need? How many faces will his prism have?

3. Mara wants to make a model of a pyramid whose base has 12 sides. She will use balls of clay for the vertices and straws for the edges. How many balls of clay will she need?

4. Look back at Exercise 3. How many straws will Mara need? How many faces will her pyramid have?

CHAPTER 11

Lesson 11.1

The Band Boosters are planning a spaghetti dinner fund raiser to purchase new uniforms. Determine whether each is a decision they will have to make. Write *yes* or *no*.

1. The location of the dinner

2. The price to charge per person

3. Whether the cafeteria is available

4. The availability of the auditorium for the choir

5. The color of the new football uniforms

6. How many people might attend the dinner

Lesson 11.2

Manuel and his committee are in charge of gathering information to determine a school mascot for the new middle school. There are 680 sixth, seventh, and eighth graders in the new school this September.

1. How many people should the committee survey?

2. Whom should Manuel and his committee survey?

3. How can Manuel get a random sample?

4. Is a random sample of the band members fair? Explain.

Lesson 11.3

Tell whether the sampling method is *biased* or *not biased*. Explain.

The 680 students want good school lunches, and a variety of foods from which to select. A survey was conducted to find out the favorite foods.

1. Randomly survey the teachers.

2. Randomly survey 1 of every 10 students in sixth, seventh, and eighth grades.

3. Ask all your friends in the sixth grade.

4. Randomly survey 10 people who get off the school bus one morning.

5. Randomly ask people in each of the grades whether their favorite fast food is hot dogs or hamburgers.

6. Randomly survey several people from each sports team.

Lesson 11.4

For Exercises 1–3, use the table at the right.

1. Copy and complete the table, by first making a frequency table and then also including the values for a cumulative frequency column.

2. How large was the sample size?

3. Could you use the data in the table to make a line plot? Explain.

Favorite Vegetable	Tally	Frequency	Cumulative Frequency
Peas	卌 I	?	?
Green Beans	卌 IIII	?	?
Potatoes	卌 卌 卌 卌 II	?	?
Carrots	卌 卌 I	?	?
Corn	卌 卌 卌	?	?

CHAPTER 12

Lesson 12.1

Choose the appropriate graph for each set of data. Then make the graph.

1.

WEIGHTS (in lbs.) OF CLASSMATES					
104	115	156	175	121	142
112	102	133	134	143	138
114	134	132	132	132	113
125	136	127	133	122	105
134	168	167	145	113	152

2.

TEAM'S PERCENTAGE OF WINS					
92-93	93-94	94-95	95-96	96-97	97-98
0.188	0.500	0.389	0.245	0.875	0.935

Lesson 12.2

Tell whether a bar graph or a histogram is more appropriate.

1. Heights to the nearest inch of all students in seventh grade.

2. Points for each starting player on the junior high basketball team.

3. Ages of all the cars in the school parking lot.

For Exercises 4–5, use the table below.

NOON TEMPERATURES FOR THIRTY DAYS IN JUNE						
Interval	75-79	80-84	85-89	90-94	95-99	100-104
Number of days	2	4	7	10	6	1

4. Make a histogram.

5. How would the number of days per interval change if instead of six intervals, only three intervals were used?

Lesson 12.3

1. Make a multiple-bar graph showing the weights of these five students before and after a six-week physical fitness program.

WEIGHTS IN POUNDS BEFORE AND AFTER					
	Jamie	Todd	Marquis	Eddie	Sammy
Before	124	132	118	148	142
After	118	118	116	128	126

2. Make a multiple-line graph showing the population changes from 1960 to 1990 for Phoenix and San Antonio.

POPULATION IN PHOENIX AND SAN ANTONIO				
	1960	1970	1980	1990
Phoenix	439,170	584,303	789,704	983,403
San Antonio	587,718	654,153	785,940	935,393

Lesson 12.4

For Exercises 1–2, use the circle graph at the right.

1. How many students were surveyed about their favorite sport to watch on TV?

2. What fraction of the circle does watching auto racing represent?

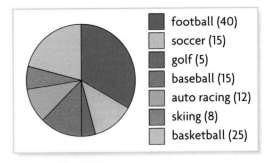

- football (40)
- soccer (15)
- golf (5)
- baseball (15)
- auto racing (12)
- skiing (8)
- basketball (25)

For Exercises 3–5, use the following data.

MONTHLY LIVING EXPENSES					
Rent	Car	Entertainment	Laundry	Food	Miscellaneous
$350.00	$120.00	$105.00	$60.00	$275.00	$90.00

3. Into how many sections would you divide a circle graph for the data?

4. For a circle graph, what angle measure would you use to represent rent?

5. Make a circle graph for the data.

CHAPTER 13

Lesson 13.1

For Exercise 1–2, use the bar graph.

1. Which activity had the most participants? Which had the least?

2. Do intramurals have more or less participants than the total participating on the baseball, basketball, and football teams combined?

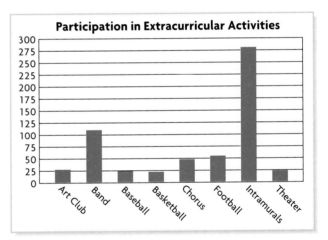

Lesson 13.2

For Exercises 1–3, use the bar graph.

1. Looking at the graph only, how many times as high is the 9th grade bar, compared to the 7th grade bar?

2. Actually, how many more hours did the 9th grade class volunteer than the 7th grade class?

3. Explain why the graph is misleading. Then make a graph that is not misleading.

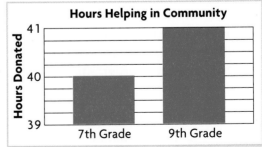

Lesson 13.3

For Exercises 1–4, use the graph at the right, which shows the average cost of a textbook over the time period from 1990 to 1997.

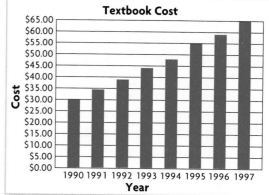

1. How much did an average textbook cost in 1990? in 1995?

2. How much did the average price increase from 1990 to 1997?

3. What has been the pattern for the cost of a textbook from one year to the next year?

4. Based on the data given, what would you predict the average price of a textbook to be in the year 1998?

Lesson 13.4

Copy and complete the table for Exercises 1–8.

	Data	Mean	Median	Mode
1.	4, 5, 7, 8, 8			
2.	6.1, 8.1, 7.4, 7.2			
3.	44, 55, 77, 88, 88			
4.	95, 95, 95, 95, 95			

	Data	Mean	Median	Mode
5.	12, 14, 18, 12, 22			
6.	1, 1, 7, 9, 3, 4, 2, 3			
7.	0.9, 0.8, 0.2, 0.5			
8.	25, 28, 21, 30, 21			

Write *true* or *false* for each.

9. Some sets of data have no mean.

10. The median is always one of the numbers in the set of data.

11. The mode is always one of the numbers in the set of data.

12. The mean, median, and mode can never all be the same number.

Lesson 13.5

For Exercises 1–7, use the table at the right.

1. What is the median?

2. What are the lower and upper quartiles?

3. What are the lower and upper extremes?

4. What is the range?

5. Make a box-and-whisker graph.

6. What fractional part of the data is less than 31? greater than 43?

7. How are the data distributed?

Weekly Sales of Different Pants Sizes					
28	44	28	44	38	42
34	40	34	32	30	44

CHAPTER 14

Lesson 14.1

Use the strategy *account for all possibilities* to solve.

1. Sarah's clothes are color coordinated. If she has three pairs of slacks (light green, light gray, and dark green), three blouses (white, tan, yellow), and two sweater vests (beige and cream), how many different choices does she have in all, if she wears one of each (slacks, blouse, and vest)?

2. On your hamburger deluxe, you may have one beef patty *or* two, provolone cheese *or* swiss cheese, *one* of these condiments (ketchup, mustard, or mayonnaise), and *one* of these health items (lettuce, tomato, or onion). How many choices do you have?

Lesson 14.2

For Exercises 1–4, use the spinner at the right.

1. How many outcomes are there for choosing 5?

2. How many outcomes are there? Name them.

3. How many favorable outcomes are there for choosing a letter?

4. What is the probability of choosing a color?

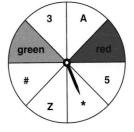

A number cube is numbered 5, 10, 25, 50, 100, 2000. Find each probability.

5. $P(25)$ 6. $P(5 \text{ or } 25)$ 7. $P(\text{a number ending in zero })$

For Exercises 8–11, use the rectangle grid at the right. Find each probability.

8. $P(\text{yellow})$ 9. $P(\text{green or blue})$

10. $P(\text{red})$ 11. $P(\text{red or blue or green})$

	A	B	C	D
1	BLUE	GREEN	GREEN	YELLOW
2	BLUE	GREEN	GREEN	GREEN
3	BLUE	GREEN	GREEN	GREEN
4	RED	RED	RED	RED
5	RED	RED	RED	RED
6	RED	RED	RED	RED

Lesson 14.3

For Exercises 1–6, use the spinner at the right. Find each probability.

1. $P(\text{Al})$ 2. $P(\text{Ed or Dee})$ 3. $P(\text{not Al or Ed})$

4. $P(\text{Garth, Al, or Cara})$ 5. $P(\text{Jim})$ 6. $P(\text{not Garth})$

Cards with team mascots are placed in a hat. There are 4 lions, 3 bears, 5 tigers, and 8 cheetahs. You choose one card without looking. Find each probability.

7. $P(\text{lion})$ 8. $P(\text{tiger or lion})$ 9. $P(\text{member of cat family})$

A bag contains some pencils: 8 blue, 12 brown, 10 red, 4 green, and 6 yellow. You choose one pencil without looking. Find each probability.

10. $P(\text{brown})$ 11. $P(\text{brown or blue})$ 12. $P(\text{orange})$

Lesson 14.4

1. Carlos spins a pointer on a circular disc with the letters A, E, I, O, and U, one letter in each of the five equally spaced regions. He spins the pointer 100 times and records his results in the table below. Find the experimental probability of each letter.

Letter	A	E	I	O	U
Times Landed On	10	30	15	5	40

2. What is the mathematical probability for each letter?

3. How many times can Carlos expect the pointer to land on *E* in the next twenty times?

4. How many times can Carlos expect the pointer to land on *U* if he spins 2,000 times?

CHAPTER 15

Lesson 15.1

Write a numerical or algebraic expression for the word expression.

1. 9 less than 12

2. 13 more than a number, x

3. 22 multiplied by 3

4. 52 divided by 2

5. the product of one half and three fourths

6. 32 increased by y

7. the sum of 9 squared and 12

8. 7 times a number, x

9. 24 decreased by $\frac{1}{5}$

10. x more than y

11. 42 times a number, y

12. 22 divided by 2

Lesson 15.2

Evaluate the numerical expression. Remember the order of operations.

1. $8 - 6 \times 5$

2. $4.8 + 6.2 - 8$

3. $9 \div 3 \times 2$

4. $(4.1 - 3.2) + 2.1$

5. $42 - 0.5$

6. $(28 - 12) + 4$

7. $32 \div 8 + 4$

8. $(16 \div 2)^2$

Evaluate the algebraic expressions for the given value of the variable.

9. $x - 6$, for $x = 12$

10. $y + 13$, for $y = 25$

11. $42 - k$, for $k = 30$

12. $38 + p$, for $p = 22$

13. $a^2 - 6$, for $a = 3$

14. $x^3 - 6$, for $x = 2$

15. $x^3 - 9 + 12$, for $x = 3$

16. $32 - a^4$, for $a = 2$

17. $4 \times p$, for $p = 13.5$

18. $7z + 12$, for $z = 3$

19. $(19 + 5) \div x$, for $x = 8$

20. $12 \times 3 - p$, for $p = 14$

21. $x^2 - x + 10$, for $x = 3$

22. $(12 \div y) \times 9$, for $y = 6$

23. $(7 \times p) \div 2$, for $p = 8$

Lesson 15.3

Compute the output for each input.

1. $x + 9$; for $x = 4, 5, 6$

2. $x + 6.4$; for $x = 9.6, 10.1, 11.6$

3. $y \times 14$; for $y = 8, 12, 10.5$

4. $y^3 - 8$; for $y = 2, 3, 4$

Make an input-output table for the algebraic expression. Evaluate the expression for 2, 3, 4, and 5.

5. $x + 12$

6. $36 - y$

7. $30 \div z$

8. $6.5 \times y$

Determine the input for the given output. Make a table if necessary.

9. $x + 9$
output = 13

10. $x - 14$
output = 9

11. $x \div 4$
output = 9

12. $y \times 8$
output = 32

Lesson 15.4

Determine whether the given value is a solution of the equation. Write *yes* or *no*.

1. $x + 8 = 12, x = 3$

2. $x \div 8 = 3, x = 24$

3. $12 \times y = 36, y = 3$

4. $8 - x = 2, x = 7$

5. $2.5 \times z = 10, z = 4$

6. $8.9 - x = 4.4, x = 3.5$

7. $x - 12.1 = 8.2, x = 20.3$

8. $9.25 \times p = 37, p = 4$

9. $x + 13 - 4 = 16, x = 6$

Use inverse operations to solve. Check your solution.

10. $x + 6 = 12$

11. $16 - r = 8$

12. $p - 12 = 4$

13. $14 = k + 8$

14. $x^2 - 9 = 7$

15. $21 = p \times 3$

16. $48 \div x = 3$

17. $x + 450 = 1{,}120$

18. $4.7 + x = 10$

19. $64 = x - 28$

20. $41.7 = x + 30.3$

21. $y + 9 = 26$

CHAPTER 16

Lesson 16.1

Determine whether the given value is a solution of the equation. Write *yes* or *no*.

1. $3x = 12; x = 4$

2. $7k = 56; k = 7$

3. $\dfrac{p}{16} = 3; p = 48$

4. $\dfrac{a}{14} = 2; a = 26$

5. $27 = 3x; x = 8$

6. $88 = 22n; n = 6$

Use inverse operations to solve. Check your solution.

7. $4x = 24$

8. $8x = 32$

9. $9 = \dfrac{p}{3}$

10. $56 = 7p$

11. $21 = \dfrac{s}{3}$

12. $45 = 9n$

13. $180 = 3d$

14. $462 = \dfrac{a}{3}$

15. $1{,}486 = \dfrac{a}{2}$

Lesson 16.2

Find the number of quarters, dimes, nickels, and pennies in the given dollar amount.

1. $13.00

2. $7.00

3. $45.00

4. $17.00

5. $3.75

6. $21.50

7. $13.25

8. $52.75

Use the formulas on page 314 to convert the English pounds and German marks to U.S. dollars. Round to the nearest penny.

9. 35 pounds

10. 235 pounds

11. 75 pounds

12. 80 marks

13. 283 marks

14. 130 marks

Lesson 16.3

Convert the temperatures from degrees Celsius to degrees Fahrenheit. Round the answer to the nearest degree.

1. 40°C

2. 2.3°C

3. 14°C

4. 20°C

5. 35°C

6. 85°C

7. 60°C

8. 41°C

Convert the temperatures from degrees Fahrenheit to degrees Celsius. Round the answer to the nearest degree.

9. 32°F

10. 47°F

11. 79°F

12. 40°F

13. 105°F

14. 72°F

15. 98°F

16. 81°F

Lesson 16.4

Use the formula $d = r \times t$ to complete.

1. $d = ?$
$r = 35$ mi per hr
$t = 4$ hr

2. $d = ?$
$r = 22$ ft per sec
$t = 50$ sec

3. $d = ?$
$r = 18.3$ km per hr
$t = 5.1$ hr

4. $d = 200$ mi
$r = ?$
$t = 5$ hr

5. $d = 1,600$ km
$r = ?$
$t = 400$ min

6. $d = 180$ ft
$r = ?$
$t = 60$ sec

7. $d = 1,300$ mi
$r = 65$ mi per hr
$t = ?$

8. $d = 4,880$ ft
$r = 20$ ft per sec
$t = ?$

9. $d = 2,100$ km
$r = 70$ km per sec
$t = ?$

10. $d = ?$
$r = 55$ mi per hr
$t = 9$ hr

11. $d = 3,500$ mi
$r = ?$
$t = 50$ hr

12. $d = 3,700$ ft
$r = 740$ ft per sec
$t = ?$

Lesson 16.5

Solve by making a table.

1. Jason contributes to a charity that helps children with serious medical problems. For every dollar he contributes, a local radio station contributes $4. He wants to make sure that a total of $50 goes to the charity. How much does he have to contribute?

2. Shawn runs 5 miles every day. His coach wants him to run 10 km. Is Shawn running far enough? Explain. (HINT: 1 mi = 1.609 km.)

CHAPTER 17

Lesson 17.1

Write the ratio in three ways.

1. four to nine
2. seven to thirteen
3. eight to three
4. seven to three

5. seven to eighteen
6. eleven to thirty-one
7. nineteen to eight
8. thirty-seven to four

For Exercises 9–10, use the figure at right.

9. Find the ratio of blue sections to red sections. Then write three equivalent ratios.

10. Find the ratio of yellow sections to all sections. Then write three equivalent ratios.

Find the missing term that makes the ratios equivalent.

11. $\frac{2}{7}, \frac{\square}{21}$
12. 7 to 3, $\square$ to 6
13. $9 : \square$, 18:8
14. $\frac{3}{7}, \frac{15}{\square}$

Lesson 17.2

Write a ratio in fraction form for each rate.

1. 247 points in 8 games
2. $8 for 2 lb
3. 150 minutes in 3 hr

4. 15¢ per page
5. 75 words in 2 minutes
6. 12 pages in 3 hr

Write the unit rate in fraction form.

7. $2.50 for 10
8. $2.70 for 12
9. 360 people for 3 sq mi

10. 240 miles in 6 hr
11. $2.75 for 25
12. 500 miles per 20 gallons

Lesson 17.3

Write the ratio as a percent.

1. $\frac{27}{100}$

2. $\frac{42}{100}$

3. 74:1

4. 19:100

5. 31:100

Write the decimal as a percent.

6. 0.3

7. 0.09

8. 0.43

9. 0.7

10. 0.0012

Write the ratio as a percent.

11. $\frac{3}{5}$

12. $\frac{1}{3}$

13. $\frac{5}{25}$

14. $\frac{3}{8}$

15. $\frac{1}{20}$

Write the percent as a decimal and as a ratio in simplest form.

16. 30%

17. 75%

18. 96%

19. 42%

Complete by using $<$, $>$, or $=$.

20. $\frac{3}{50}$ ☐ 5%

21. 40% ☐ $\frac{7}{20}$

22. 50% ☐ $\frac{13}{20}$

23. 0.35 ☐ $\frac{1}{3}$

24. 0.015 ☐ 1.5%

25. $\frac{5}{16}$ ☐ 6.25%

26. 25% ☐ $\frac{1}{2}$

27. 20% ☐ $\frac{1}{5}$

Lesson 17.4

Tell what percent of the figure is shaded.

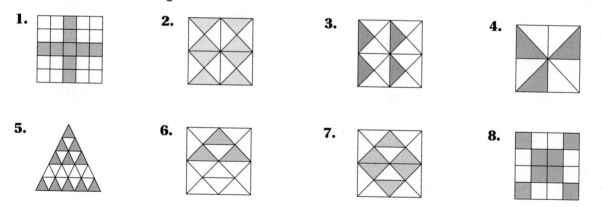

1.

2.

3.

4.

5.

6.

7.

8.

Lesson 17.5

Write a proportion to solve.

1. 7 bags of potatoes weigh 28 lbs. How much do 3 bags weigh?

2. Theresa's car uses 5 gallons of gasoline on a 250-mile trip. How much gasoline does she need to travel 800 miles?

3. The ratio of rolls of film to exposures is 1 to 36. How many exposures are in 5 rolls of film?

CHAPTER 18

Lesson 18.1

Solve the problem by acting it out.

1. Bill reviewed his spending for the past year. His expenditures totaled $15,000. Of these expenditures, 60% went to necessities. How much did he spend on necessities?

2. Samantha and her husband paid an interest rate of 9% last year on their $200,000 home loan. How much did they spend in interest for the year?

Lesson 18.2

Use a ratio in simplest form to find the percent of the number.

1. 20% of 8   **2.** 30% of 90   **3.** 45% of 75   **4.** 50% of 58

5. 25% of 60   **6.** 85% of 220   **7.** 40% of 20   **8.** 10% of 90

Use a decimal to find the percent of the number.

9. 14% of 20   **10.** 35% of 90   **11.** 47% of 71   **12.** 40% of 60

13. 91% of 37   **14.** 38% of 42   **15.** 15% of 62   **16.** 32% of 50

Use the method of your choice to find the percent of the number.

17. 20% of 15   **18.** 150% of 300   **19.** 60% of 120   **20.** 80% of 78

21. 75% of 90   **22.** 300% of 85   **23.** 6.5% of 70   **24.** 42% of 20

25. 53% of 106   **26.** 82% of 82   **27.** 3% of 24   **28.** 120% of 90

Find the sales tax. Round to the nearest cent when necessary.

29. price: $30 <br> tax rate: 6%   **30.** price: $29.99 <br> tax rate: 7.5%   **31.** price: $9.28 <br> tax rate: 8%   **32.** price: $89.50 <br> tax rate: 6.5%

Lesson 18.3

1. Find the angles you would use to make a circle graph.

2. Make a circle graph of the data in Exercise 1.

ICE CREAM		
Item	Percent	Angle
Mint	10%	
Vanilla	35%	
Chocolate	37.5%	
Strawberry	17.5%	

Lesson 18.4

Find the amount of discount.

1. regular price: $31.00
25% off

2. regular price: $65.00
50% off

3. regular price: $42.00
75% off

4. regular price: $116.50
30% off

5. regular price: $87.00
50% off

6. regular price: $97.00
15% off

7. regular price: $38.20
35% off

8. regular price: $137.50
20% off

9. regular price: $95.00
40% off

10. regular price: $325.00
85% off

11. regular price: $180.00
60% off

12. regular price: $225.20
25% off

13. regular price: $37.00
75% off

14. regular price: $223.60
45% off

15. regular price: $14.40
50% off

Lesson 18.5

Find the simple interest.

	Principal	Yearly Rate	Intrest for 1 Year	Intrest for 2Years
1.	$65.00	4%	?	?
2.	$210.00	2.1%	?	?
3.	$735.00	7%	?	?
4.	$1,300.00	3.9%	?	?
5.	$2,250.00	2.3%	?	?
6.	$7,200.00	9%	?	?

Chapter 19

Lesson 19.1

Look at each figure. Tell whether each pair of shapes appear to be *similar, congruent, both,* or *neither.*

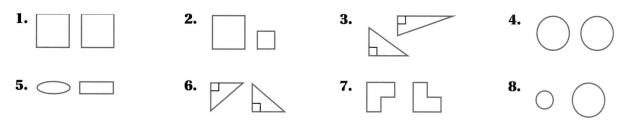

9. Draw two figures that are congruent. Then draw two figures that are not congruent but are similar.

Lesson 19.2

Name the corresponding sides and angles. Write the ratio of the corresponding sides in simplest form.

1.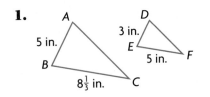

2.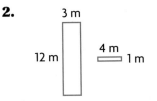

Tell whether the figures in each pair are similar. Write *yes* or *no*.

3.

4.

Lesson 19.3

The figures in each pair are similar. Find *n*.

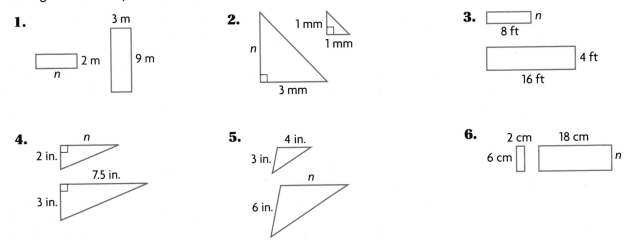

1. 3 m 2 m 9 m *n*

2. 1 mm 1 mm *n* 3 mm

3. *n* 8 ft 16 ft 4 ft

4. *n* 2 in. 7.5 in. 3 in.

5. 4 in. 3 in. *n* 6 in.

6. 2 cm 18 cm 6 cm *n*

Lesson 19.4

Use the two similar right triangles to write a proportion. Then solve.

1.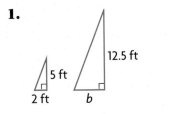
12.5 ft 5 ft 2 ft *b*

2.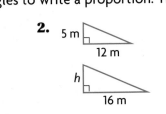
5 m 12 m *h* 16 m

3.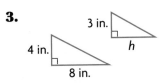
3 in. 4 in. *h* 8 in.

CHAPTER 20

Lesson 20.1

Find the ratio of length to width.

1. $l = 6, w = 4$

2. $l = 5, w = 15$

3. $l = 2.7, w = 0.9$

4. $l = 12, w = 2$

5. $l = 28, w = 27$

6. $l = 8.1, w = 0.9$

7. $l = 72, w = 9$

8. $l = 17, w = 13$

9. $l = 27, w = 5$

Find the missing dimension.

10. scale: 1 in. : 5 ft
drawing length : 7 in.
actual length : ☐ ft

11. scale: 1 in. : 5 ft
drawing length : 4 in.
actual length : ☐ ft

12. scale: 1 in. : 5 ft
drawing length : ☐ in.
actual length : 15 ft

13. scale: 1 in. : 8 ft
drawing length : 2 in.
actual length : ☐ ft

14. scale: 1 in. : 8 ft
drawing length : ☐ in.
actual length : 72 ft

15. scale: 5 cm : 1 mm
drawing length : ☐ cm
actual length : 8 mm

16. scale: 5 cm : 1 mm
drawing length : 15 cm
actual length : ☐ mm

17. scale : 5 m : 1 km
drawing length : 20 m
actual length : ☐ km

18. scale: 5 m : 1 km
drawing length : ☐ m
actual length : 10 km

Lesson 20.2

Write and solve a proportion to find the actual miles. Use a map scale of
1 inch = 25 miles.

1. map distance: $1\frac{1}{2}$ in.

2. map distance: 3 in.

3. map distance: $5\frac{1}{2}$ in.

4. map distance: 7 in.

5. map distance: 9 in.

6. map distance: 18 in.

Lesson 20.3

Draw a diagram to solve.

1. Enya is on a 5-day hiking trip. The first day she hikes 4 miles north. The second day she hikes 3 miles east. The third day she hikes 7 miles south. The fourth day she hikes 3 miles west. In what direction and how many miles does she go on a fifth day to return to the starting point?

2. Tom sails 200 yards north, and then goes 300 yards east. Next he goes north again 500 yards and west for 300 yards. In what direction and how far must he go from that point to meet the first part of his path? How far has it been to his starting point?

3. To get to the library, Michele walks 3 blocks north, and then 1 block west. Draw a map of her walk, using the scale 1 in. = 1 block.

4. To get to a gas station, Dave walked 4 blocks south, and 3 blocks east. In what directions and what distances must Dave walk to get back to his car?

Lesson 20.4

In Exercises 1–6, tell whether the rectangle is a golden rectangle.

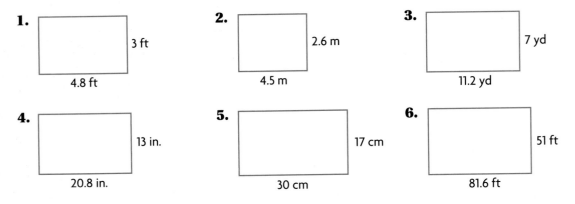

1. 3 ft / 4.8 ft

2. 2.6 m / 4.5 m

3. 7 yd / 11.2 yd

4. 13 in. / 20.8 in.

5. 17 cm / 30 cm

6. 51 ft / 81.6 ft

CHAPTER 21

Lesson 21.1

Change to the given unit.

1. 40 ft = $\underline{?}$ in.

2. 50 yd = $\underline{?}$ ft

3. 20 c = $\underline{?}$ fl oz

4. 16 lbs = $\underline{?}$ oz

5. 13 weeks = $\underline{?}$ days

6. 600 ft = $\underline{?}$ yds

Use a proportion to change to the given unit.

7. 6 years = $\underline{?}$ mo

8. 6 tons = $\underline{?}$ lbs

9. 3 yds = $\underline{?}$ in.

10. 10 gal = $\underline{?}$ qt

11. 72 pt = $\underline{?}$ gal

12. 4 mi = $\underline{?}$ ft

13. 10,560 ft = $\underline{?}$ mi

14. 272 ft = $\underline{?}$ yd $\underline{?}$ ft

15. 248 min = $\underline{?}$ hours $\underline{?}$ min

16. $7\frac{1}{2}$ ft = $\underline{?}$ yd $\underline{?}$ ft $\underline{?}$ in.

Lesson 21.2

Change to the given unit.

1. 300 kL = $\underline{?}$ L

2. 50 dm = $\underline{?}$ cm

3. 600 m = $\underline{?}$ km

4. 20 g = $\underline{?}$ kg

5. 4,000 mm = $\underline{?}$ km

6. 0.004 mm = $\underline{?}$ m

Use a proportion to change to the given units.

7. 12 L = $\underline{?}$ mL

8. 22 kL = $\underline{?}$ L

9. 0.22 m = $\underline{?}$ km

10. 650,000 mg = $\underline{?}$ g

11. 450 mm = $\underline{?}$ cm

12. 0.042 m = $\underline{?}$ mm

13. 0.057 L = $\underline{?}$ mL

14. 200 g = $\underline{?}$ kg

15. 0.0030 kL = $\underline{?}$ L

Lesson 21.3

Tell which measurement is more precise.

1. 8 ft or 97 in.

2. 58 mm or 5 cm

3. 95 mm or 9 cm

4. 73 in. or 2 yds

5. 28 in. or 2 ft

6. 22 mm or 2 cm

7. 482 in. or 40 ft

8. 9.3 cm or 93.3 mm

9. 600 m or 0.6 km

10. 29 yd or 89 ft

11. 5,282 ft or 1 mile

12. 50 cm or 500 mm

13. 85 mm or 8 cm

14. 37 ft or 12 yds

15. 9.1 cm or 9 cm

16. 4 m or 403 cm

17. 532 mm or 53 cm

18. 19 yds or 57 ft

Lesson 21.4

Use the network to find the shorter route. Give the distance. Distances are in kilometers.

1. *BDCE* or *BFEC*

2. *ECBF* or *DBCF*

3. *CFBC* or *CBFE*

4. *EFCD* or *DEFC*

5. *FEDB* or *CFBD*

6. *DCEF* or *EDCB*

7. *BFCD* or *CEDB*

8. *CBDE* or *BDCF*

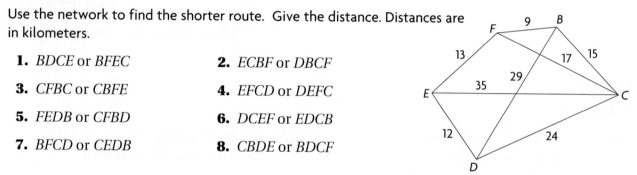

Lesson 21.5

Find the perimeter.

1.

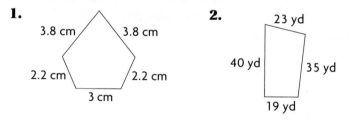

3.8 cm   3.8 cm

2.2 cm   2.2 cm

3 cm

2.

23 yd

40 yd   35 yd

19 yd

3.

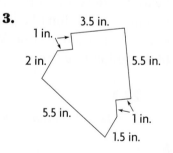

3.5 in.

1 in.

2 in.   5.5 in.

5.5 in.   1 in.

1.5 in.

The perimeter is given. Find the missing length.

4. *P* = 230 cm

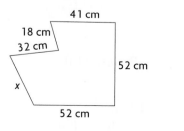

41 cm

18 cm
32 cm

52 cm

x

52 cm

5. *P* = 124 yd

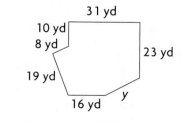

31 yd

10 yd
8 yd

23 yd

19 yd

16 yd   *y*

6. *P* = 401 cm

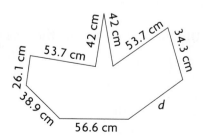

26.1 cm   53.7 cm   42 cm   42 cm   53.7 cm   34.3 cm

38.9 cm   56.6 cm   *d*

CHAPTER 22

Lesson 22.1

Estimate the area of the figure. Each square is 1 in².

1.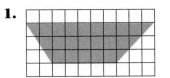

2.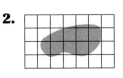

3.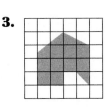

4.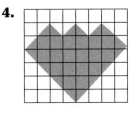

5.

6.

7.

8.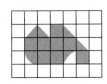

Lesson 22.2

Use a formula to solve.

1. A patio is built out of poured blocks. Each block is 15 in. on each side. 244 blocks were used to build the patio. At a cost of $3.25 for each block, find the area of the patio and its cost.

2. A garden measures 18 ft on each side. What is the area of the garden?

3. Evan marked a 215 ft by 170 ft rectangular section of a parking lot to make an unloading zone. What is the area of the unloading zone?

4. Dean has an oriental rug that is 6 ft long and 12 ft wide. What is the area of the rug?

Lesson 22.3

Use the formula to find the area of the parallelogram.

1. $b = 9$ in.
   $h = 5$ in.

2. $b = 6.2$ m
   $h = 3.1$ m

3. $b = 13\frac{1}{2}$ ft
   $h = 10$ ft

4. $b = 7$ yd
   $h = 4$ yd

5. $b = 5$ ft
   $h = 2$ ft

6. $b = 12$ yd
   $h = 6$ yd

Use the formula to find the area of the triangle.

7. $b = 40$cm
   $h = 20$ cm

8. $b = 4$ ft
   $h = 1\frac{1}{2}$ ft

9. $b = 21$ ft
   $h = 4$ ft

10. $b = 6$ cm
    $h = 0.6$ cm

11. $b = 16$ m
    $h = 1.6$ m

12. $b = 32$ in.
    $h = 0.8$ in.

Lesson 22.4

Find the perimeter and area of each figure. Then double the dimensions and find the new perimeter and area.

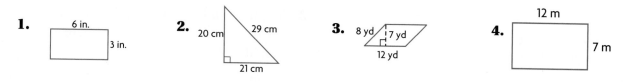

1. 6 in. / 3 in.

2. 20 cm, 29 cm, 21 cm

3. 8 yd, 7 yd, 12 yd

4. 12 m, 7 m

Find the perimeter and area of each figure. Then halve the dimensions and find the new perimeter and area.

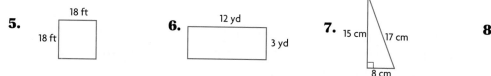

5. 18 ft, 18 ft

6. 12 yd, 3 yd

7. 15 cm, 17 cm, 8 cm

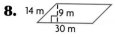

8. 14 m, 9 m, 30 m

Lesson 22.5

Find the area of each circle. Round to the nearest tenth.

1. $r = 9$ yd   **2.** $d = 12$ mm   **3.** $r = 7$ ft   **4.** $r = 4.2$ in.

5. $d = 24$ cm   **6.** $r = 9.2$ mm   **7.** $d = 8$ yd   **8.** $d = 7\frac{1}{2}$ ft

9. $r = 45$ mm   **10.** $r = 16$ in.   **11.** $r = 27$ cm   **12.** $r = 3.5$ cm

CHAPTER 23

Lesson 23.1

Find the volume of each rectangular prism.

1. length = 6 in.
width = 3 in.
height = 4 in.

2. length = 8 ft
width = 5 ft
height = 10 ft

3. length = 10 m
width = 2 m
height = 5 m

4. length = 7 cm
width = 9 cm
height = 3 cm

Find the volume of each triangular prism.

5. length = 12 cm
width = 2 cm
height = 1 cm

6. length = 10 in.
width = 3 in.
height = 2 in.

7. length = 10 ft
width = 2 ft
height = 2 ft

8. length = 12 m
width = 13 m
height = 5 m

9. length = 8 ft
width = 2 ft
height = 2 ft

10. length = 20 m
width = 10 m
height = 10 m

11. length = 20 in.
width = 5 in.
height = 10 in.

12. length = 15 cm
width = 12 cm
height = 8 cm

Lesson 23.2

Find the volume of each prism. Then double the dimensions and find the new volume.

1. length = 4 cm
width = 2 cm
height = 1 cm

2. length = 8 in.
width = 2 in.
height = 2 in.

3. length = 10 ft
width = 5 ft
height = 2 ft

4. length = 7 cm
width = 5 cm
height = 4 cm

5. length = 8 ft
width = 8 ft
height = 8 ft

6. length = 4 m
width = 2 m
height = 2 m

7. length = 6 in.
width = 6 in.
height = 3 in.

8. length = 9 ft
width = 8 ft
height = 6 ft

Find the volume of each prism. Then halve the dimensions and find the new volume.

9. length = 8 cm
width = 4 cm
height = 2 cm

10. length = 8 in.
width = 8 in.
height = 1 in.

11. length = 16 ft
width = 3 ft
height = 2 ft

12. length = 12 cm
width = 13 cm
height = 3 cm

13. length = 12 ft
width = 4 ft
height = 2 ft

14. length = 5 m
width = 4 m
height = 4 m

15. length = 7 in.
width = 4 in.
height = 2 in.

16. length = 5 cm
width = 6 cm
height = 10 cm

Lesson 23.3

Find the volume of each cylinder. Round to the nearest whole number.

1. height = 6 in.
radius = 3 in.

2. height = 5 ft
radius = 2 ft

3. height = 9 m
radius = 1 m

4. height = 9 cm
radius = 2 cm

5. height = 5 in.
radius = 7 in.

6. height = 6 cm
radius = 9 cm

7. height = 2 in.
radius = 12 in.

8. height = 18 ft
radius = 4 ft

9. height = 7 ft
radius = 7 ft

10. height = 6 in.
radius = 8 in.

11. height = 3 ft
radius = 4 ft

12. height = 27 in.
radius = 9 in.

13. height = 7 m
radius = 2 m

14. height = 10 in.
radius = 14 in.

15. height = 2 ft
radius = 6 ft

16. height = 14 cm
radius = 11 cm

17. height = 6 m
radius = 16 m

18. height = 7 in.
radius = 14 in.

19. height = 5 cm
radius = 8 cm

20. height = 22 ft
radius = 5 ft

Lesson 23.4

Find the surface area of each prism.

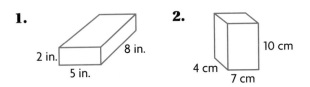

1. 2 in. 8 in. 5 in.

2. 10 cm 4 cm 7 cm

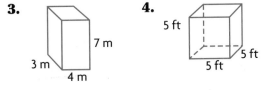

3. 7 m 3 m 4 m

4. 5 ft 5 ft 5 ft

CHAPTER 24

Lesson 24.1

Write an integer to represent each situation. Then describe the opposite situation, and write an integer to represent it.

1. a temperature increase of 5 degrees

2. the wind speed decreases by 12 mph

3. depositing $510 into a savings account

4. a temperature decrease of 7 degrees

Write the absolute value.

5. $|{}^-3|$

6. $|{}^-12|$

7. $|+36|$

8. $|{}^-160|$

9. $|+62|$

10. $|+4|$

11. $|+400|$

12. $|{}^-4|$

13. $|+1|$

14. $|{}^-7|$

15. $|+27|$

16. $|{}^-271|$

Lesson 24.2

Write each rational number in the form $\frac{a}{b}$.

1. $3\frac{1}{6}$

2. 0.5

3. 0.27

4. 13.4

5. $2\frac{2}{5}$

6. 3.18

7. 10.02

8. 300

9. 13

10. 0.04

11. $5\frac{2}{5}$

12. $7\frac{2}{3}$

13. 0.36

14. $5\frac{1}{3}$

15. 312

16. $4\frac{1}{4}$

Lesson 24.3

Find the terminating decimal for the fraction.

1. $\frac{3}{10}$

2. $\frac{2}{5}$

3. $\frac{3}{8}$

4. $\frac{9}{25}$

5. $\frac{3}{20}$

6. $\frac{3}{16}$

7. $\frac{3}{25}$

8. $\frac{17}{200}$

9. $\frac{4}{5}$

10. $\frac{19}{25}$

11. $\frac{9}{16}$

12. $\frac{4}{8}$

Find the repeating decimal for the fraction.

13. $\frac{3}{11}$

14. $\frac{7}{3}$

15. $\frac{2}{9}$

16. $\frac{7}{9}$

17. $\frac{4}{45}$

18. $\frac{17}{9}$

19. $\frac{7}{60}$

20. $\frac{5}{11}$

21. $\frac{4}{15}$

22. $\frac{7}{11}$

23. $\frac{8}{9}$

24. $\frac{7}{6}$

Write the fraction as a decimal.

25. $\frac{7}{50}$

26. $\frac{19}{20}$

27. $\frac{12}{15}$

28. $\frac{10}{9}$

29. $\frac{13}{3}$

30. $\frac{3}{40}$

31. $\frac{7}{4}$

32. $\frac{17}{25}$

Lesson 24.4

Find a rational number between the two given numbers.

1. $\frac{3}{8}$ and $\frac{5}{6}$

2. $\frac{1}{8}$ and $\frac{1}{4}$

3. $\frac{5}{9}$ and $\frac{11}{15}$

4. $1\frac{1}{8}$ and $1\frac{1}{3}$

5. $\frac{-1}{4}$ and $\frac{-1}{8}$

6. $3\frac{5}{6}$ and $4\frac{1}{8}$

7. 1.8 and 1.9

8. −1.5 and −1.3

9. 4.23 and 4.235

10. −3.55 and −3.5

11. 3.02 and 3.03

12. 0.89 and 0.9

13. −5.16 and −5.15

14. $\frac{1}{3}$ and 0.4

15. −6.3 and −6.2

16. $\frac{1}{4}$ and 0.28

17. $\frac{2}{3}$ and 0.8

18. $\frac{1}{9}$ and $\frac{2}{8}$

19. −6.01 and −6

20. 1.4 and 1.41

21. 3 and 3.1

22. −7.08 and −7.07

23. $\frac{1}{7}$ and $\frac{1}{4}$

24. $2\frac{4}{7}$ and $2\frac{7}{8}$

25. $\frac{1}{8}$ and $\frac{2}{6}$

26. 2.5 and 2.6

27. −5.2 and −5

28. 3.8 and 3.82

29. −5 and −4.9

30. $-12\frac{1}{4}$ and $-12\frac{1}{3}$

Lesson 24.5

Compare. Write $<$, $>$, or $=$.

1. $0.4 \square 0.38$

2. $\frac{2}{7} \square 0.25$

3. $-0.6 \square \frac{-2}{5}$

4. $\frac{7}{9} \square 0.8$

5. $0.28 \square \frac{2}{7}$

6. $\frac{3}{13} \square 0.23$

7. $\frac{-4}{9} \square \frac{-2}{5}$

8. $-3 \square -3.1$

9. $2\frac{1}{5} \square 2\frac{4}{13}$

10. $0.87 \square 0.868$

11. $\frac{-5}{8} \square \frac{-5}{9}$

12. $2\frac{1}{16} \square 2\frac{1}{10}$

Compare the rational numbers and order them from least to greatest.

13. $\frac{1}{3}, \frac{2}{5}, 0.38, \frac{1}{2}$

14. $\frac{2}{7}, \frac{1}{3}, \frac{1}{4}, 0.26$

15. $0.92, \frac{9}{8}, \frac{8}{9}, 0.924$

16. $0.23, \frac{2}{9}, \frac{1}{4}, \frac{1}{5}$

17. $\frac{-2}{5}, \frac{-1}{3}, \frac{-1}{2}, \frac{-3}{8}$

18. $\frac{1}{2}, \frac{1}{6}, \frac{1}{9}, 0.1$

CHAPTER 25

Lesson 25.1

Find the sum.

1. $+2 + {}^-6$

2. $^-5 + {}^+5$

3. $^-82 + {}^-8$

4. $^-4 + {}^+8$

5. $^-33 + {}^-18$

6. $^-9 + {}^-4$

7. $+4 + {}^+12$

8. $+64 + {}^-31$

9. $^-82 + {}^+71$

10. $^-21 + {}^-7$

11. $+120 + {}^-42$

12. $^-19 + {}^+19$

13. $^-67 + {}^+58$

14. $^-14 + {}^-57$

15. $+45 + {}^-33$

16. $^-90 + {}^-111$

17. $+47 + {}^-12$

18. $+83 + {}^-15$

Lesson 25.2

Find the difference.

1. $^-2 - {}^+6$
2. $^-3 - {}^-5$
3. $^+5 - {}^+3$
4. $^+5 - {}^-3$
5. $^-8 - {}^-3$
6. $^-8 - {}^+3$
7. $^-7 - {}^-7$
8. $^-12 - {}^-10$
9. $^+3 - {}^+6$
10. $^+7 - {}^-3$
11. $^-13 - {}^+6$
12. $^-2 - {}^-2$
13. $^-7 - {}^+7$
14. $^+23 - {}^+31$
15. $0 - {}^-4$

Lesson 25.3

Complete the pattern.

1. $^-4 \times {}^+2 = {}^-8$
   $^-4 \times {}^+1 = {}^-4$
   $^-4 \times 0 = 0$
   $^-4 \times {}^-1 = \square$
   $^-4 \times {}^-2 = \square$
   $^-4 \times {}^-3 = \square$

2. $^-5 \times {}^+3 = {}^-15$
   $^-5 \times {}^+2 = {}^-10$
   $^-5 \times {}^+1 = {}^-5$
   $^-5 \times 0 = \square$
   $^-5 \times {}^-1 = \square$
   $^-5 \times {}^-2 = \square$
   $^-5 \times {}^-3 = \square$

3. $^-8 \times {}^+3 = {}^-24$
   $^-8 \times {}^+2 = {}^-16$
   $^-8 \times {}^+1 = {}^-8$
   $^-8 \times 0 = 0$
   $^-8 \times {}^-1 = \square$
   $^-8 \times {}^-2 = \square$
   $^-8 \times {}^-3 = \square$

Find the product.

4. $^-3 \times {}^-2$
5. $^-8 \times {}^+6$
6. $^-3 \times {}^-12$
7. $^-8 \times {}^+9$
8. $^+7 \times {}^-12$
9. $^-30 \times {}^-20$
10. $^-31 \times {}^+6$
11. $^-12 \times {}^-5$
12. $^-7 \times {}^-3$
13. $^-10 \times {}^-8$
14. $^+17 \times {}^-9$
15. $^-23 \times {}^-5$

Lesson 25.4

Find the quotient.

1. $^-48 \div {}^-8$
2. $^-72 \div {}^+9$
3. $^+115 \div {}^-23$
4. $^-126 \div {}^+7$
5. $^-36 \div {}^+3$
6. $^-36 \div {}^-3$
7. $^-60 \div {}^-6$
8. $^-84 \div {}^-12$
9. $^+21 \div {}^-7$
10. $^+6 \div {}^-2$
11. $^+186 \div {}^-6$
12. $^+126 \div {}^-7$
13. $^-260 \div {}^-65$
14. $^+84 \div {}^-6$
15. $^+176 \div {}^-11$

CHAPTER 26

Lesson 26.1

Evaluate the numerical expression. Remember the order of operations.

1. $8 + 1 \times 2$

2. $9 - 6 \times 4$

3. $^-4 \times ^-2 + 2$

4. $(^-20 - 14) + 7$

5. $62 \times ^-2$

6. $^-8 \times (^-28 + 26)$

7. $^-4 \times \frac{6}{4}$

8. $9 \times ^-2 + 10$

9. $12 + (^-8 \times ^-4)$

Evaluate the algebraic expression for the given value of the variable.

10. $x - 6$, for $x = 12$

11. $x - 32$, for $x = 11$

12. $9x$, for $x = 9$

13. $42 + k$, for $k = 18$

14. $21 + a$, for $a = ^-14$

15. $y - 9.3$, for $y = 12$

16. $k^2 - ^-8$, for $k = ^-4$

17. $x^3 - ^-3$, for $x = 3$

18. $k^2 + 7$, for $k = 5$

19. $5 + k - 16$, for $k = ^-4$

20. $^-21 + a^3$, for $a = 3$

21. $20.1 - m$, for $m = 3.4$

22. $8z$, for $z = ^-3.5$

23. $y - ^-18$, for $y = ^-4$

24. $\frac{x}{5}$, for $x = 125$

25. $\frac{8}{j}$, for $j = ^-64$

26. $-\frac{72}{y}$, for $y = ^-8$

27. $16k$, for $k = 9$

Lesson 26.2

Solve and check.

1. $x + 8 = 4$

2. $13 = y + 20$

3. $k + 6 = ^-2$

4. $^-9 = n + 8$

5. $c - 10 = ^-4$

6. $^-9 = w - 13$

7. $r - 9 = ^-6$

8. $^-21 = x - 13$

9. $x + 120 = 14$

10. $^-32 = y + 147$

11. $p - 243 = ^-182$

12. $^-847 = y - 391$

13. $^-4y = 36$

14. $^-6x = 72$

15. $^-7x = 42$

16. $^-3m = ^-69$

17. $10y = ^-40$

18. $8d = ^-56$

Lesson 26.3

Find the whole-number solutions to the inequality.

1. $x < 4$

2. $x < 9$

3. $x \leq 3$

4. $x \leq 6$

Write the algebraic inequality represented by the number line.

5.

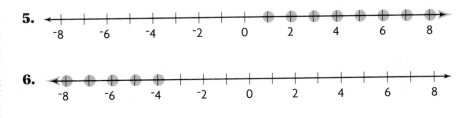

6.

Lesson 26.4

Write the ordered pair for the point on the coordinate plane.

1. point *A*
2. point *B*
3. point *C*
4. point *D*
5. point *E*
6. point *F*
7. point *G*
8. point *H*
9. point *I*
10. point *J*
11. point *K*
12. point *L*
13. point *M*
14. point *N*
15. point *P*
16. point *R*

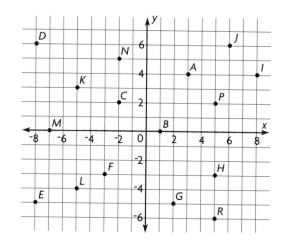

Lesson 26.5

Use the first three values of *x* and *y* to complete the table.

1.

input (*x*)	1	2	3	4	5
output (*y*)	3	4	5	?	?

2.

input (*x*)	1	2	3	4	5
output (*y*)	6	12	18	?	?

3.

input (*x*)	3	4	5	6	7
output (*y*)	1	2	3	?	?

4.

input (*x*)	2	1	0	⁻1	⁻2
output (*y*)	8	4	0	?	?

5. What expression can you write using *x* to get the value of *y* in Exercise 1?

6. What expression can you write using *x* to get the value of *y* in Exercise 2?

CHAPTER 27

Lesson 27.1

Copy each figure onto a coordinate grid. Perform the indicated transformation. Give the new coordinates of the vertices.

1.

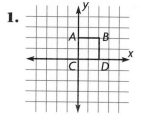

2 units to the right

2.

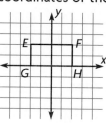

4 units down

3.

reflect across
the *y*-axis

Lesson 27.2

Find a pattern and solve.

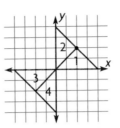

1. Sandra made the design at right. What pattern of transformations did she use?

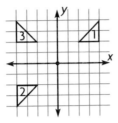

2. Sherry made the design at right. What pattern of transformations did she use?

Lesson 27.3

Draw the next two figures in each geometric pattern.

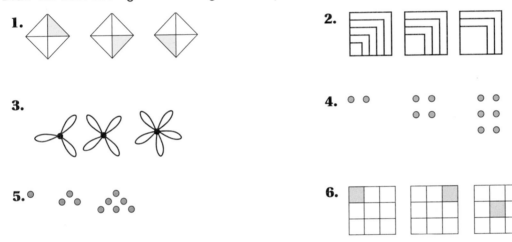

1.

2.

3.

4.

5.

6.

Lesson 27.4

Make the tessellation shape described by each pattern. Then form two rows of a tessellation.

1.   2. 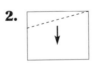   3.   4.

Write *yes* or *no* to tell whether the figure forms a tessellation.

5.   6.   7.   8.

CHAPTER 28

Lesson 28.1

Find the next three terms in the sequence.

1. 7, 10, 13, 16, ...

2. 23, 230, 2,300, 23,000, ...

3. 11, 23, 35, 47, ...

4. 2, 6, 18, ...

5. 7, 17, 37, 67, ...

6. 34, 35, 37, 40, ...

7. 95, 89, 83, 77, ...

8. 0.76, 0.73, 0.70, 0.67, ...

9. 440, 44, 4.4, 0.44, ...

10. 8,000, 4,000, 2,000, 1,000, ...

11. 2.3, 3.9, 5.5, ...

12. 62.4, 62.2, 62.0, 61.8, ...

Lesson 28.2

Find the next three terms in the sequence.

1. $\frac{2}{5}, \frac{4}{5}, 1\frac{1}{5}, ...$

2. $1\frac{3}{5}, 2\frac{1}{5}, 2\frac{4}{5}, ...$

3. $\frac{1}{10}, \frac{1}{5}, \frac{3}{10}, ...$

4. $\frac{3}{14}, \frac{3}{7}, \frac{9}{14}, \frac{6}{7}, ...$

5. $2, 3\frac{1}{2}, 5, ...$

6. $20, 18\frac{2}{3}, 17\frac{1}{3}, ...$

7. $\frac{3}{8}, \frac{7}{8}, \frac{11}{8}, ...$

8. $5, 4\frac{11}{12}, 4\frac{5}{6}, ...$

Lesson 28.3

Find the next three terms in the sequence.

1. $\frac{1}{2}, \frac{3}{2}, \frac{9}{2}, \frac{27}{2}, ...$

2. $\frac{1}{8}, \frac{4}{8}, \frac{16}{8}, ...$

3. $\frac{1}{4}, \frac{1}{8}, \frac{1}{16}, \frac{1}{32}, ...$

4. $\frac{7}{8}, \frac{7}{16}, \frac{7}{32}, ...$

5. $\frac{1}{27}, \frac{1}{9}, \frac{1}{3}, 1, ...$

6. $\frac{1}{3}, \frac{2}{9}, \frac{4}{27}, ...$

Write the first four terms of the sequence.

7. pattern: multiply by 4
first term: $\frac{1}{2}$

8. pattern: multiply by $\frac{1}{3}$
first term: $\frac{1}{6}$

9. pattern: multiply by $\frac{1}{2}$
first term: $\frac{1}{2}$

10. pattern: multiply by $\frac{2}{5}$
first term: $\frac{1}{2}$

11. pattern: multiply by $\frac{3}{8}$
first term: $\frac{1}{2}$

12. pattern: multiply by $\frac{1}{12}$
first term: $\frac{1}{3}$

Lesson 28.4

Find the next three terms in the sequence.

1. −13, −9, −5, −1, ...

2. 4, 1, −2, −5, ...

3. 3, −6, 12, −24, ...

4. 3,000, −300, 30, −3, ...

5. 14, 7, 0 , −7, ...

6. −33, −22, −11, 0, ...

7. 4, −8, 16, −32, ...

8. 27, −9, 3, −1, ...

9. −35, −20, −5, 10, ...

10. −64, −53, −42, −31, ...

11. 2, −10, 50, −250, ...

12. −5,000, 1,000, −200, 40, ...

13. 20, −2, 0.2, −0.02, ...

14. 0.4, −1.2, 3.6, −7.2, ...

15. 250, −50, 10, −2, ...

Be a Good TEST TAKER

A test is one way you show what you have learned. Almost every day in math class you answer questions, use manipulatives, or solve problems on paper. All of these activities are not very different from the tests you will take. So, you are getting ready for tests every day.

Some tests are used to give you a grade. Some are used to see how much you have learned over a long period of time. Some require that you choose from several possible answers. Some require that you explain how you got an answer.

THE TIPS ON THESE PAGES WILL HELP YOU BECOME A BETTER TEST TAKER.

GETTING READY FOR A TEST

What you do at home the night before the test and the morning of the test is very important. Studying hard at the last minute can make you so tired on the day of the test that you will not be able to do your best. But spending some time thinking over the topics and reminding yourself of what you have learned are important. Follow these tips:

- RELAX AND GET A GOOD NIGHT'S SLEEP.

- EAT A GOOD BREAKFAST ON THE MORNING OF THE TEST.

- TELL YOURSELF THAT YOU WILL DO THE VERY BEST YOU CAN DURING THE TEST.

- PROMISE YOURSELF THAT YOU WILL NOT WORRY AND GET UPSET.

TAKING THE TEST

Understand the Directions

If your teacher gives directions orally, look at him or her and pay attention. Ask yourself these questions:

- Can I write on the test itself?
- How many items are on the test?
- How long can I work on the test?

If you don't understand printed directions, reread them. Follow all test directions carefully.

Answer the Questions

Read each question slowly. Be sure that you understsand what you are asked to do. Read the question a second time if necessary.

If you are stopped by an unfamiliar word, try reading on to the end of the sentence or problem to figure it out. Or, try using the first part of the word as a clue, and think of what makes sense in the sentence or problem.

On some tests you are asked to explain your thinking. The questions require that you write an explanation of how you got an answer or why your answer is reasonable. Organize your thoughts before you write!

Choose the Correct Answer

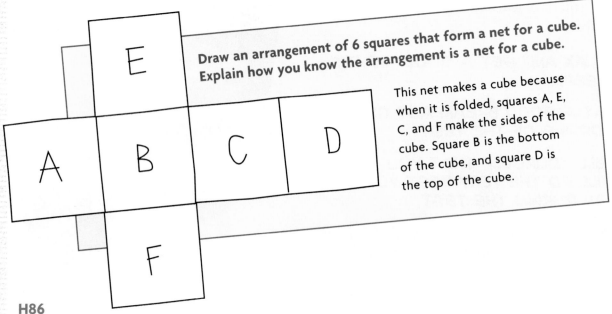

Draw an arrangement of 6 squares that form a net for a cube. Explain how you know the arrangement is a net for a cube.

This net makes a cube because when it is folded, squares A, E, C, and F make the sides of the cube. Square B is the bottom of the cube, and square D is the top of the cube.

Multiple-choice questions have answer choices. Look at each choice carefully.

- Eliminate the ones that look wrong to you. Sometimes estimating can help you eliminate unreasonable choices.

- Then concentrate on the ones that look like reasonable answers.

- If you are still not sure which choice is right, make your best guess.

- Skip the question if all the choices seem correct. Come back to the question later if you have time.

- Some multiple-choice questions give you the answer choice of *None* or *Not Here*. Before you choose one of these or any other answer, reread the problem and recheck your work to be sure your choice is correct.

1. Write an algebraic expression for the following.

6 less than the product of x and 5.

A $6 - x(5)$

B $6 - x + 5$

C $5x - 6$

D $6 < x5$

Think: I need to subtract 6 from the product of x and 5.

- Answer A can't be correct because the product is being subtracted from 6.

- Answer B can't be correct because 5 is being added to x.

- Answer C is correct because 6 is being subtracted from 5x.

Mark Your Answers

Sometimes you will do all your work and show your answers on a piece of notebook paper. At other times you will use a separate answer sheet that may be scored by a machine. A machine can't tell the difference between your answers and other stray marks, so you need to mark carefully.

Study the answer sheet before you begin. Find out which way the numbers go. On some answer sheets you only have to mark the letter of the answer you choose. On other answer sheets you might have to make a mark for each digit of your answer in the right order.

Keep your place on the answer sheet. Make sure you mark the answer for each question on a choice for the right question number.

If you need to change an answer on an answer sheet, erase your mark cleanly. The machine will score an answer that was not erased cleanly as a wrong answer.

Keep Track of Time

If you have a set amount of time to complete a test or a section of a test, look at the clock to find out how much time you have left. Sometimes, your teacher will write on the chalkboard the number of minutes remaining. Glance quickly at the clock or the chalkboard to keep track of the time.

Don't waste a single minute while you are working. Move quickly from one question to another. But don't work so fast that you make careless mistakes.

If you are running out of time and still have many questions to do, try to work faster. If you have only a minute or two left, glance at the remaining questions to find the easiest ones. Answer those and skip the rest.

25 minutes left

EVALUATING HOW YOU DID

When you finish the test, look over your answer sheet. Erase any stray marks you see. Use any time you have left to check your answers and be sure that you have recorded each answer in the correct place on the answer sheet.

If your graded test is returned to you, look carefully at the questions you missed. Try to determine what errors you made. Rework the problems if you can. Don't hesitate to ask your teacher for help in understanding what you did wrong.

YOU CAN BE A GOOD TEST TAKER. THE KEY TO SUCCESS IS DEVELOPING AND USING GOOD TEST-TAKING STRATEGIES.

TABLE OF MEASURES

METRIC UNITS	CUSTOMARY UNITS

Length

1 millimeter (mm) = 0.001 meter (m)
1 centimeter (cm) = 0.01 meter
1 decimeter (dm) = 0.1 meter
1 kilometer (km) = 1,000 meters

1 foot (ft) = 12 inches (in.)
1 yard (yd) = 36 inches
1 yard = 3 feet
1 mile (mi) = 5,280 feet
1 mile = 1,760 yards
1 nautical mile = 6,076.115 feet

Capacity

1 milliliter (mL) = 0.001 liter (L)
1 centiliter (cL) = 0.01 liter
1 deciliter (dL) = 0.1 liter
1 kiloliter (kL) = 1,000 liters

1 teaspoon (tsp) = $\frac{1}{6}$ fluid ounce (fl oz)
1 tablespoon (tbsp) = $\frac{1}{2}$ fluid ounce
1 cup (c) = 8 fluid ounces
1 pint (pt) = 2 cups
1 quart (qt) = 2 pints
1 quart = 4 cups
1 gallon (gal) = 4 quarts

Mass/Weight

1 milligram (mg) = 0.001 gram (g)
1 centigram (cg) = 0.01 gram
1 decigram (dg) = 0.1 gram
1 kilogram (kg) = 1,000 grams
1 metric ton (t) = 1,000 kilograms

1 pound (lb) = 16 ounces (oz)
1 ton (T) = 2,000 pounds

Volume/Capacity/Mass for Water

1 cubic centimeter (cm³) → 1 milliliter → 1 gram
1,000 cubic centimeters → 1 liter → 1 kilogram

TIME

1 minute (min) = 60 seconds (sec)
1 hour (hr) = 60 minutes
1 day = 24 hours
1 week (wk) = 7 days

1 year (yr) = 12 months (mo)
1 year = 52 weeks
1 year = 365 days

FORMULAS

Perimeter

Polygon	$P = $ sum of the lengths of the sides
Rectangle	$P = 2(l + w)$
Square	$P = 4s$

Circumference

Circle	$C = 2\pi r$, or $C = \pi d$

Area

Circle	$A = \pi r^2$
Parallelogram	$A = bh$
Rectangle	$A = lw$
Square	$A = s^2$
Trapezoid	$A = \frac{1}{2} h(b_1 + b_2)$
Triangle	$A = \frac{1}{2} bh$

Surface Area

Rectangular Prism	$S = 2(lh + lw + wh)$

Volume

Cylinder	$V = \pi r^2 h$
Pyramid	$V = \frac{1}{3} Bh$
Rectangular Prism	$V = lwh$
Triangular Prism	$V = \frac{1}{2} lwh$

Other

Celsius (°C)	$C = \frac{5}{9} \times (F - 32)$
Diameter	$d = 2r$
Fahrenheit (°F)	$F = (\frac{9}{5} \times C) + 32$
Pythagorean Property	$c^2 = a^2 + b^2$

Consumer

Distance traveled	$d = rt$
Interest (simple)	$I = prt$

SYMBOLS

$<$	is less than		1:2	ratio of 1 to 2
$>$	is greater than		%	percent
$\leq$	is less than or equal to		$\cong$	is congruent to
$\geq$	is greater than or equal to		$\approx$	is approximately equal to
$=$	is equal to		$\perp$	is perpendicular to
$\neq$	is not equal to		$\parallel$	is parallel to
10^2	ten squared		$\overleftrightarrow{AB}$	line AB
10^3	ten cubed		$\overrightarrow{AB}$	ray AB
10^4	the fourth power of 10		$\overline{AB}$	line segment AB
2^3	the third power of 2		$\angle ABC$	angle ABC
$2.\overline{6}$	repeating decimal 2.666 ...		$m\angle A$	measure of $\angle A$
$^+7$	positive 7		$\triangle ABC$	triangle ABC
$^-7$	negative 7		°	degree (angle or temperature)
(4,7)	the ordered pair 4,7		π	pi (about 3.14)
$5/hr	the rate $5 per hour		P(4)	the probability of the outcome 4

GLOSSARY

A

absolute value The distance from a point on the number line to zero *(page 466)*

acute angle An angle whose measure is greater than 0° and less than 90° *(page 164)*
Example:

acute triangle A triangle whose three angles are acute *(page H21)*
Example:

algebraic expression An expression that is written using one or more variables *(page 292)*
Examples: $x - 4$; $2a + 5$

algebraic operating system (AOS) The set of procedures some calculators use to automatically follow the order of operations *(page 49)*

angle A geometric figure formed by two rays that have a common endpoint *(page 164)*
Example:

area The number of square units needed to cover a given surface *(pages 422, H25)*

Associative Property Addends can be grouped differently; the sum is always the same. Factors can be grouped differently; the product is always the same. *(pages 36, 41)*
Examples: $(8 + 7) + 4 = 8 + (7 + 4)$
$(5 \times 2) \times 6 = 5 \times (2 \times 6)$

average The number obtained by dividing the sum of a set of numbers by the number of addends *(page 258)*

axes The horizontal line (*x*-axis) and the vertical line (*y*-axis) on the coordinate plane *(page 507)*

B

bar graph A graph that uses separate bars (rectangles) of different heights (lengths) to show and compare data *(page 236)*
Example:

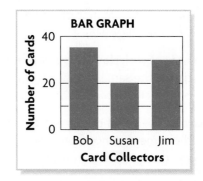

base A number used as a repeated factor *(page 26)*
Example: $8^3 = 8 \times 8 \times 8$
The base is 8. It is used as a factor three times.

base A side of a polygon or a face of a solid figure by which the figure is measured or named *(page 196)*
Examples:

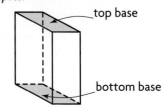

biased sample A sample that is not representative of the population *(page 224)*

bisect To divide into two equal parts *(page 170)*

box-and-whisker graph A graph that shows how far apart and how evenly data are distributed *(page 262)*
Example:

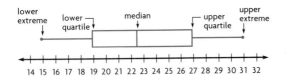

C

capacity The amount of liquid a container can hold *(page 448)*

center of a circle The point inside a circle that is the same distance from each point on the circle *(page H20)*

circle The set of points in a plane that are the same distance from a given point called the center of the circle *(page H20)*

circle graph A graph using a circle that is divided into pie-shaped sections showing percents or parts of the whole *(pages 244, 246, 356)*
Example:

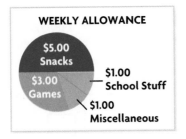

WEEKLY ALLOWANCE

$5.00 Snacks
$3.00 Games
$1.00 School Stuff
$1.00 Miscellaneous

circumference The distance around a circle *(page 416)*

clustering A method used in estimation when all addends are about the same *(page 50)*

Commutative Property The property which states that numbers can be added in any order or can be multiplied in any order without changing the sum or the product. *(pages 36, 41)*
Examples: $9 + 4 = 4 + 9$
$6 \times 3 = 3 \times 6$

compatible numbers Pairs of numbers that are easy to compute mentally *(page 52)*

compensation An estimation strategy in which you change one addend to a multiple of ten and then adjust the other addend to keep the balance *(page 36)*
Example: $16 + 9$
$(16 - 1) + (9 + 1)$
$15 + 10 = 25$

composite number A whole number greater than 1 with more than two whole-number factors *(page 84)*

cone A solid figure with a circular base and one vertex *(page 197)*
Example:

←vertex

congruent Having the same size and shape *(pages 166, 167)*

congruent figures Figures that have the same size and shape *(page H23)*
Example:

congruent polygons Polygons that have all sides and all angles congruent *(pages 370, H23)*
Example:

coordinate plane A plane formed by a horizontal line (*x*-axis) that intersects a vertical line (*y*-axis) at a point called the origin *(page 507)*
Example:

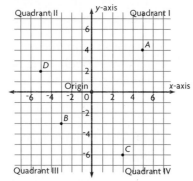

cross products Two equal products obtained by multiplying the second term of each ratio by the first term of the other ratio in a proportion *(page 343)*

cube A rectangular solid figure with six congruent faces *(page 203)*
Example:

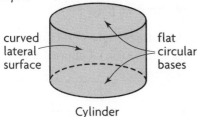

cumulative frequency A column in a table that keeps a running total *(page 229)*

customary measurement system A measurement system that measures length in inches, feet, yards, and miles; capacity in cups, pints, quarts, and gallons; weight in ounces, pounds, and tons; and temperature in degrees Fahrenheit *(pages 406, H24, H89)*

cylinder A solid figure with two parallel bases that are congruent circles *(page 197)*
Example:

curved lateral surface — flat circular bases

Cylinder

D

data A set of information *(page 228)*

decimal A number that uses place value and a decimal point to show tenths, hundredths, thousandths, and so on *(pages 16, H10)*

decimal system A numeration system based on grouping by tens *(page 16)*

degree A unit for measuring angles *(page H90)*

degree Celsius (°C) A metric unit for measuring temperature *(page H90)*

degree Fahrenheit (°F) A customary unit for measuring temperature *(page H90)*

denominator The bottom part of a fraction *(pages 22, H14)*
Example: $\frac{3}{8}$ ← denominator

diameter A line segment through the center of a circle, with endpoints on the circle *(page H20)*
Example:

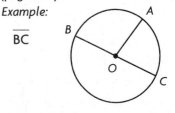

$\overline{BC}$

difference The answer in a subtraction problem *(page H7)*

discount The amount by which the original price is reduced *(page 358)*

Distributive Property Multiplying the sum by a number is the same as multiplying each addend by the number and then adding the products. *(page 40)*
Example: $4 \times (3 + 5) = 32$
$(4 \times 3) + (4 \times 5) = 32$

dividend The number to be divided in a division problem *(page H8)*

divisible A number is divisible by another number if the quotient is a whole number and the remainder is zero. *(page H4)*

divisor The number by which a dividend is divided in a division problem *(page H8)*

E

edge The line segment along which two faces of a solid figure meet *(page 200)*
Example:

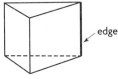

edge

equation An algebraic or numerical sentence that shows two quantities are equal *(page 298)*

equilateral triangle A triangle with three congruent sides *(page H21)*
Example:

4 cm 4 cm

4 cm

equivalent decimals Decimals that name the same number or amount *(pages 18, H11)*
Example: 0.4 = 0.40 = 0.400

equivalent fractions Fractions that name the same amount or part *(page 92)*
Example: $\frac{1}{2} = \frac{4}{8} = \frac{5}{10}$

equivalent ratios Ratios that make the same comparisons *(page 332)*

estimate An answer that is close to the exact answer and is found by rounding, by clustering, or by using compatible numbers *(pages 50, 112, 130, H9, H10)*

evaluate In a numerical expression, perform the operations and write the expression as one number. In an algebraic expression, replace the variable with a number and perform the operation in the expression. *(page 294)*

experimental probability The ratio of the number of times the event occurs to the total number of trials or times the activity is performed *(page 280)*

exponent A number that tells how many times a base is to be used as a factor *(page 26)*
Example: $2^3 = 2 \times 2 \times 2 = 8$
The exponent is 3, indicating that 2 is used as a factor 3 times.

expression A mathematical phrase that combines operations, numerals, and/or variables to name a number *(page 292)*

F

face One of the polygons of a solid figure *(pages 200, H23)*
Example:

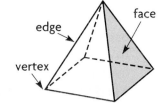

face

edge

vertex

factor A number that is multiplied by another number to find a product *(page H3)*

factor tree A diagram that shows the prime factors of a number *(pages 86, H4)*

Fibonacci sequence The infinite sequence of numbers formed by adding two previous numbers to get the next number *(page 552)*
Example: 1, 1, 2, 3, 5, 8, 13, . . .

fractal-like A mathematical figure that appears to have self-similarity *(page 533)*

fraction A number that names part of a group or part of a whole *(page H14)*

frequency table A table that organizes the total for each category or group *(page 229)*

G

Golden Ratio A ratio equivalent to the value of about 1.6 *(page 394)*

Golden Rectangle A rectangle with a length-to-width ratio of about 1.6 to 1 *(page 394)*
Example:

1

1.6

greater than (>) More than in size, quantity, or amount; the symbol > stands for *is greater than*. *(pages H7, H90)*
Example:
Read 7 > 5 as seven is greater than five.

greatest common factor (GCF) The largest number that is a factor of two or more numbers *(page 89)*

hexagon A six-sided polygon *(page H21)*
Example:

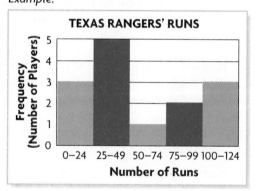

histogram A bar graph that shows the number of times data occur within certain ranges or intervals *(page 240)*
Example:

TEXAS RANGERS' RUNS

Frequency (Number of Players) vs Number of Runs

Number of Runs	Frequency
0–24	3
25–49	5
50–74	1
75–99	2
100–124	3

Identity Property of One The property which states that the product of 1 and any factor is the factor *(pages 42, H6)*

Identity Property of Zero The property which states that the sum of any number and zero is that number *(pages 42, H6)*

indirect measurement The technique of using similar figures and proportions to find a measure *(page 378)*

inequality A mathematical sentence containing <, >, ≤, ≥, or ≠ to show that two expressions do not represent the same quantity *(page 504)*
Examples:
$2 \times 3 < 8$; $6 + 5 > 9$

integers The set of whole numbers and their opposites *(pages 30, 466)*
Example:

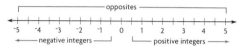

intersecting lines Lines that cross at exactly one point *(page 162)*
Example:

interval A set of numbers such as 7-10, that includes the first and last numbers and the numbers between *(page 230)*

inverse operations Operations that undo each other; addition and subtraction are inverse operations; multiplication and division are inverse operations. *(page 300)*

isosceles triangle A triangle with two congruent sides and two congruent angles *(page H21)*
Example:

8 ft   8 ft
5 ft

lateral face In a prism or a pyramid, a face that is not a base *(page 196)*

least common denominator (LCD) The smallest number, other than zero, that is a multiple of two or more denominators *(page 108)*

least common multiple (LCM) The smallest number, other than zero, that is a multiple of two or more given numbers *(page 88)*

less than (<) Smaller in size, quantity, or amount; the symbol < stands for *is less than*. *(page H90)*
Example: Read 5 < 7 as five is less than seven.

line A straight path that goes on forever in opposite directions *(page 160)*

line graph A graph in which line segments are used to show changes over time *(page 237)*
Example:

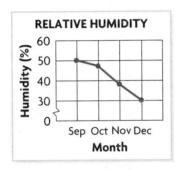

line plot A number line with dots or other marks to show frequency *(page 228)*
Example:

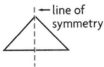

line of symmetry A line that divides a figure into two congruent parts *(page 178)*
Example:

line of symmetry

line segment Part of a line; it has two endpoints. *(page 160)*
Example:

R •————————• S

line symmetry A figure has line symmetry if a line can separate the figure into two congruent parts *(page 178)*

liter A metric unit for measuring capacity *(pages 408, H89)*

lower extreme The least number in a set of data *(page 262)*

lower quartile The median of the lower half of a set of data *(page 262)*

map scale A ratio that compares the distance on a map with the actual distance *(page 390)*

mathematical probability The number of favorable outcomes divided by the number of possible outcomes *(page 272)*

mean The average of a group of numbers *(pages 258, H26)*

measure of central tendency A measure used to describe data; the mean, median, and mode are measures of central tendency. *(pages 258, H26)*

median The middle number or the average of the two middle numbers in an ordered set of data *(pages 258, H26)*

meter A metric unit for measuring length *(pages 408, H24, H89)*

metric system A measurement system that measures length in millimeters, centimeters, meters, and kilometers; capacity in liters and milliliters; mass in grams and kilograms; and temperature in degrees Celsius *(pages 408, H24, H89)*

mixed number A number that is made up of a whole number and a fraction *(page 96)*
Examples:
$$3\frac{3}{4}, \ 1\frac{7}{8}, \ 2\frac{1}{6}$$

mode The number or numbers that occur most often in a collection of data; there can be more than one mode or none at all. *(pages 258, H26)*

multiple The product of a given number and a whole number *(pages 84, H3)*

multiple-bar graph A bar graph showing two or more sets of data at once *(page 242)*
Example:

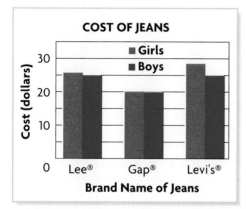

multiple-line graph A line graph showing two or more sets of data at once *(page 242)*
Example:

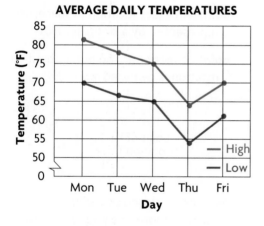

negative integers Integers less than 0 *(pages 30, 466)*

net An arrangement of two-dimensional figures that folds to form a three-dimensional figure *(page 202)*
Example:

network A graph with vertices and edges *(page 412)*
Example:

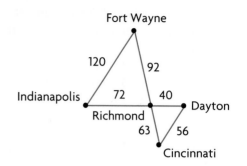

numerator The top part of a fraction *(pages 22, H14)*
Example: $\frac{3}{8}$ ← numerator

numerical expression A mathematical phrase that includes only numbers and operation symbols *(page 292)*

obtuse angle An angle whose measure is greater than 90° and less than 180° *(page 164)*
Example:

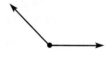

obtuse triangle A triangle that has one obtuse angle *(page H21)*
Example:

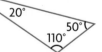

octagon An eight-sided polygon *(page 172)*
Example:

opposites Two numbers that are an equal distance from 0 on the number line *(page 30)*

ordered pair A pair of numbers used to locate a point on a coordinate plane *(page 506)*

order of operations The order in which operations are done; first, do the operations within parentheses; next, clear exponents; then, multiply and divide from left to right; and last, add and subtract from left to right. *(page 48)*

Example: $3 \times (8 - 3) + 2^2$
$3 \times (5) + 2^2$
$3 \times (5) + 4$
$15 + 4 = 19$

origin The point on the coordinate plane where the *x*-axis and the *y*-axis intersect, (0,0) *(page 507)*

outcome A possible result in a probability experiment *(page 272)*

parallel lines Lines in a plane that do not intersect *(page 162)*
Example:

$\overleftrightarrow{AB} \parallel \overleftrightarrow{CD}$
Read: Line *AB* is parallel to line *CD*.

parallelogram A quadrilateral whose opposite sides are parallel and congruent *(page H22)*

pentagon A five-sided polygon *(page 172)*
Example:

5 sides, 5 angles

percent The ratio of a number to 100; *percent* means "per hundred." *(page 336)*
Example:
$25\% = \frac{25}{100}$

perimeter The distance around a polygon *(page 414)*

period Each group of three digits in a number *(pages 16, H6, H10)*

perpendicular lines Lines that intersect to form 90°, or right, angles *(page 162)*
Example:

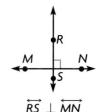

$\overleftrightarrow{RS} \perp \overleftrightarrow{MN}$
Read: Line *RS* is perpendicular to line *MN*.

pi (π) The ratio of the circumference of a circle to its diameter; $\pi \approx 3.14$ or $\frac{22}{7}$ *(page 417)*

place value The value of a digit as determined by its position in a number *(page 16)*

plane A flat surface that goes on forever in all directions *(page 160)*

point An exact location in space, usually represented by a dot *(page 160)*

point of rotation The point about which a rotation is centered *(page 179)*
Example:

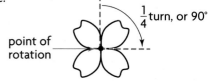

polygon A closed plane figure formed by three or more line segments *(page 172)*

polyhedron A solid figure with flat faces that are polygons *(page 196)*

population A particular group of people, such as sixth graders *(page 222)*

positive integer A whole number greater than zero *(pages 30, 466)*

precision A property of measurement that is related to the unit of measure used; the smaller the unit of measure used, the more precise the measurement is. *(page 410)*

prediction An estimate made by looking at a trend over time and then extending that trend to describe a future event *(page 256)*

prime factorization A number written as the product of all its prime factors *(page 86)*
Example:

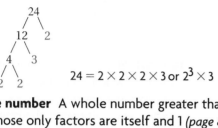

$24 = 2 \times 2 \times 2 \times 3$ or $2^3 \times 3$

prime number A whole number greater than 1 whose only factors are itself and 1 *(page 84)*

principal The amount of money borrowed or saved *(page 360)*

prism A solid figure whose bases are congruent, parallel polygons and whose other faces are parallelograms *(page 196)*
Examples:

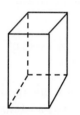

rectangular prism   triangular prism

probability (P) The chance that an event will occur expressed as the ratio of the number of favorable outcomes to the number of possible outcomes *(page 272)*

product The answer in a multiplication problem *(page H8)*

Property of Zero The property which states that the product of 0 and any number is 0 *(pages 42, H6)*

proportion A number sentence or an equation that states that two ratios are equivalent *(page 342)*
Example: $\frac{2}{3} = \frac{4}{6}$

pyramid A solid figure whose base is a polygon and whose other faces are triangles with a common vertex *(page 198)*
Examples:

rectangular pyramid   triangular pyramid

quadrant One of the four regions of the coordinate plane *(page 507)*

quadrilateral A four-sided polygon *(page 172)*
Examples:

quotient The answer in a division problem *(page H8)*

radius A line segment with one endpoint at the center of a circle and the other endpoint on the circle *(page H20)*

random sample A sample in which every individual or item in the population has an equal chance of being selected *(page 222)*

range The difference between the greatest and least numbers in a set of numbers *(page 230)*

rate A ratio that compares two quantities having different units of measure *(page 334)*

ratio A comparison of two numbers *(page 332)*
Example: 3 to 5, or 3:5, or $\frac{3}{5}$

rational number Any number that can be expressed as a ratio in the form of $\frac{a}{b}$ where a and b are integers and $b \neq 0$ *(page 468)*

ray A part of a line that has one endpoint and goes on forever in only one direction *(page 160)*
Example:

ray JK, $\overrightarrow{JK}$

reciprocal One of two numbers whose product is 1 *(page 146)*

rectangle A parallelogram with four right angles *(page H22)*

reflection The figure formed by flipping a geometric figure over a line of reflection to obtain a mirror image *(page 184)*
Example:

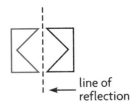

line of reflection

regular polygon A polygon in which all sides and all angles are congruent *(page 172)*

relation A set of ordered pairs *(page 510)*

repeating decimal A decimal in which one or more digits repeat endlessly *(page 471)*
Example: 0.333 . . . , or 0.3̄

rhombus A parallelogram whose four sides are congruent and whose opposite angles are congruent *(page H22)*
Example:

right angle An angle whose measure is 90°
(page 164)
Example:

right triangle A triangle with exactly one right angle *(page H21)*
Example:

60° 90° 30°

rotation A turning of a figure about a fixed point *(page 184)*
Example:

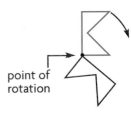

point of rotation

rotational symmetry A figure has rotational symmetry when it can be rotated less than 360° around a central point, or point of rotation and still match the original figure. *(page 179)*

S

sample A group of people or objects chosen from a larger group to provide data to make predictions about the larger group *(page 222)*

sample space The set of all possible outcomes *(page 270)*

scale The ratio between two sets of measurements *(page 385)*

scale drawing A reduced or enlarged drawing whose shape is the same as an actual object and whose size is determined by the scale *(page 384)*

scalene triangle A triangle with no congruent sides *(page H21)*
Example:

self-similarity A figure has self-similarity if it contains a repeating pattern of smaller and smaller parts that are like the whole, but different in size. *(page 532)*
Example:

sequence An ordered set of numbers *(page 538)*

similar figures Geometric figures that have the same shape and angles of the same size *(page 368)*

simple interest The amount obtained by multiplying the principal by the rate by the time *(page 360)*

simplest form A fraction is in simplest form when the numerator and denominator have no common factors other than 1. *(page 94)*

simulation A model of an experiment that would be too difficult or too time-consuming to actually perform *(page 278)*

solid figure A three-dimensional figure *(pages 196, H22)*

solution A value that, when substituted for the variable, makes an equation true *(page 299)*

solve To find the value of a variable that makes an equation true *(page 299)*

sphere A solid with all points an equal distance from the center *(page H22)*
Example:

square To square a number means to multiply it by itself *(page 28)*
Example: $3^2 = 3 \times 3 = 9$

square A rectangle with four congruent sides *(page H22)*

square root One of the two equal factors of a number *(page 29)*
Example:
5 is the square root of 25 because $5^2 = 25$.

standard form The form in which numerals are usually written, with digits 0 through 9, separated into periods by commas *(page 16)*
Example: 634,578,910

stem-and-leaf plot A method of organizing data in order to make comparisons; the ones digits appear horizontally as leaves, and the tens digits and greater appear vertically as stems. *(page 238)*
Example:

Card-Stacking Competition

Stem	Leaves
1	1 2 6 8 8 9
2	1 1 1 9
3	1 2 3 3
4	2 7
5	0 4

Key: 3 | 2 = 32

straight angle An angle whose measure is 180° *(page 164)*
Example:

sum The answer in an addition problem *(page 50)*

surface area The sum of the areas of all the faces, or surfaces, of a solid figure *(page 452)*

T

tally table A table with categories for recording each piece of data as it is collected *(page 228)*

tangram A puzzle consisting of seven polygon-shaped pieces that can be rearranged to make various figures or shapes *(page 182)*

term One of the numbers in a ratio *(page 332)*

term Each of the numbers in a sequence *(page 538)*

terminating decimal A decimal that ends; a decimal for which the division operation results in a remainder of zero *(page 470)*
Examples: $\frac{1}{2} = 0.5$
$\frac{5}{8} = 0.625$

tessellation An arrangement of shapes that completely covers a plane, with no gaps and no overlaps *(page 188)*

transformation A movement that doesn't change the size or shape of a figure is a rigid transformation. *(page 184)*

translation A movement of a geometric figure to a new position without turning or flipping it *(page 184)*
Example:

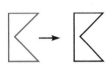

trapezoid A quadrilateral with only two parallel sides *(page H22)*
Example:

tree diagram A diagram that shows all the possible outcomes of an event *(page 270)*

triangle A three-sided polygon *(page 172)*

triangular number A number that can be represented by a triangular array *(page 538)*
Example: 6

unit rate A rate in which the second term is 1 *(page 334)*

unlike fractions Fractions whose denominators are not the same *(page 104)*

upper extreme The greatest number in a set of data *(page 262)*

upper quartile The median of the upper half of a set of data *(page 262)*

variable A letter or symbol that stands for one or more numbers *(page 292)*

Venn diagram A display which shows relationships among sets by using ovals or other shapes to represent individual sets *(page 468)*
Example:

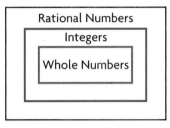

| Rational Numbers |
| Integers |
| Whole Numbers |

vertex The point where two or more rays meet; the point of intersection of two sides of a polygon; the point of intersection of three or more edges of a solid figure; the top point of a cone *(pages 164, 197)*

volume The number of cubic units needed to occupy a given space *(page 440)*

x-axis The horizontal axis on the coordinate plane *(page 507)*

x-coordinate The first number in an ordered pair; tells whether to move right or left along the x-axis of the coordinate plane *(page 507)*

y-axis The vertical axis on the coordinate plane *(page 507)*

y-coordinate The second number in an ordered pair; tells whether to move up or down along the y-axis of the coordinate plane *(page 507)*

Answer

Chapter 1
Page 17
1. 4 ones 3. 4 thousandths 5. two tens
7. eight hundred thousands 9. forty six
11. one thousand, five hundred, and one
tenth 13. one million, thirty-seven
thousand, eight hundred four 15. 0.15
17. 850,247.56 19. 9,456,302

Pages 20–21
1. < 3. > 5. < 7. > 9. > 11. =
13. > 15. > 17. $290 < 295 < 298 < 299$
19. $1.124 < 1.214 < 1.412 < 1.421$ 21. 0.405
$< 1.05 < 1.125 < 1.25 < 1.405 < 1.45$
23. $\$90.22 > \$90.12 > \$90$
25. $1,623 > 1,263 > 1,260 > 1,203 > 1,063$
27. 1.51, 1.55
29. 1.5342, 1.56, 1.507
31. $0.280 > 0.268 > 0.265$
33. 6 ten > 1 ten, so $5,361 > 5,316$

Pages 24–25
1. $\frac{2}{10}$ 3. $\frac{17}{100}$ 5. $\frac{99}{100}$ 7. $\frac{25}{100}$ 9. $\frac{325}{1,000}$
11. $\frac{3,525}{10,000}$ 13. 0.2 15. 0.6 17. 0.9375
19. 0.0087 21. 0.25 23. 0.4375 25. <
27. > 29. > 31. < 33. > 35. 0.125
37. 0.75 lb 39. The 3 keeps repeating; the
remainder is not 0. 41. 16 43. 216
45. 5 in.^{2} 47. G

Page 27
1. 12^3 3. 4^4 5. 7^2 7. 2^5 9. 1,024 11. 1
13. 8 15. 14 17. 100,000,000 19. 1,024
21. 8^2 23. 10^3 25. greater than;
$18,000,000 > 10,000,000$
27. Take 2 steps backward. 29. Jump down.
31. 200,024 33. 70,072 35. G

Page 31
1. < 3. > 5. > 7. < 9. > 11. <

33.

1. C 3.

Chapter 2
Page 37
1. 36 3. 45 5.
13. 41 15. 41 17.
23. 23 25. 21 27. 14
33. 72 CDs 35. 13 CDs

Page 39
1. 31 grape, 19 orange
3. 21 red tickets, 15 blue tickets 5.
7. 165 mi 9. 77 stamps 11. 6 weeks

Pages 42–43
1. 8 3. 55 5. 32 7. 153 9. 192 11. 168
13. 495 15. 180 17. 294
23. 37; Commutative 25. 1; One
27. 9; Commutative 29. 90 31. 63 33. 210
35. 72 37. 52 39. 224 41. 240 items
43. Possible response:
$24 \times 14 = (24 \times 10) + (24 \times 4),$
$26 \times 14 = (26 \times 10) + (26 \times 4),$
$28 \times 14 = (28 \times 10) + (28 \times 4)$

Pages 46–47
1. 297 3. 10,920 5. 26,522 7. 1,245,454
9. 1,312,800 11. 14 13. 12 15. 26 r2
17. 503 r2 19. 301 21. 125 23. 89
25. 14 27. $20\frac{5}{12}$ 29. $241\frac{11}{14}$ 31. $25\frac{11}{15}$
33. $317\frac{7}{9}$ 35. $50\frac{4}{9}$ 37. \$8,700 39. $49\frac{4}{6}$
43. 13,000 45. $^-3 < ^-2 < 4 < 5$
47. $^-6 < ^-3 < 6 < 7$ 49. G

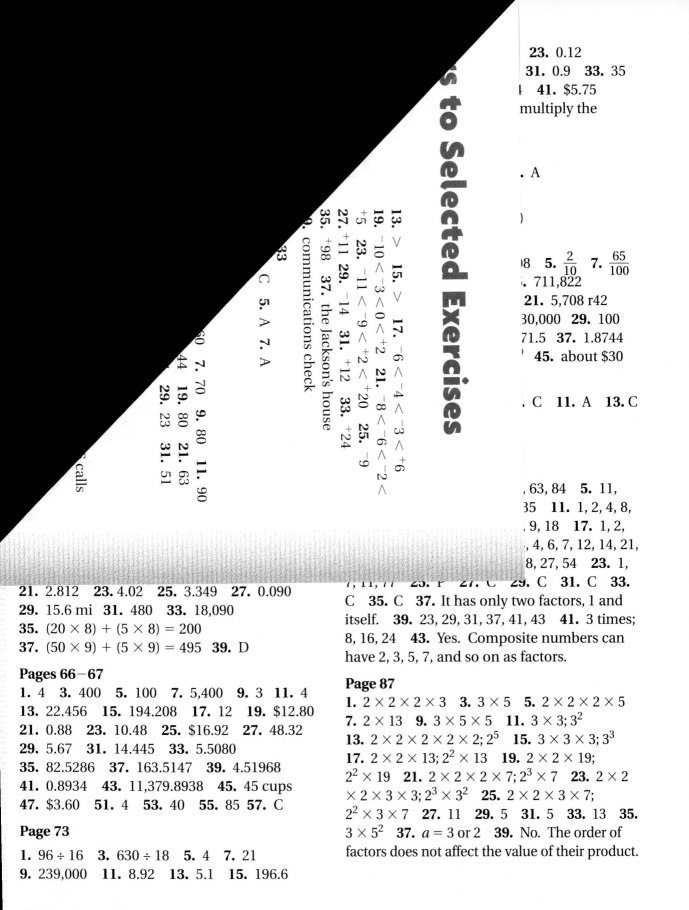

23. 0.12
31. 0.9 **33.** 35
41. $5.75
multiply the

. A

5. $\frac{2}{10}$ **7.** $\frac{65}{100}$
. 711,822
21. 5,708 r42
30,000 **29.** 100
71.5 **37.** 1.8744
45. about $30

. C **11.** A **13.** C

13. > **15.** > **17.** $^-6 < ^-4 < ^-3 < ^+6$
19. $^-10 < ^-3 < 0 < ^+2$ **21.** $^-8 < ^-6 < ^-2 <$
23. $^-11 < ^-9 < ^+2 < ^+20$ **25.** $^-9$
27. $^+11$ **29.** $^-14$ **31.** $^+12$ **33.** $^+24$
35. $^+98$ **37.** the Jackson's house
. communications check

C **5.** A **7.** A

7. 70 **9.** 80 **11.** 90
19. 80 **21.** 63
29. 23 **31.** 51

calls

, 63, 84 **5.** 11,
35 **11.** 1, 2, 4, 8,
, 9, 18 **17.** 1, 2,
, 4, 6, 7, 12, 14, 21,
8, 27, 54 **23.** 1,
, 11, 77 **25.** P **27.** C **29.** C **31.** C **33.**
C **35.** C **37.** It has only two factors, 1 and
itself. **39.** 23, 29, 31, 37, 41, 43 **41.** 3 times;
8, 16, 24 **43.** Yes. Composite numbers can
have 2, 3, 5, 7, and so on as factors.

21. 2.812 **23.** 4.02 **25.** 3.349 **27.** 0.090
29. 15.6 mi **31.** 480 **33.** 18,090
35. $(20 \times 8) + (5 \times 8) = 200$
37. $(50 \times 9) + (5 \times 9) = 495$ **39.** D

Pages 66−67
1. 4 **3.** 400 **5.** 100 **7.** 5,400 **9.** 3 **11.** 4
13. 22.456 **15.** 194.208 **17.** 12 **19.** $12.80
21. 0.88 **23.** 10.48 **25.** $16.92 **27.** 48.32
29. 5.67 **31.** 14.445 **33.** 5.5080
35. 82.5286 **37.** 163.5147 **39.** 4.51968
41. 0.8934 **43.** 11,379.8938 **45.** 45 cups
47. $3.60 **51.** 4 **53.** 40 **55.** 85 **57.** C

Page 73
1. $96 \div 16$ **3.** $630 \div 18$ **5.** 4 **7.** 21
9. 239,000 **11.** 8.92 **13.** 5.1 **15.** 196.6

Page 87
1. $2 \times 2 \times 2 \times 3$ **3.** 3×5 **5.** $2 \times 2 \times 2 \times 5$
7. 2×13 **9.** $3 \times 5 \times 5$ **11.** $3 \times 3; 3^2$
13. $2 \times 2 \times 2 \times 2 \times 2; 2^5$ **15.** $3 \times 3 \times 3; 3^3$
17. $2 \times 2 \times 13; 2^2 \times 13$ **19.** $2 \times 2 \times 19;$
$2^2 \times 19$ **21.** $2 \times 2 \times 2 \times 7; 2^3 \times 7$ **23.** 2×2
$\times 2 \times 3 \times 3; 2^3 \times 3^2$ **25.** $2 \times 2 \times 3 \times 7;$
$2^2 \times 3 \times 7$ **27.** 11 **29.** 5 **31.** 5 **33.** 13 **35.**
3×5^2 **37.** $a = 3$ or 2 **39.** No. The order of
factors does not affect the value of their product.

Page 91

1. 12 **3.** 6 **5.** 20 **7.** 28 **9.** 24 **11.** 60
13. $2 \times 2 \times 2$; 8 **15.** 2×2; 4 **17.** 2×3; 6
19. 3; 3 **21.** 2×2; 4 **23.** 3 **25.** 3 **27.** 2
29. 3 **31.** 1 **33.** 3 **35.** 60 of each **37.** 12
packages **39.** 5 **41.** 4 **43.** 7 **45.** 7.82
47. 20 **49.** H

Page 95

1. 1;1 **3.** 1, 3; 3 **5.** 1, 5; 5 **7.** $\frac{1}{6}$ **9.** $\frac{1}{8}$
11. $\frac{5}{9}$ **13.** $\frac{1}{5}$ **15.** $\frac{7}{11}$ **17.** $\frac{7}{10}$ **19.** 6 **21.** 8
23. 12 **25.** any multiple of 4
27. when 1 is the GCF of the numerator and
the denominator **29.** 3 **31.** 0.035 **33.** 0.6
35. J

Page 97

1. $1\frac{3}{4}$ **3.** $5\frac{1}{2}$ **5.** 9 **7.** $5\frac{1}{6}$ **9.** $4\frac{1}{7}$ **11.** $4\frac{11}{16}$
13. $3\frac{4}{9}$ **15.** $1\frac{1}{5}$ **17.** $5\frac{1}{2}$ **19.** $\frac{11}{3}$ **21.** $\frac{16}{3}$
23. $\frac{37}{9}$ **25.** $\frac{37}{4}$ **27.** $\frac{53}{11}$ **29.** $\frac{33}{5}$ **31.** $\frac{29}{4}$ **33.** $\frac{15}{2}$
35. $\frac{25}{12}$ **37.** $9\frac{1}{2}$ and $\frac{19}{2}$ **39.** $\frac{28}{3}$ and $9\frac{1}{3}$
41. $\frac{17}{8}$ and $2\frac{1}{8}$ **43.** 4 **45.** 55
47. Yes. The pillow cover requires only $1\frac{1}{2}$ yd
of fabric.
49. Possible answer: change whole number to
a fraction; find the sum of fractions.

Page 99

1. C **3.** C **5.** E **7.** C

Chapter 5
Page 103

1. 5 **3.** 11 **5.** 1 **7.** $\frac{2}{3}$ **9.** 1 **11.** $\frac{3}{7}$ **13.** $\frac{1}{5}$
15. $\frac{1}{8}$ **17.** $\frac{1}{10}$ **19.** $\frac{1}{3}$ **21.** $\frac{14}{17}$ **23.** $1\frac{1}{4}$ **25.** $\frac{3}{5}$
27. $\frac{1}{3}$ **29.** $\frac{1}{5}$ **31.** $\frac{1}{6}$ **33.** $1\frac{1}{2}$ **35.** $\frac{1}{2}$
37. $\frac{4}{5}$ of the lumber **39.** $\frac{2}{4}$, or $\frac{1}{2}$ of the album

Page 107

1. $\frac{1}{4} + \frac{1}{12} = \frac{4}{12}$; or $\frac{1}{3}$ **3.** $\frac{2}{5} + \frac{1}{10} = \frac{5}{10}$, or $\frac{1}{2}$
5. $\frac{4}{6}$, or $\frac{2}{3}$ **7.** $\frac{10}{24}$, or $\frac{5}{12}$ **9.** $\frac{1}{6}$ **11.** $\frac{5}{8}$

13. $\frac{3}{10}$ **15.** $\frac{47}{60}$ **17.** $\frac{3}{8}$ **19.** $\frac{47}{60}$ **21.** $\frac{11}{12}$ yd

23. Possible word problem: Jane has $\frac{5}{12}$ yd of
fabric. She uses $\frac{1}{4}$ yd for a vest. How much
fabric does she have left?

Page 109

1. $\frac{5}{6} + \frac{4}{6}$ **3.** $\frac{3}{15} + \frac{7}{15}$ **5.** $\frac{2}{8} + \frac{5}{8}$ **7.** $\frac{15}{20} + \frac{16}{20}$
9. $\frac{3}{8} + \frac{4}{8}$ **11.** $\frac{9}{12} + \frac{2}{12}$ **13.** $\frac{5}{6}$ **15.** $\frac{4}{5}$
17. $\frac{11}{8}$, or $1\frac{3}{8}$ **19.** $\frac{7}{12}$ **21.** $\frac{13}{12}$, or $1\frac{1}{12}$
23. $\frac{9}{10}$ **25.** $\frac{9}{10}$ **27.** $\frac{14}{15}$ **29.** $\frac{5}{6}$ can
31. $\frac{17}{12}$ hr, or $1\frac{5}{12}$ hr

Page 111

1. $\frac{3}{8} - \frac{2}{8}$ **3.** $\frac{20}{36} - \frac{9}{36}$ **5.** $\frac{35}{60} - \frac{12}{60}$ **7.** $\frac{1}{10}$
9. $\frac{2}{9}$ **11.** $\frac{3}{14}$ **13.** $\frac{7}{15}$ **15.** $\frac{1}{8}$ **17.** $\frac{1}{2}$ **19.** $\frac{4}{21}$
21. $\frac{5}{18}$ **23.** $\frac{7}{12}$ **25.** $\frac{5}{24}$ **27.** $\frac{7}{8}$ tank
29. much less than **31.** about $\frac{1}{2}$ **33.** $\frac{1}{2}$
35. $\frac{3}{4}$ **37.** B

Pages 114–115

1. about 1 **3.** about $\frac{1}{2}$ **5.** about 0 **7.** $1 + 1$
9. $\frac{1}{2} + 1$ **11.** $1 - \frac{1}{2}$ **13.** 0 **15.** 1 **17.** 0
19. $\frac{1}{2}$ **21.** $1\frac{1}{2}$ **23.** 2 **25.** $\frac{1}{2}$ **27.** $1\frac{1}{2}$ **29.** 1
31. $1\frac{1}{2}$ **33.** 1 **35.** 0 **37.** about 6 c
39. about 2 mi **41.** Compare the numerator
with the denominator to decide whether the
fraction is closer to 0, $\frac{1}{2}$, or 1. **43.** $\frac{9}{4}$ **45.** $\frac{11}{3}$
47. $\frac{26}{5}$ **49.** 8 **51.** 6 **53.** B

Page 117

1. D **3.** C **5.** A **7.** C

Chapter 6
Page 121

1. $1\frac{1}{5} + 1\frac{3}{5} = 2\frac{4}{5}$ **3.** $2\frac{6}{8}$, or $2\frac{3}{4}$ **5.** $3\frac{4}{8}$, or $3\frac{1}{2}$
7. $3\frac{1}{2}$ hr **9.** $6\frac{5}{8}$

Page 125

1. $1\frac{3}{2}$ 3. $2\frac{5}{3}$ 5. $4\frac{13}{8}$ 7. $1\frac{2}{5}$ 9. $1\frac{1}{2}$
11. $\frac{1}{6}$ 13. $2\frac{2}{3}$ 15. $1\frac{1}{2}$ qt   17. $1\frac{11}{12}$ mi

Pages 128–129

1. $1\frac{2}{8}+1\frac{3}{8}$ 3. $2\frac{5}{10}+3\frac{4}{10}$ 5. $2\frac{5}{20}+3\frac{8}{20}$
7. $1\frac{3}{4};2\frac{3}{4}$ 9. $1\frac{5}{6};5\frac{5}{6}$ 11. $1\frac{3}{10};6\frac{3}{10}$
13. yes   15. no   17. $2\frac{3}{8}$ 19. $8\frac{3}{10}$
21. $11\frac{3}{20}$ 23. $13\frac{4}{9}$ 25. $10\frac{1}{18}$ 27. $3\frac{3}{10}$
29. $2\frac{1}{12}$ 31. $2\frac{13}{20}$ 33. $1\frac{4}{5}$ 35. $1\frac{3}{8}$
37. $28\frac{1}{8}$ dollars   39. $41\frac{7}{9}$ mi   41. 0
43. $\frac{1}{2}$ 45. 1   47. $\frac{1}{2}$ 49. $\frac{4}{9}$ 51. D

Page 131

1. 6   3. $3\frac{1}{2}$ 5. $2\frac{1}{2}$ 7. 13   9. 17   11. 4
13. $5\frac{1}{2}$ 15. 2   17. about 20 gal
19. Round the fractions to 0, $\frac{1}{2}$, or 1. Rewrite the problem. Then add or subtract.   21. 0.15
23. 0.025   25. $1\frac{4}{15}$ 27. $\frac{13}{24}$ 29. H

Page 133

1. B   3. D   5. B   7. B

Chapter 7

Page 139

1. $\frac{3}{8}$ 3. $\frac{3}{10}$ 5. $\frac{7}{2}$, or $3\frac{1}{2}$ 7. $\frac{1}{4}$ 9. $\frac{2}{15}$
11. $\frac{7}{10}$ 13. $\frac{2}{7}$ 15. $\frac{1}{6}$ 17. $\frac{3}{4}$ 19. 2   21. $\frac{25}{6}$, or $4\frac{1}{6}$ 23. $\frac{5}{9}$ 25. $\frac{28}{5}$, or $5\frac{3}{5}$ 27. 3
29. 2, 3, or 3, 2   31. $93\frac{3}{4}$, or \$93.75   33. Use a model; multiply the numerators and the denominators, and write the answer in simplest form.   35. 4   37. 5   39. $\frac{11}{12}$
41. $3\frac{9}{10}$ 43. G

Page 141

1. 2   3. 7, 3   5. 4   7. 2, 9   9. $\frac{3}{1}\times\frac{1}{7}$

Page 143

11. $1\times\frac{1}{2}$ 13. $\frac{1}{2}\times\frac{3}{5}$ 15. $\frac{1}{5}\times\frac{1}{2}$ 17. $\frac{1}{9}$
19. $\frac{1}{2}$ 21. $\frac{2}{3}$ 23. $\frac{3}{7}$ 25. $\frac{1}{6}$ 27. $\frac{7}{12}$ 29. 7
31. 20   33. 13 teens   35. 245 apartments
37. You do not have to simplify the product; multiplying the factors is easier.

Page 143

1. $\frac{19}{4}\times\frac{1}{8}$ 3. $\frac{4}{3}\times\frac{15}{4}$ 5. $\frac{14}{5}$, or $2\frac{4}{5}$ 7. $\frac{143}{24}$, or $5\frac{23}{24}$ 9. $\frac{49}{6}$, or $8\frac{1}{6}$ 11. $\frac{11}{12}$ 13. -15
15. 85   17. $\frac{50}{7}$, or $7\frac{1}{7}$ 19. $\frac{13}{2}$, or $6\frac{1}{2}$
21. 4 mi   23. $40\frac{1}{4}$ mi   25. $\frac{14}{12}$

Page 147

1. $\frac{3}{4}$ 3. $2\frac{5}{8}$ 5. 20   7. $\frac{20}{27}$ 9. 8   11. $\frac{5}{32}$
13. $\frac{1}{40}$ 15. $\frac{7}{8}$ 17. 48 hamburgers
19. 80 times   21. $\frac{7}{3}$ 23. $\frac{9}{2}$ 25. $\frac{14}{5}$ 27. $\frac{4}{9}$
29. $4\frac{7}{18}$ 31. B

Page 149

1. $4\frac{3}{4}$ mi   3. $1\frac{1}{4}$ yd   5. 16 books   7. 2 times
9. 1 hr   11. 189 in.^{2}

Page 151

1. C   3. A   5. D   7. B   9. B

Page 153

1. =A2*4   3. 42   5. 24

Pages 154–155

1. prime numbers   3. least common denominator (LCD)   5. 10   7. 12   9. 5
11. 5   13. $\frac{3}{7}$ 15. $\frac{1}{2}$ 17. $\frac{1}{3}$ 19. $\frac{2}{3}$ 21. $\frac{1}{4}$
23. $\frac{22}{7}$ 25. $\frac{11}{4}$ 27. $\frac{13}{3}$ 29. $1\frac{4}{5}$ 31. $1\frac{5}{12}$
33. $3\frac{1}{2}$ 35. $\frac{2}{3}$ 37. $\frac{25}{18}$, or $1\frac{7}{8}$ 39. $\frac{1}{10}$ 41. $10\frac{1}{2}$
43. $\frac{7}{8}$ 45. $2\frac{11}{12}$ 47. $\frac{1}{3}$ 49. $8\frac{21}{25}$ 51. 8
53. $\frac{4}{3}$, or $1\frac{1}{3}$ 55. $\frac{2}{7}$ 57. $\frac{8}{15}$ 59. 2 hr

Page 157

1. B   3. D   5. C   7. A   9. D   11. C
13. A

Chapter 8

Page 161
1. $\overline{PQ}$, $\overline{QR}$, $\overline{PR}$ 3. $\overleftrightarrow{PQ}$, $\overrightarrow{QR}$, $\overrightarrow{QP}$, $\overleftrightarrow{RQ}$ 5. lines
7. point 9. points 11. line segment

Page 163
1. $\overleftrightarrow{BD}$, $\overleftrightarrow{EG}$, $\overleftrightarrow{FH}$ 3. $\overleftrightarrow{AB}$, $\overleftrightarrow{CD}$, $\overleftrightarrow{BF}$, $\overleftrightarrow{DH}$ 7. Both
lines cross at one point; perpendicular lines
form right angles, but intersecting lines might
not 9. greater than 11. less than 13. $\frac{21}{4}$,
or $5\frac{1}{4}$ 15. D

Page 165
1. acute 3. acute 5. obtuse
7. 180°, straight 9. 90°, right 11. 18°,
acute 13. right 15. acute 17. acute
19. acute 21. No; the measure of an angle is
the measurement from one ray to another ray
with a common vertex.

Pages 168–169
1. no 3. yes 5. 114°, 145°; no 13. 3;
triangle 15. 6; hexagon 17. $4\frac{2}{3}$
19. $1\frac{1}{9}$ 21. C

Page 173
1. parallelogram; yes 3. trapezoid; no
5. Figure 4 7. scalene 9. 12 triangles; 11
quadrilaterals 11. hexagon 13. Regular:
all sides and angles are congruent; not
regular: no sides or angles need to be
congruent.

Page 175
1. B 3. D 5. C 7. B 9. A

Chapter 9

Pages 180–181
5. yes 7. no 9. no 11. yes 13. yes
15. no 17. $\frac{1}{2}$, 180° 19. $\frac{1}{2}$, 180°
23. 8 pieces 29. parallel 31. perpendicular
and intersecting 33. H

Pages 186–187
1. translation 3. rotation 9. false; rotation
11. true 13. rotation, translation, reflection
or rotation 15. reflection, rotation, rotation,
reflection 17. right 19. pop 21. 90° 23.
45° 25. obtuse 27. right 29. B

Page 189
1. yes 3. no 5. each angle: 90°; sum: 360°
7. each angle: 90°; sum: 360° 9. 360°
11. two squares, a triangle, and a hexagon
13. no 17. They cover a plane with no gaps
or overlaps, and the angles at each vertex
have a sum of 360°.

Page 191
1. yes 3. yes 5. Heather is left of Howie.
7. 4 months 9. yes

Page 193
1. D 3. C 5. B 7. A 9. D

Chapter 10

Pages 198–199
1. base 3. lateral face 5. rectangular
prism 7. triangular prism 9. cone
11. true 13. true 15. true 17. False; a
cylinder is not a polyhedron 19. six sides
23. 6, 6; hexagon 25. 4, 4; rectangle
27. yes 29. no 31. H

Page 201
1. N, M, R, P, S 3. *MRS, MSP, MPN, MNR,
RSPN* 5. polyhedron 7. 4, 6, 5, 8, 6, 10
9. 4 colors 11. pentagonal prism,
hexagonal pyramid

Page 203
3. yes 5. yes 7. 5 9. 3 in. × 2 in. 17. A

Page 205
1. cone 3. hexagonal prism 5. triangular
pyramid 7. cone and cylinder 13. top:
circle; side: rectangle 15. Possible answer:
cereal box

Page 209
1. 20 balls of clay 3. 11 balls of clay
5. $30.00 7. $\frac{1}{8}$ of the cake 9. 24 cookies
11. 8 sides

Page 211

1. B **3.** C **5.** B **7.** A

Page 213

1. The shape does not change. It only moves to the right.

Pages 214– 215

1. perpendicular **3.** vertex **5.** 1 **7.** 120°; obtuse **9.** 46°; acute **11.** $\frac{1}{2}$, or 180° **13.** No. **15.** cylinder; not a polyhedron **17.** pentagonal prism; polyhedron **19.** 4 faces, 6 edges, 4 vertices **21.** ray; yes: any two points of the ray can be the endpoints of a line segment **23.** No; octagons will not tessellate a plane.

Page 217

1. A **3.** B **5.** C **7.** C **9.** C **11.** D

Chapter 11

Page 221

1. a and b **5.** 3 **7.** 32 **9.** cone
11. cylinder **13.** D

Page 223

1. 57 people **3.** Possible response: by randomly choosing students at lunch **5.** Yes; all voters had equal chances of selection.
7. 100 students **9.** about 56 students

Page 225

1. biased; excludes female teachers **3.** biased, sample not large enough **5.** biased, sample not large enough **7.** biased, excludes men **9.** not biased **11.** No; boys and girls are randomly selected **13.** This sample excludes voters that are not homeowners.

Pages 230–231

1. 20; 46; 80 **7.** 9 heights **11.** A frequency table shows totals. You don't have to count.
13. bar graph **15.** 4 faces **17.** G

Page 233

1. B **3.** D **5.** C **7.** E

Chapter 12

Page 239

1. bar graph **3.** stem-and-leaf plot
11. 10 **13.** 62 **15.** D

Page 241

1. histogram **3.** bar graph **5.** The number of members in some of the different age groups would increase.

Page 243

3. multiple-line graph; changes over time
5. 3 sets; 1 set for each year
7. multiple-line graph

Page 247

1. 100 people **3.** 5 sections **11.** change over time **13.** yes **15.** B

Page 249

1. D **3.** B **5.** C **7.** C

Chapter 13

Page 253

1. October; July; there were 4.5 times as many sales in October as in July. **3.** Sales increased. **5.** 400 more gallons, or 5 times as many gallons **7.** No longer make Stellar Kiwi; make more Coco Delight.

Page 255

1. Lucy **3.** No. Lucy sold 5 more boxes of cookies, not 15 more boxes. **5.** about 4 times **7.** about 40,000; no

Page 257

1. 10; 15 **3.** 5 CDs a yr **5.** 40 CDs
7. $2,400 **9.** $160 **11.** about $2,000

Pages 260– 261

1. 8, 7, 7 **3.** 27, 28, 32 **5.** 46.4, 45.8, 45.7
7. 4.2, 3.4, 7.8 **9.** true **11.** 85; 100 **13.** 85; 100 **15.** No. You have to find the sum of the data and divide by 10. **17.** 7; 7.5; 8 **19.** 9
21. 15.2 **23.** B

Page 265

1. 22 **3.** 15; 32 5. 1; 9 **9.** $\frac{7}{8}$ **13.** F

Page 267

1. C **3.** D **5.** B **7.** A

Chapter 14

Page 271

1. 6 choices **3.** 9 choices **5.** 3 hr
7. 4 groups of 10, 3 groups of 8 **9.** $12 each week **11.** Friday

Pages 274–275

1. 1 **3.** $\frac{1}{4}$ **5.** $\frac{5}{6}$ **7.** $\frac{1}{2}$ **9.** $\frac{2}{9}$ **11.** $\frac{2}{3}$
13. $\frac{1}{4}$ **15.** $\frac{4}{5}$ **17.** $\frac{2}{5}$ **19.** $\frac{1}{6}$ **21.** $\frac{1}{2}$

23. 12 sections **25.** 1 **27.** Favorable; a selected event; possible: any event; examples will vary. **29.** $\frac{1}{4}$ **31.** $\frac{1}{8}$ **33.** 4 **35.** C

Page 277

1. $\frac{1}{4}$ **3.** 0 **5.** $\frac{3}{8}$ **7.** $\frac{1}{10}$ **9.** $\frac{1}{2}$ **11.** $\frac{1}{4}$
13. $\frac{5}{8}$ **15.** $\frac{3}{4}$ **17.** 0 **19.** $\frac{8}{35}$ **21.** $\frac{9}{35}$

23. 1 **25.** $\frac{1}{2}$

Page 281

1. $\frac{2}{5}$ **3.** $\frac{6}{25}$ **5.** $\frac{1}{4}$ **7.** 24 times **9.** number of times 4 lands ÷ total rolls **11.** 23 **13.** 9
15. a larger amount **17.** F

Page 283

1. A **3.** B **5.** E **7.** A **9.** C

Page 285

1. E10;=mean (E2:E9)

Pages 286–287

1. range **3.** probability **9.** $200 **11.** 6; 6; none **13.** 92; 90; 90 **15.** $\frac{1}{10}$ **17.** $\frac{9}{10}$
19. $\frac{16}{30} = \frac{8}{15}$ **21.** $\frac{18}{30} = \frac{3}{5}$ **23.** 800

Page 289

1. A **3.** C **5.** C **7.** B **9.** C **11.** C

Chapter 15

Page 293

1 $\frac{1}{2} \times \frac{2}{5}$ **3.** $87 - k$ **5.** $32.7 \times p$
7. the sum of 42 and 2.5 **9.** twelve less than a number, c **11.** The sum of 52 and x
13. 98 divided by 4.1 **15.** $s - 2$
17. Algebraic expressions contain a variable; numerical expressions do not. **19.** 10
21. 35 **23.** 0 **25.** $\frac{1}{3}$ **27.** B

Page 295

1. 17 **3.** 40 **5.** 6.1 **7.** $\frac{3}{4}$ **9.** 8 **11.** 5 **13.** 52
15. 17.5 **17.** 14 **19.** 19 **21.** $30.00 − $12.00; $18.00 **23.** 45 ÷ f; 4.5 pieces

Page 297

1. 6 − 6, 0; 7 − 6, 1; 8 − 6, 2 **3.** 100 ÷ 4, 25; 102 ÷ 4, 25.5; 104 ÷ 4, 26 **5.** 12, 13, 14, 15
7. 10, $6\frac{2}{3}$, 5, 4 **9.** 5 **11.** 8 **13.** $5.25 \times h$; 23 hr. **15.** 50

Pages 302–303

1. yes **3.** no **5.** no **7.** no **9.** yes
11. $a = 2$ **13.** $r = 35$ **15.** $b = 55$
17. $c = 54$ **19.** $m = 902$ **21.** $y = 1{,}574$
23. $x = 14.6$ **25.** $s = 103\frac{2}{5}$ **27.** $y - 9 = 17$
29. m = Mary's height; $m - 60 = 6$
31. x = amount in savings; $x + $125 = 324; $x = $199 **33.** 42 **35.** 7 **37.** $\frac{3}{7}$ **39.** $\frac{2}{7}$
41. $\frac{6}{7}$ **43.** J

Page 305

1. D **3.** D **5.** B **7.** A

Chapter 16

Page 313

1. yes **3.** no **5.** yes **7.** no **9.** $x = 2$
11. $y = 35$ **13.** $m = 12$ **15.** $s = 75$
17. $d = 10$ **19.** $b = 80$ **21.** $11x = 66$
23. g = games won; $5g = 120$; $g = 24$; 24 games **25.** Use the inverse operation.
27. 60; 150; 300; 1,500 **29.** 170; 425; 850; 4,250 **31.** $y = 22$ **33.** $d = 20$ **35.** G

Page 315

1. 32; 80; 160; 800 3. 108; 270; 540; 2,700
5. 24; 60; 120; 600 7. 78; 195; 390; 1,950
9. 10; 25; 50; 250 11. $40.65 13. $37.20
15. $133.34 17. 50 pennies and 50 dimes
19. 10 pounds; they are worth more in U.S.
dollars 21. 7 23. 23 25. 56 27. 22
29. 85 31. H

Page 317

1. 68°F 3. 140°F 5. 36°F 7. 59°F
9. 212°F 11. 194°F 13. 190°F 15. 208°F
17. 174°F 19. 30°C 21. 70°C 23. 31°C
25. 1°C 27. 9°C 29. 33°C 31. 39°C
33. 32°C 35. 31°C 37. Yes; the
temperature was about 13°C. 39. 57°C

Page 319

1. $d = 52$ ft 3. $t = 4$ sec. 5. 855 ft
7. 25 mi per hr 9. 6 ft per sec 11. 12 sec
13. 376 mi 15. 125 sec 17. 24 ft per min
19. It would increase; the distance would
decrease; as time increases or decreases, so
does distance.

Page 321

1. 10.6 lb 3. 75 times 5. 12 onions
7. 62, 87, and 117 tickets 9. 24 hr

Page 323

1. B 3. D 5. C 7. C

Page 325

1. $= A6/2$ 3. 3, 5, 7 5. 1.5, 2, 2.5
7. 0, 2, 4 9. 1.2, 1.6, 2 11. 6.5, 8.5, 10.5
13. 4.65, 6.65, 8.65

Pages 326–327

1. numerical expression 3. numerical
5. algebraic 7. $x + 7$ 9. 5 11. 3
13. 10 + 7, 17 17. yes 19. $a = 46$
21. $x = 9$ 23. $x = 20$ 25. $y = 240$
27. $y = 12$ 29. 28; 140 31. 140; 700
33. 125; 1,250 35. 104°F 37. 54°F 39. 0°C
41. 210 mi 43. 429 sec
45. Karen; 2 mi ≈ 3.2 Km

Chapter 17

Page 333

1. 5 to 9, 5:9, $\frac{5}{9}$ 3. 10 to 1, 10:1, $\frac{10}{1}$ 5. 9 to
20, 9:20, $\frac{9}{20}$ 7. 21 to 10, 21:10, $\frac{21}{10}$ 9. 4:2;
possible answers: 2:1, 8:4, 12:6 11. 6
13. 25 15. 6 17. 12 19. 7 21. 30
25. 12:4, or 3:1 27. Possible answer: A ratio
compares two numbers.

Page 335

1. $\frac{12 \text{ eggs}}{\$1.10}$ 3. $\frac{90 \text{ words}}{2 \text{ min}}$ 5. $\frac{60 \text{ mi}}{3 \text{ gal}}$ 7. $\frac{\$1.89}{3 \text{ pens}}$
9. $\frac{\$15}{5 \text{ tapes}}$ 11. $\frac{\$0.15}{1}$ 13. $\frac{50 \text{ mi}}{1 \text{ hr}}$ 15. $\frac{20 \text{ mi}}{1 \text{ gal}}$
17. $\frac{\$52}{1 \text{ tire}}$ 19. $0.35 21. $0.37
23. P.O.: $0.32; drugstore: $0.35
25. the rate when the second term is 1

Pages 338–339

1. 35% 3. 84% 5. 40% 7. 10% 9. 38%
11. 80% 13. 80% 15. 24% 17. $66\frac{2}{3}$%
19. 0.2; $\frac{1}{5}$ 21. 0.9; $\frac{9}{10}$ 23. 0.48; $\frac{12}{25}$
25. 0.17; $\frac{17}{100}$ 27. 0.09; $\frac{9}{100}$ 29. 0.125; $\frac{1}{8}$
31. 100 33. 15 35. 29 37. <; $\frac{2}{50} = 4$%
39. <; 0.0018 = 0.18%
41. 36% want supreme
43. 20%; 20% of 35 = 7 and $\frac{1}{5}$ of 35 = 7
45. $\frac{1}{4}$ 47. $\frac{1}{2}$ 49. $n = 5$ 51. $t = 9$ 53. C

Page 341

1. 40% 3. 25% 5. 60% 7. 12.5%
9. 20%; 40% 11. 50%

Page 345

1. 18 sec 3. 25 gal 5. 2 quarters, 18 nickels
7. 30°F 9. 90 11. $24,000 13. 9%
15. 40% 17. 8% 19. 41°F 21. 68°F
23. 150°C 25. A

Page 347

1. B 3. C 5. B 7. C

Chapter 18

Page 351

1. 600 bags **3.** 600 shows **5.** $10,000
7. 8 in the front, 9 in the back **9.** $59
11. the first floor

Page 355

1. $\frac{3}{5}$ **3.** $4\frac{1}{2}$ **5.** 108 **7.** 5.6 **9.** 59.64
11. 3 **13.** 32.2 **15.** 0.24 **17.** 98.01
19. 32 **21.** 3.85 **23.** 200 **25.** 225
27. $1.30 **29.** $6.25 **31.** $24.80
33. Change the percent to a ratio or decimal and multiply by the number. **35.** 36 **37.** 45
39. $\frac{1}{10}$, 0.1 **41.** $\frac{9}{20}$, 0.45 **43.** $\frac{8}{25}$, 0.32 **45.** H

Page 357

1. 30% = 108°, 25% = 90°, 12.5% = 45°, 32.5% = 117° **3.** 175 people. **5.** 150 people. **7.** $100 **9.** Multiply the percent by 360°.

Page 359

1. $42.00 **3.** $3.60 **5.** $8.75 **7.** $37.50
9. $11.99; $35.96 **11.** $12.97; $38.92
13. $15 **15.** $112.50; $337.50

Page 361

1. 1.00, 2.00 **3.** 11.50, 23.00 **5.** 82.80, 165.60 **7.** $109 **9.** 1 **11.** 3.6 **13.** 25%
15. 33.3% **17.** A

Page 365

1. B **3.** A **5.** D **7.** D

Chapter 19

Page 371

1. both **3.** neither **5.** neither **7.** both
9. 15 **11.** 15 **13.** yes; because they are the same shape but not the same size

Page 375

3. No. Angles are not congruent, and ratios are not equivalent. **5.** yes **7.** Check that the corresponding angles are congruent and that the ratios of the lengths of the corresponding sides are equivalent.

9. $x = 24$ **11.** $9 **13.** $44.25 **15.** F

Page 377

1. $n = 12$ ft **3.** $n = 2.5$ cm **5.** $n = 3$ mi
7. No. Their ratios are not equivalent.
9. 2 in. wide

Page 379

1. $\frac{2}{n} = \frac{4}{10}$; $n = 5$ yd **3.** 8 m **5.** 6 **7.** 6
9. $0.50 **11.** $2.19 **13.** B

Page 381

1. C **3.** E **5.** E **7.** A

Chapter 20

Page 387

1. $\frac{4}{3}$ **3.** $\frac{2.4}{1.2}$, or $\frac{2}{1}$ **5.** 15 **7.** 36 **9.** 48
11. $\frac{1}{2} = \frac{3}{n}$; $n = 6$ ft **13.** $x = 10$ **15.** $x = 20$
17. yes **19.** H

Page 391

1. $\frac{1}{50} = \frac{1.5}{n}$; 75 mi **3.** $\frac{1}{50} = \frac{4.5}{n}$; 225 mi
5. $\frac{1}{50} = \frac{12}{n}$; 600 mi
7. $\frac{1}{50} = \frac{8.25}{n}$; $n = 412.5$ mi
9. $\frac{1}{50} = \frac{n}{7}$; $n = 350$ mi
11. 103.5 mi; 115 mi **13.** $1\frac{1}{2}$ in.; 138 mi
15. 20 in.

Page 393

3. north 4 mi. **5.** 2 four-person boats
7. 11 classes **9.** 15 ft **11.** 6 onions, 15 oz of cheese

Page 395

1. yes; $\frac{4}{2.5} = 1.6$ **3.** no; $\frac{12}{4.5} \approx 2.67$ **5.** no; $\frac{12}{10} = 1.2$ **7.** yes; $\frac{101}{60} \approx 1.6$ **9.** If its $\frac{l}{w} \approx 1.6$, then it is a Golden Rectangle. **11.** 3
13. $x = 12\frac{1}{2}$ ft **15.** C

Page 397

1. B **3.** B **5.** D **7.** E

Page 399

1. = A3*B3   **3.** = A3*B3*5   **5.** $2.40, $12.00
7. $22.00, $110.00   **9.** $76.80, $384.00

Pages 400–401

1. ratio   **3.** Golden Ratio   **5.** $\frac{240\text{mi}}{4\text{hr}}; \frac{60\text{mi}}{1\text{hr}}$
7. $\frac{630\text{ words}}{9\text{ min}}; \frac{70\text{ words}}{1\text{ min}}$ **9.** 50 people
11. 30 people   **13.** $8.40   **15.** $38.50   **17.**
$6.30   **19.** $36.16   **21.** $200.00   **23.** $88.60
25. $y = 10^{\text{cm}}$ **27.** $y = 10$ ft   **29.** $a = 15$ ft
31. 300 mi   **33.** 525 mi   **35.** 8 in.   **37.** $18

Chapter 21

Page 407

1. 240   **3.** 90   **5.** 12   **7.** 60   **9.** 4   **11.** 20
13. 15 yd   **15.** 6T   **17.** centimeter
19. liter   **21.** 6   **23.** 42.5   **25.** C

Page 409

1. divide by 1,000   **3.** divide by 1,000
5. divide by 1,000   **7.** multiply by 1,000
9. 10; 1   **11.** 0.1; 0.01; 0.001   **13.** 4,000
15. 0.01   **17.** 0.001   **19.** 9,000   **21.** 0.00034
23. 32   **25.** 2.512 km   **27.** 2.5 L

Page 411

1. 1 in.; $1\frac{1}{2}$ in.   **3.** 85 in.   **5.** 71 mm

7. 111 in.   **9.** 395 in.   **11.** 400 m
13. 10,520 ft   **15.** 65 mm   **17.** 62 mm
19. 211 mm   **21.** 4,000 m   **23.** 2,650 yd
25. 30 ft

Page 413

1. BCDE, 52 km   **3.** FCBE, 101 km
5. EFDC, 96 km   **7.** CBFE, 73 km   **9.** HSBS,
28 mi   **11.** by listing the different routes,
finding the distance for each route, and
identifying the shortest route   **13.** 208.4
15. 6   **17.** 6   **19.** B

Page 415

1. $x = 12$ m; 43 m   **3.** $x = 27$ ft, $y = 58$ ft;
610 ft   **5.** $y = 34$ yd   **7.** 20 km   **9.** Possible
answers: $P = s + s + s + s$, or $P = 4s$; 460 m

Page 419

1. B   **3.** A   **5.** D   **7.** C

Chapter 22

Page 423

1. about 27 in.^{2} **3.** about 7 in.^{2} **5.** about
30 in.^{2} **7.** about 18 in.^{2} **9.** about 18 ft^2
11. Count full squares, almost-full squares,
and half-full squares   **13.** 12   **15.** 5   **17.** 30
19. H

Page 425

1. 7.5 ft^2 **3.** 150 ft^2; $750   **5.** $200   **7.** 12
baskets **9.** $3\frac{5}{8}$ in.   **11.** $4.50   **13.** 255 women

Pages 428–429

1. 10 ft^2 **3.** 102.3 m^2 **5.** 112 in.^{2} **7.** 92 ft^2
9. 0.2 m^2 **11.** 14.175 m^2 **13.** 100 cm^2
15. 45 in.^{2} **17.** 9 cm^2; 4.5 cm^2 **19.** 60 ft^2
21. 32   **23.** 6   **25.** 0.01   **27.** B

Page 431

1. 20 in., 16 in.^{2}; 40 in., 64 in.^{2} **3.** 36 yd,
48 yd^2; 72 yd, 192 yd^2 **5.** 80 ft, 400 ft^2; 40 ft,
100 ft^2 **7.** 56 cm, 84 cm^2; 28 cm, 21 cm^2 **9.**
44 ft   **11.** No. The area only doubled,
because the width didn't change.

Page 435

1. 201 yd^2 **3.** 260 cm^2 **5.** 113 yd^2
7. 7.1 m^2 **9.** 88.2 cm^2 **11.** 1,962.5 yd^2
13. 38.5 m^2 **15.** 60.8 cm^2 **17.** 379.9 ft^2
19. 642.1 m^2 **21.** 907.5 in.^{2} **23.** 66.5 cm^2
25. The one with the 9 in. radius; its area is
about 254 in.^{2} **27.** about 3.44 cm^2 **29.**
Area is the number of square units needed to
cover the circle; circumference is the distance
around.

Page 437

1. D   **3.** B   **5.** C   **7.** A

Chapter 23

Page 443

1. 24 in.^{3} **3.** 75 cm^3 **5.** 324 ft^3 **7.** 900 cm^3
9. 162 cm^3 **11.** $12   **13.** It has three
dimensions.   **15.** 1,280   **17.** 720   **19.** 90 ft^2
21. 14.4 cm^2 **23.** H

Page 447

1. 24 cm^3; 192 cm^3 **3.** 60 m^3; 480 m^3

5. 60 m³, 7.5 m³   **7.** 840 ft³, 105 ft³
9. 120 in.³   **11.** 13 ft²   **13.** 79 m²   **15.** 3 in.²
17. 40 m²   **19.** B

Page 451
1. about 50 m³   **3.** about 1,413 ft³   **5.** about
1,690 cm³   **7.** about 603 cm³ or 603 mL
9. about 16 hr   **11.** πr^2 is the area of the
base, and h is the height.

Page 453
1. 198 in.²   **3.** 148 m²   **5.** 416 ft²   **7.** 440 ft²
9. 142 cm²

Page 455
1. C   **3.** B   **5.** D   **7.** C   **9.** A

Page 457
1. = 2*A3 + 2*B3   **3.** = A3*B3   **5.** 26, 40
7. 44, 120   **9.** 40, 99

Pages 458–459
1. perimeter   **3.** 240   **5.** 0.2   **7.** 10,000
9. 18,000   **11.** 46.4 mi   **13.** 50 mi
15. 38.5 m²   **17.** 102 in.²   **19.** 201 m²
21. 60 ft³; 480 ft³   **23.** 64 m³; 8 m³   **25.** 402
cm³   **27.** 80 ft²; 80 ft²

Page 461
1. B   **3.** C   **5.** D   **7.** A   **9.** B   **11.** D

Chapter 24
Page 467
1. $^+2$; losing 2 pounds; $^-2$ **3.** $^-5$; gaining
5 yd; $^+5$ **5.** 1   **7.** 5   **9.** 12   **11.** 28   **13.** 80
15. 295   **17.** $^-630$ **19.** It is its distance from
zero on a number line.   **21.** $\frac{2}{1}$ **23.** $\frac{6}{5}$
25. 120 ft³   **27.** 320 cm³   **29.** J

Page 469
1. $\frac{21}{4}$ **3.** $\frac{32}{100}$, or $\frac{8}{25}$ **5.** $\frac{10}{7}$ **7.** $\frac{100}{1}$ **9.** $\frac{5}{2}$
11. $\frac{260}{1}$ **13.** R   **15.** I and R   **17.** R   **19.** R
21. R   **23.** R   **25.** all   **27.** R
29. true; $\frac{^-10}{1} = ^-10$ **31.** $1\frac{1}{2}$, or 1.5
33. $\frac{20}{100}$ or $\frac{1}{5}$ **35.** $\frac{123}{2}$

Pages 472–473
1. 0.1   **3.** 0.625   **5.** 0.$\overline{7}$ **7.** 0.$\overline{54}$ **9.** 1.$\overline{3}$
11. 0.$\overline{1}$ **13.** 0.$\overline{4}$ **15.** 0.$\overline{6}$ **17.** 0.$\overline{8}$ **19.** 0.06
21. 0.875   **23.** 0.3   **25.** 0.91$\overline{6}$ **27.** 0.82
29. 0.025   **31.** 0.5 in.; yes; he recorded 0.15
in. more   **33.** Yes. $9\frac{3}{4}$ is equal to 9.75.
35. terminating: no remainder; repeating:
remainder repeats   **37.** 10   **39.** 62.8 in.³
41. C

Page 475
1. $^-1\frac{3}{4}$ **3.** $^-1\frac{3}{10}$ **5.** $\frac{5}{24}$ **7.** $1\frac{5}{16}$ **9.** 2.38
11. $^-7.44$ **13.** $1\frac{5}{8}$ **15.** $\frac{3}{20}$, or 0.15   **17.** 1.8
19. $^-3.251$ **21.** $^-2\frac{3}{4}$ **23.** $\frac{4}{5}$
25. Yes; $\frac{5}{8}$ is between $\frac{1}{2}$ and $\frac{3}{4}$.   **27.** Possible
answer: Between 0.25 and 0.50; you don't
have to find a common denominator.

Page 477
1. 2.5   **3.** >   **5.** <   **7.** $^-1.6 < \frac{4}{5} < \frac{6}{3} < 3.8$
9. $^-1.25 < \frac{^-1}{2} < 0 < \frac{1}{10} < 0.3$ **11.** 0.64 <
$6.02 < 6\frac{1}{8} < 6\frac{3}{4} < 6.8$ **13.** 3.3 > 2.4 > 1.5 >
$^-1.9 > ^-2$ **15.** John   **17.** Sheila

Page 479
1. D   **3.** C   **5.** D   **7.** A

Chapter 25
Page 485
1. $^+3 + ^+2 = ^+5$ **3.** $^+12$ **5.** 0   **7.** $^-1$
9. $^-60$ **11.** $^-116$ **13.** $^+2$ **15.** $^-17$
17. $^+51$ **19.** $^+64$ **21.** $^+59$ **23.** $^+63$
25. $^+35$ **27.** $^+5 + ^-2 = ^+3$; 3rd floor
29. $^+4°$F

Page 489
1. $^-3 + ^-5$; $^-8$ **3.** $^-5$ **5.** $^+9$ **7.** $^-15$ **9.** 0
11. $^+20$ **13.** $^-16$ **15.** $^+5$ **17.** $^-54$ **19.**
$^-25°$ **21.** Change subtraction to addition,
and use the opposite of the number to be
subtracted.   **23.** 60   **25.** 504   **27.** 0.$\overline{3}$ **29.**
0.$\overline{2}$ **31.** 0.3   **33.** J

Page 491

1. $^+4$; $^+6$ **3.** $^-9$; $^-6$ **5.** $^-48$ **7.** $^-20$
9. $^-21$ **11.** $^-96$ **13.** $^-60$ **15.** $^-30$
17. $^+75$ **19.** $^-180$ **21.** $^-480$ **23.** $^+1{,}000$
25. $^-105$ **27.** $^+154$ **29.** $^-192$ **31.** $^-24°$

Page 493

1. $^+3$ **3.** $^+6$ **5.** $^+2$ **7.** $^-16$ **9.** $^+5$
11. $^+10$ **13.** $^+11$ **15.** $^-20$ **17.** $^-30$
19. $^-27$ **21.** $^-4°$ **23.** $^-40 \div ^-10 = ^+4$; a
positive number is greater than a negative
number. **25.** 8 **27.** 5 **29.** $^-6.5 < ^-5.1 <$
$6.2 < 7.1 < 7.7 < 8.0$ **31.** C

Page 495

1. C **3.** D **5.** B **7.** D

Chapter 26

Page 499

1. 13 **3.** 21 **5.** $^-192$ **7.** $^-12$ **9.** 7
11. 47 **13.** 29 **15.** $^-20$ **17.** $^-40.5$
19. $^-7$ **21.** $^-4$ **23.** $^-147$ **25.** $^-$$9
29. $a = 40$ **31.** $x = 15$ **33.** $^-11$ **35.** $^-14$
37. H

Page 501

1. $x = ^-3$ **3.** $k = ^-6$ **5.** $c = 5$ **7.** $r = ^-3$
9. $x = ^-113$ **11.** $k = 55$ **13.** $v = ^-8$
15. $p = 8$ **17.** $y = ^-5$ **19.** $y = ^-24$
21. $y = ^-30$ **23.** $m = 15$ **25.** $p = ^-72$
27. $x = 19{,}270$ **29.** $^-7x = 63$ **31.** $x + 7 = $
$^-34$ **33.** $d - 30 = ^-67$; $^-37$ ft is her original
depth. **35.** Multiply each side by $^-7$.
Replace x with the solution and divide by $^-7$.

Page 505

1. 0, 1 **3.** 0, 1, 2, 3, 4, 5, 6, 7 **5.** 0, 1, 2, 3, 4, 5
17. $x \le 3$, or $x < 4$ **19.** $x < ^-2$, or $x \le ^-3$
21. $y \ge 12$ **23.** $y > 9$ **25.** $g > j$ or $j < g$

Page 509

1. (2, 6) **3.** (2, $^-2$) **5.** (0, 7) **7.** (3, 0)
9. (6, $^-4$) **11.** ($^-7$, $^-6$) **13.** (4, $^-7$)
27. triangle **29.** They are located in the first
quadrant. They are located in the third
quadrant. **31.** $y = 15$ **33.** $y = ^-7$
35. $^-72$ **37.** 0 **39.** J

Page 511

1. 5; 6 **3.** 6; 7 **7.** $x + 1$ **9.** (0, 5), (1, 6),
(2, 7), (3, 8), (4, 9) **11.** (7, 35), (8, 40),
(9, 45), (10, 50)

Page 513

1. D **3.** B **5.** C **7.** D

Page 515

1. ($^-1$, 3), (2, 0), (3, 4), (5, 2), (2, 2)
3. 2 units to the left and 3 units up

Pages 516–517

1. integers **3.** origin **5.** R **7.** all **9.** R
11. $0.\overline{6}$ **13.** 0.55 **15.** 0.8 **17.** < **19.** >
21. $^-17$ **23.** $^+88$ **25.** $^-10$ **27.** $^+4$
29. $^+72$ **31.** $^+8$ **33.** $^-13$ **41.** $^-1$x, or $^-$x
43. yes; $\frac{3}{5}$ is between $\frac{1}{2}$ and $\frac{3}{4}$ **45.** a square

Page 519

1. B **3.** C **5.** A **7.** A **9.** C **11.** A

Chapter 27

Page 525

1. E' (0, $^-1$), F' (3, $^-1$), G' (3, $^-4$), H' (0, $^-4$)
3. S' (2, 6), T' (2, 2), U' (4, 2) **5.** A' (0, 0),
B' ($^-6$, 0), C' ($^-4$, 2), D' ($^-2$, 2) **7.** E' (1, 3),
F' (4, 3), G' (4, 1), H' (1, 1) **9.** A' (0, 0),
B' (0, $^-2$), C' ($^-3$, 0) **11.** 10 **13.** $^-5$
15. $x = 3$ **17.** $x = 6$ **19.** C

Page 527

1. translation, rotation, translation
3. translation **5.** 3 ft **7.** 6 ft **9.** $18

Page 529

1. right, row 2; left, row 3; right, row 3 **3.** 4^2,
5^2, 6^2 **5.** Add 2 more cubes in each row, and
start a new row of 2 cubes on top. **7.** 25
rectangular prisms; 36 rectangular prisms
9. yes **11.** no **13.** Quadrant I
15. Quadrant IV **17.** Quadrant I **19.** C

Page 531

5. yes **7.** no

Page 535

1. D **3.** B **5.** D **7.** A

Chapter 28

Page 541

1. add four to each term **3.** subtract 12 from each term **5.** add 0.89 to each term **7.** 28, 32, 36 **9.** 24, 30, 37 **11.** 335, 485, 665 **13.** 16, 8, 0 **15.** 0.03, 0.003, 0.0003 **17.** 7.2, 7.95, 8.7 **19.** $200, $220, $240, . . .; $300 **21.** $55, $59, $63, $67, . . .; 6 weeks **23.** 6 **25.** 15 **27.** 9 **31.** J

Page 545

1. add $\frac{1}{16}$ to each term **3.** add $\frac{4}{27}$ to each term **5.** subtract $\frac{1}{5}$ from each term **7.** subtract $\frac{1}{12}$ from each term **9.** add $\frac{12}{10}$ to each term **11.** $\frac{1}{4}, \frac{5}{16}, \frac{3}{8}$ **13.** 17, $18\frac{1}{2}$, 20 **15.** $8\frac{1}{9}, 9\frac{1}{3}, 10\frac{5}{9}$ **17.** $17\frac{1}{2}, 16\frac{1}{4}, 15$ **19.** $7\frac{1}{3}, 6\frac{1}{6}, 5$ **21.** $7\frac{1}{2}, 6\frac{4}{9}, 5\frac{7}{18}$ **23.** $\frac{1}{2}, 1\frac{1}{4}, 2, . . .; 2\frac{3}{4}$ hr **25.** Possible answer: Find the common denominator, and then find the pattern.

Page 547

1. $\frac{16}{3}, \frac{32}{3}, \frac{64}{3}$ **3.** 2, 4, 8 **5.** $\frac{1}{8}, \frac{5}{8}, \frac{25}{8}, \frac{125}{8}$ **7.** $\frac{7}{2}, \frac{21}{20}, \frac{63}{200}; \frac{189}{2,000}$ **9.** $\frac{1}{16}$ **11.** -5 **13.** $^+2$ **15.** $^-5$ **17.** yes **19.** no **21.** D

Page 549

1. $+5, ^-7, ^-19, ^-31, . . .$ **3.** $+3, ^-6, ^-15, ^-24, . . .$ **5.** subtract 21 **7.** add $^-2$ **9.** multiply by $^-4$ **11.** $^-11, ^-14, ^-17$ **13.** $^-0.02, ^+0.002, ^-0.0002$ **15.** $^+12, ^+18, ^+24$ **17.** $^-12, ^+6, ^-3$ **19.** $^-117, ^-96, ^-75$ **21.** $^-64, ^-16, ^-4$ **23.** $^+0.009, ^-0.0009, ^+0.00009$ **25.** 80°F, 76°F, 72°F, . . .; 68°F **27.** $^-386,561$; multiply 23 by $^-7$ and so on until you get to the first number.

Page 551

1. D **3.** D **5.** A **7.** C

Page 553

1. to keep the balance by adding to one side what is taken out on another side

Pages 554–555

1. sequence **3.** $A'\ (^-3, 2), B'\ (^-3, 0), C'\ (^-1, 0)$ **5.** $D'\ (^-1, ^-2), E'\ (^-1, ^-1), F'\ (0, 0)\ G'\ (0, ^-3)$ **13.** $\frac{5}{12}, \frac{1}{4}, \frac{1}{12}$ **15.** $\frac{5}{8}, \frac{3}{4}, \frac{7}{8}$ **17.** $4, 2\frac{2}{3}, 1\frac{1}{3}$ **19.** $\frac{2}{135}, \frac{2}{405}; \frac{2}{1,215}$ **21.** $\frac{13}{48}, \frac{13}{96}, \frac{13}{192}$ **23.** 8, 32, 128 **25.** $^-0.2, ^-0.02, ^-0.002$ **27.** $^-11, ^-14, ^-17$ **29.** $^-9, ^+90, ^-900$ **31.** $^-1, \frac{^+1}{4}, \frac{^-1}{16}$ **33.** 165 books

Page 557

1. A **3.** C **5.** A **7.** B **9.** B **11.** B **13.** A

INDEX

area of triangle, 427–429
Celsius to Fahrenheit, 140, 316–317
circumference, 416–417
distance, 318–319
Fahrenheit to Celsius, 140, 316–317
generate, 424–425
volume of cylinder, 448–449, 450–451
volume of rectangular prism, 440–443
volume of triangular prism, 442–443
See also Tables of measures

Fractals, 532–533
Fractions. *Adding and subtracting decimals and fractions is a focal point at Grade 6 and is found throughout this book.*
adding, 102–103, 104–105, 106–107, 108–109, H17
calculators, H34, H35
comparing, 24–25, H15
to decimals, 23–25, H19
decimals to, 22–25
dividing, 144–145, 146–147
estimating differences, 114–115
estimating sums, 112–115
equivalent, 92–93, H14
like, 102–103, H17
mixed numbers, 96–97, 126, H16
multiplying, 136–139
ordering, H16
to percents, 336–339
percents to, 336–339
rounding, 112–115
simplest form of, 94–95, 102, 140–141, H15
subtracting, 102–103, 104–105, 106–107, 110–111, H17
understanding, H14
unlike, 104–105, 106–107, 108–109, 110–111
using, 331, 463
Frequency tables, 229–231

G

Gardening, 431
See also Connections
Geoboards, 172, 369
Geography, 31, 409, 415, 467
See also Connections
Geography Links, 106, 113, 318, 334, 507

Geometric probability, 273–275
Geometry. *The use of concepts of geometry and spatial reasoning is a focal point at Grade 6 and is found throughout this book.*
angles, 164–165
area, 273, 422–423, 424–425, 426–429, 430–431, 432–433, 434–435, H25, H74
circle graphs, 244–245
circles, 416–417
circumference, 416–417
cones, 197, H22
congruent figures, 166–169, 370–371, H23
constructions, 166–169, 170–171
coordinate plane, 506–509, 522–525, 526–527
cylinders, 197, H22
Golden Rectangles, 394–395
lines, 160–161
line segments, 160–161
networks, 412–413
parallelograms, 426–429, H22
perimeter, 414–415, 430–431, H25
planes, 160–161
points, 160–161
polygons, 172–173, H21
prisms, 196–197, H22
pyramids, 198, H22
quadrilaterals, 172–173, H21, H22
rays, 160–161
rectangles, 172–173, H22
reflections, 184–187, 522–525, H20
rhombuses, H22
rotations, 184–187, 522–527, H20
shrinking, 389
similar figures, 368–369, 370–371, 372–375, 376–377, 378–379
squares, 172–173, H22
stretching, 388
surface area, 452–453
symmetry, 178–181
tangrams, 182–183
tessellations, 188–189, 190–191, 530–531
translations, 184–187, 522–527, H20
trapezoids, 172–173, H22
triangles, 172–173, 426–429, H21
volume, 440–443, 444–447, 448–449, 450–451
Glossary, H91–H102

Golden
Ratio, 394–395
Rectangle, 394–395
Graphs
analyzing, 252–253, 254–255, 256–257
bar, 236–239, 242–243, 244–245, 252–253, 254–255, 256–257
box-and-whisker, 262–263, 264–265
choosing appropriate, 239, 241
circle, 244–245, 246–247, 252–253, 356–357
coordinate grid, H29
coordinate plane, 506–509
histograms, 240–241
inequalities, 504–505
interpreting, 252–265
line, 237, 239, 242–243, 253, 254–255, 256–257
line plots, 228–231, 259–261
misleading, 254–255
predictions from, 256–257
of relations, 510–511
stem-and-leaf plots, 238–239, 259–261
Greatest common factors (GCF), 89–91, H5
Guess and Check, 5, 38–39

H

Health, 115, 293
See also Connections
Health Links, 290, 404, 414, 492
Hexagonal prisms, 197
Hexagonal pyramids, 198
Hexagons, 172–173, H21
Histograms, 240–241
History, 395
See also Connections
History Links, 34, 118, 268, 406, 427, 524
Hobbies, 53, 107, 131, 163, 169, 199, 313, 341, 345, 377, 411, 447, 541, 545
See also Connections

I

Identity Property
of One, 42–43, H6

PHOTO CREDITS *Page Placement Key:* (t)-top (c)-center (b)-bottom (l)-left (r)-right (fg)-foreground (bg)-background (i)-inset